VACUUM ELECTRON AND ION TECHNOLOGIES

VACUUM ELECTRON AND ION TECHNOLOGIES

D.S. KARPUZOV (EDITOR)

NOVA SCIENCE PUBLISHERS, INC.
COMMACK

Art Director: Maria Ester Hawrys
Assistant: Director: Elenor Kallberg
Graphics: Denise Dieterich and Kerri Pfister
Manuscript Coordinator: Roseann Pena
Book Production: Tammy Sauter, and Benjamin Fung
Circulation: Irene Kwartiroff and Annette Hellinger

Library of Congress Cataloging-in-Publication Data

Vacuum Electron and Ion Technologies / D.S. Karpuzov (Ed.)
p. cm.
Includes bibliographical references and index.
ISBN 1-56072-242-8 :
1. Electron beams--Industrial applications. 2. Ion beams--Industrial applications. 3. Manufacturing processes. 4. Vaccum technology. 5. Thin films. I. Karpuzov, D.S.

TK7835.V23 1995 95-21123
621.381--dc20 CIP

6080 Jericho Turnpike, Suite 207
Commack, New York 11725
Tele. 516-499-3103 Fax 516-499-3146
E Mail Novasci1@aol.com

Printed in the United States of America

CONTENTS

PART VI. SURFACE ANALYSIS

PREFACE

This volume contains a selection of the invited and contributed papers presented at the 8th International School on Vacuum Electron and Ion Technologies held in Varna, Bulgaria, from September 26th to October 1st, 1993. The meeting was organized by the Institute of Electronics, Bulgarian Academy of Sciences and was sponsored by the European Community Commission, Bulgarian Ministry of Science and Education, "Evrika" and "Cyril & Methodius" Foundations and the Union of Scientists in Bulgaria. The event continued the 15-years-old tradition of bringing together physicists, chemists, material scientists and engineers from Eastern and Western European countries, involved in studies of vacuum technologies, thin films and surface science. This time it attracted nearly one hundred participants from 18 countries, both young researchers and leading specialists.

The scientific program emphasized on the physical and the engineering aspects of the field, incorporating fundamental and applied research studies oriented toward processing/analysis of new materials. It covered traditional as well as new topics such as:

(a) Electron & Ion Sources and Beam Optics;
(b) Electron-Beam Processing;
(c) Electrical Discharges and Plasma Treatments;
(d) Deposition and Growth of Thin Films;
(e) Ion Beam Modification of Materials;
(d) High-Temperature Superconducting Thin Films;
(f) Surface Analysis

The School's activities involved 44 oral presentations of invited lectures and selected contributions in 8 plenary sessions, and 87 contributed presentations in 3 poster sessions. The peer reviewing of the manuscripts resulted in the selection covered in this volume.

The organizers want to thank everybody who contributed in making the 8th International VEIT School a success. Special gratitude is due to our sponsors whose generosity made the preparation of the meeting possible. The editors of NOVA Science Publishers who contributed to the fast appearance of this volume is also acknowledged.

D. S. Karpuzov,
Editor

Vacuum Electron and Ion Technologies
D. S. Karpuzov (Ed.)

ISBN 1-56072-242-8

SHOCK TUBE EXPERIMENTS ON PULVERIZED COAL COMBUSTION AND ENERGY CONVERSION.

A. Veefkind, V.E. Banin and V.S. Bajovic
Eindhoven University of Technology
V.A. Ivanov and V.A. Bityurin
Institute for High Temperatures,
Russian Academy of Sciences, Moscow
C.A. Borghi
Institute of Electrotechnics, University of Bologna

Abstract

A number of shock tube experiments will be presented concerning coal combustion and energy conversion. Ignition and subsequent combustion of small coal particles (6 and 70 μm) have been realized by shock heating of pulverized coal suspensions in different gas mixtures. Temperature development, burn-out and combustion products formation can be investigated with simple techniques. The determination of ignition delay times will be discussed and possibilities to determine activation energies and reaction rates will be indicated. Energy conversion experiments are carried out to demonstrate and investigate the possibility to employ hot flow parts in a combustion products gas flow as gaseous current conductors, in order to generate electrical energy from the flow when it is passed through a magnetic field.

Introduction

It is expected that in the coming decennia pulverized coal will be frequently employed for electrical power production. Direct combustion as well as gasification systems will be used to provide heat to be converted in combined cycles. The conditions in the topping cycles require combustion or gasification to take place at high pressures and high temperatures. Fundamental data for combustion and gasification have to be collected for these conditions. They are important because the designs of advanced combustors and gasifiers require accurate matching between the history of the coal particles and the flow field of the reacting medium. It will be demonstrated in this presentation that shock tube experiments can contribute to produce such data. The determination of ignition delay times, activation energies and reaction rates will be discussed.

In order to increase efficiencies of future power plants energy conversion methods are developed which can operate at high temperatures. Magnetohydrodynamic (MHD) generators are particularly suitable to satisfy this requirement. An intrinsic disadvantage of combustion MHD generators in that they cannot be operated at temperatures below 2100 K because then the electrical conductivity of the seeded combustion products becomes too low. In this presentation a new method is proposed: energy conversion with gaseous conductors (ECGC). This method employs an alternating flow in which large relatively cold flow parts are alternated with small hot flow parts. Shock tube experiments are used for the validation of the method.

Shock tube facility

The shock tube is schematically shown in Fig. 1. It has a total length of 12 m and a diameter of 224 mm and consists of a driver section and a test section separated by a double diaphragm formed by two aluminium plates. Prior to the experimental run the driver section is filled with helium upto a pressure of 11 bar and the test section with test gas to a typical pressure of 0.1 bar. The composition of the test gas depends on the aim of the experiment. When pulverized coal is studied it is dispersed as a suspension in the test gas. In Table 1 the compositions are listed for several experimental objectives. When the shock tube is used for energy conversion experiments the objective is to generate as much thermal energy as possible within a short time. For this reason a gaseous mixture is chosen which guarantees a good mixture between fuel and oxydizer. Moreover for energy conversion experiments the test gas is seeded with an alkali compound to increase the electrical conductivity. The rupture of the double diaphragm is described in [1]. After the rupture the helium expands into to test section resulting in a shock wave travelling in the test gas towards the end plate. When the shock reaches the end plate it reflects. After the passage of the reflected shock wave the test gas obtains the stagnation conditions. For the experiments considered here the stagnation pressure is around 8 bar. The stagnation temperature can be varied by the initial pressure ratio between driver gas and test gas and by the admixture of argon in the driver gas. The range of stagnation temperatures which can be obtained is between 700 and 1700 K for molecular gases. For argon the maximum temperature is around 3500 K. The mentioned temperatures have to be considered as pre-combustion temperatures. After ignition and subsequent combustion temperatures around 4000 K can be reached in molecular gas mixtures. The pre-combustion conditions can be calculated with simple shock tube theory [2] and appear to be strongly dependent on the speed of the incident shock. For this reason the shock velocity is determined from pressure transducer signals. When the test gas is chemically reacting the simple shock tube theory does no hold anymore. Numerical simulation is then necessary.

Table 1. Test gas mixtures.

Object of study	Mixture
Pyrolysis	Coal + nitrogen or argon
Combustion	Coal + x% oxygen + y% nitrogen
Gasification	Coal + x% CO_2 + y% nitrogen
Energy conversion	CO + x% oxygen + y% nitrogen + seed (cesium or potassium compounds).

Numerical Simulations

The experimental studies are accompanied by a numerical simulation of the main physical processes taking place in the shock tube and in the energy conversion channel which will be described later. The mathematical modelling of these processes is based on a one-dimensional and time dependent gasdynamical description. The nonsteady state of the gas mixture is described by the one-dimensional Euler's equations including body forces and energy sources [3]. The considered gas components are CO, CO_2, He, H_2O O and O_2. Coal particles have to be considered separately. The thermodynamical characteristics are defined with the approximations from [4]. The chemical evolution of this gas mixture is described with two reactions:

$$CO + \frac{1}{2} O_2 = CO_2$$

(3)

$$2O + M = O_2 + M$$

Here M is a non-specified molecule. The first reaction describes the complex chain mechanism of CO combustion in O_2. The reaction rate has been taken from [5]:

$$K_{CO} = 2.85 \times 10^{12} \cdot T \cdot [CO]\,[O_2]^{0.25}\,[H_2O]^{0.5} \times \exp(-15000/T) \tag{4}$$

The oxygen atoms recombination reaction rate has been taken from [6]:

$$K_o = 10^5/T \tag{5}$$

The reverse reaction rates are derived using the chemical equilibrium constants from [4].

The numerical simulation of the stagnation pressures at 0.1 m from the end plate is compared with a number of experiments with a mixture of CO and oxygen as test gas. The result is given in Fig. 2. showing a good agreement. The two bumps are caused by two detonation waves which propagate in opposite directions with respect to the gas from the point of ignition and which are separated by a rarefaction region.

Coal combustion experiments

- Ignition delay times

The experimental arrangement is schematically presented in Fig. 3. Coal particles are dispersed in the test section by an aspiration system. The main measurements are carried out in a cross section at 0.04 m from the end plate. There optical windows are used for infrared diagnostics on different wave lengths and for temperature measurements. A pressure transducer is also placed in that position. The experiments have been performed under the conditions obtained after the passage of the reflected shock, when the gas velocity is zero. The pre-combustion gas properties (gas density and temperature) are calculated from the measured velocity of the incident shock and the initial shock tube conditions. The typical test time available for experiments is 5 ms after which period the conditions are terminated by the arrival of rarifaction waves. The experiments were performed with two different coals: a polish coal with 6 μm particles and an australian coal with 70 μm particles. Typical pre-combustion conditions were 500 or 1000 mg coal particles in an oxygen-nitrogen mixture with a density around 3 kg/m^3 and temperatures between 1000 and 1200 K. Two radiation intensity signals on the wavelengths 1.36 μm and 2.2 μm have been chosen for infrared pyrometry of the particle surfaces. The signals of the two pressure transducers (see Fig. 3) have been used to define the starting moment of the measurements and to determine the shock velocity.

Ignition delay times are determined from the particle surface temperature behaviour as presented in Fig. 4. The initial time (- 1 ms on the figure) is the moment when the reflected shock passes the observation ports. It can be seen that practically immediately the particles obtain the pre-combustion gas temperature, in this case around 1100 K. Then it takes about 3 ms (the ignition delay time) before the ignition causes an further temperature increase to about 2100 K. Figure 5 shows the ignition delay time as a function of the partial oxygen concentration under constant gas temperature in the oxygen-nitrogen mixture for 6 μm particles. The total density of the gas mixture is constant. Three sets of experiments with different gas temperatures (about 1050 K and 1120 K) and different amounts of coal injected (500 mg and 1000 mg) have been performed. In both cases of higher temperature or higher coal particle concentration ignition is faster. The former is obvious and the latter can be explained by mutual influence of the coal particles on each other by accumulation of the volatile matter in the gas volume or by heating up the gas during the process of oxidation before the ignition. Figure 5 also shows a characterisitic step behaviour of the particle ignition as function of the partial oxygen concentration. At first when there is oxygen enough the particle ignites. By decreasing the amount of oxygen available it is possible to reach the point when no ignition occurs. Late ignition of the particles can be explained by a typical shock tube effect. It is given in Fig. 6. A secondary compression occurs 2-3 ms after the passage of the reflected shock. This compression wave results from the encounter of the reflected shock with the medium interface. Although the effect is in reality much weaker than expected from the calculations it still accounts for an additional temperature increase of 40-50 K, enough to cause ignition. This situation gives us the possibility to use the Frank-Kamenetsky criterion [7] for the ignition to find the activation energy and pre-exponential factor in the Arrhenius type of equation for chemical reaction on the surface supposing a linear dependence of the reaction rate on the oxygen concentration ($w = A \cdot Q \cdot \rho \cdot \exp(-E/T)$ with w the specific surface heat production and ρ the oxygen mass density). The effective activation energy E calculated in the case presented in Fig. 5 is equal to 10100 K and the pre-exponential factor A = 220 m/s with the reaction heat Q taken equal to 30 MJ/kg. It should be mentioned that a possible influence of devolatilisation has not been taken into account. The long period (1 - 2.5 ms) of constant temperature before ignition (see Fig. 4) corresponds to a situation of energy equilibrium at the particle near the point of ignition. This situation is characterised by the relatively low resultive heat flux to the particle surface which can be balanced by the endothermical process of devolatilisation. This situation ends by the ignition if the devolatalisation stops or the gas volume is heated up in the processes mentioned above.

The two step behaviour does not appear in the experiments with 70 μm coal particles (Fig. 7) because the ignition gas temperature for big particles is lower than for small ones. Figure 7 also shows the influence of the kind of coal on the process of the ignition.

- **Other possibilities**

Fig. 8 represents the time behaviour of the continuum radiation and of the radiation at a wavelength 4.3 μm at the band of CO_2.

The continuum radiation (the lower curve in the main part of the figure) is proportional to the total radiating particle surface. The decrease of the intentsity with time represents the burn-out of particles. Quantitative results can be obtained when continuum radiation is measured at two wavelenths so that particle surface temperatures and total radiating surface area can be determined simultaneously. In such a way reaction rates can be determined. Especially because they are obtained at high temperatures and fairly high pressures they are of interest for advanced applications.

The radiation at a wavelength of 4.3 μm (the middle curve in the main part of the figure) has to be corrected for the continuum radiation contribution to find the characteristics of CO_2 production. The corrected curve is the third curve of the figure. When the radiation is measured at a wavelength of 4.7 μm similar information can be obtained for the production of CO. Systematic measurements of CO_2 and CO production are a subject of future work.

Energy Conversion

The here presented shock tube experiments on energy conversion concern the method of direct energy conversion with gaseous conductors (ECGC). With this method electrical energy is generated from an alternating plasma flow. In our case the plasma consists of alkali seeded combustion products. In the channel flow hot and well conducting plasma is alternated with cooler and poorly conducting plasma as schematically presented in Fig. 9. The good conducting plasma typically represents 10 volume percents. As in magnetohydrodynamic generators the plasma flow is directed perpendicular to a magnetic field. The induced current will be concentrated in the hot plasma flow parts. The Lorentz force is counteracted by the local pressure gradient mainly generated by the pushing action of the colder gas. In this way the expansion work is delivered by the colder gas. The Ohmic dissipation inside the hot nonuniformities provides for a self-sustaining mechanism to keep the local gas temperature at a high enough level.
ECGC allows a substantial energy conversion at average temperatures which are considerably lower than in uniform plasma flows (e.g. 1600 K instead of 2100 K). The thermodynamic advantage is presented in Fig. 10 for a combined ECGC-steam cycle.

A tri-lateral co-operation research project has been set up with the aim to study ECGC. The participating institutes are the Institute for High Temperatures of the Russian Academy of Sciences (IVTAN), the University of Bologna and the Eindhoven University of Technology. This presentation reports about experiments with the Eindhoven Shock Tunnel Facility. The experimental program is directed to the demonstration of the practical feasability of the method.

For the experiments the test gas consists of a mixture of CO and O_2 (partial pressures before the rupture of the diaphragms equal to 74 and 37 mbar resp.). The gas is seeded with Cs by injecting 10 ml of a saturated solution of Cs-acetate in isopropyl alcohol as a liquid spray in the test section a few seconds before the rupture of the diaphragms. Through a hole in the end plate the gas can flow from the stagnation region in the shock tube to the conversion channel and from there into a dump tank. (Fig. 11, insert). The channel (Fig. 11) diverges in two directions from 58.4 x 58.4 mm^2 in the throat to 120.1 x 120.1 mm^2 at the channel exit over a length of 875 mm. The first electrode pair is situated at 75 mm downstream of the throat. There are 81 electrode pairs with a width of 1 mm. The segmentation length is 10 mm. The magnetic field is constant over the channel length and has a maximum value of 2.5 T. The standard load is 5 Ω for each electrode pair.

A hot flow part is formed by means of a discharge pulse, obtained from a capacitor of 1 μF, charged to 6000 V prior to the experiment. With a spark gap switch the voltage is applied to the connected electrode pair at a well defined time, resulting in a pulsed arc in the channel (typical duration 10 μs).
The responses of several downstream electrode pairs to the passage of the hot flow part has been measured. The results are presented in Fig. 12. Apparently nonuniformities can be formed by the employed technique and a finite life time can be observed. Three aspects of the nonuniformities are considered more systematically as a function of the channel position: the current increment, their

axial dimension and their motion (Figs 13, 14 and 15). The qualitative conclusion on the current increment is that the dissipation in the hot flow parts is not yet large enough to maintain the initial increment. This can be explained by the large electrode voltage drops which appear due to the cold walls and have therefore to be considered as a typically experimental effect. The motion of the hot flow parts indicates a velocity practically equal to the main flow velocity. Apparently there is a good momentum transfer between the hot flow parts and the main flow. A growth of the axial size of the hot flow parts is concluded from the time periods during which they pass along the electrode pairs, in combination with the velocities determined from Fig. 14. The increase of the axial dimension requires further study.

Similar flow simulations as carried out for the shock tube have been performed for the channel flow. Also the introduction of a hot flow part at a certain position is simulated. Experimental and calculated electrode current responses to the passage of a hot flow part are given in Fig. 16. It can be seen that also according to the calculations the internal disipation is not large enough to obtain the self-sustaining temperature conditions. The calculations suggest the earlier mentioned conclusion that the large voltage drops are probably responsable. Differences between the calculations and the experiment are mainly caused by the modelling of the electrical conductivity.

Concluding remarks

With a shock tube facility valuable fundamental data can be produced concerning coal combustion and gasification processes. They are of particular interest for application in advanced power stations because they are obtained at high pressures and high temperaturs. The pressures of the shock tube experiments presented in this paper are often still lower than the practical values but experiments with higher pressures are possible. Ignition delay times, activation energies and combustion reaction rates are presented as possible process characterizations which can be obtained from shock tube experiments with simple tools. In the near future gasification processes will be investigated in a similar way. More advanced optical diagnostics can make it possible to examine more closely combustion products formation and the production of contaminating substances.

Energy conversion with gaseous conducters (ECGC) makes it possible to employ a temperature range from 3000 K down to 1600 K to convert heat from a combustion product gas into electrical energy. In a combined cycle an overall efficiency of 60% is possible with this technique. Shock tube experiments are suitable to investigate fundamental aspects of the conversion method. The possibility to create gaseous conducters in a laboratory experiment has been presented. Further experiments must prove that the dissipation inside the conducters can be made large enough to sustain the necessary high temperatures.

Acknowledgement

The presented work has been carried out as a part of the research program "Coal and Electricity" at the Eindhoven University of Technology. It is incorporated in the national program of the Foundation for Fundamental Research on Matter (FOM) and has been partially supported by the Netherlands Technology Foundation (STW). W.F.H. Merck and M.A. van Buuren are kindly acknowledged for their scientific contributions, Mrs. M.H.A.J. van Rixtel-Kerkhoff for the preparation of the manuscript and A.P.C. Holten, A.W.M. van Iersel and H.F. Koolmees for their technical assistance.

References

1. A. Veefkind, "Inert Gas MHD Generator Experiments with a Shock Tunnel Facility", Magnetohydrodynamics: An International Journal, Vol. 2, No's 2-3, pp. 149-161, 1989.

2. H. Oertel, "Stossrohre", Springer Verlag, Vienna 1966.

3. V.A. Ivanov, "Calculations of Two-Phase Flow under Strong MHD Interaction in One-Dimensional Monodispers Approximation", Teplophysica Vysokikh Temperature, Vol. 29, No. 2, 1991 (in Russian).

4. L.V. Gurvitch et al., "Thermodynamic Properties of Individual Substances, Nauka Moscow, 1978 (in Russian).

5. T.V. Vilenshii and D.M. Khzmalian, "Dynamics of Combustion of Test Fuels", Moscow, 1979.

6. W.C. Gardiner et al., "Combustion Chemistry", Springer Verlag, 1984.

7. D.A. Frank-Kamenetskii, "Diffusion and heat transfer in chemical kinetics", Plenum 1969.

8. V.A. Bityurin, "MHD Electrical Power Generation with Plasma Nonuniformities Gas Interaction", Proc. 30th Symp. on Engineering Asp. of MHD, P.6.1, Inner Harbor Baltimore, 1992.

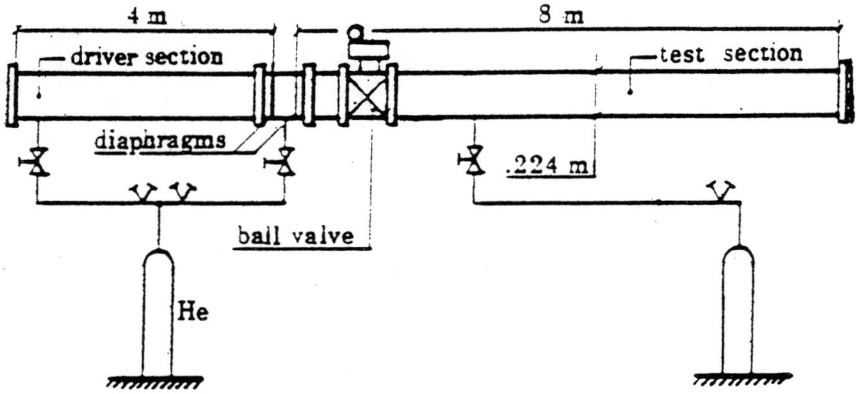

Fig. 1. Diagram of the shock tube facility.

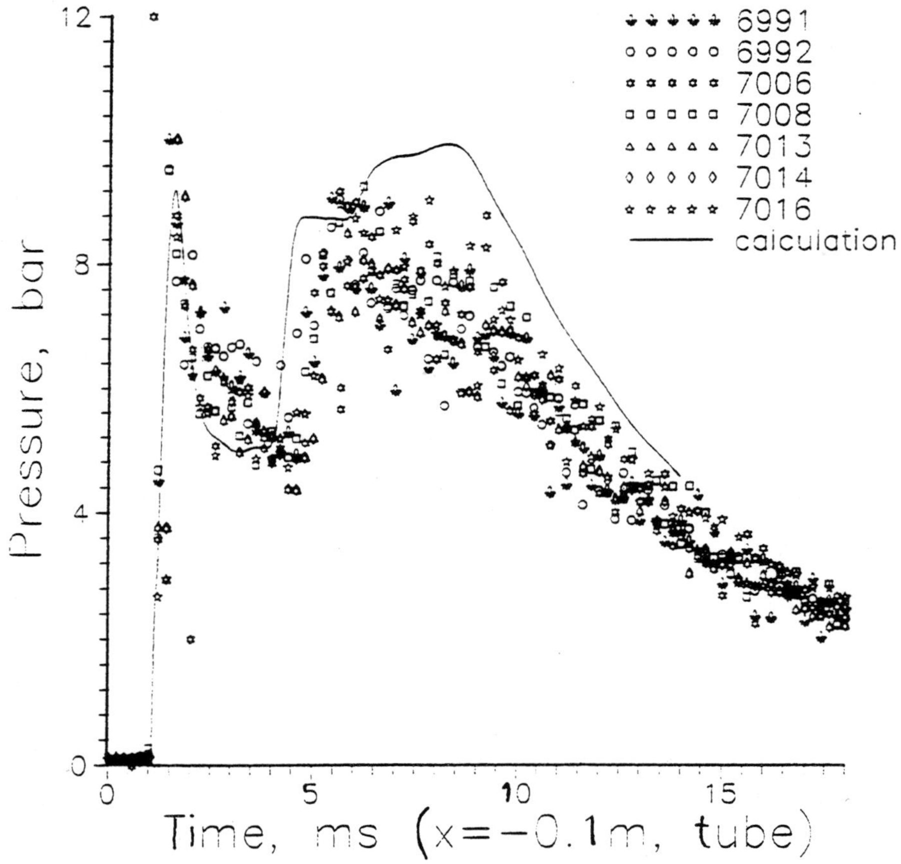

Fig. 2. Calculated and measured pressures in the stagnation region of the shock tube, o.1 m upstream of the end plate for a reacting test gas mixture consisting of CO and oxygen.

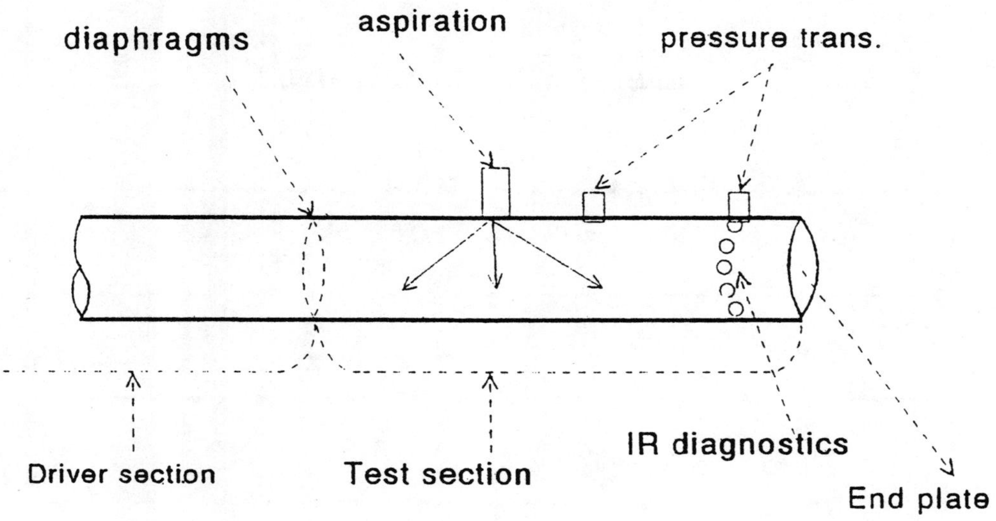

Fig. 3. Experimental set-up for coal combustion experiments.

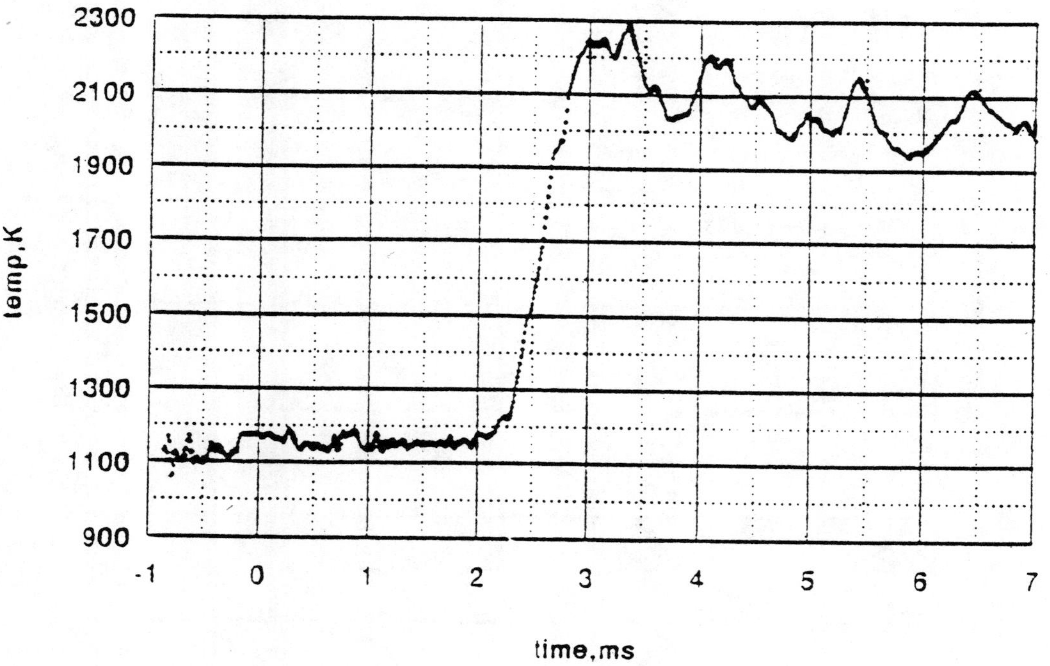

Fig. 4. Particle surface temperature as a function of time for 6 μm particles.

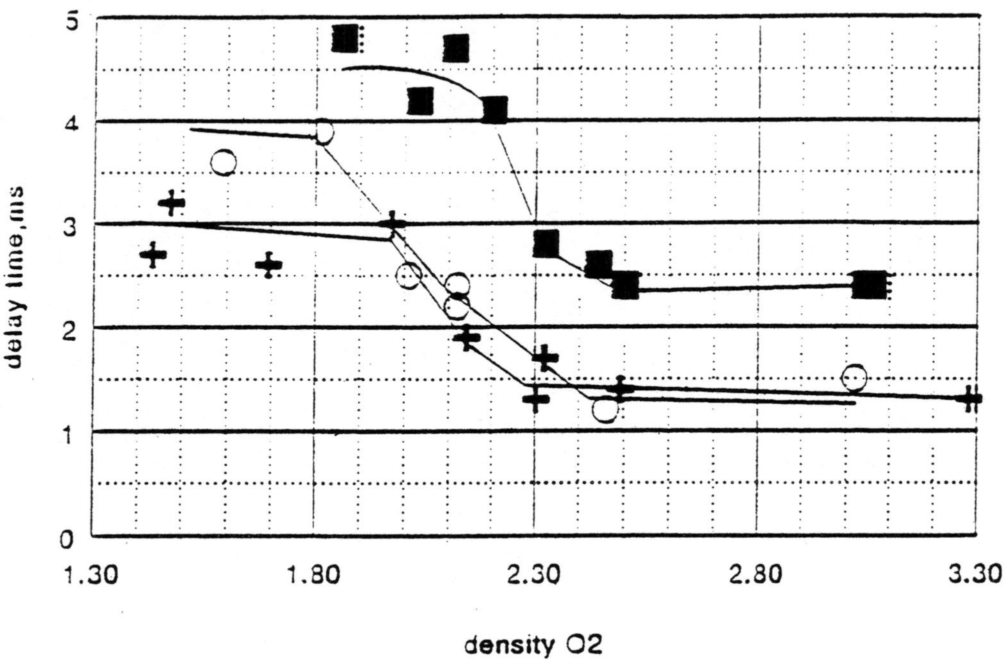

Fig. 5. Ignition delay time as a function of the partial density of oxygen for 6 μm particles. + corresponds to a gas temperature of 1120 K and an amount of 500 mg; 0 corresponds to a gas temperature of 1050 K and 1000 mg of coal and ■ to 1050 K and 500 mg.

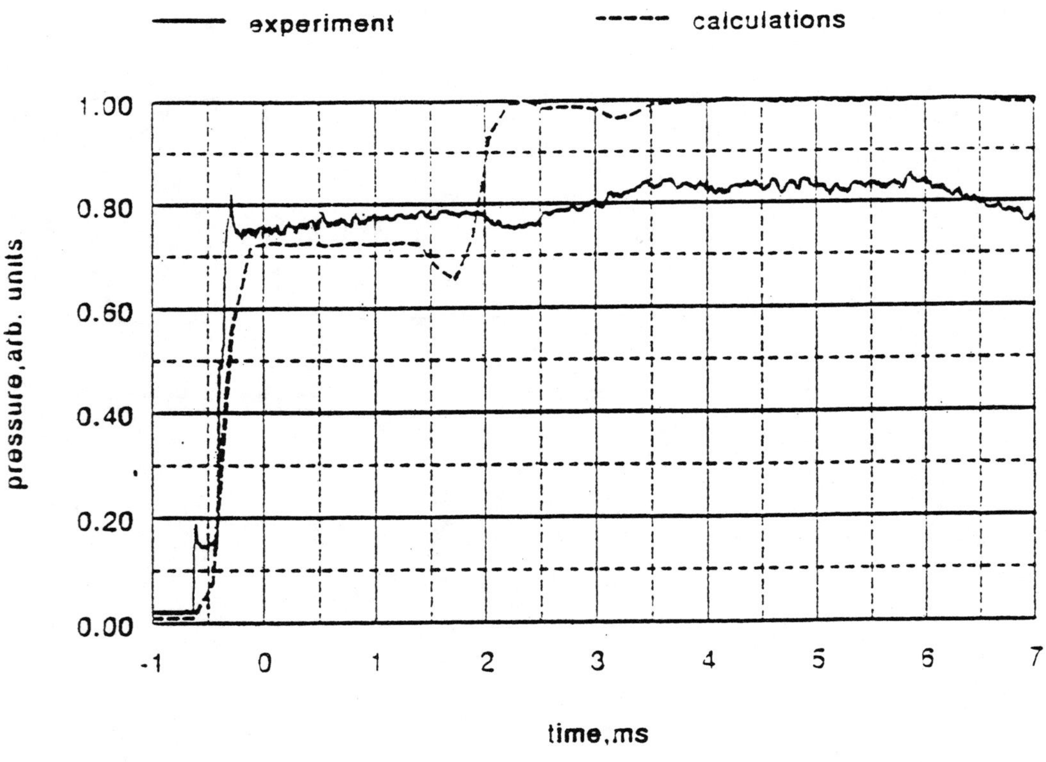

Fig. 6. Calculated and measured pressureis for an ignition delay time experiment.

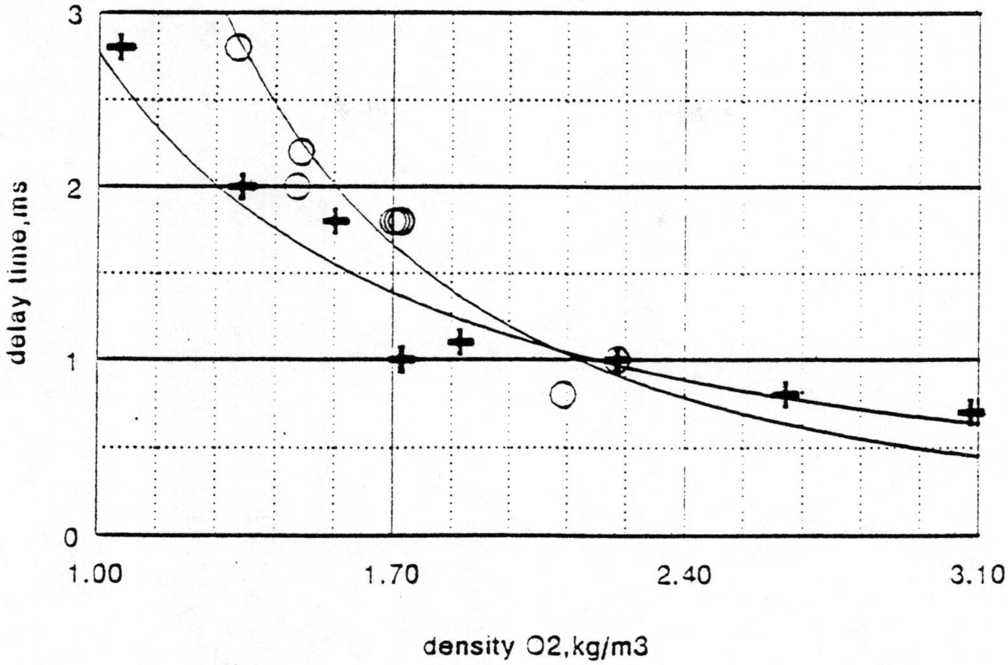

Fig. 7. Ignition delay time as a function of the partial oxygen density for polish (0) and australian (+) coals.

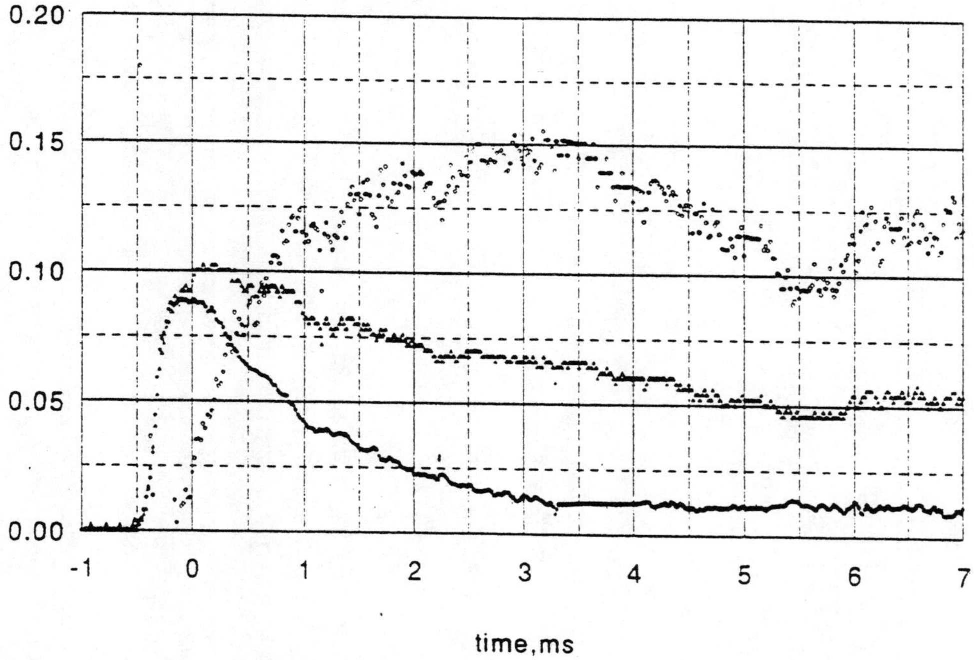

Fig. 8. Radiation intensities (in arbitrary units) from particle surface radiation (lower curve in the main part of the figure), radiation at 4.3 μm wavelength (middle curve) and resulting CO_2 radiation obtained after correction for the continuum contribution (top curve).

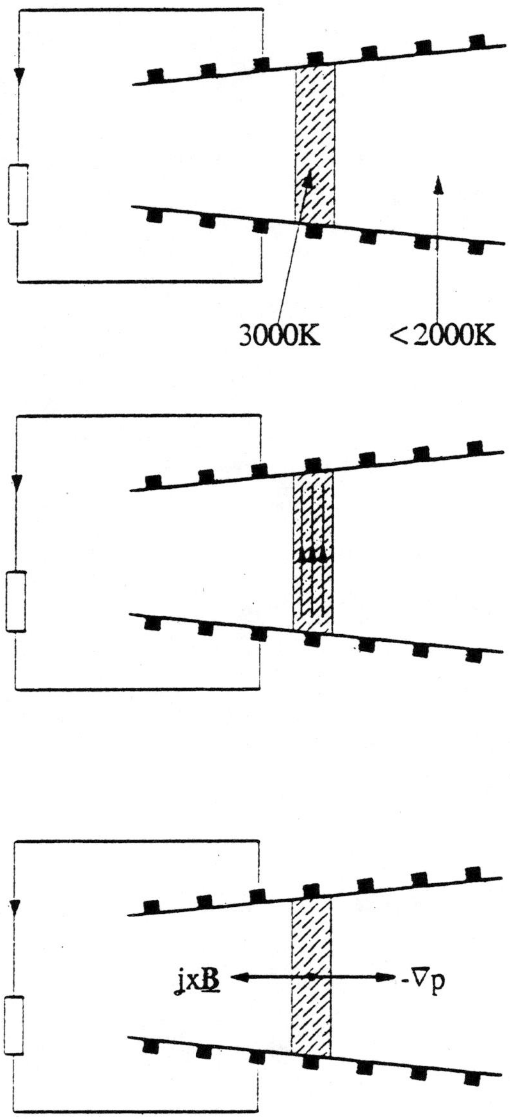

Fig. 9. Principle of energy conversion with gaseous conducters.

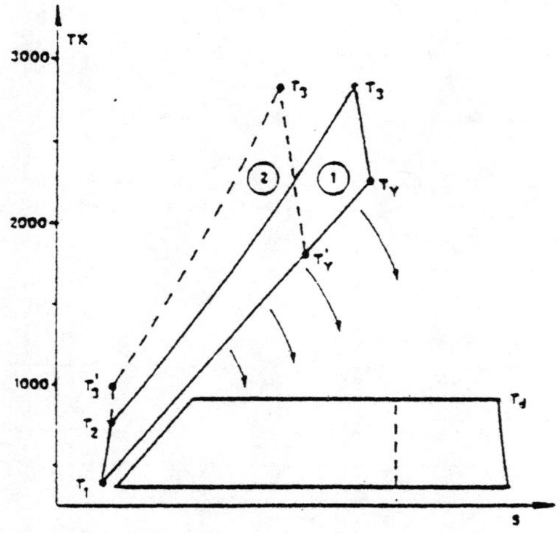

Fig. 10. From [8]. Thermodynamic diagram of a combined MHD/steam cycle with low (1) and high (2) enthalpy extraction of the topping cycle.

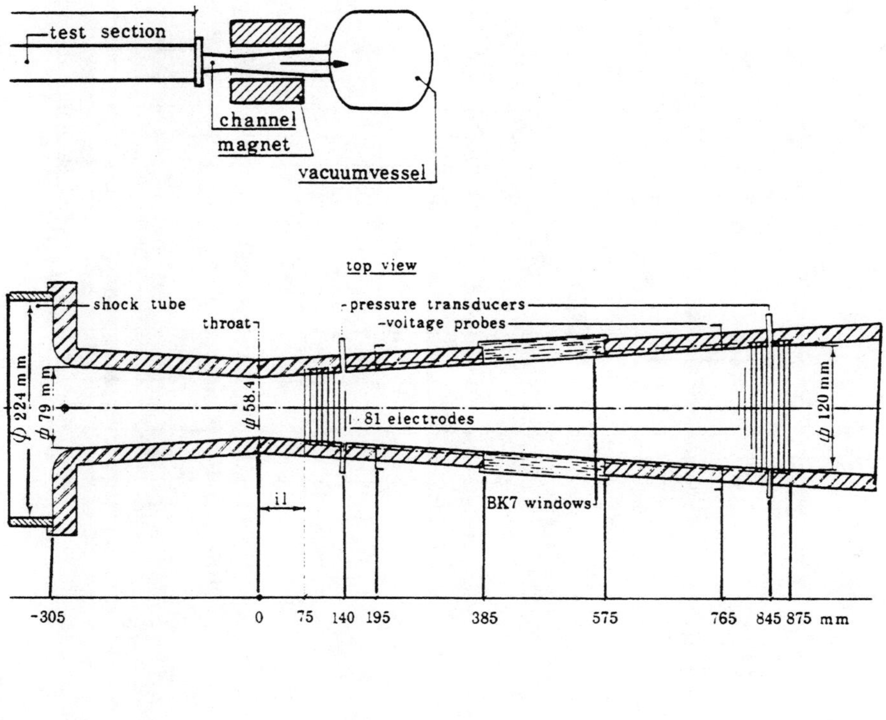

Fig. 11. Schematic of the energy conversion channel and connection to the shock tube (insert).

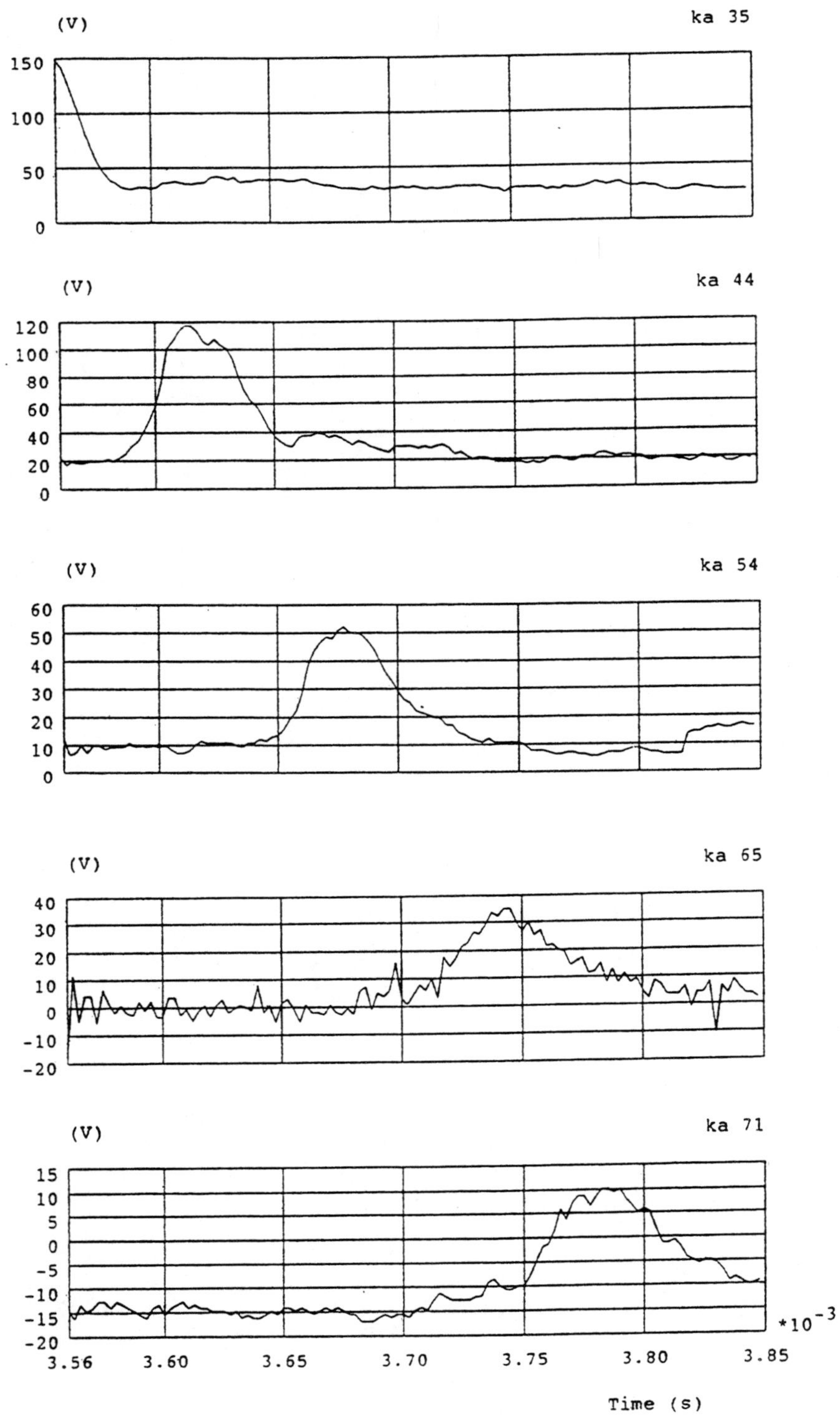

Fig. 12. Electrode voltages as a function of time for the time interval during which the hot flow part is passing.

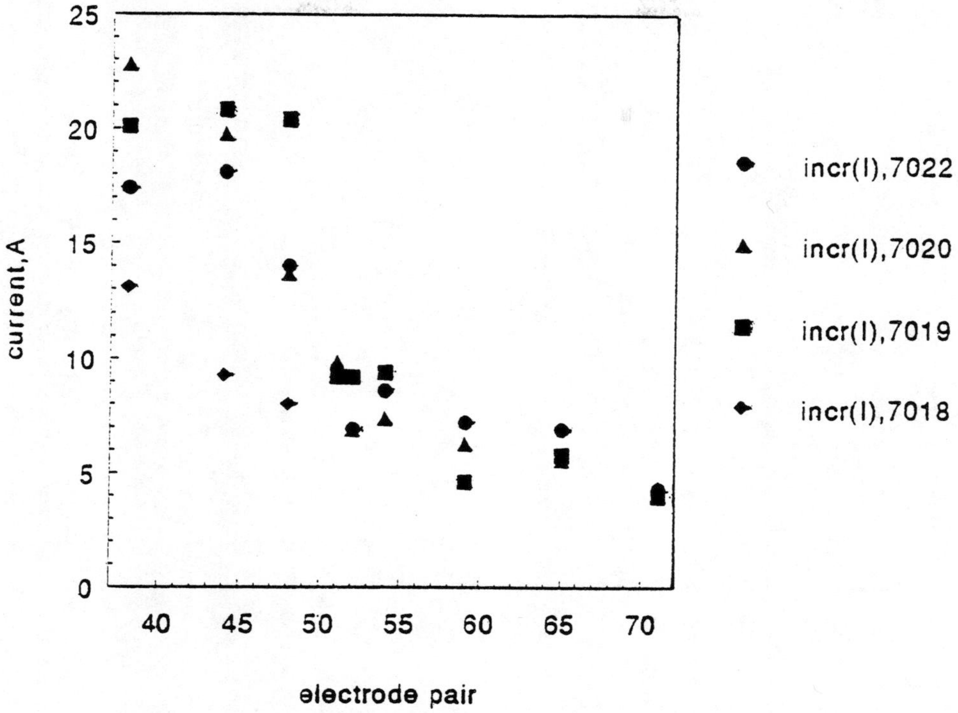

Fig. 13. Maximum current increase in the hot flow part for several runs with approximately the same conditions.

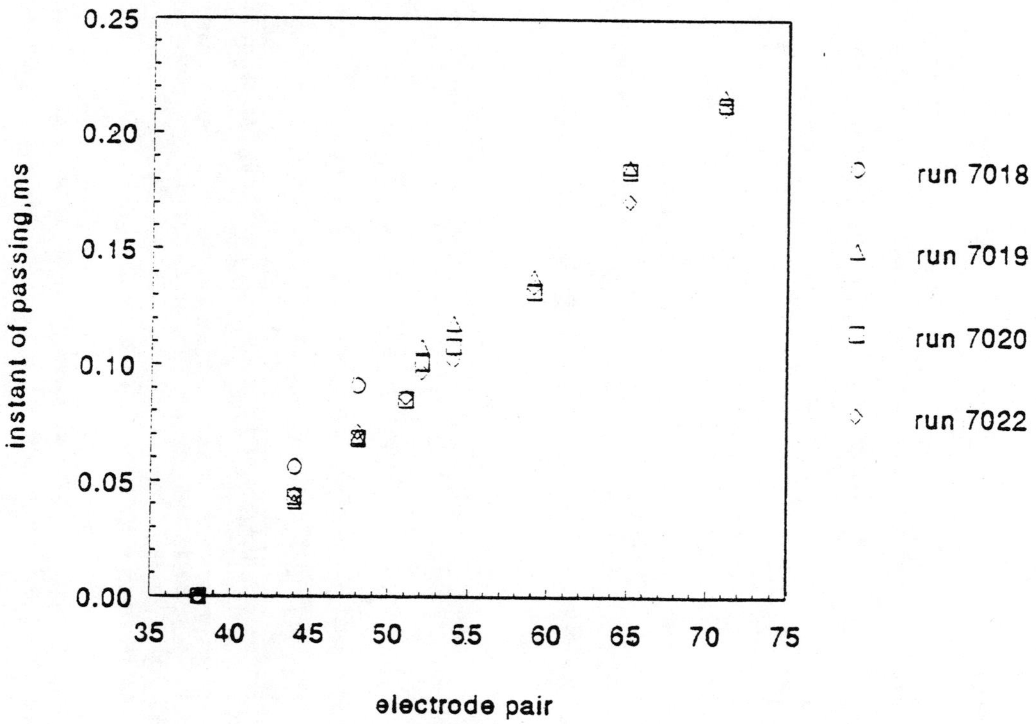

Fig. 14. Instants of passing of the hot flow part for the same runs as in Fig. 13.

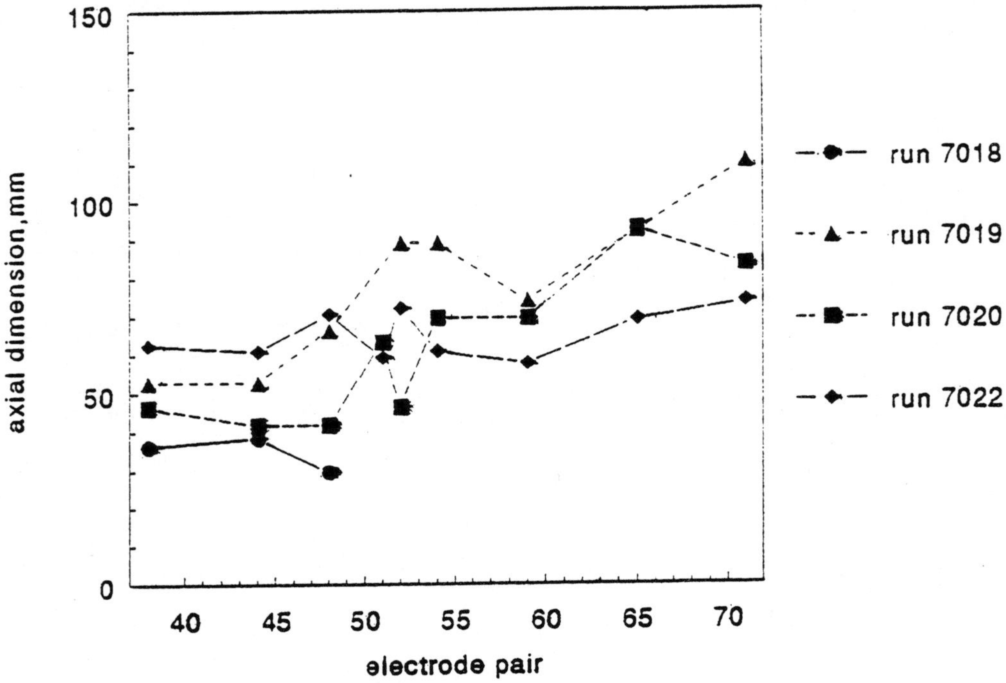

Fig. 15. Axial dimensions of the hot flow part for the same runs as in Fig. 13.

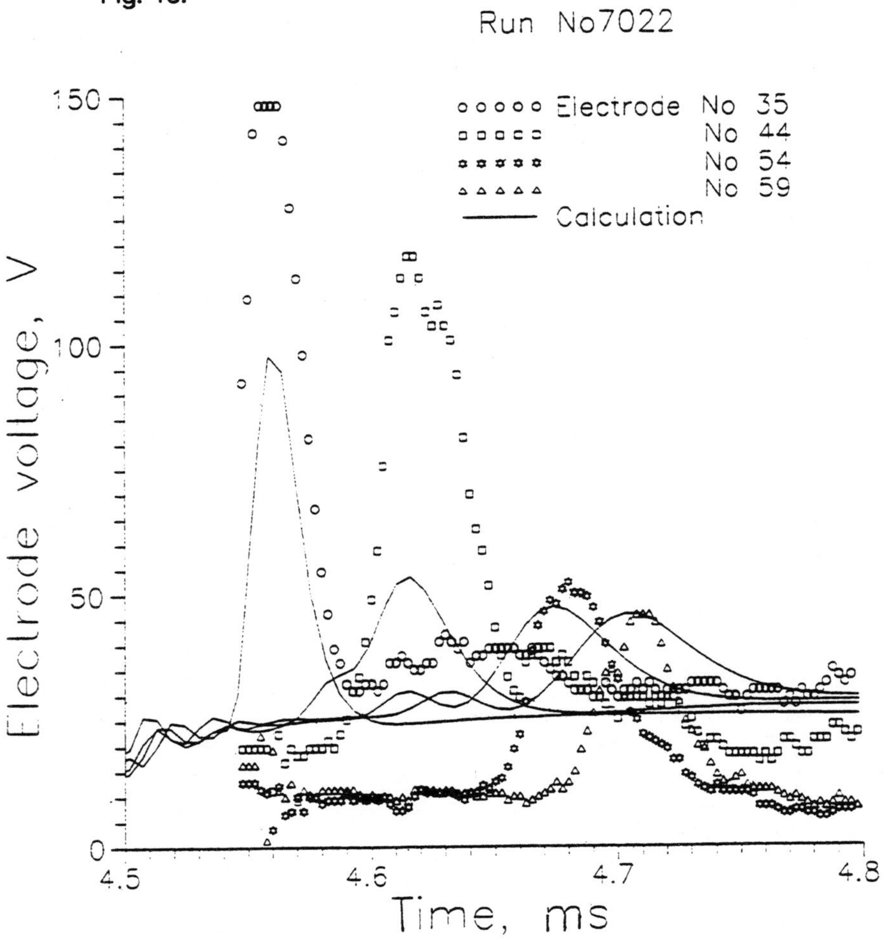

Fig. 16. Calculated and measured voltages for a particular run (the same as in Fig. 12) during a time interval in which the hot flow part passes.

Vacuum Electron and Ion Technologies
D. S. Karpuzov (Ed.)

ISBN 1-56072-242-8

INTERMOLECULAR POTENTIALS OF GLOBULAR MOLECULES

B. Stefanov and L. Zrkova
Institute of Electronics, Bd. Tzarigradsko
Chaussee 72, Sofia 1784, Bulgaria

Abstract. The model of vibrationally excited states of the molecules (VESM) was developed and motivated by Stefanov (1992). It was successively used for simultaneous approximation and prediction of the second virial coefficient B and the viscosity η of the heavy F or Cl containing molecules such as CF_4 (Stefanov 1992), SF_6, $\mathbf{BF_3}$ and SiF_4 (Stefanov and Zarkova 1993) and CCl_4 (Zarkova 1993) in a wide temperature range (200 to 1000K).

The approximation of B and they η for gaseous SF_6 was chosen as an example in order to justify the applicability of the VESM method for other globular molecules. The paper is concluded with a comparison of our calculations with the results of the recently published experiments (Strehlow and Vogel 1989), approximations (Boushehri et al 1987, Trengove and Wakeham 1980, Aziz et al 1991 and reference data (Dymond and Smith 1980,) , as well as with theresults of other investigations (Lichtenthaler and Schafer 1969, Kumar et al 1985)

1 Introduction.

The gases BF_3 CF_4, SiF_4, CCl_4, SF_6, etc. are prominent materials widely used in different areas of application (plasma processing, gaseous dielectrics, laser technique, pure material production). Usually the available experimental data are not enough and not consistent and this calls for tables with reliable recommended data. That is why .the efforts of many researchers were directed toward developing different models of interactions between two globular (quasi sphericlally symmetric) molecules.

Different approaches could be used: for restoring the parameters of the potential function: one can simply derive them from measured equilibrium and/or transport properties or one can develop more or less sophisticated many parametric potentials. As early as in 1758 the first model potential function was proposed by Boscovich. From that time up to now in all models the authors assume the globular molecules in the gas as a ground state and try to describe interactions between them with the use of one single potential. It is widely spread practice to derive the parameters of a such potential from one measured, for example equilibrium (B), property. The fit in this case is perfect (the experiments and calculations coincide within less than 1 percent discrepancy) but one can not use this potential to approximate transport properties (viscosity, thermal conductivity or diffusion). To obviate this effect one has to proceed simultaneously the transport and equilibrium properties. This procedure produces good results in the case of light (such as CH_4 or SiH_4) globular molecules or rare gases. but is not good for the heavy F or CL-containing globular molecules Although the spherically symmetric potential functions are simple and convenient to use, they fail to describe even the qualitative aspects in this case: discrepancy is about 15 percent and the measured $B_{exp}(T)$ and calculated $B_{calc}(T)$ even cross. Therefore authors try different tricks: including a fourth parameter (hexadecapole or dipole moment) or constructing more sophisticated anisotropic models (Aziz et al 1991).

The VESM model (Stefanov 1992) is quite different: the gas of globular molecules is considered as a mixture of a ground state and different vibrationally excited states with slightly different sizes. Therefore the interaction between every pair of molecules could be described by a simple three parameter potential functions $\varphi_{kl}(r)$ with slightly different parameters.

2 Basic considerations

A full detailed description of the VESM model is given elsewhere (Stefanov 1992). Briefly, it is based on the well known fact of the existence of the vibrationally excited states of the molecule and corresponding to it enlargement of the molecular radii (see Herzberg 1945). Since the normal frequencies ν_k and the degeneracy $g_{\upsilon i}$ of the molecules are known, one can calculate the relative populations x_{kl} of the states through the vibrational partition function

$$Z = 1 + \sum_{n=1}^{\infty} \sum_{\upsilon_1=0}^{n} \cdots \sum_{\upsilon_N=0}^{n-\upsilon_1-\ldots-\upsilon_{N-1}} \prod_{i=1}^{N} g_{\upsilon i} \exp\left(-\upsilon_i\theta_i/T\right)$$

where N is the total number of normal vibrations, υ_i is the number of the vibrationally excited level of the i-th normal vibration, and $\theta_i = h\nu_i / k$, ν_i being the i-th normal frequency. The relative population of the ground vibrational state is $x_0 = Z^{-1}$; the sum of the relative populations of all other vibrational states is $1-x_0$.

To consider the real situation in the gas, one has to take into account all vibrational modes i=0 to N and states υ_i, i.e. to consider many potentials

$$\varphi_{kl} = \varepsilon_{kl}\left(1-6/n\right)\left\{\left(6/n\right)\left(r_m^{(kl)}/r\right)^n - \left(r_m^{(kl)}/r\right)^6\right\} \quad (1)$$

with parameters: n (repulsive power), ε_{kl} (depth of the potential) and $r_m^{(kl)}$ (collisional distance). The analysis showed that in the case of SF the number of the excited states with the substantial impact to the interaction is $N_S = 16$ (including the ground state). These level are specified by individual parameters and the remainder is assumed as one state (k=17) with average parameters. The relative population of the ground state x_0 decreases from 0.67 to 0.14 when the temperature increases from 200 to 800 K. The ratio $\sum_{k=1}^{16} x_k / x_{17}$ changes from 34.8 to 0.03 at T =200 and 800 K respectively. It is obvious that even small differences between potentials lead to substantial changes of the overall interaction when the temperature is changed. Therefore, to describe interactions between SF_6 molecules we have to choose between three alternative approaches :(1) a sophisticated many parameter potential describing different temperature regions in different manner; (2) a potential function with temperature dependent parameters and (3) a set of simple potentials with temperature dependent contribution to the interaction. The advantage of the third one besides the simplicity is the clear physical background.

In the adopted model $\varepsilon_{kl} = \varepsilon\left[r_m / \left(r_m + \Delta_{kl}\right)\right]^6$, where Δ_{kl} is the distance between states k and l, r_m, ε and n are parameters of the equilibrium (ground) state. Since n remains the same for all φ_{kl} , the family of these seventeen simple three-parameter ((6-n) or (exp-6)) potential functions reflexes in fact the anisotropy of the short range forces while the long range forces remain unchanged.

Although the value of enlargements of the molecular radii δ_V is less than 1 percent of the equilibrium value (r_m), it has a substantial contribution to the temperature dependencies of B and η. In accordance with the mixture rule (Hirschfelder et al 1954)

$$1/\eta^{0.5} = \sum_{k,l} x_k x_l / \eta_{k,l}^{0.5}(T); \qquad B = \sum_{k,l} x_k x_l B_{k,l} \tag{2}$$

where x_k and x_l are the relative populations of levels (including the ground state) and modes $\sum_{k,l} x_{k,l} = 1$, and η_{kl}, B_{kl} are values of the viscosity and the second virial coefficients calculated by means of the corresponding potential functions $\varphi_{kl}(r)$, k,l=0 or υ_i, υ_j;

$$B_{kl} = -2\pi N_0 \int_0^\infty r^2 \exp\left(-\varphi_{kl}/kT\right) dr \tag{3}$$

N_0 being Avogadro's number, k - the Boltzmann constant and r - the distance between two excited molecules;

3 Results

The model was proved for SF_6. By solving the ill-posed problem of deriving the parameters (n, r_m, ε) from 29 measurements of B and η (178 experimental points). The best fit for the available experimental data (B and η) by means of eqs.(2) with the use of seven (or less) independent variables ε, r_m, n δ_4,δ_5, was achieved by minimization of the sum of weighted relative deviations

(a_B and a_η are the relative errors of the individual experiments or experimental points) following the next procedure:

(a) assuming the number P of the independent variables and their values in the frame of the accepted shape of the potential;

(b) minimizing the sum of squared deviations F until the best solution is achieved and the parameters are defined;

(c) checking the sensitivity of the solution to the number of parameters, to the number of the excited states N_s taken into account and to the specific experiment (so called moving control);

(d) repeating the minimization with optimised P and N_s;

(e) statistical analysis of the results;

(f) comparing the results with the information from other investigations.

The results of the mimimization procedure with different number of parameters (P=7 to 5) are presented in Table 1. Comparing different versions we found that the increasing of F and the change of ε, n, r_m, and δ were not significant, whereas the computing time was diminished proportionally to the number of variables. When the *rms* for two versions with a different number of variables is almost the same in the bounds of the acceptable error, it is preferable (see Tikhonov and Arsenin 1977) to choose the solution occupying less dimensional phase space (less number of variables P). Furthermore, in this case the computer time decreases substantially. Therefore, we assumed $\delta_4 \equiv \delta_5 \equiv \delta_6$ and considered only 5 parameters ε, n, r_m, δ_4 and δ_0

Table 1. Results of fitting the experimental data by VESM potentials using a different number of variables P in the fitting procedure (N_s=17, ε in K, r_m, δ_4, δ_5, δ_6, and δ_0 - in 10^{-10} m)

No	P		n	ε	r_m	δ_4	δ_5	δ_6	δ_0	F
1	7		32.03	423.8	5.02	0.0526	0.0400	0.0257	0.0145	60.20
2	6	$\delta_5 = \delta_4$	32.07	420.9	5.02	0.0589	0.0589	0.0115	0.0140	60.27
3	6	$\delta_6 = \delta_4$	31.88	422.1	5.02	0.0175	0.0805	0.0175	0.0143	60.25
4	6	$\delta_6 = \delta_5$	32.07	422.6	5.02	0.0228	0.0228	0.0228	0.0143	60.26
5	5	$\delta_6 \equiv \delta_5 \equiv \delta_6$	32.0	425.4	5.01	0.0350	0.0350	0.0350	0.0149	60.32

The final solution was obtained in this manner (version 5) with ε =425.4 K, n = 32, r_m =5.02x10^{-10} m, $\delta_4 \equiv \delta_5 \equiv \delta_6$ = 0.035x10^{-10} m, and δ_0 = 0.015x10^{-10} m, F = 60.32, i.e. the *rms* = 0.582 in units of the accepted errors a_B and a_η. The variance-covariance matrix is given in Table 2.

Table 2. A covariance-variance matrix (ε in K, r_m ,$\delta_4 \equiv \delta_5 \equiv \delta_6$ in 10^{-10} m)

n	ε	r_m^2	$\delta_6 \equiv \delta_5 \equiv \delta_6$	δ_0
-0.286x10^{-7}	-0.392x10^{-6}	0.809x10^{-8}	-0.532x10^{-9}	0.527x10^{-6}
0.119x10^{-6}	0.163x10^{-5}	0.337x10^{-7}	0.110x10^{-5}	
0.181x10^{-5}	0.247x10^{-4}	0.329x10^{-7}		
0.876x10^{-4}	0.120x10^{-2}			
0.392x10^{-1}				

The statistical check showed that the experiments are quite satisfactorily fitted over the whole temperature range. The deviations R_i are less than the accepted experimental errors (a_B and a_η) and the scattering from the reference line (our calculations) is practically equal to zero (the mean value of samples $\langle R_i \rangle = \sum R_i / M$ is equal to 0.09 for B and to 0,037 for η). This result is important for the temperature dependence of the second virial coefficient since viscosity could be approximated successfully with the use of many different potentials. The same holds for the high temperature measurements of B, but there is a drastic discrepancy between the approximations and the experiments at low temperatures (T<325 K) increasing when T decreases (Aziz et al 1991). In our work we succeeded to achieve the best fit for this area. The results of statistical check of the individual experiments confirmed that even for the experiments at lowest temperatures (200- 413 K, Hahn et al 1974) *rms* =0.45a_B, which means R_i = -8.49 cm^3 mol^{-1}. This result is not surprising as far as the calculations with the use of the adopted VESM model are the most accurate at low temperatures, where the relative populations of the ground state + fifteen considered vibrationally excited states are dominating.

4 Comparison with approximations from the set for validation

A limited number of papers propose a method of calculation of B and η for SF_6. A comparison is made with recently published experimental (Strehlow and Vogel 1989) and calculated (Mears et al 1969, Dymond and Smith 1980, Trengove and Wakeham 1986, Boushehri et al 1987). There is a satisfactory coincidence over the whole temperature range (the maximal deviations in the temperature range 200 - 800 K are ± 8 cm^3 mol^{-1} for B and from -0.7 to +1.3 % - for η) regardless of the fact that the authors use different experimental methods or different potential functions valid only for the considered property. Articles which consider simultaneously the equilibrium and transport properties of SF_6 (Boushehri et al 1987, Aziz et al 1991) over a wide temperature range were published recently. Despite the different procedures, the coincidence is satisfactory at temperatures 250 to 900 K (maximum 8 cm^3 mol^{-1} for B and maximum 0.8% for η). The experimental results of (Strehlow and Vogel 1989) coincide with the set of approximation (R_i =-0.26% , rms = 0.361).

To prove the reliability of the VESM model, we compare the results of the most recent work with those predicted by the VESM model using the same data for validation. The substantial discrepancy was observed at low temperature (T = 200 to 300 K). This area is usually difficult for predicting the second virial coefficients. With the use of VESM model we succeed to diminish substantially this discrepancy between calculations and experiments. For example, at 200 K the measured B (Hahn et al 1974) is -693 cm^3 mol^{-1} while the calculated B is -670.9 in our approximation and -646.6 cm^3 mol^{-1} - as calculated by Aziz et al (1991) respectively. The explanation of the better fit of the experimental data achieved in this work is possibly implied in the comparison of the proposed potential functions (see Figure 1). Our estimation of the value of the dispersion constant derived from the parameters of the best solution (ε =425.4 K, n = 32 and r_m = 5.01x10^{-10} m), is C_6 =596.5 a.u., which is in very good coincidence with the accurate calculations of Kumar et al (1985) - 585.4 ± 11.7 a.u. Almost the same figure for C_6 was chosen in the fitting procedure of Aziz et al (1991), but their attractive potential includes also the terms of a higher orders. The estimations of the dispersion energy (- C_6R^{-6}) in our model compared to (- C_6R^{-6}- C_8R^{-8} - $C_{10}R^{-10}$) Aziz et al (1992) at R = 1.475 (r_m = 4.705x10^{-10} m) showed adiscrepancy between calculated values in terms of r of about 7% . The two potentials differ mainly in the area of low-range forces ($r < r_m$).

When the experiments were satisfactorily fitted by means of only five parameters (version 5 of Table 1) we looked for additional information for intermolecular interactions in order to verify the physical background of the model. The brief look over the potentials proposed for SF_6 (Table III of Aziz et al, 1991) shows that r_m changes from 4.705 to 6.720 x10^{-10} m; ε - from 155 to 494.7. At the same time the value of r_m determined from crystallographic data (Lichtenthaler et al, 1968) is 5.02 x10^{-10} m which is equal to the estimated value in this work. This surprisingly good agreement without value of r_m deserves a special attention, since the determination from crystallographic data does not depend on the shape of the potential.

5. Conclusion

The approximation of the equilibrium and transport properties made with the use of the VESM model approved its validity for SF_6 in a wide temperature range. The best fit with less independent variables compared to the earlier published procedures was achieved at low temperatures, where the VESM calculations are the most accurate. Two facts speak in favour of the VESM model and are to be considered when the quality of the model is discussed: (a) the value of the collisional distance r_m estimated from crystallographic data (5.03x10^{-10} m) is almost equal to the derived in this work and (b) the good coincidence of the van der Vaals constant (C_6 = 596.5 ±0.5 a.u.) with the accurate quantum mechanic calculations.

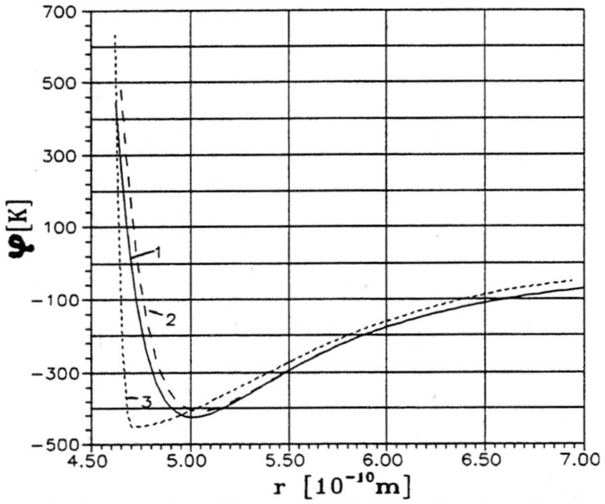

Figure 1. Potential curves for interactions between different vibrationally excited levels: 1 (solid line) - (0-0) interaction between two ground state molecules ; 2 (----) - (0-1) interaction and 3 (....) - MMSV potential (Aziz et al 1991).

References:

Aziz R A , Slaman M J, Taylor W L and Hurly J J 1991 *J. Chem. Phys.* **94** 10344

Boscovich R"*Theoria Philosophica Naturalis reducta ad unicam legem virum in natura existentium*", Vienna, 1758

Boushehri A, Wzowski J, Kestin J and Mason E A 1987 *J. Chem. Ref. Data* **16** 445

Dymond J H and Smith E B 1980 *The virial coefficients of pure gases* (Oxford: Clarendon Press)

Herzberg (1945)G. ,"*Molecular Spectra and Molecular Structure. II. Infrared and Raman Spectraof Polyatomic Molecules*" (New York: Van Nostrand)

Hirschfelder J D, Curtiss Ch F and Bird R B 1954 *Molecular Theory of Gases and Liquids* (London: John Willey)

Klein M and Smith E J 1968 *Res. NBSA* **72** 359

Kumar A, Fairley G R G,Meath W J 1985 *J. Chem. Phys.* **83** 70

Lichtenthaler R N and Schafer K 1969 *Ber. Bunsenges. Phys. Chem.* **73** 42

Stefanov B 1992), J. Phys. B: At. Mol. & Opt. Phys. **25 B** , 4519

Stefanov B and Zarkova L 1993) 13 ECTP, Lisbon, Portugal.

Strehlow T and Vogel E 1989 *Physica A* **161** 101

Tikhonov A N and Arsenin V Y 1977 *Solution of Ill-Posed Problems* (London: John Willey)

Trengove R D and Wakeham W A 1987 *J. Chem. Ref. Data* **16** 175

Zarkova L1993, 8 VEIT'93, Varna, Bulgaria

Vacuum Electron and Ion Technologies
D. S. Karpuzov (Ed.)

ISBN 1-56072-242-8

RESTORING OF ELECTRON - SYMMETRICAL TOP MOLECULE CROSS SECTIONS

B. Stefanov and P. Pirgov
Institute of Electronics, bd. Tzarigradsko
72, Sofia 1784 Bulgaria

Abstract

A semiempirical method is described for calculating the electron - molecule interaction potential, electron density, elastic and vibrational inelastic cross sections using experimental data. CF_4, SiF_4 and BF_3 are considered as examples. The experimental data from IR and Raman spectra and the molecule properties calculations (dipole matrix elements, polarizability, the distance between atoms in the molecule), beam measurements of total momentum transfer and differential cross sections and swarm data (diffusion and drift velocity) are used for extracting the interaction potential, the molecule electron density and the differential cross sections. Electron energy distribution functions, drift velocity and characteristic energy are calculated with obtained differential electron - molecule cross sections.

1. Introduction

The need of the plasma physics and chemistry requires knowledge of the electron energy distribution function, drift velocity, diffusion tensor, different reaction rates, etc. Generally different cross sections - differential, total and momentum transfer are the molecular parameters when the above quantities are calculated.

At present the theoretical ab initio computation of cross sections, the electron transport (swarm) data determination of cross sections and the cross sections measurement using electron beams, comprises three independent means of cross section determination. They represent independent complementary approaches to the problem. Each one has its advantages and limitations and all of them are capable of providing useful information. Errors may occur when beam measured cross sections are included in swarm calculations, especially at low values of the reduced electric field (E/N).

The cross sections extracting from swarm data is an ill- posed inverse problem which has been solved successfully for many atoms and molecules. This procedure can be formalized by considering the deviations between experimental and calculated values of some swarm characteristics (obtained by solution of the Boltzmann equation) as functions of parameters defining cross sections and then minimizing these deviations until best fitting is obtained [1-3]. Supplementary information is needed to specify a unique set of cross sections [4]. Usually there are not enough experimental data, so only integral cross sections can be restored. Besides, one can observe a great difference between cross sections proposed from various authors [10] and beam derived cross sections [5,6,7].

Ab-initio theory using highly accurate interaction potentials and target wave functions with a very tight control on numerical accuracy have been developed recently [8, 9]. Nevertheless, the cross sections produced are incompatible with the results of transport analysis via the Boltzmann equation especially at low energies [5,6,7,10,11].

In this paper we made an attempt to combine electron- molecule cross section calculations based on a parametrized model interaction potential with Boltzmann analysis of transport coefficients at low energy ($\leq$ 10 eV). This study allows the unification and justification of various properties: molecular parameters (such as interatomic distance, effective charge of atoms etc.); IR and Raman spectra; beam measurements and swarm data.

2. Method

The method of processing the data is described elsewhere [12,13,14,15], briefly it consists of:

1. Approximating the molecular parameters by means of some constants:

a. molecular electron density $\rho(r) = \sum |\varphi_i(r)\, \varphi_j(r)|$ - a few screening factors α_i corresponding to i-th molecular wave function $\varphi_i(r)$ and one exchange constant α_x;

$$\varphi_i(r) = N\exp(-\alpha_i r)$$

$$V_{ex}(r) = -(\tfrac{3}{2}\pi)\alpha_x \sqrt[3]{3\pi^2\rho(r)}F(r)$$

b. dipole matrix elements for each vibrational transition d_i.

2. Calculating the scattering **K** matrix via the Schrodinger equation;

3 Calculating the momentum transfer elastic and the total vibrational inelastic cross sections from the **K** matrix;

4 Calculating the electron energy distribution function via the Boltzmann equation;

5 Comparing between the calculated swarm values and/or cross sections W_{calc} and the experimental data W_{exp} by forming terms:

$$F = \sum\left[\ln\left(W_{calc}/W_{exp}\right)\right]$$

6 Minimisation the function $F(\alpha_i, \alpha_x, d_i)$

3. Results and discussion

The rotational cross sections corresponding to octopole transition and the electron attachment cross section are very small ($Q \leq 0.06 \times 10^{-20} m^2$) [417,418] and do not affect the electron energy distribution function (EEDF). The electron excitation cross sections have thresholds ~ 8eV and up to 10eV are also small compared to momentum - transfer and vibrational inelastic cross sections [2,10]. The cross sections of CF_4 and SiF_4 corresponding to ν_1 (ν_2) are approximately 10 times less then those of ν_3 (ν_4) [6,7,10]. That is why we considered only ν_3 and ν_4 vibrational cross sections. For the same reason we neglect the ν_1 vibrational mode for BF_3. The molecular data [10,16-19] are shown in Table 1.

A comparison between the experimental data and our best solutions are made in table 2. The root mean square deviations *(RMS)* between calculation and experiments as well as quoted experimental errors and discrepancies between the various authors are represented there. Our deviations from measured values of drift velocity, characteristic energy and cross sections are consistent with the experimental errors and discrepancies between the experiments themselves (one experiment was taken as a base).

Table 1. Molecular properties of CF_4, SiF_4 and BF_3.

Substance	CF_4	SiF_4	BF_3
Rotational constant, $x10^{-7}$ a.u.	8.6767	6.2810	12.478
Polarizability, a.u.	25.0	58.9	32.93
Distance between central and F atoms, a.u.	2.4982	2.9328	2.4755
Dipole moment matrix elements for			
ν_2 vibration	-	-	0.408-0.501
ν_3 vibration	0.132-0.167	0.109	0.121-0.137
ν_4 vibration	0.020-0.047	-	0.195-0.303

Table **2**. A comparison between the experimental data used in the minimization procedure and our best solution. In the column "Property" *Q* MERT means the Modified effective range theory of beam measured differential cross sections; *Q* - total and/or momentum transfer cross section; *I* - beam measured differential cross sections; *W* - measured drift velocity in pure substance; *W*(%)- (in parentheses the percentage to *Ar* in *Ar* mixtures is given); and D/μ - measured transversal diffusion to mobility ratio (characteristic energy).

Reference	Property	Number of experimental points	Quoted experimental error	Discrepancy between the experimental data in %	RMS deviations
		CF_4			
5,6	I	109	30-60	-	57
5,6	Q	14	30-60	-	48
5,6	Q MERT	7	30-60	-	53
7	I	103	25-50	30	61
7	Q	14	25-50	30	44
20	Q	32	30-40	100	58
21	W	23	8	9	11
21	W (0.1, 0.5, 2, 10, 30,)	97	8	9	12
22	W	15	12	12	10
22	W (1, 2, 10, 20)	52	12	12	11
23	W	39	5	-	10
24	W	8	15	30	12
23	D/μ	1	15	-	33
24	D/μ	8	5	-	36
25	D/μ	16	25	-	28
		SiF_4			
26	W	30	8	-	12
		BF_3			
26	W	45	8	-	11

The obtained model potential (full line) and the model electron density (broken line) as functions of the distance from the molecular centre are shown in Figure 1. In the vicinity of $r = 0$, the molecular potential is merely the Coulomb potential of the C nucleus while at large r - dipole polarizability potential.

Figures 2 represent the obtained integral cross sections (the differential cross sections are also available). Figures 3 and 4 give the calculated values of W and D/μ. It can be seen that even a very small addition of fluorine containing substance changes notably transport properties.

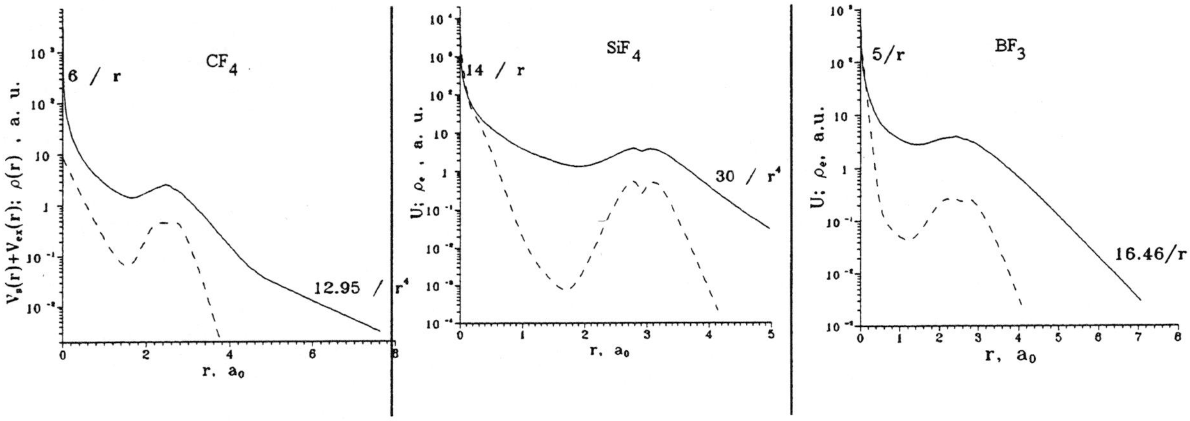

Figure 1. Electron- molecule interaction potential (full line) and electron density as a function of the distance from the molecular centre in atomic units.

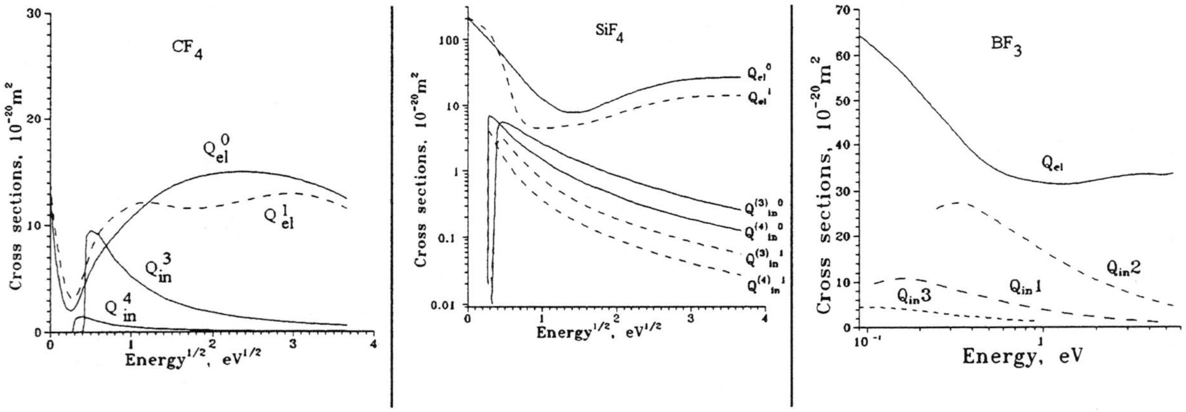

Figure 2. The obtained integral cross sections

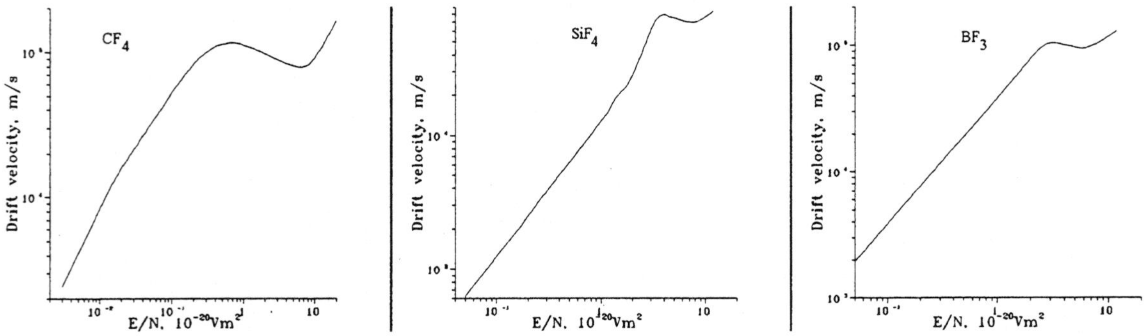

Figure 3. The obtained drift velocity as a function of reduced electric field

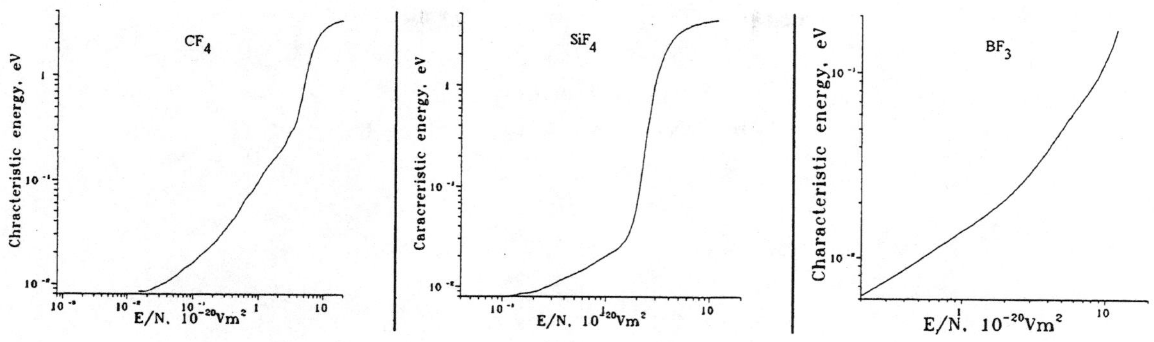

Figure 4. The obtained diffusion to mobility ratio (characteristic energy) as a function of reduced electric field

4. Conclusion

It is preferable to use ab initio calculations of the cross sections when possible. But for a molecule containing many atoms the theory gives accurate cross sections only for relatively large energies ≥ 5 eV. Often there simply are no calculations of the electron - molecule cross sections. The restoring of cross sections using only swarm experimental data demonstrates a discrepancy within a factor of 5. This can be explained with the peculiarities of the ill-posed inverse problems. As a rule at least 4 - 6 parameters are used for an approximation of a particular cross section. Often the minimization function has 12- 18 dimensions and it is difficult to find a unique set of cross sections. Besides, the error in just one cross section can introduce errors in other cross sections derived from a Boltzmann-type analysis.

The method described above allows obtaining much more information at the same time: model interaction potential, model electron density and differential cross sections. The ill- posed nature of the problem can be substantially reduced by using experimental data from quite different areas of physics (molecular structure data, IR and Raman spectra beam measured differential cross sections, etc.).

References

1. L. C. Frost and A. V. Phelps, Phys. Rev. 127, 1621 (1962).
2. M. Hayashi, *Swarm Studies and Inelastic ElectronMolecule Collisions*, Springer-Verlag, New York (1985).
3. M. G. Curtis, I. C. Walker and K. J. Mathieson, *J. Phys. D* **21**, 1271 (1988).
4. A. N. Tichonov and V. Y. Arsenin, *Solution of ill-posed problems*, Wiley, London (1977).
5. A. Mann and F. Linder, *J. Phys. B* **25**, 533 (1992).
6. A. Mann and F. Linder, *J. Phys. B* **25**, 545 (1992).
7. L. Boesten, H. Tanaka, A. Kobayashi, M. A. Dillon and M. Kimura, *J. Phys. B* **25**, 1607 (1992).
8.R. W. Crompton and A. Morrison, *Swarm Studies and Inelastic Electron - Molecule Collisions*, Springer-Verlag, New York (1986).

9.N. F. Lane, *Rev. Mod. Phys.* **52**, 29 (1980).
10. W. L. Morgan, *JILA Data Center Report No 34*, University of Colorado, National Institute of Standards and Technology Boulder, Colorado, (1991).
11.S. J. Buckman, M. J. Brunger, D. S. Newman, G. Snitchler, S. Alson, D. W. Norcross, M. A. Morrison, B. C. Saha B C, G. Danby and W. K. Trail, *Phys. Rev. Lett.* **65**, 3253 (1990).
12.B. Stefanov, N. Popkirova and L. Zarkova, *J. Phys. B* **21**, 3989 (1988).
13.P. Pirgov P and B. Stefanov, *J. Phys. B* **23**, 2879 (1990).
14.P. Pirgov and B. Stefanov, *Plasma Chem. Plasma Proc.* **11**, 287 (1991).
15.P. Pirgov and B. Stefanov, *Plasma Chem. Plasma Proc.* **13**, No4 (1993).
16. Miller C K, Orr B J and Ward J F 1981, J. Chem. Phys. 74 4858
17. Kim K and Park C W 1987, J. Mol. Str. 161, 297
18. Heijmen B, Bizzarri A, Stolte S, and Reuss J 1989 Chem. Phys. 132 331
19. Kim K 1984 J. Phys. Chem. 88 2394
20. R. K. Jones, *J. Chem. Phys.* **84**, 813 (1986).
21.S. R. Hunter, J. G. Carter and L. G. Christophorou, *J. Appl. Phys.* **58**, 3001 (1985).
22.L. G. Christophorou, D. L. McCorkle, D. V. Maxey and J. G. Carter, *Nucl. Instr. Meth.* **163**, 141 (1979).
23.S. R. Hunter, J. G. Carter and L. G. Christophorou, *Phys. Rev. A* **38**, 58 (1988).
24.M. S. Naidu and A. N. Prasad, *J. Phys. D* **5**, 983 (1972).
25.C. S. Lakshminarasimha, J. Lucas and D. A. Prince, *Proc. IEE* **120**, 1044 (1973).
26. Hunter S R, Carter J G and Christophorou L G, 1989 J. Appl. Phys. **65** 1858

Vacuum Electron and Ion Technologies
D. S. Karpuzov (Ed.)

ISBN 1-56072-242-8

LANGMUIRE PROBE MEASUREMENTS OF ELECTRON ENERGY DISTRIBUTION FUNCTION IN GAS MIXTURES CONTAINING CF4

Tc. Popov, P. Pirgov and D. Bossev
Institute of Electronics, 72 Tzarigradsko
Chaussee , Sofia 1784, Bulgaria

Abstract

Electron energy distribution functions (EEDF) of Langmuire probe measurements in $Ar+4\%CF_4$ DC gas discharge in the range E/N=19-22 Td at the pressure 1,8 Torr are reported. The result satisfactorily agree with the solution of the Boltzmann equation. The specific details of the construction of the electrodes and the probe as well as their influence on the experimental results are discussed.

Introduction

The use of CF_4 in many different contemporary technologies such as plasma chemistry etching, thin layers dielectric deposition (plasma polymerisation) etc. requires reliable knowledge of the electron energy distribution function (EEDF) in a wide range of reduced electric field E/N and different gas mixtures. Simulations and calculations are based on solving the Boltzmann equation with the use of measured or calculated electron - atom or electron - molecule cross-section and EEDF. Measured data are of great importance as a base for justification in these cases. For CF_4 such measurements are far from enough and accurate because of the experimental difficulties arising due to the aggressive nature of the substance.

As far as we know, no experimental EEDF data are available in the literature. Below we report on EEDF Langmuire probe measurements. The EEDF can be determined with the help of the Druyvesteyn formula [1]:

$$f(\varepsilon) = \frac{2^{3/2} m_e^{1/2}}{e^3 n_e S} e^{1/2} \frac{d^2 I}{dV^2} \tag{1}$$

where m_e and e are the electron mass and charge, S is the probe area and $\varepsilon=eV$, where V is the probe potential with respect to the space potential. The second derivative of the probe characteristic I" can be obtained with the electrical modulation method [2]. This method does not directly measure I", but a convolution of I" with the apparatus function $\varphi(V-V')$:

$$\tilde{I}''(V) = \int_{-\infty}^{\infty} I''(V')\varphi(V-V')dV' \tag{2}$$

To correctly obtain $f(\varepsilon)$, a deconvolution procedure with known $\varphi(V-V')$ must be done.

Description of the experimental device

The quartz discharge tube (L = 0.8 m, r = 0.018 m) is shown in fig. 1. Since the components of the gas discharge (fluorine and carbon-fluorine radicals) are extremely aggressive, a continuous flow of gas was maintained with a minimum flow rate $5x10^5$ mol/l at a pressure p = 240 Pa. Special attention was paid to the choice of the materials. Using quartz for the tube and electrical insulation inside the device and graphite electrodes and reference probe we succeeded to diminish substantially the impurities in the discharge. By means of a the movable anode (5) the distance between electrodes was varied. It was used to measure the mean longitudinal electrical field E(x) and to define the value of the reduced field E/N (N being the particle density in the discharge).

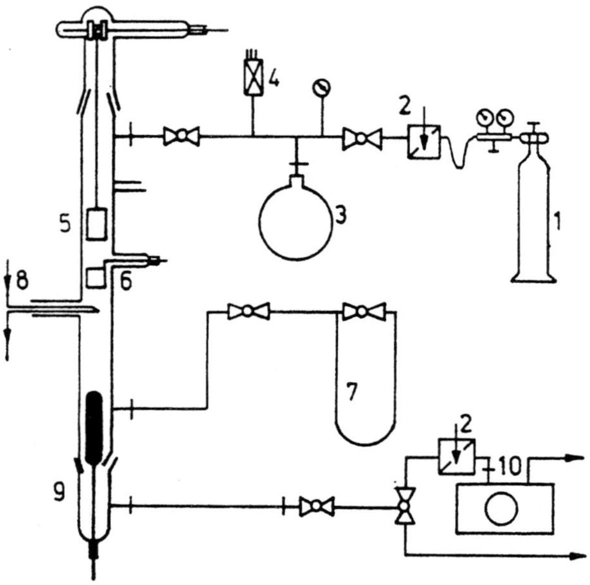

Figure 1.:1- gas bottle, 2- inlet valve, 3- buffer, 4- thermocouple manometer, 5- anode, 6- reference probe, 7- manometer, 8- probe 9- cathode and 10- vacuum pump.

The most difficult problem was the construction of the measuring probe. At low temperature of the probe (same as the temperature in the discharge) the chemical products from the discharge are covered the surface of the probe. The hypothesis of the existence of a polymerized film on the surface was proposed by some authors [3,4] to explain why the signals can not be registered when the probe is not heated. In the same time, heating the probe at a temperature higher than 600K resulted in a rapid destruction of the material of the probe even when platinum was used. To obviate this controversy we used an electrically heated tungsten wire (ϕ = 4.4x10^{-5} m) which was axially moved through the discharge in such a way that the time of the exposure of its effective part to the influence of the chemical reactions in the central zone of the discharge was kept short. This uninisulated part of the wire was long 3x10^{-3} m and it moved with velocity ~10^{-3} m/s.

First we tried to measure in pure CF_4 but the ratio noise/signal was too high. Therefore we added Ar as a buffer gas and operation in the mixture Ar+4%CF4 was acceptable.

A gas chromatography technique was used to check the concentration of the components and their purity.

The second derivative of the probe characteristic I" was measured by superimposing a low frequency (ω= 3kHz) voltage $\Delta V = V_0 \sin(\omega t)$ on the DC probe potential V. Then the magnitude of the circuit current with a frequency 2ω is proportional to I"(V). The signal was detected by means of a selective amplifier, lock-in-amplifier and recorded by computer.

In our case the apparatus function is [5,6]:

$$\varphi(V-V')=\frac{4}{3}\frac{e^2}{m^2}V_0\left[1-\left(\frac{V-V'}{V_0}\right)^2\right]^{3/2} \qquad |V-V'|\le V_0$$
$$\varphi(V-V')=0 \qquad |V-V'|>V_0 \tag{3}$$

It is known that fluctuations of the plasma potentials lead to a distortion and broadening of the apparatus function. The experimental check of any presence of such effects was made using the following procedure. The second derivative of an ideal diode [7] I(V) - chart was recorded while the plasma noise signal was applied upon it. Fig. 2 shows that the theoretical curve (expressions (3) is a good fit of the experimental points.

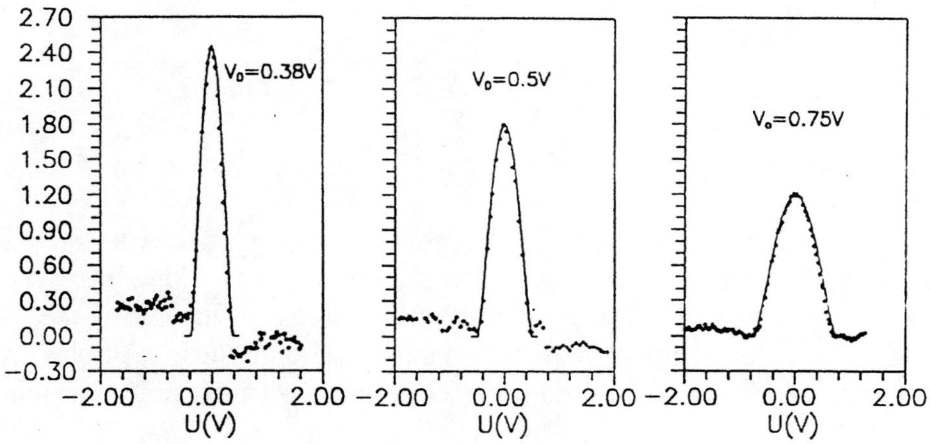

Figure 2. :Apparatus function for different magnitudes of the low frequency signal V_0.

Results and discussion

Results of the measurements performed at a gas pressure p=240 Pa, discharge currents I=30 mA (E/N=22 Td), and 50 mA (E/N=19 Td) are shown on Fig. 3. One can observe a big distance (in volts) between zero and maximum of the second derivative of the electron probe current as well as the local minimum in the vicinity of 1.5V. Similar results are obtained for Ne [8] when not sufficiently heating the probe. It is shown that this influences mostly the curve of the second derivative of the electron probe current in the area from zero to maximum of the signal while the rest of the curve is not distorted. In our case we think that the chemical reactions on the surface on the probe during the measurements are the actual reason of the distortion of the signal .

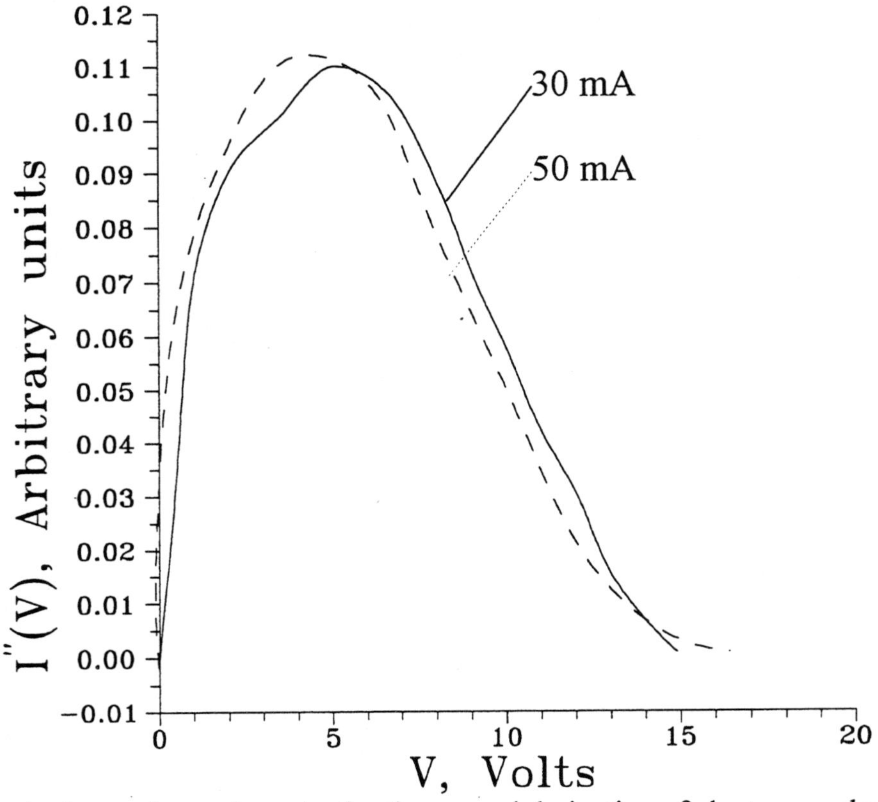

Figure 3. Typical experimental results for the second derivative of electron probe current

In the calculations the experimentally measured noise was eliminated in the following manner: Several knots were chosen on a appropriate places along the energy axis. Then the EEDF were approximated by a cubic spline (ln f_0 as a function of the energy). Then the sum of squared deviations f_0^{exp}/f_0^{calc} was minimised until the best fit was obtained

The Boltzmann equation was solved using a two-term hydrodynamic approximation. In the collision integral the following terms were considered:

1. Effective elastic electron - molecule collisions (the sum of the elastic and rotational inelastic scattering;
2. Vibrationall and electronic excitation of the molecule by electron impact.
3. Electron molecule ionization and attachment (including dissociative attachment);
4. Electron - electron and electron - ion collisions.

The Ar and CF_4 cross sections were taken from [9].

Up to now we had not measured the density populations of the excited levels, (an input parameter in the Boltzmann equation calculations). That is why we solved two boundary cases:

a) There are no excited particles in the discharge (see curve a) on Fig. 4).

b) Thermodynamic equilibrium - the excited levels are populated according Boltzmann's law (see curve b) on Fig. 4).

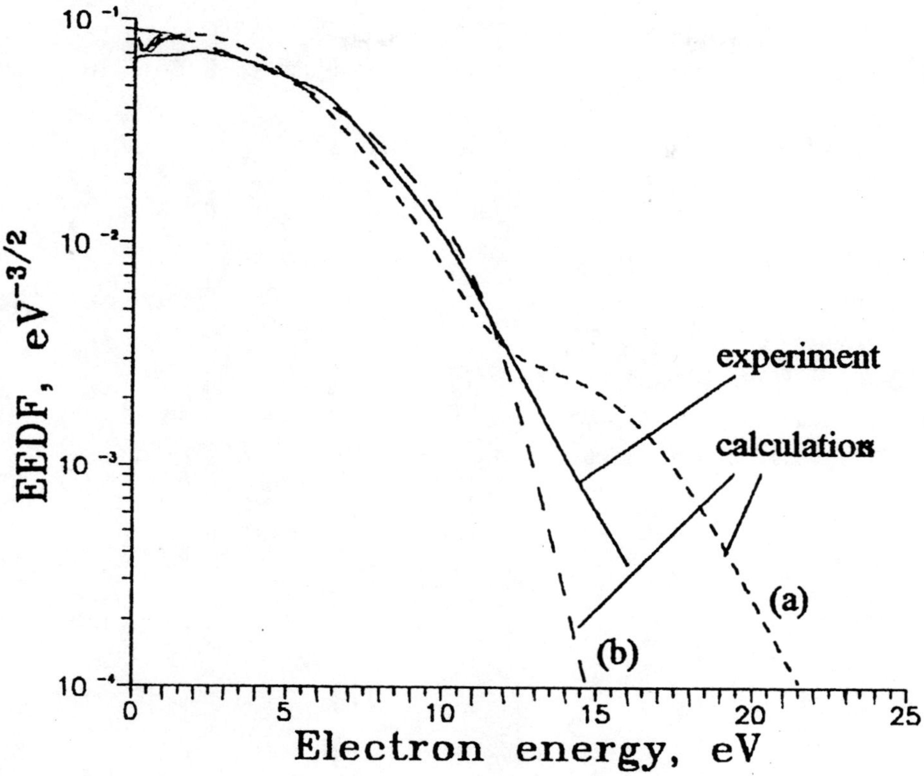

Figure 4. Comparison between the experimental curve and the calculations for E/N = 21 Td.

Since there are a few possible mechanisms changing the zero position of the EEDF (see the discussion above), we can not take into account the measured data at low energy. In order to compare the experiment with the calculations (Fig. 4) we used the following procedure:

1. Replacement the experimental zero position to some positive value.
2. Normalization the obtained EEDF $\left[\int f_0 \sqrt{\varepsilon} d\varepsilon = 1\right]$
3. Comparing this EEDF with the ones calculated (see curves a) and b) on Fig.4))
4. Repeat steps 1. to 3. until best coincidence is achieved.

Conclusion

The final result satisfactorily agrees with the solution of the Boltzmann equation in the range 0 to 12 V.

The shape of the EEDF at high energies (11 to 17 eV) is a source of information concerning chemical reactions in gas discharges. The results will be published in the future.

This work was initiated by Prof. B. Stefanov and supported by Bulgarian National Research Foundation (Contract F-7).

References

1. Druyvestein M.J. Der Niedervoltbogen. Ztschr. Phys., **64**, S.781, 1930.
2. Boid R.L.F., Twiddy N.D., Electron energy distribution in plasmas. I. Proc. Roy. Soc., **A250**, 53, 1959.
3. Thomas T.L., Battle E.L. J. Appl. Phys., **41**, p 3428, (1970).
4. Klagge S., Beitr. Plasmaphys. **15**, S309-317, (1975).
5. Blagoev A.B. et al. Sov. Phys. Tech. Phys. **45**, p 579, (1975).
6. Blagoev A.B. et al. Sov. Phys. Tech. Phys. **51**, p 2022, (1981).
7. Rutkowski G., Handbook of integrated-circuit operational amplifiers, Prentice-Hall Inc., Englewood Cliffs, New Jersey, USA, (1975).
8. Herrmann D., Klagge S. Beitr. Plasmaphys. **12**, S 17-25, (1972).
9. W. L. Morgan, *JILA Data Center Report No 34*, University of Colorado, National Institute of Standards and Technology Boulder, Colorado, (1991).

Vacuum Electron and Ion Technologies
D. S. Karpuzov (Ed.)

ISBN 1-56072-242-8

INVESTIGATION OF MULTI-FREQUENCY DISCHARGES BY OPTICAL AND ELECTRICAL METHODS

S. Hascík, J. Huran
Institute of Electrical Engineering, Slovak Academy of Sciences,
842 39 Bratislava, Dúbravská cesta 9, Slovakia

I. INTRODUCTION

Reactive ion etching of semiconductor materials has been expected to play an important role in the fabrication of eleectronic and opto-electronic integrated circuits. In order to accurately transfer patterns with minimum damage on the substrate, the accurate control of the both the ion bombardment energy and productions of etching species is extremely important. A multi-frequency discharge reactor [1] made by combining radio frequency (rf) system and low frequency (lf) system provides semiindependetly control of radical production and ion bombardment energy.

The purpose of this of this article is to present the efects of lf power on the self-bias voltage of substrate electrode and relative concentration of etching particles.

II. EXPERIMENTAL

The present experiments were caried out in a commercial parallel plate reactor XPL01.fy XENNON, reported previously [2], modificated in a multifrequency system. Fig. 1 shows a block diagram of the reactor approach used to apply two frequencies at the same time. Rf 13.56 MHz voltage is applied to the upper electrode and coupled to the lower electrode via high frequency ground . Additionaly lf (20,45,100kHz) voltage is applied to the lower electrode and coupled via a low frequency ground. Two lenses optical system mounted on table constructed for lift together with a 0.3 m monochromator BENTHAM M300 EA and a photomultiplier Hamamatsu R 955 selected light from chosen axis gap position and collected it onto an entrance slit of the monochromator (fig. 2). Measurements were performed at pressure 10 Pa with 30 sccm of CCl_2F_2 and 4% Ar for actinometry. The rf

(13.56 MHz) power was nominally 100 W. The lf power was varied from 0 to 100 W. The Cl 837.6 nm line and Ar 826.4 nm line were used for determining of relative ground concentration of Cl [3] and the F 685.6 nm line and Ar 706.7 line for F[4].

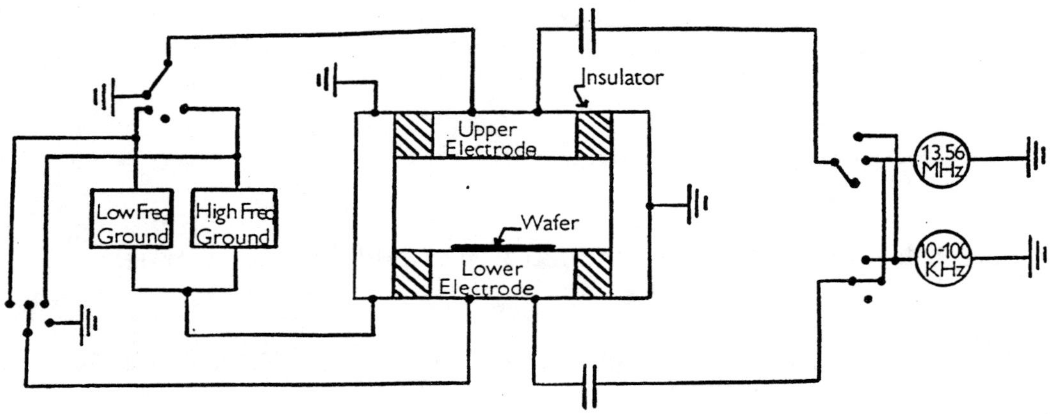

Fig.1 A block diagram of multifrequency discharge reactor

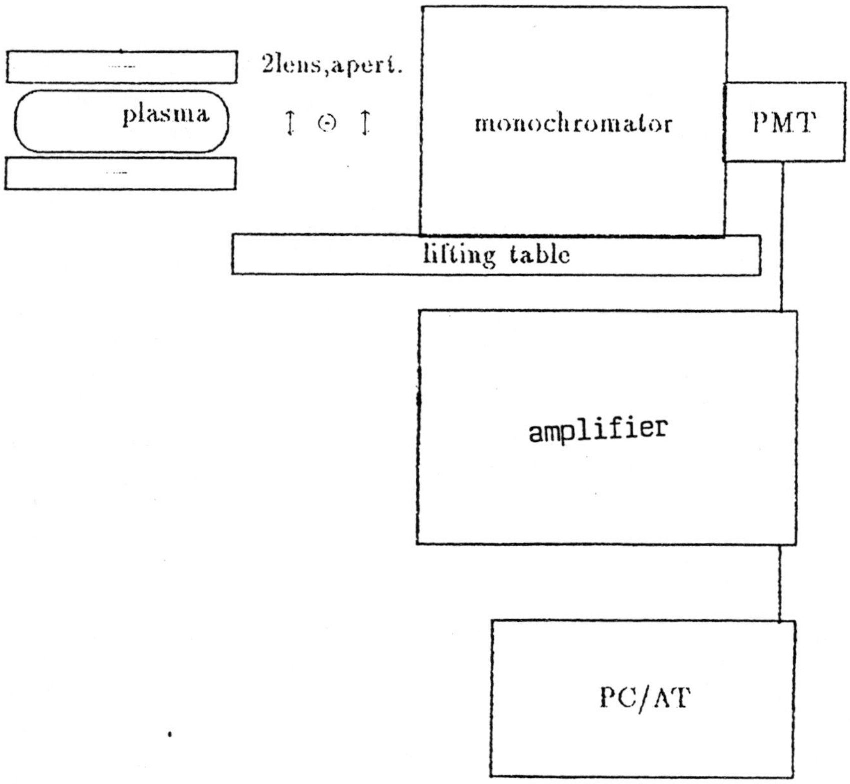

Fig.2 Experimental arrangement for actinometric measurements

III. RESULTS AND DISCUSION

Figures 3(a), 3(b) and 3(c) show the efects of varying the plasma generating lf power on the self-biases. When the plasma is generating by 13.56 MHz from

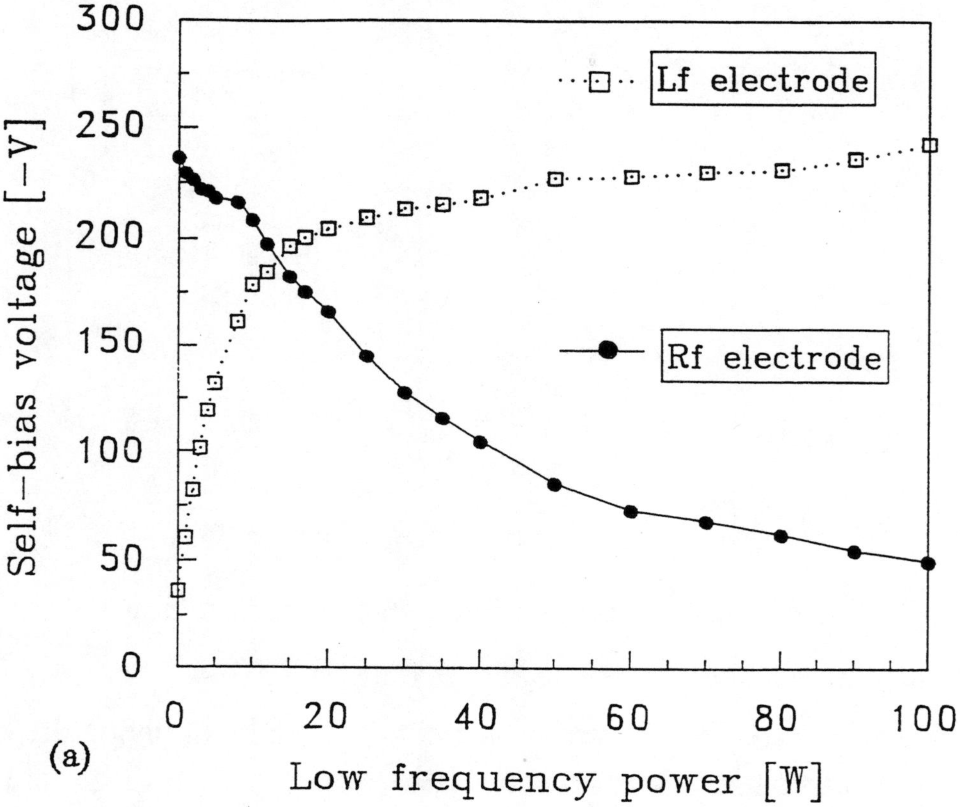

(a)

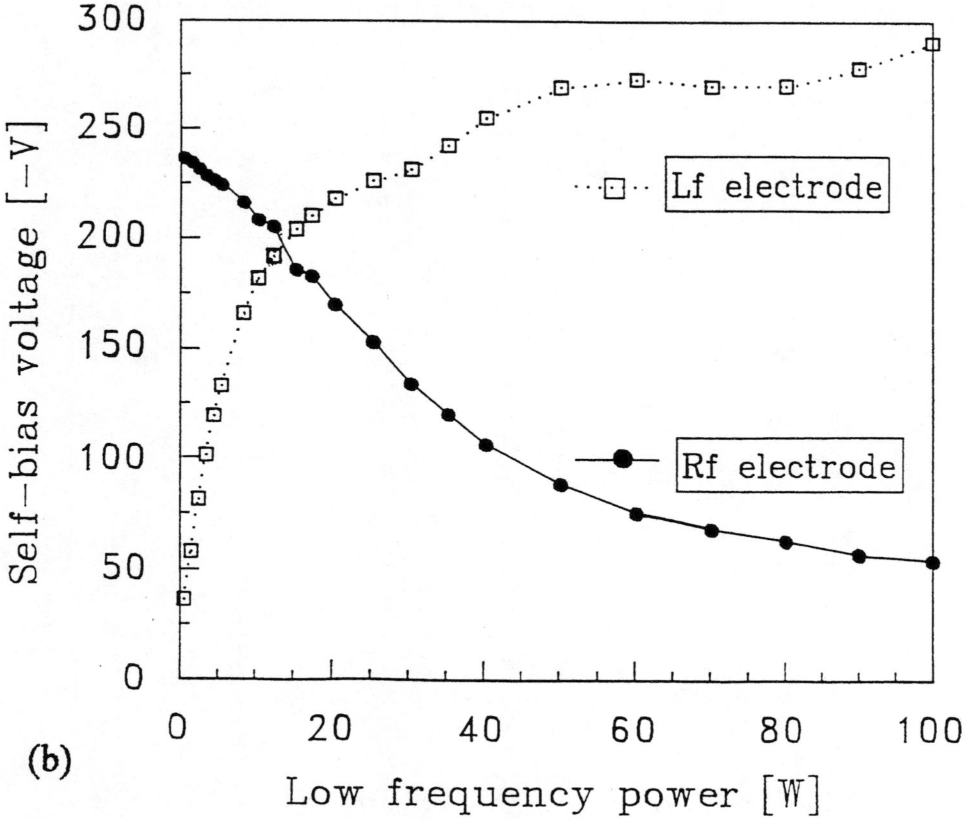

(b)

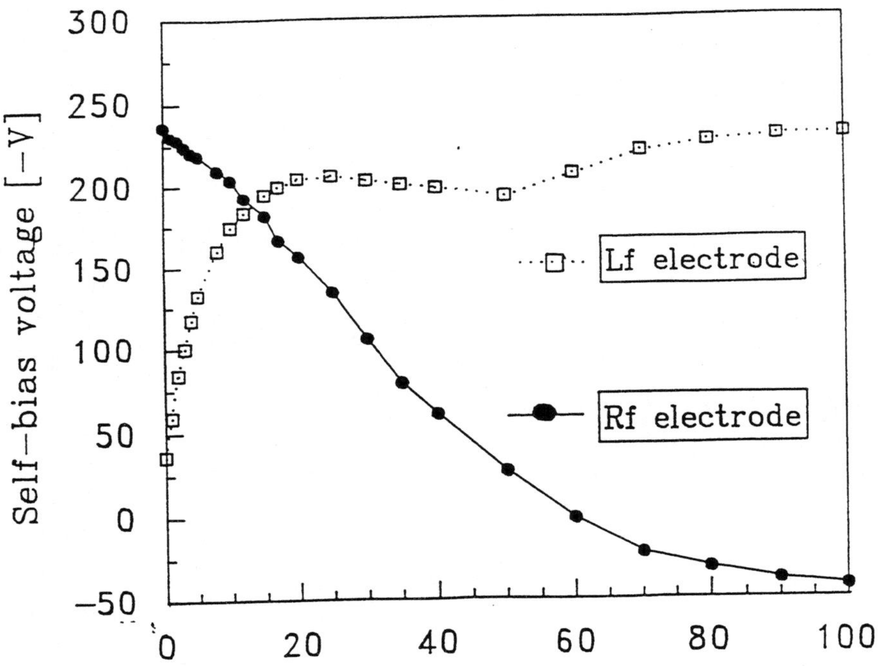

(c)

Fig.3. Self bias voltage as a function of lowfrequency power
(a) 20 kHz, (b) 45 kHz, (c)100 kHz.

the top electrode, the substrate self-bias can be varied by applying the lf power. The effects of varying lf power on the relative concentration of ground state Cl and F in the multifrequency CCl_2F_2 discharge are shown in figs. 4 and 5. At the

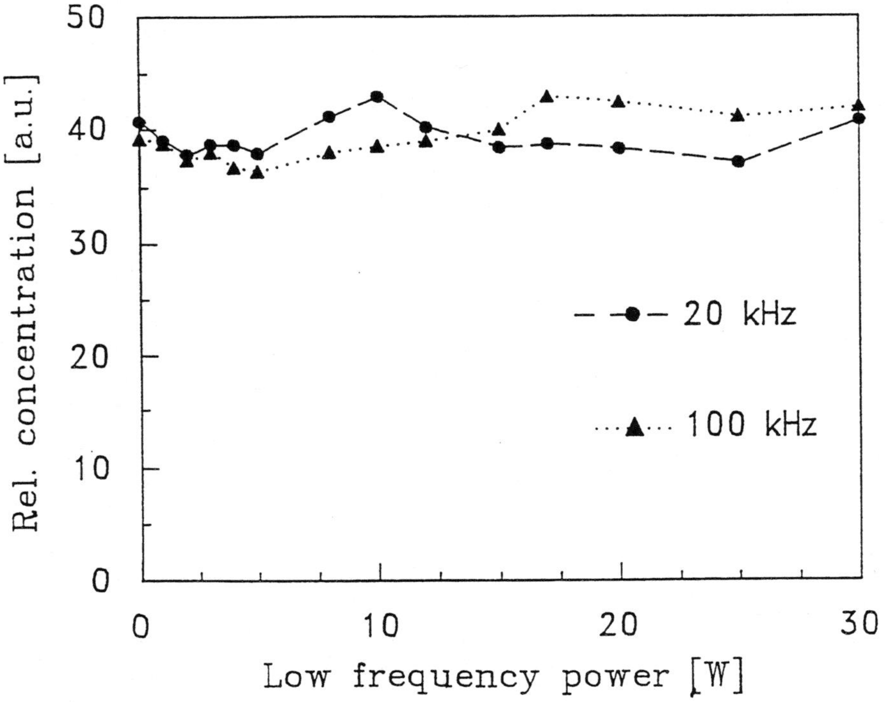

Fig. 4. Dependence of the rel.conc.of Cl on the lf power.

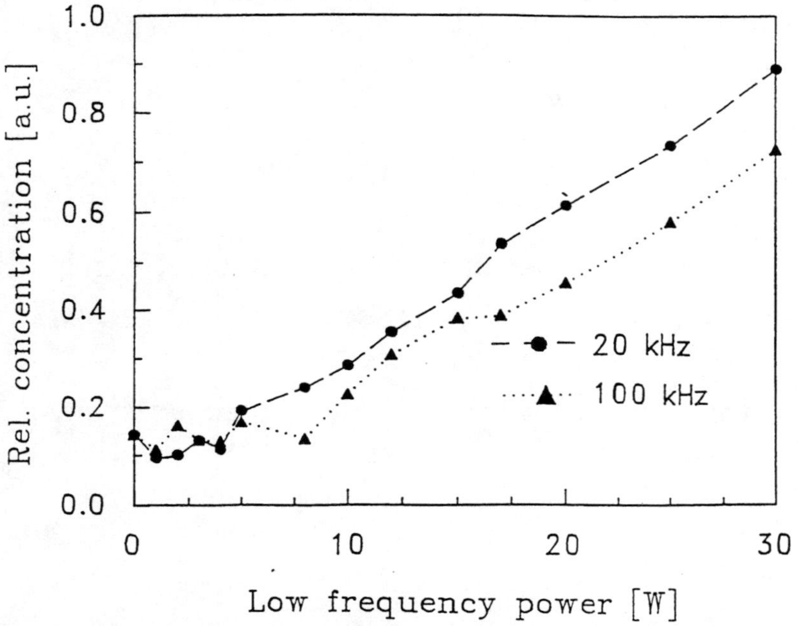

Fig.5. Dependence of the rel.conc.of F on the lf power

constant rf power an the Cl relative concentration is relatively independendt on the lowfrequency power . The relative concentration of F increases aproximately linearly with the power.

In order to achieve the indenpendent control of the ion energy bombardment energy and the ion flux ,the substrate lf biasing power should be kept as low as posible[5]. For example, the substrate self -bias voltage increases from -30V to -200V for changing of lf power from 0 to 30W (30% of rf power)(fig. 6).

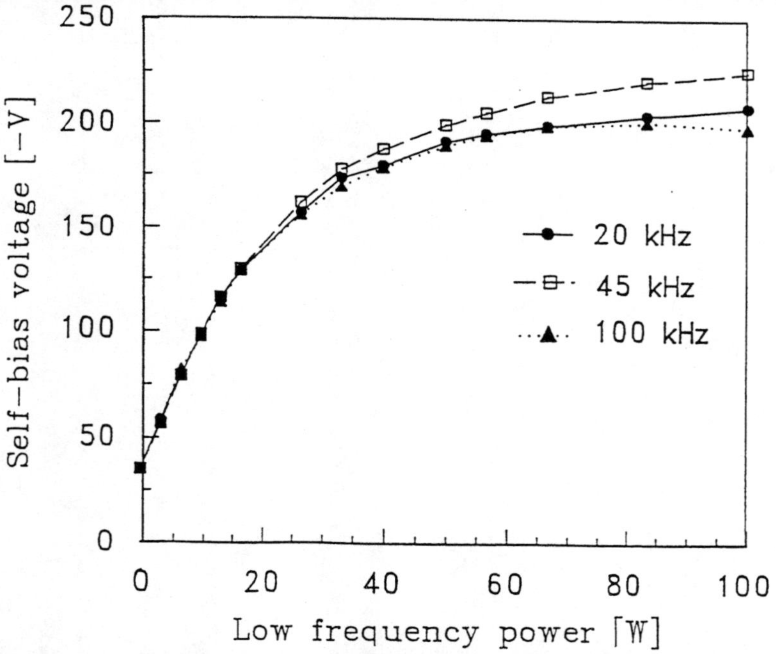

Fig. 6. Effects of lf power on self-bias voltages of substrate electrode.

The relative concentration of F increases about eight times at the same condition.

IV. CONCLUSIONS

The multi-frequency discharge reactor provides the control of ion energy bombardment and concentration of etching species in relatively wide range.

The authors are grateful to the Slovak Grant Agency for science (grant No 2/999056/93) for partial supporting of this grant.

REFERENCES:

[1] Ornellas S.P., De Ocampo J.P.: The Etching of Silicon in a Multi-Frequency Discharge Reactor. In: Proceedings of the Sixth Symposium on Plasma Processing (The Electrochemical Society) **6** , Pennington 1987 ,166

[2] Haščík Š., Horniaková A., Huran J.: Acta Phys. Slov. **41** (1991), 374

[3] Gottsho R.A., Donelly,V.M.:J. Appl. Phys **56** (1984), 245

[4] Heinrich F., Hofmann P., Muller K.P.: Microelect. Engn **13** (1991), 433

[5] Lowe H.D., Goto H.H., Ohmi, T.: J. Vac. Sci. Technol. **A9,** (1991), 3090

Vacuum Electron and Ion Technologies
D. S. Karpuzov (Ed.)

ISBN 1-56072-242-8

HIGH CHEMICAL ACTIVITY AND PASSIVATION OF ULTRADISPERSED PARTICLES SYNTHESIZED UNDER THE CONDITIONS OF ELECTRIC ARC LOW-TEMPERATURE PLASMA

G. P. Vissokov
Institute of Electronics, Bulgarian Academy of Sciences,
72 Zarigradsko Chosse boul., Sofia 1784, Bulgaria

1. Introduction

The high chemical activity, expressed by either intensive oxidation or spontaneous ignition in contact with atmosphere, represents a technological problem related to the preparation of ultradispersed particles (UDP). Plasmachemically obtained powders of tungsten [1-6], molybdenum [7-9[, rhenium, manganese [9-12], iron [9] and some catalysts [9,13-16] are pyrophores; they are synthesized by reduction of the respective oxygen compounds in low-temperature (LTP).

The present paper aims to explain the high chemical activity and, in some cases, the pyrophoric properties of plasmachemically synthesized UDP on the basis of experimental results and literature data as well as to establish effective methods for their passivation (stabilization).

2. Experimental procedure

Our extensive studies were directed to the obtaining of ultradispersed manganese by subjecting a manganese oxide concentrate to reduction by hydrogen, ammonia or a propane-butane gaseous mixture under the conditions of electric arc LTP [9-12]. The content of manganese oxides - MnO_2, Mn_2O_3 and Mn_3O_4 (Mn^{2+} - 43.1%, Mn^{4+} - 23%) in the starting concentrate was up to 99.9%, with particle size distribution (mass %) as follows: below 1μm - 72.7; 1-2 μm - 21.1; 2-3 μm - 3.0; 3-5 μm - 1.9; 5-7 μm - 0.6; 7-10μm - 0.2. The specific surface of the starting concentrate was 5.7 m^2/g. Investigations were carried out by a plasmachemical installation described in detail in [9,11]. A plasma jet with necessary power of up to 5 kW was generated by a direct current plasmatron. Ar, H_2, N_2 or mixtures thereof were used as plasma-forming gases. The plasma jet and the reactants were mixed in a mixing chamber. The plasmachemical reactor was 150 mm long, with channel diameter of 12 mm. The residence time of the reactants in the channel was about 10^{-3} s. The products of the plasmachemical process were fixed by an abrupt decrease of the temperature in a quenching device where an inert gas or a reactant were fed radially through three nozzles; the quenching rate was 10^6 K/s. The products of the plasmachemical reaction were cooled to room temperature in a dust-chamber where about 70 % of the ultradispersed solid phase was deposited. The remaining amount of the powder was deposited in the bag filter at the outlet of the heat exchanger-dust collector. Two modifications of the plasmachemical reactor were used. In the first case the reactor was made of copper and its walls were water-cooled (wall temperature - 300 K) while in the second case a refractory material was used without cooling of the walls; their temperature was in the range 1000-1500 K depending on the process regime. The average temperature of the plasmachemical resctor was determened calorimetrically.

Investigations for the estabilishment of reductions of a molybdenite concentrate (Mo - 34%, S - 30%, Cu - 3%, Fe - 5 %, Re - 0.05%, SiO_2 and Al_2O_3 up to 100 % by mass, specific surface 4 m^2/g) were also carried out in view of the application of the product in powder metallurgy [7-9]. The installation used comprised an electric arc plasmatrom, plasmachemical reactor, quenching device and dust-chamber. Argon (technical grade) was used as plasma-forming gas while butane, ammonia or hydrogen served as reducing agents and dust carriers. Quenching was effected by the cold walls of the dust-chamber as well as by argon blowing.

A similar installation was used for the study of the reduction of iron oxides, and more particularly Fe_2O_3, by hydrogen under the conditions of electric arc LTP as well as for the synthesis and/or regeneration of numerous catalysts finding application in ammonia production. There are catalysts for steam conversion (reforming) of natural gas [9,16,18], for low-temperature steam conversion of carbon oxide [9,16,17,19,20] as well as for ammonia synthesis [9,13,15-17,20-22].

The plasmachemical reactor with "cold" or "hot" walls served for the synthesis of UDP of AlN [9,13,23,24] and Mg_3N_2 [9,13,25] using elemental aluminium or manganese and nitrogen (99.99%) as raw materials.

The synthesized UDP were characterized specific surface using the method of Klyachko-Gurvich [26], their particle size was determined by electron microscopy, their phase and structural composition was studied by X-ray diffraction (Mössbauer spectroscopy [21,22] as well as by derivatographic and elemental analyses. Reduction and activation of catalysts were studied by the flow method.

3. Results and discussion

The experimental results on the reduction of the manganese oxide as raw material by H_2, NH_3 or CH_4 in a cold-walled reactor were unsatisfactory since reduction rate was low. This is basically due to high radial temperature gradients in the plasmachemical reactor. Furthermore, intensive oxidation of UDP was observed upon its removal from the dust-chamber. The attempts to passivate the final product by means of nitrogen containing 0.5 % O_2 or by CO or CO_2 do not give better results. The presence of nitrogen in the gaseous phase (an argon/nitrogen mixture is used as plasma-forming gas) prevents the spontaneous ignition of the particles due to surface formation of manganese nitrides Mn_3N_2, Mn_4N, Mn_xN) on the ultradispersed manganese (Table 1)

Best results are obtained with the hot-walled reactor using nitrogen as a passivation agent. The reducing agent-hydrogen is used also as a powder carrier. The results obtained [9,12] show that the process efficiency, expressed as reduction rate of the manganese oxide, is higher than 50 % and a highly dispersed product is obtained; its particle size is below 100 nm while its specific surface is of about 80 m^2/g (Table 1).

The basic reasons for the decrease of the process efficiency (the reduction rate) are the spontaneous ignition and the low passivation rate of UDP. In our opinion, due to effective quenching ($dT/d\tau \sim 10^6$ K/s), a complete reduction of the oxides to elemental manganese takes place in the plasmachemical reactor,

Table. 1. Some technological and physicochemical properties of chemically active ultradispersed powders synthesized under the conditions of electric-arc, low-temperature plasma

Ultradispersed substance	Basic reaction of plasma-chemical synthesis	Temperature range in the plasma-chemical reactor, K	Particle size, nm	Specific surface, m^2/g	Passivation agent	Composition of the protective layer	Ref.
Mn	$Mn_xO_y + yH_2 \rightleftarrows xMn + yH_2O$	2000–4000	<100	up to 80	N_2	Mn_3N_2, Mn_4N, Mn_xN	9–12
Mo	$MoS_2 + 2H_2 \rightleftarrows Mo + 2H_2S$	2000–4000	<100	20–380	CO	$Mo(CO)_6$	7–9
α-Fe, ϑ-Fe	$Fe_xO_y + yH_2 \rightleftarrows xFe + yH_2O$	2000–3000	10–100	up to 160	N_2 (0.5% O_2) CO	Fe_xO_y, $Fe(CO)_5$ $Fe_2(CO)_9$, $Fe_3(CO)_{12}$	9
CA-1 type catalyst for ammonia synthesis	$3Fe + 2O_2 \rightleftarrows Fe_3O_4$	1100–3400	20–60	10–40	N_2 (0.5–2% O_2) CO_2 (up to 2% O_2)	Fe_xO_y Fe_xO_y	9,13,15 16,17 20–22
Catalyst for low-temperature conversion of CO with H_2O	$(Cu,Zn,Al) + O_2 \rightleftarrows$ $\rightleftarrows CuO(Cu_2O) + ZnO + Al_2O_3$	up to 5100	10–40	45–51	N_2 (1% O_2)	CuO, Cu_2O	9,16,17 19,20
Catalyst for natural gas reforming	$(Ni,Al) + O_2 \rightleftarrows NiO + Al_2O_3$	2000–3000	10–30	up to 110	H_2 (1–2% O_2)	NiO	9,16,17 18
AlN	$2Al + N_2 \rightleftarrows 2AlN$	3300–3800	50–70	60–100	annealing to 1000 K	–	9,13,23, 24
Mg_3N_2	$3Mg + N_2 \rightleftarrows Mg_3N_2$	2000–2500	10–60	up to 180	annealing to 900 K		9,13,25

which in turn fixes a non-equilibrium crystal structure of UDP. The elimination of the pyrophoric properties of plasmachemically ultradispersed manganese could be achieved by (i) annealing of UDP in an inert atmosphere (Ar, N_2) leading to the occurrence of recrystallization processes which reduce to a great extent the non-equilibrium states in the crystal lattice and the structural defects of manganese, (ii) storage of the reaction product in an inert atmosphere or in an organic liquid and (iii) complete surface nitration or oxidant of the ultradispersed manganese particles.

Treatment of a molybdenite concentrate with butane in a cold-walled reactor results in a reduction rate of 83.5 % while the specific surface of the product varies from 20 to 73 m^2/g. The use of hydrogen as a reducing agent leads to a rate of 90 % and the specific surface amounts to 380 m^2/g (Table 1, [7-9]).

In our opnion, complete reduction of the concentrate under the conditions of electric arc LTP could be achieved by (i) effective quenching, i.e. abrupt decrease of the temperature of the reaction products below 500 K so as the reaction $Mo + 2H_2S \rightleftarrows MoS_2 + 2H_2$ does not take place. Experimental results [7-9] under the same conditions show that the reduction rate is not influenced by using gas quenching. Due to the high chemical activity of ultradispersed molybdenum, the above reaction takes place in the dust-chamber. (ii) Maintainance of a temperature higher than 1200 K in the dust-chamber when the reaction equlibrium is shifted to Mo and H_2S. A maximal reduction rate of 93 % is achieved in this case. It should be noted that the amount of the reducting agent should be always much higher than the stoichiometrical one.

The high chemical activity (pyrophoric properties) of the product obtained by reduction of molybdenite concentrate with hydrogen is undesirable since ultradispersed molybdenum is oxidized to MoO_3 by atmospheric oxygen and the specific surface decreases abruptly. These unfavourable conditions are created mainly in the dust-chamber. In order to overcome the spontaneous ignition of ultradispersed molybdenum, passivation with CO is carried out, resulting in the formation of a thin surface layer of $Mo(CO)_6$. The latter prevents further diffusion of oxyden molecular in the volume of the particles. The molybdenum carbonyl thus obtained sublimated at 310-320 K and decomposed within the temperature range of 420-670 K. It should be noted that bulk molybdenum reacts with carbon monoxide at 420 K and a pressure of 20 - 25 MPa.

Our studies [9] of the reduction of a mixture of iron oxides, and more particularly of Fe_2O_3 with hydrogen under conditions of the electric arc LTP show that a reduction rate of almost 100% is achieved; the specific surface of the extremely pyrophoric iron amounts to 160 m^2/g. X-ray structural anallysis and Mössbauer spectra show the presente of α-Fe and of the ultraparamagnetic ϑ-Fe in UDP. By the rise of the quenching rate from 10^5 to 10^8 K/s, the ϑ-Fe content increases from about 50% to almost 100%. Thus the plasmachemical mehod could be applied to the oreparation of ultradispersed iron in the high-temperature ϑ-modification. The extremely high chemical activity of the latter is prevented by treating it in a flow of nitrogen containing 0.5% O_2, resulting in the formation of a thin surface layer of a complex mixture of iron oxides (Fe_xO_y, Table 1). A passivation effect is observed by treating the ultradispersed iron particles with carbon monoxide, at which surface layers of several molecules of $Fe(CO)_5$, $Fe_2(CO)_9$ and even $Fe_3(CO)_{12}$ are formed. In order to obtain $Fe(CO)_5$ by heating of an iron powder with particle size in the micron range, a temperature of about 420-470 K and pressure of about 10 MPa are required. It is clear that the high chemical activity of the plasmachemically obtained ultradispersed iron with particle size of 10-100 nm (Table 1) is the reason for the formation of iron carbonyls at room temperature and atmopheric pressure.

The plasmachemically synthesized and/or renerated catalysts of the CA-1 type for ammonia synthesis show a catalytic activity that is higher by 15-25% than that of their coventional analogues; furthermore, their reduction is 3 to 5 times faster [9,13,15-17,20-22]. Regardless of the reduction method (isothermal or stepwise), they show an extremely high chemical activity (pyrophoric properties) in contact with air. The most common method of passivation of reduced catalysts for ammonia synthesis is their treatment with nitrogen containing 0.5 to 2% (by volume) of oxygen; elemental iron is thus oxidized to Fe_xO_y (Table 1). According to A.V. Krilova [27], three forms of oxygen can participate in the passivation of reduced conventional catalysts for ammonia synthesis, i.e. slightly chemisorbed oxygen (layer of 0.5 to 1 molecule) which is separated by heating of the catalyst to 423 K regardless of the nature of the carrier gas; strongly chemisorbed oxygen (a layer of 2 to 4 molecules) which leaves the sorption centres at 500-800 K in helium by diffusion in the bulk; oxygen in the form of oxides (4-6 molecular layers) which separates at 500-800 K only by hydrogenation in H_2 atmosphere. The particular form of sorbed oxygen could be determined by the activation energy of hydrogenation; this value is E = 77.8±6 kJ/mol in the case of plasmachemically synthesized and/or regenerated catalyst for ammonia synthesis and it corresponds to the heat of iron oxidation, i.e. chemisorption occurs during passivation. Carbon dioxide (technical grade) containing up to 2% (by volume) of oxygen is used as a passivation agent. Similar to the case of passivation with nitrogen containing up to 0.5 vol. % oxygen, complex oxide mixtures are formed on the surface of the catalyst particles (Fe_xO_y, Table 1).

The derivatographic study of passivated plasmachemically synthesized and/or regenerated catalysts after exploitation shows separation of water upon heating (endothermal effect below 373 K) and initation of oxidation of the CA-1 type catalyst in air at 380 K (appearance of an exothermal effect). Maximum oxidation rate is observed at about 450 K.

It should be noted that cooling in air down to the evaporation temperature of liquid nitrogen (77 K) results in the removal of the pyrophoric properties of the plasmachemically synthesized and/or regenerated reduced catalysts for ammonia synthesis, low-temperature steam conversion of carbon monoxide or for methane reforming while the initial temperature of oxidation is increased by about 20 K.

Plasmachemically synthesized ultradispersed aluminium nitride (Table 1, [9]) is more chemically active than the bulky one but it does not show pyrophoric properties. In humid atmosphere it undergoes slow hydrolysis according to the reaction $AlN + 3H_2O \rightleftarrows Al(OH)_3 + NH_3$ and diffraction maxima of $Al(OH)_3$ are observed in the X-ray diffractograms in this case. After annealing to about 1000 K in pure nitrogen the chemical activity of the nitride is somewhat decreased. However, plasmachemically synthesized ultradispersed aluminium nitride should be stored in tightly closed containers.

Plasmachemically synthesized magnesium nitride [25] shows a high reactivity and it is rapidly hydrolyzed in humid atmosphere: $Mg_3N_2 + 6H_2O \rightleftarrows 3Mg(OH)_2 + 2NH_3$. Its gradual heating to 900 K followed by cooling to room temperature decreases to a great extent the chemical affinity to water vapours. Nevertheless, plasmachemically synthesized magnesium nitride sould be stored in inert atmosphere (Ar, He or Ne free of oxygen and water vapours).

The high activity of plasmachemicalli synthesized UDP with respect to atmospheric oxygen and water vapours can be explained by the peculiarities of of their crystal structure formed under the conditions of LTP [28]. The abrupt cooling upon quenching of the plasma jet leads to the formation of a less perfect crystal lattice with a large number of vacancies. Migration of the metal atoms in the defect nodes is hampered by the low-temperature. As a rule, defective crystal structures are characterized by an enhanced surface activity [28] and for this reason oxygen is actively chemisorbed on the ultradispersed metal powders. There are no conditions of rapid chemisorption head transfer and for this reason the metal particles are heated to the temperature of rapid oxidation; increased dispersity enhances the occurrences of these processes.

An equation is derived from the heat balance of powder oxidation [29].

$$T \geq \frac{E}{2R\ln\dfrac{A^{1/2}\Delta H m^{2/3}\theta^{1/3}}{K\lambda p d \tau^{1/2}}}$$

relating the temperature of spontaneous ignition with the physicochemical parameters of the specimen. This equation satisfactorily coincides our experimental results; it show that the temperature of self-igni tion T increases with the rise of the activation energy of oxidation E, the heat conductivity of the powder λ, the density p and diameter d of the particles as well as with the heating duration τ and decreases with the rise of the preexponential multiplier A, the thermal effect of oxidation ΔH, the mass of the specimen m and bulk mass of the powder θ. Experimental data show that the impacts of the particle diameter, the specimen and the duration of heating qualitatively reply to the above equation. The exact quantitative check of the equation is however hampered by the absence of precise values of E, A and K.

4. Conclusion

A high chemical activity and even pyrophoric properties (in some cases) are observed with plasmachemically synthesized UDP of metals (Mn, Mo, Fe), oxide catalysts after reduction to an active form and some nitrides. Nitrogen, carbon monoxide or dioxide, inert gases containing small amounts of oxygen, metal halides, hydrocarbons, etc. can be used in each particular case as sorbates protecting plasmachemically synthesized UDP from an active interaction with oxyde. Unfortunately, the sorption of gases on the UDP surface can only slower the oxidation rate rather than to completely prevent this process. The high reactivity of the powders is decreased in numerous cases by thermal treatment in an inert gas flow or in vacuum. Chemisorption of gases on the surfaces of somemetals offers some advantages with respect to passivation since surface metal complexes can be formed, with composition similar to those of chemical compounds. As an additional requirement,these complexes should rapidly decompose during the further treatment of the powders, e.g. sintering. The carbonyls of numerous metals (Fe, Mo, etc.) prove appropriate in this respect. General theoretical concepts defining apriori the most efficient passivation agents for the various highly active plasmachemically synthesized UDP are still not formulated.

Acknowledgement
The X-ray structural and phase analysis performed by Dr. R. Radanov on a TUR-M 62 X-ray apparatus is gratefully acknoweledged

References

1. British Pat. No 1098246 (1968).
2. D.M. TCHIZIKOV, S.S. DEJNEKA, I.K. TAGIROV, I.Y BASIEVA and Yu.V. TZVETKOV in Issledovanie protzessov v metalurgii tzvetnoych i redkich metallov" (Nauka, Moscow, 1969) p. 41.
3. D.M. TCHIZIKOV and Yu.V. TZVETKOV, ibid. (NAUKA, Mosckow, 1969) p.54.
4. M.G. CHARTON and G. DAVIS, Nature 175 131 (1955).
5. T. MILLNER and A. HUGEDUS, Z. Anorg. Allgem. Chem 289 288 (1957).
6. J. NEUGEBAUER, in Proc. Int. Symp. Reinstoffe Wiss. Techn. Tanugber. 2nd (1965) p.755.
7. G.P. VISSOKOV and N.B. VIDENOV, "Sb. Dokladi ot symposiuma Fizikata v himitcheskata promichlenost" (MNP, Sofia, 1978) p.159.
8. Idem, Himiya I industriya 50 389 (1978).
9. G.P. VISSOOV, "Prilozhna plasmahimiya. Prilozhenie na niskotemperaturnata plasma v organitchata himitchna tehnologiya i metallurgiyata", Part 2 (Tehnika, Sofia, 1987 p.325.
10. G.P. VISSOKOV and N.B. VIDENOV, Godichnik na VHTI-Sofia 25 49 (1978).
11. G.P. VISSOKOV, N.B. VIDENOV and N.TZ. GERASSIMOV, ibid. 25 52 (1978).
12. G.P. VISSOKOV, N.Tz. GERASSIMOV and D.H. OLIVER, Himiya i industriya 52 408 (1980).
13. G.P. VISSIKOV, "Prilozhna plasmohimiya. Prilozhenie na niskotemperaturnata plasma v neorganitchnata himitchna tehnologya", Part 1 (Tehnika, Sofia, 1984) p. 294.
14. Idem, Godichnik na VHTI-Sofia 29 181 (1984).
15. G.P.VISSOKOV and G.P. GEORGIEV, ibid. 29 186 (1984).
16. G.P. VOSSOKOV, in "Plasmohimiya-89", edited by L.S. Polak, Vol. 1 (INHS USSR Academy of Sciences, Moscow, 1989) p.33.
17. Idem, Izvestyia po himiya BAN 16 275 (1983).

18. Idem, Godichnik na VHTI-Sofia 29 131 (1984).
19. G.P. VISSOKOV and O.V. JANKOV, "Sb. Geterogennii kataliz. Trudi V Mezhdunarodnogo symposiuma", Vol. 2 (BAN, Sofia, 1983) p. 219.
20. G.P. VISSOKOV in Proc. 6th Int. Symp. Heterogeneous catalysis (BAN, Sofia, 1987) Part 2 p.303.
21. T.M. PEEV, G.P. VISSOKOV, I. CZAKO-NAGY and A. VERTES, Appl. Catal. 19 301 (1985).
22. G.P. VISSOKOV, T.M. PEEV, I. CZAKO-NAGY and A. VERTES, ibid. 27 257 (1986).
23. G.P. VISSOKOV and L.B. BRAKALOV, J. Mater. Sci. 18 2011 (1983)
24. G.P. VISSOKOV, Silikattechnik 40 263 (1989).
25. G.P. VISSOKOV, L.B. BRAKALOV and K.D. MANOLOVA, Godichnik na VHTI-Sofia 29 190 (1984).
26. A.L. KLYACHKO-GUREVICH, Izv. USSR Acad. Sci. OHN 10 1884 (1961).
27. A.V. KRILOVA, D. Sc.Thesis (Moscow, 1982) p.40.
28. G.P. VISSOKOV, J. Mater. Sci. 27 5561 (1992).
29. B.G. Ioffe, Porochkovaya Metallurgiya 35 102 (1965).

Vacuum Electron and Ion Technologies
D. S. Karpuzov (Ed.)

ISBN 1-56072-242-8

INFLUENCE OF SiO_2-MONOMER GENERATION ON ULTRAFINE SILICA PRODUCTION IN FLOW PLASMA REACTOR

E. Balabanova, D. Oliver
Institute of Electronics, Bulg. Acad. Sci. boul. Tzarigradsko
Chaussee 72, Sofia 1784, Bulgaria

ABSTRACT

The production of ultrafine silica powder by a rapid cooling of silica sand vapour in a flow plasma reactor is considered. The particle formation and growth is described as a pure free-molecular coagulation taking place under non isothermal conditions in the reactor. The formation of ultrafine particles during the cooling of the vapour is accompanied by generation of a monomer (precursor of the particles), since vaporised silica is decomposed to SiO and O_2. The monomer, SiO_2 molecule, is obtained by SiO oxidation. The influence of the monomer generation rate on the process of coagulation is investigated. The study is performed by simulating the processes in the reactor.

INTRODUCTION

The silica sand can be used as a raw material for ultrafine silica production. The sand is vaporised by using thermal plasma. The obtained vapour at atmosphere pressure is mixed with oxygen and cooled in a flow plasma reactor. It is well known that SiO_2 is decomposed during the vaporisation to SiO and O_2 [1]. The cooling of the vapour leads to the monomer generation. The cooling agent (O_2) also plays the role of an oxidising agent. The kinetics of SiO oxidation is not well known. For this reason two models of the monomer generation are used. The effect of SiO_2 monomer generation on the particles' production, respectively on particle size distribution function (PSDF) is studied.

MODEL

The model is based on our experimental data on silica sand processing [2]. The vapour with mean mass temperature 3000 K is cooled rapidly to the temperature of SiO_2 condensation (2000 K), in a flow plasma reactor. A constant cooling rate of

10^6 K/s is accepted. The estimation of this value is in agreement with the concept about cooling rate given in [3]. The mixing of the hot vapour and the cooling agent is assumed to be instantaneous. At such cooling rate, the cooling time is 10^{-3} s. The vapour supersaturates at such a rapid cooling and coagulates. The free-molecular coagulation is the process responsible for particle formation and growth.

The discrete-sectional method of Rosinsky-Snow [4] is used to describe the coagulation. In this method particles are included in discrete groups (sections) with masses m, 2m, 2^2 m, ...,2^n m (m is the mass of the monomer). The Rosinsky-Snow equation of coagulation is

$$\frac{dz_n}{dt} = \sum_{l<n-1} 2^{l-n} C_{l,n} z_n z_l + \tfrac{3}{4} C_{n-1,n-2} z_{n-1} z_{n-2} + \tfrac{1}{2} C_{n-1,n-1} z_{n-1}^2 - \sum_{l=n-1}^{M} C_{n,l} z_n z_l \quad (1)$$

where z_n , z_l are the concentrations of particles in n-group and l-group (cm^{-3}), M is the number of groups and $C_{n,l}$ is the collision frequency function for collisions between particles of groups n and l,

$$C_{n,l} = \left(2^{n/3} + 2^{l/3}\right) \frac{1+2^{n-l}}{2^n} F(m,\rho,T) \quad (2)$$

$$F(m,\rho,T) = 2^{1/6} 3^{2/3} m^{1/6} \pi^{1/6} \rho^{-4/6} (kT)^{1/2} \quad (3)$$

where ρ is the mass density of the particles, (2.2 g/cm^3 for silica particles), k is Boltzmann's constant, T is absolute temperature.

The evolution of the monomer concentration is given by:

$$\frac{dz_0}{dt} = S - z_0 \sum_{l=1}^{M} C_{o,l} z_l - \lambda z_0 \quad (4)$$

Here S is a term presenting the source of the monomers; $C_{0,l}$ is the collision frequency function for collisions between the monomer and the other particles. On order to describe the variation of the particle concentrations with the variation of the volume during cooling (the pressure being assumed constant), we incorporated the term λz_n in equations (1), (4). The coefficient λ is determined from the cooling rate α, $\lambda = -\alpha/T$.

The accurate mechanism of the SiO oxidation (monomer generation) is not known. For this reason we use (i) one smooth monotonously increasing function for presentation of the monomer source, given by

$$S^{(1)} = \frac{[SiO_2]_0}{\theta}\left(1 + \cos\frac{\pi t}{\theta}\right) \quad (5)$$

where $[SiO_2]_0$ is the total initial monomer concentration, θ is the conversion (oxidising) time; (ii) the following parallel chemical reactions to present the monomer generation:

$$SiO + O_2 \xrightarrow{k_1} SiO_2 + O \tag{6}$$

$$SiO + O + O_2 \xrightarrow{k_2} SiO_2 + O_2 \tag{7}$$

$$O + O + O_2 \xrightarrow{k_3} O_2 + O_2 \tag{8}$$

The formulae for the chemical reaction rates are $w_1=k_1[SiO][O_2]$, $w_2=k_2[SiO][O][O_2]$, $w_3=k_3[O]^2[O_2]$. The term of the monomer source in equ. (4) is now $S^{(2)}=w_1+w_2$ .The evolution of the concentrations of the different chemical components is given by

$$\frac{d[SiO]}{dt} = -w_1 - w_2 - \lambda\,[SiO] \tag{9}$$

$$\frac{d[O]}{dt} = w_1 - w_2 - 2w_3 - \lambda\,[O] \tag{10}$$

$$\frac{d[O_2]}{dt} = -w_1 + w_3 - \lambda\,[O_2] \tag{11}$$

Since the chemical kinetics data related to SiO oxidation are not known, the estimation of the rate coefficients k_1 and k_2 on the basis of the rate coefficients of the corresponding reactions of CO oxidation to CO_2 is made. Such an estimate can be justified due to the same physic-chemical properties of gaseous silicon and carbon oxides and also due to the same linear structure of SiO_2 and CO_2 molecules [5]. The bond length of Si-O atoms [5] is greater than that of C-O atoms [6] and the binding energy for Si-O atoms in SiO [7] is lower than that of C-O in CO [6]. One can thus assume that the activation energy of SiO oxidation to SiO_2 is lower compared with that of CO oxidation.

The reaction (6) is more important reaction for SiO_2 production than (7), because w_1 is larger than w_2. Unfortunately, various expressions for rate coefficient of the reaction $CO + O_2 \xrightarrow{k} CO_2 + O$ corresponding to (6) are recommended. Sultzmann et al. [8] gave $k' = 5.62\times10^{-12}$ exp(-51000/RT) cm^3/s, Clark et al. [9] obtained higher rate coefficient which can be represented by $k'' = 4.88\times10^{-12}$ exp(-28630/RT) cm^3/s.

Our calculations are run both with k_1=k' and k_1 =k". The coefficients k_2 and k_3 are taken to be $1.55x10^{-34}$ cm^6/s [10] and $5.14x10^{-33}$ cm^6/s [11], respectively.

RESULTS AND DISCUSSION

Calculations of PSDF for both cases, $S=S^{(1)}$ and $S=S^{(2)}$, are performed and their self-preservation is checked. The self-preserving form of the functions is investigated by the evolution of the calculated mean diameters. Since the obtained PSDF are of a log-normal character their integral form are presented by probability function (erf-function). The calculated mean diameters are mean median diameters d_m, corresponding to the 50% probability of the PSDF.

At $S=S^{(1)}$ for the conversion time (θ) we used the values 0, 10^{-5}, 10^{-4}, 10^{-3} s. Zero conversion time corresponds to the case of instantaneous monomer generation, the conversion time 10^{-3} s corresponds to the case of slowest monomer generation. The total monomer concentration, $[SiO_2]_0$ is 10^{18} cm^{-3}, which corresponds to the concentration of the hot vapour.

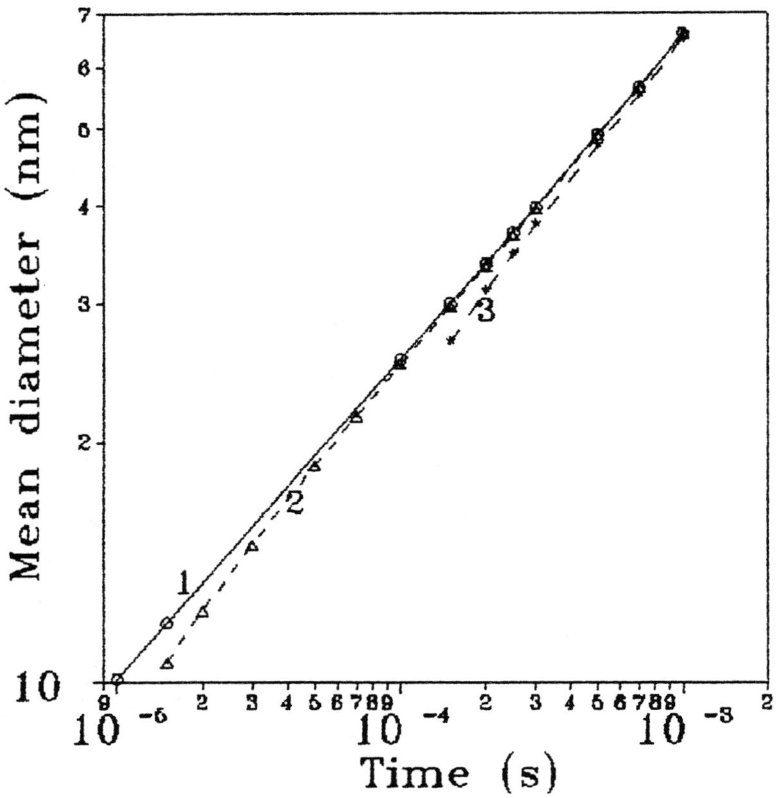

Figure 1. Evolution of the mean diameter for instantaneous monomer generation θ=0 (line 1), for non-instantaneous monomer generation ($S^{(1)}$) at $\theta=10^{-5}$ s (line 2), at $\theta=10^{-4}$ s (line 3)

Figure 1 shows the evolution of the mean diameters (d_m) of the particles accepting $S=S^{(1)}$. For $\theta=0$, d_m follows the relation $d_m \sim t^{0.4}$ presenting the Lai self-preserving law [12], nearly from the start of coagulation. For $\theta=10^{-5}$, 10^{-4} s, d_m deviates from the self-preserving line until times of about 5θ. When $\theta=10^{-3}$ s PSDF can not reach self-preserving form to the end of coagulation. The obtained results show that the monomer source influences the self-preservation of PSDF.

When $S=S^{(2)}$ the conversion rate of SiO to SiO_2 is calculated. It is determined by the ratio of the SiO concentration to the sum of the concentrations of all Si containing products. It is presumed that the oxidation is completed when the conversion rate drops below 1%. Its duration defines the oxidation time. The evolution of the conversion rates calculated for all initial ratios $SiO:O_2$ at both rate coefficients k' and k" is presented in Figure 2. The oxidation of SiO is not completed for the coagulation time for each of the $SiO:O_2$ ratios at k_1=k' as well as for ratio 1:1 at k_1=k". For the other ratios the oxidation is completed before the end of coagulation (the oxidising time is in the range of $4x10^{-5}$ to 10^{-4} s.

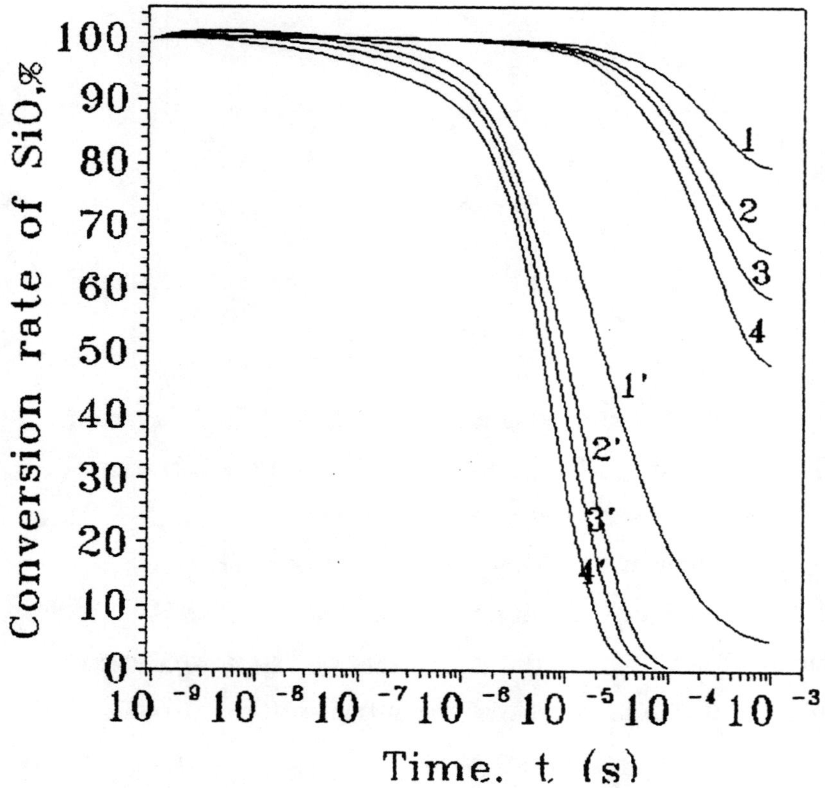

Figure 2. Conversion rate of SiO for $SiO:O_2$ = 1:1; 1:3; 1:5; 1:10 at $k_1 = k'$ (lines 1,2,3,4) and $k_1 = k''$ (lines 1',2',3',4')

From our experiments, as well as the experiments described in [13] which were carried out at nearly the same conditions, it is known that the products in the end of the reactor do not contain SiO. Therefore fast chemical monomer generation seems plausible.

The effect of the monomer generation rate on the evolution of PSDF is studied and it is obtained that PSDF can not reach self-preserving form when the monomer generation is slow (the oxidising time is longer than the cooling time). It is observed for all ratios $SiO:O_2$ at k_1=k' and for ratio 1:1 at k_1=k". PSDF are of self-preserving form when monomer generation is instantaneous ($k_1 \to \infty$) or fast (k_1=k" and $SiO:O_2$ >1:1), see Figure 3.

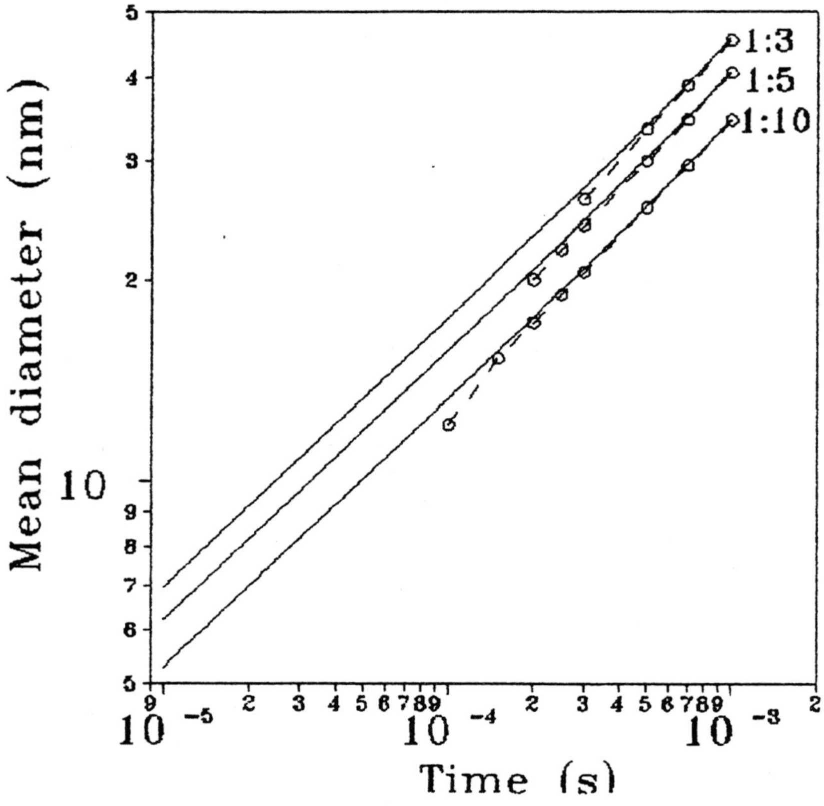

Figure 3. Evolution of the mean diameter for instantaneous monomer generation $k_1 \to \infty$ (solid lines), for $k_1 = k''$ (dashed lines).

The Lai self-preserving relation is followed nearly from the start of coagulation ($t=10^{-5}$ s) in the cases with instantaneous monomer generation. (see solid lines in Figure 3). A delay of self-preservation is observed when k_1=k". The smaller the ratio $SiO:O_2$, the longer is the delay(see dashed lines in Figure 3).

When instantaneous monomer generation was accepted earlier [14], it was shown that the final mean size of the particles depends on the initial monomer concentration. From Figure 3 it can be seen that the use of different initial ratios $SiO:O_2$ leads to different final mean sizes of the particles. The mean size is larger at smaller $SiO:O_2$.ratios. The final mean sizes calculated at $S=S^{(1)}$ are larger compared with the mean sizes calculated at $S=S^{(2)}$. The reason is the smaller $SiO:O_2$ ratio (1:0.5) corresponding to the oxidation at stoichiometric ratio of the components.

The experimental final mean size obtained at ratio $SiO:O_2$=1:10, d_m=43 nm [14] shows a satisfactory agreement with the calculated one.

CONCLUSIONS

The rate of SiO_2-monomer generation affects the evolution of PSDF. The slower is the generation, the longer is the delay of establishing of self-preservation. The calculations carried out for different oxidation times at $S=S^{(1)}$ as well as for different values of rate coefficients and ratios $SiO:O_2$ at $S=S^{(2)}$ show that the SiO_2 monomer generation seems to be a process faster than the coagulation of the particles.

REFERENCES

1 Schick H.L.: Chem. Rev. 60 (1960) 331

2 Oliver D.H., Balabanova E. G., Vrantchev A.P, Vissokov G.P., EnikovR.Z.: 14th Czechoslovak seminar in plasma physics and technology, Research reports part 1 p. 118, July 1987

3 Sundstrow D.W., De Michiell R.L.: Int. Eng. Chem. Process Des. Dev. 10 (1971) p.114

4 Rosinsky J., Snow R.: J. Meteorology 18 (1961) 1736

5 Toropov N.A., Barzakovsky V.P.: High temperature chemistry of silica and others oxide systems, (in Russian), Publ. Acad. Sci. USSR, Moscow, 1963

6 Moelwyn-Hughes E.A., Physical chemistry, Pergamon press, London-New York-Paris (1961) p.906

7 Huczko A., Meabus P.: ISPC-8, Tokyo, 1987, 2099

8 Sultzmann K.G.P., Myers B.F., Bartle E.R.: J. Chem. Phys. 42 (1965) 3969

9 Clark T.C., Garnett S.H., Kistiakowsky G.B., ibid. 51 (1969) 2885

10 Kondratev V.N. Intesarova E.I.:J. Chem. Kinetics 1 (1969) 105

11 Solonkhin R.I.: Combustion and Flame 11 (1967) 489

12 Gans I., Gauvin W.H., ISPC-8, Tokyo, 1987, 2387

13 Lai F.S., Friedlander S.K., Pich J., Hidy G.M.: J. Colloid Interface Sci. 39 (1972) 395

14 Balabanova E.G., Oliver D.H., Levitsky A.A.: 11 ESCAMPIG, St. Petersburg, Aug. 25-28, 1992, 410

Vacuum Electron and Ion Technologies
D. S. Karpuzov (Ed.)

ISBN 1-56072-242-8

STATISTICAL STUDIES OF THE ION THERMOEMISSION FLUCTUATIONS

Longin Gladyszewski
Institute of Physics, Maria Curie
Chaussee 72, Sofia 1784, Bulgaria

Abstract

This work concerns research on fluctuations (noises) of the ion thermoemission currents of the alkali metals and rare earth atoms emitted from the tungsten and rhenium surfaces. These noises are generated as a result of adsorbate density fluctuations.

The spectral density functions and their temperature dependence are discussed in terms of the surface diffusion noise model at adsorption-desorption equilibrium.

Method

The method is based on the relationship between the number n of atoms adsorbed on the emitter and the ion current i emitted from this emitter. Due to random migration and desorption of adatoms, n undergoes fluctuations dn around its mean value. These fluctuations are reflected in work function $d\phi$ and finally - in ion current fluctuations di, $di = i(t) - i_o$.

The mean square of the ion current fluctuation $\langle di^2 \rangle$ is given by:

$$\langle di^2 \rangle = a\, i_o^2\, (\pi\, e\, \mu/\varepsilon_o kT)^2\, n\, D\, \tau\,, \qquad (1)$$

where a is a coefficient depending on the geometry of the ion source, i_o is DC ion current intensity, μ the dipole moment of the adsorbed atoms, ε_0 the permittivity of vacuum, n the concentration of adsorbed atoms, D the diffusion coefficient and τ is the mean residence time of adatom [1].

Experimental

The measurements were performed in a stainless steel vacuum chamber with a high vacuum of 10^{-9} Torr.

The procedure of depositing atoms on the emitter was described earlier [1, 2] by the author and details concerning the ion source construction can be also found there. Details of the Faraday collector and measurement system can be found in ref. [2].

After the pressure in the vacuum chamber had reached 10^{-9} Torr, the temperature of the ions emitter was increased gradually and the DC and noise components of the ion current were recorded. The ion thermoemission noise was amplified by a wide-band DC electrometer. The frequency response of the amplifier system was tested by modulating the focusing electrode voltage applied to the ion source by means of a sine-generator.

To describe the ion current fluctuations the following statistical functions are used:

(1) the spectral density function $S(f) = \langle di^2 \rangle / df$, where df is the frequency bandwidth,

(2) the probability density function:

$$P(di) = \lim_{\Delta i \to 0} \text{Prob.}[di < di(t) < di + \Delta i]/\Delta i.$$

The spectral density function S(f) was investigated with a selective nanovoltmeter Unipan-237. The probability density function P(di) was tested using an analog-digital converter and multichannel pulseheight analyser. The analog-digital converter is a specially constructed linear gate, triggered by 5 μs pulses from a 5 kHz square-pulse generator.

Results

1).Spectral density function.

The fact that the adsorption particle system has itself a relaxation time τ, causes that the wide-range diffusion spectrum is modified by this relaxation time, and the current fluctuations have a frequency spectrum with spectral density function:

$$S(f) = S_o / [1 + (\omega\tau)^2] \quad (2)$$

where $\omega = 2\pi f$ and S_o is asymptotic value of S(f) for $f \to 0$.

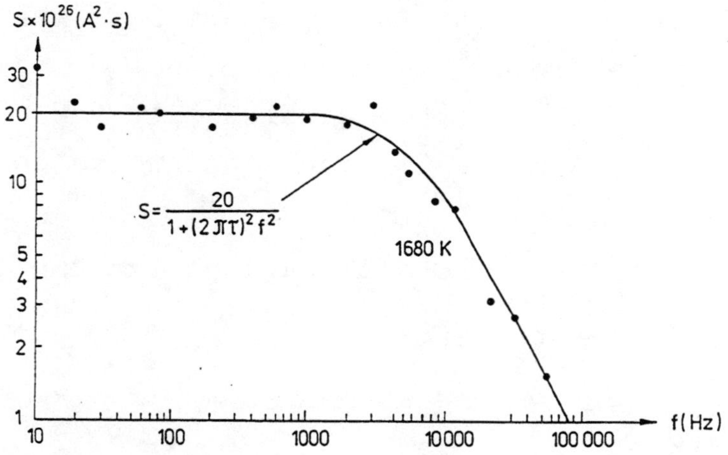

Fig.1.Spectral density function for Li on Rhenium.

In this form of the function S(f), the low frequency spectrum is flat for $f \ll 1/2\pi\tau$, and for $f \to 0$, $S(f) \to S_o$. The high frequency asymptote is proportional to f^{-2}. The turnover frequency $f^* = 1/2\pi\tau$ and $S^* = 1/2 \cdot S_o$ .Fitting the S(f) curves (2) to the experimental data we determine τ.

2).Probability density function.

Special graph paper was used to illustrate the probability density function, such that a normal distribution law resulted in two straight lines (see Fig.2.a.).

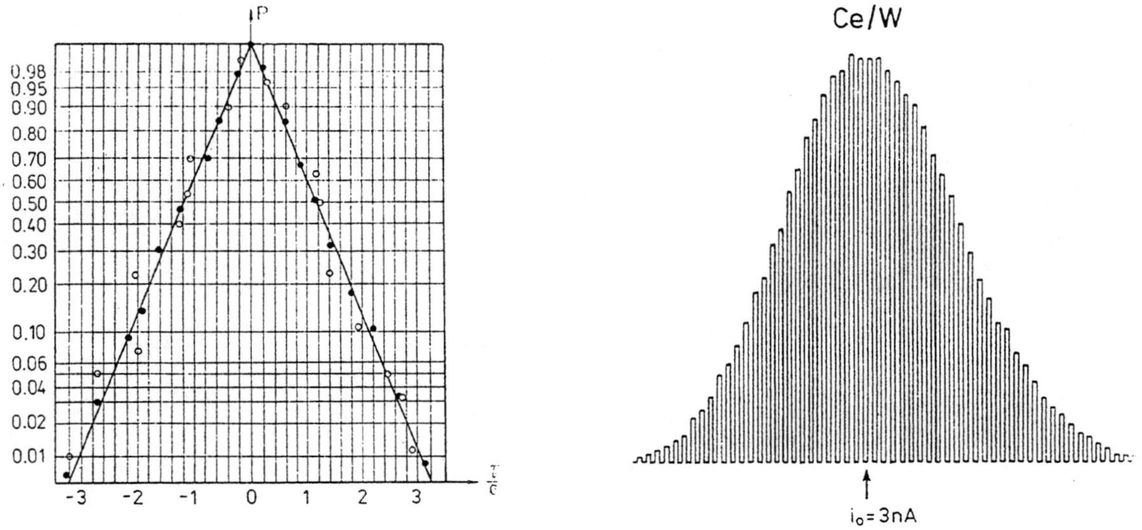

Fig.2.a).Experimental relative distribution of instantaneous amplitude fluctuations for Eu^{+} plotted on a special scale and compared to a normal distribution (two straight lines). $\gamma = di/i_o$ is the relative fluctuation, σ is the standard deviation, P is the relative probability.
b).Amplitude distribution of the Ce^{+} current fluctuations.

To describe the amplitude probability density function, the following central moments were used:

(1).the mean value of the relative fluctuation: $\gamma = di/i_o$,

(2). variance σ^2, where $\sigma^2 = \langle di^2 \rangle$,

(3),(4). third μ_3 and fourth, μ_4, central moments: $\mu_K = \langle di^k \rangle$. It is more convenient to use the dimensionless parameters:"skewness" Sk and "kurtosis" Ex instead of μ_3 and μ_4 moments.

These values are defined by $Sk = \mu_3 / \sigma^3$ and $Ex = \mu_4 / \sigma^4$, (for normal density Sk = 0 and Ex = 3).

Sk and Ex parameters define the asymmetry and slenderness of the experimental distribution, respectively.

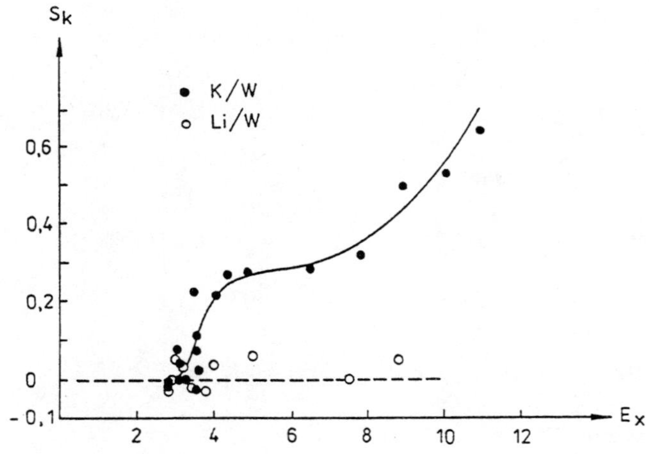

Fig.3.Relationship between the skewnees Sk and kurtosis Ex of the amplitude probability density of the K^+ and Li^+ ion thermoemission noises.The anode temperature increases by 50K from point to point in the range 1000–1800K for K^+ and 1550–2100K for Li^+.

3). The activation energy for surface diffusion.

To investigate the surface diffusion , the normalized mean square of the fluctuations was applied:

$$m = \langle di^2 \rangle T^2/i_c^2 \, \tau \approx D_0 \exp(-E/kT), \qquad (1)$$

where E is the activation energy for surface diffusion. The plot of log m versus 1/T represents a straight line with slope E.

It was established that the diffusion energy E is in good agreement with a linear function of the ionic radius r_i of the alkali metals: $E = 0.88 - 0.33\, r_i$.

Conclusions

(1). The spectral densities may always be represented by a Lorentzian function: $S(f) = S_0 / [1 + (\omega\tau)^2]$.Fitting the S(f) curves to the experimental data we determine the mean residence time τ of the adatoms on the surface.

(2) For relatively low temperature (T < 1300K) the amplitude probability density appears Gaussian.

(3). For T > 1300K, large values of Sk were observed for potassium emission. This fact may be explained by the increase of the asymmetry when the pulsed emission of potassium impurities appear [4].

(4). The method used made it possible to determine the surface diffusion energy [5].

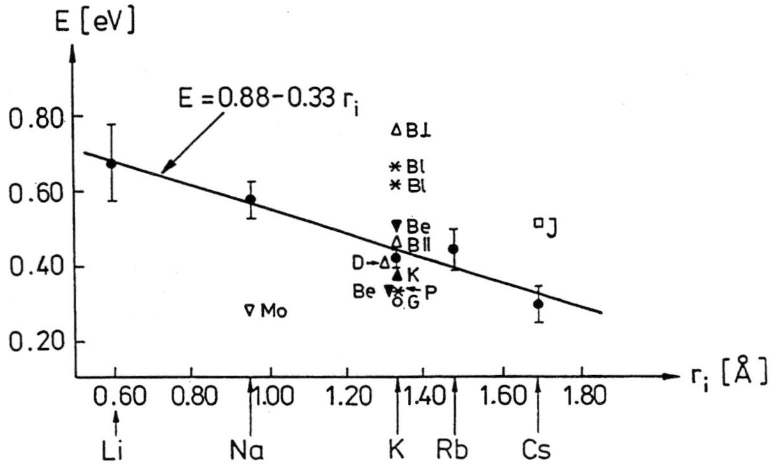

Fig.4.Surface diffusion energy of alkali atoms on polycrystalline tungsten as a function of ionic radius r_i . (Mo-Morin [6], B-Bayat [7], Bl-Błaszczyszyn [8], Be-Bęben [9], D-Dąbrowski [10], K-Kleint [11], G-Gomer [12], P-Popp [13], J-Jamba [14]).

Acknowledgement.

This work was supported by the 2P302/080/05 KBN Research Program.

References.

[1] Gładyszewski L., Surface Sci. 200 (1988) 386.

[2] Gładyszewski L., Isotopenpraxis, 22, 6 (1986) 212.

[3] Gładyszewski L.,Gładyszewski G., Surface Sci. 247 (1991) 274.

[4] Minturn R.,Datz S.,Taylor E., J.Appl.Phys.,31(1960)876

[5] Gładyszewski L.,Surface Sci., 231 (1990) 120.

[6] Morin R., Surface Sci., 155 (1985) 187.

[7] Bayat B., Wassmuth H.K., Surface Sci., 133 (1983) 1.

[8] Błaszczyszyn R.,Kleint Ch., J.de Physique C2,3,47 (1986) 393.

[9] Bęben J.,Kleint Ch., Męclewski R.,J.de Physique C9(1984) 45.

[10] Dąbrowski A.,Kleint Ch., Surface Sci.,119 (1982) 118.

[11] Kleint Ch., Surface Sci., 25 (1971) 394, 411.

[12] Schmidt L., Gomer R., J,Chem.Phys., 45 (1966) 1605.

[13] Popp G., Ann.der Physik, B11 (1964) 115.

[14] Jamba D.M., Husmann O.K., J.Appl.Phys. 38 (1976) 2630.

Vacuum Electron and Ion Technologies
D. S. Karpuzov (Ed.)

ISBN 1-56072-242-8

CURRENT MODULATION IN LIQUID METAL ION SOURCES

G.L.R. Mair
Department of Physics, University of Athens
Section of Solid State Physics
Panepistimiopolis, Zografos
GR 157 84, Greece.

Abstract

The rapid modulation of current from an LMIS has important uses in ion microprobe applications. In this paper, the phenomenon is successfully modelled for the case where the current is modulated using a thermal process. Some, possible, undesirable, side effects of the method are also discussed.

1. Introduction

The ion-emitting cone in a liquid metal ion source (LMIS) can form at the apex of a solid needle covered by a liquid metal film; or, it can form at the nozzle of a capillary filled with liquid metal. The cone is maintained in dynamic equilibrium under the competing influences of electric and surface tension forces, and liquid flow. The apex region of the cone is in the form of a jet. It is from the apex (~10 Å radius) of this jet that the metal ions are field-emitted.

2. The current - voltage law in LMIS

Because of the large space charge infront of the apex, arising in turn from the smallness of the emission area, the mechanism that controls the current is different to that by which the ions are emitted. The latter mechanism is, essentially, some form of field-evaporation. The space-charge model of Mair[1] gives the current-voltage (i-V) characteristics of low flow impedance LMIS as:

$$i=\frac{3\pi\,(2e/m)^{1/2}R\gamma\,\cos\varphi}{V_x^{1/2}}\,(V/V_x-1) \tag{1}$$

Here, (e/m) is the charge-to-mass ratio of the ions, R is the capillary radius or needle apex radius, γ is the surface tension of the liquid metal, φ is the cone half-angle, taken as 49.3°

after Taylor[2]; finally, V_x is the source extinction voltage, for which the cone is barely sustainable and i->0. V_x is given by:

$$V_x = (2\ KR\gamma \cos\varphi/\varepsilon_o)^{1/2} \qquad (2)$$

where K is a geometric factor and ε_o the electric constant.

3. Current modulation in LMIS

According to eq. (1), altering the value of the surface tension γ, should alter the current i, and γ can be altered by heating the liquid metal.

Therefore, by focusing a chopped laser beam on the liquid metal cone one should be able to modulate the current. Infact, this is what Jousten et al[3] did, motivated by a desire to study the feasibity of a rapidly modulated ion column. Such a column would by useful for the ion-milling of complicated shapes: instead of changing the sweep rate, almost pixel-by-pixel, one can keep it constant and modulate the current, i.e. alter the dose.
The surface tension γ is related to its melting point (T_m) value γ_m as:

$$\gamma=\gamma-(d\gamma/dT)\ (T-T_m) \qquad (3)$$

and for gallium[4] γ=0.718 Nm^{-1} and $d\gamma/\ dT=10^{-4}\ Nm^{-1}K^{-1}$.

For a small decrease ($\delta\gamma$) in γ, arising form a temperature increase ΔT above the melting point, eqs (1) - (3) yield a corresponding increase (Δi) in current

$$\Delta i=i_m-i=i_o\frac{d\gamma/dT}{2\gamma_m}\Delta T \qquad (4)$$

where i and i_m are, respectively, the current at temperature T and T_m. The constant i_o is defined as:

$$i_o=3\pi\ (2e/m)^{1/2}R\gamma_m \cos\varphi/V_{xm}^{1/2}$$

where the source extinction voltage, V_{xm}, under melting point operating conditions is related to its value, V_x at temperature T as:

$$V_x=V_{xm}\left[1-\frac{d\gamma/dT}{2\gamma_m}\Delta T\right], \qquad (5)$$

We now write for the power density , I(t), of the chopped laser beam:

$$I(t)=I_o\ (1-\cos\omega t)/2 \qquad (6)$$

I_o being a constant and $\omega=2\pi\nu$ (ν is the laser light modulation frequency).

We will next assume that heat losses from the liquid cone are proportional to the excess energy content q(t) of the cone, with proportionality constant λ. If we denote by Q_o the energy content of the target at time t=0, then the energy content at time t will be Q(t)= Q_o+q(t). Therefore, the energy balance equations becomes:

$$dq/dt= - \lambda q+\alpha A_c I_o (1-\cos \omega t)/2 \tag{7}$$

A_c being the effective area irradiated and α some dimensionless net energy absorption coefficient between 0 and 1.

$$A_c=Rh \tag{8}$$

Since A_c is taken as the projection of the cone on the plane perpendicular to the laser beam. Also,

$$I_o=P/(\pi d^2/4) \tag{9}$$

where P is the laser power and d the diameter of the laser spot. The "heat loss constant" λ which has the dimensions of s^{-1} , and can also be interpreted as a characteristic inverse thermal response time of the target, is given dimensionally, as:

$$\lambda \sim \frac{k}{\varrho c L^2} \tag{10}$$

k being the thermal conductivity, ϱ the density, c the specific heat and L a characteristic length of the target. For gallium, with L~1-10μm, $k=30 Wm^{-1}K^{-1}$, $\varrho= 6.09x10^3 kgm^{-3}$ and $c=398 Jkg^{-1}K^{-1}$, $\lambda \sim 10^5$-107 s^{-1}.

Equation (6) has a steady-state solution that reads:

$$q(t)= \frac{2\alpha P A_c}{\lambda \pi d^2}\left[1-\frac{\cos (\omega t-\theta)}{\sqrt{1+(\omega/\lambda)^2}}\right] \tag{11}$$

The phase difference (θ) between q(t) and the incident radiation I(t) is a measure of the thermal inertia of the target . It is given by:

$$\theta=\tan^{-1} (\omega/\lambda) \tag{12}$$

If M is the mass of the target, they

$$q(t)=Mc\Delta T \tag{13}$$

Combining, therefore, equationς (4), (5), (7), (10) and (12), and with $M=\pi\varrho R^2 h/3$, we have

$$\Delta i = \overline{\Delta i}\left[1 - \frac{\cos(\omega t - \theta)}{\sqrt{1 + (\omega/\lambda)^2}}\right] \tag{14}$$

where

$$\overline{\Delta i} = \frac{9\alpha\,(d\gamma/dT)\,(2e/m)^{1/2}\cos\varphi}{\lambda\pi\, d^2\, \varrho c\, V_{xm}^{1/2}}\, P \tag{15}$$

$\overline{\Delta i}$ is the mean increase of the current from its value under melting point source operating conditions.

The oscillatory part of the current is:

$$\tilde{\Delta i}\,(t) = -\Delta i_o \cos(\omega t - \theta) \tag{16}$$

where the amplitude Δi_o is given by:

$$\Delta i_o = \frac{\overline{\Delta i}}{\sqrt{1 + (\omega/\lambda)^2}} \tag{17}$$

In Fig. 1 we compare the experimental results of Jousten et al for $\Delta i_o/\overline{\Delta i}$ with the predictions of this model. The fit is within a value for λ, which is different for the three laser spot sizes used: the larger the laser spot size the larger the volume of target irradiated and, therefore, the longer its thermal response time ($1/\lambda$).

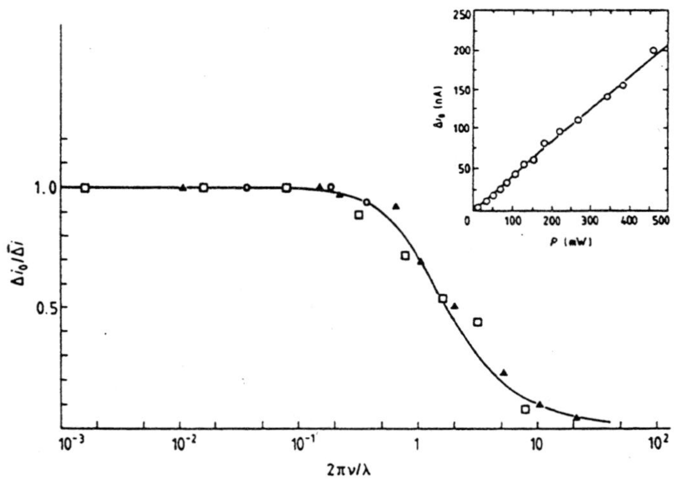

Figure 1. Relative amplitude of modulated ion current against normalized laser chopping frequency, for three different laser spot sizes. ▲, □, ○: experiment; continuous curve: theory. ▲: $d = 70\,\mu$m, $\lambda/2\pi = 0.955 \times 10^4\,\text{s}^{-1}$; □: $d = 22\,\mu$m, $\lambda/2\pi = 6.37 \times 10^4\,\text{s}^{-1}$; ○: $d = 15\,\mu$m, $\lambda/2\pi = 2.73 \times 10^5\,\text{s}^{-1}$. Inset: experimental current modulation amplitude against laser power, $d = 15\,\mu$m, $\nu = 100$ Hz. All results refer to gallium.

The inset in Fig. 1 shows the amplitude of modulation (Δi_o) of the current vs laser power

(P); d=15μm and $\omega=2\pi$ x100 kHz. The slope of the curve is about 4×10^{-7} AW^{-1}. For a spot size of d=15μm, $\omega=2\pi\times100$ kHz, the value of λ inferred for this particular spot size and V_{xm}=8 kV, we have:

$$\Delta i_o/P=3.6\times10^{-7}\ AW^{-1} \tag{18}$$

which is within 10% of the experimental value, assuming[5] $\alpha=1$.

4. Discussion

It is quite obvious that a thermal process can rapidly modulate the ion current in a LMIS, with modulation frequencies of, possibly, up to $\sim10^8 s^{-1}$. However, there might be a problem associated with the enhancement of undesirable droplet emission, when such a modulated ion column is directed on to a target:

Am approximate analytical formula for the jet length in a LMIS reads as[6]:

$$\ell=\frac{2[m/(2e)]^{1/2}V_x^{1/2}\,i}{3\pi\gamma} \tag{19}$$

The periodic heating of the cone with the resultant changes in surface tension would cause a corresponding periodic oscillation in the jet length ($\delta\ell$); viz:

$$\widetilde{\delta\ell}=-\delta\ell_o\ \cos(\omega t-\theta) \tag{20}$$

$$\delta\ell_o=\frac{\overline{\delta\ell}}{\sqrt{1+(\omega/\lambda)^2}} \tag{21}$$

$$\overline{\delta\ell}=\frac{6\alpha(d\gamma/dT)\cos\varphi\ P}{\pi^2\lambda\gamma\varrho c d^2} \tag{22}$$

For typical values, for gallium, $\delta\ell$ is of the order of 10 Å.

Now, characteristic inverse thermal response times, $1/\lambda$, are at most $\sim10^8 s^{-1}$, and this is about the upper limit for the laser chopping frequency . At the same time instability growth rates[7] for the jet are $\sim10^{10} s^{-1}$, so that an oscillation, or disturbance, of the jet is likely to lead to a growing instability resulting in the emission of a droplet.

5. Conclusions

We have presented a successful model for the current modulation in LMIS using a thermal process. However, the method may have some limitations in its applicability for the construction of a rapidly modulated ion column, in that it may precipitate the emission of undesirable droplets.

References

1. Mair G. L. R. , J. Phys. D, **17** (1984) 2323
2. Taylor G. I, Proc. Roy. Soc., **A280** (1984) 313
3. Jousten K., Holmes J.F. and Orloff J., J. Phys. D., 24 (1991) 458
4. Smithell's Metal Reference Book, 1976, 53[d] edn.
5. Gray DE (ed) American Institute of Physics Handbook 3[d] edn (New York; Mc Graw Hill)
6. Mair G.L.R. and Forbes RG, J. Phys. D, **24** (1991) 2217
7. Basset A.B., Am. J. Math, 16 (1894) 93

Vacuum Electron and Ion Technologies
D. S. Karpuzov (Ed.)

ISBN 1-56072-242-8

COMPUTER SIMULATION OF SOME EMISSION CHARACTERISTICS OF ALLOY LIQUID METAL ION SOURCES

R.G. Vichev, S. Georgieva and N. Drandarov
Institute of Electronics, Bulgarian Academy of Sciences,
72 Blvd. Tzarigradsko Chaussee, 1784 Sofia, Bulgaria

A Monte-Carlo computer code simulating the emission process of ions in liquid metal ion sources (LMIS) is developed. The electric field of the tip is taken to be that created by a sphere-on-orthogonal cone. A Poisson distribution is used to describe the time dependence of the emission process. The Coulomb interaction between the ions is taken into account. The virtual source size, energy and angular distributions of ion species emitted from AuSi LMIS are evaluated for different reference planes perpendicular to the emitter axis.

Liquid metal ion sources (LMIS) have many applications in surface analysis and in industry as devices capable of producing submicron ion beams. The final diameter of the beam depends on the virtual source size, the energy and angular distributions of the emitted ions. Knowledge of these parameters is important for better designing and optimization of ion focusing systems.

MODEL

A Monte-Carlo computer code was used to simulate the ion trajectories in the diode region of a AuSi LMIS. The computer code described in Ref. 1 was further extended to treat the emission process of alloy LMIS where ion species with different mass and charge are emitted. Electron microscopy pictures of the frozen AuSi emitter showed that it can be reasonably approximated by a sphere of radius R_o joined to an orthogonal cone with half angleof 15° [1]. The potential distribution in the vicinity of such an emitter tip is given by the so called sphere-on-orthogonal cone (SOC) model, which adequately describes the electric field in the diode region [2]:

$$V(r,\vartheta)=V_r\left(\frac{R_o}{2R}\right)^n\left[\left(\frac{2r}{R_o}\right)^n - \left(\frac{R_o}{2r}\right)^{(n+1)}\right] P_n(\cos\vartheta)$$

where R and V_r are the distance and the potential difference between the emitter and the extractor, respectively, P_n is the Legendre function, R_o is the emitter radius, ϑ is the polar angle with respect to the emitter axis (denoted as Z). The value of n (0.24) was obtained from Ref. 3. For R_o we adopted the value of 10 nm [4]. The model parameters were chosen the same as in Ref. 1. The values of R and V_r were 1.5 mm and 5500 V, respectively. The Au^+ ions constituted 90 % and Si^{2+} ions 10 % of the emission current. All ions were emitted normal to the surface with an energy of 1 eV within a polar angle ϑ = 0.1 rad. Random numbers were used to define the initial ion position. The emission process was described by a Poisson distribution. The mean number of ions generated in a time interval 10^{-13} s was 1.03 corresponding to the measured ion current (25 μA) corrected for the solid angle of emission taken for the calculations and the angular distribution of the extracted ions [1]. For each time interval, the exact number of emitted particles and the exact time of emission were obtained using random numbers. The ion trajectories were followed by a standard integration routine. Pairwise Coulomb interaction between the ions could be taken into account. The points of intersection of the ion trajectories with reference planes chosen normal to the Z axis were recorded together with the ion energy and angle ϑ. The tangential lines drawn through these points crossed the plane containing the paraxial image to form the virtual source. The virtual source size, the energy and angular distributions of the ions were obtained for reference planes situated at different distances from 3 R_o to 100 R_o from the centre of the emitting sphere. Ions were followed until they cross a plane at a distance 150 R_o from the centre of the emitting sphere to avoid edge effects. The computer time needed for calculating the virtual source size at greater distances increased significantly as the number of ions in the diode region increased. Thus, calculations for reference planes at greater distances were not carried out due to restrictions in computer time. A total number of about 3000 and 5000 ions were followed for the case with and without Coulomb interaction, respectively. The virtual source size D was determined as follows. The number of points per unit area was found for rings centered around the Z axis. We took the full width at half maximum (FWHM) of the obtained histograms for the diameter of the virtual source.

RESULTS AND DISCUSSION

The model described above was used to calculate the virtual source size, the energy and angular distributions of Au^{+} and Si^{2+} ions for three reference planes situated at $Z=3\ R_o$; $10\ R_o$ and $100\ R_o$ from the centre of emitting sphere. The obtained results are shown in Table 1.

Table 1. Virtual source size (D), energy (ΔE) and angular spread ($\Delta\vartheta$) calculated for three reference planes at $Z=3\ R_o$, $10\ R_o$ and $100\ R_o$. The subscript "i" denotes that the corresponding values were obtained taking into account the Coulomb interaction between ions. ΔE and $\Delta\vartheta$ are the FWHM of the energy and angular distributions.

Reference plane	Ion	Virtual size [A]		Energy spread [eV]		Angular spread [deg]	
		D	D_i	ΔE	ΔE_i	ϑ	$\Delta\vartheta_i$
$3R_o$	Au^{+}	60	110	0.6	17	5.5	9
	Si^{2+}	60	110	1.5	17	5.5	9
$10R_o$	Au^{+}	170	340	0.3	20	5	9
	Si^{2+}	170	340	0.7	20	5	9
$100R_o$	Au^{+}	1500	3000	0.3	24	4	8
	Si^{2+}	1500	3000	0.6	24	4	8

The normalized angular distributions of Au^{+} and Si^{2+} ions calculated for the first ($Z=3\ R_o$) and third ($Z=100\ R_o$) reference planes are shown in Fig. 1. It is seen that the full width at half maximum (FWHM) of the angular distributions $\Delta\vartheta$ computed without Coulomb interaction between ions is reduced from 5.5^o to 4° as the distance from the emitter grows from $3\ R_o$ to $100\ R_o$ (Figs. 1a, 1c). The same is also valid for the angular ditributions calculated taking into account the Coulomb interaction which are much broader ($\Delta\vartheta=9^o$ for $Z=3\ R_o$ and $\Delta\vartheta=8^o$ for $Z=100\ R_o$ .Figs. 1b, 1d). Both Au^{+} and Si^{2+} ions have similar angular distributions with equal $\Delta\vartheta$ for a given reference plane.

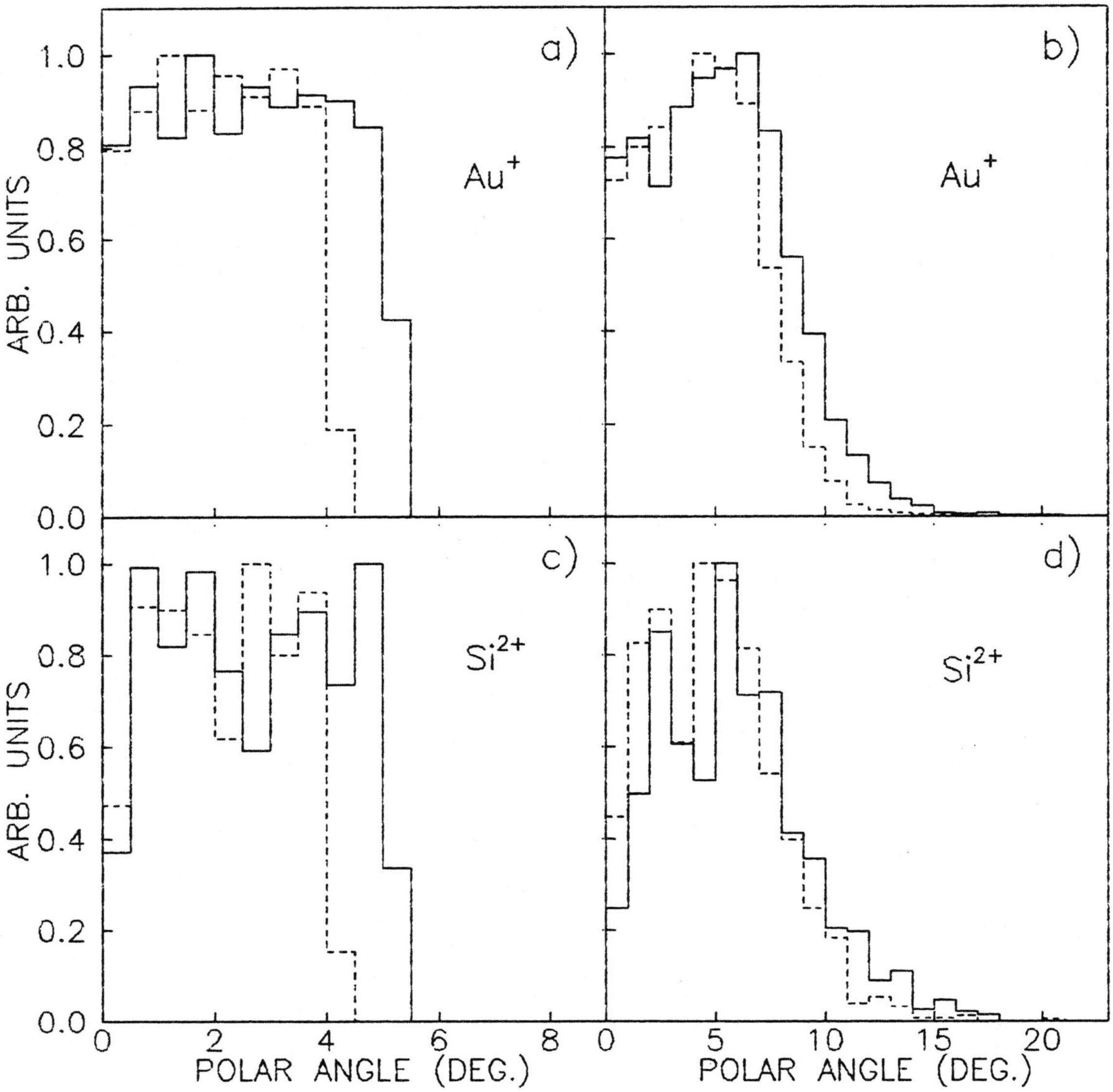

Fig. 1. Angular distributions of Au^+ and Si^{2+} ions for two reference planes: 3 R_o (full lines) and 100 R_o (broken lines) calculated with (Figs. 1b, 1d) and without the Coulomb interaction between ions (Figs. 1a, 1c).

The virtual source diameters D and D_i computed for the three reference planes are presented in Table 1. The values of D obtained without ion interaction are about two times smaller than those of D_i calculated with ion interaction. D_i increases significantly with the distance from the emitter (from 120 A for Z=3 R_o to 3000 A for Z=100 R_o).

The radial density distributions of Au^+ and Si^{2+} ions calculated for Z=100 R_o with ion interaction are shown in Fig. 2a,c. Both distributions are similar in shape and have

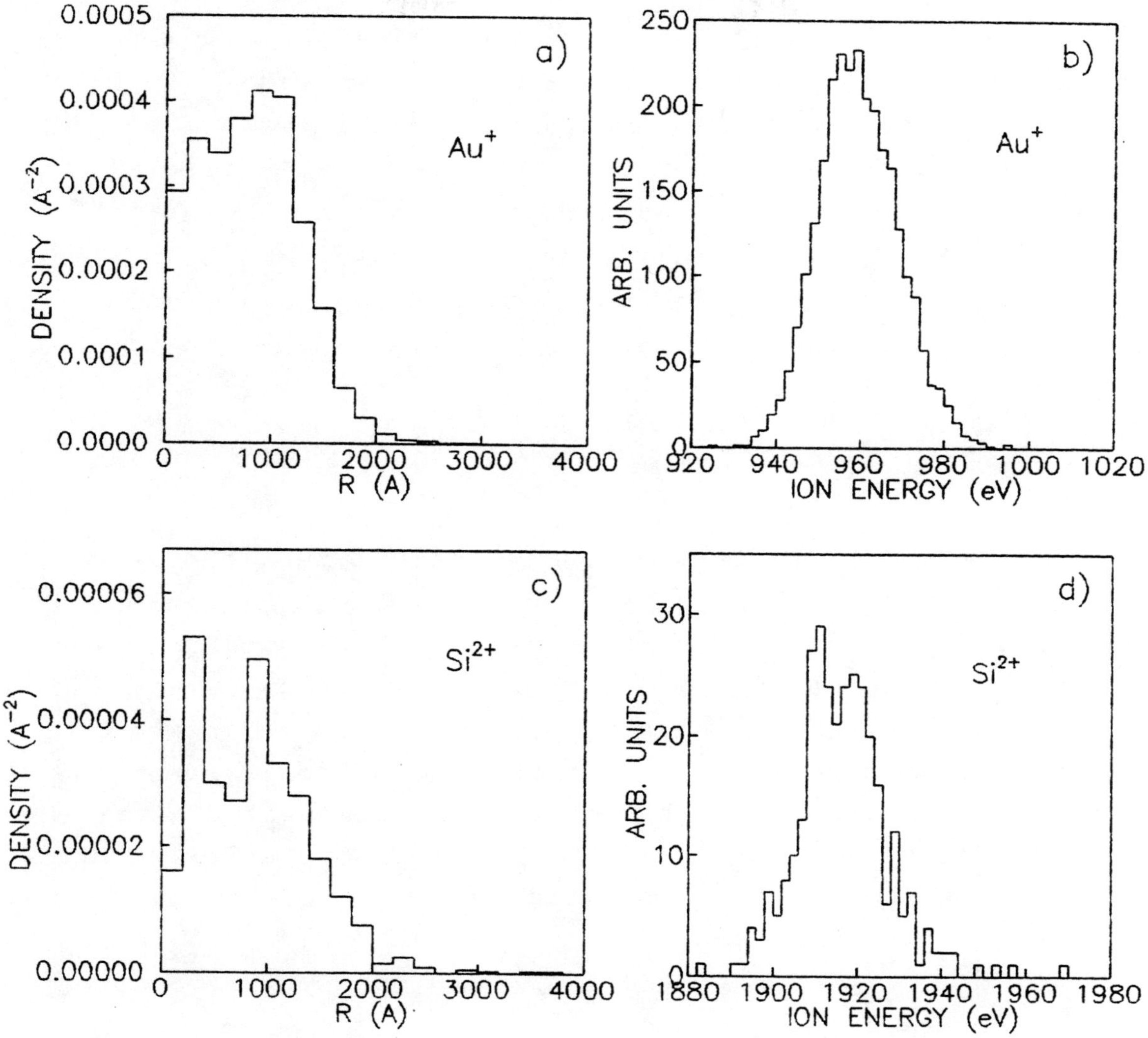

Fig. 2. Radial density distributions (Figs 2a, 2c) and energy distributions (Figs. 2b, 2d) of Au^+ and Si^{2+} ions for Z=100 R_o. The Coulomb interaction between ions has been taken into account.

approximately the same FWHM. (D_i=3000 A). The experimentally measured value for the virtual source size of an AuSi LMIS is 6 μm [1], i.e. an order of magnitude larger. One reason for this discrepancy is that the reference plane is situated at a distance which is very small compared to the distance emitter-extractor R. Another factor contributing to the small virtual source diameter obtained in the simulations is the adopted small emission angle ϑ.

The FWHM of the Au^+ and Si^{2+} energy spectra are presented in Table 1. The values of ΔE for Si^{2+} are two times greater than those computed for Au^+. The energy spectra obtained without interaction are very narrow with ΔE being within 1–2 eV. The corresponding values ΔE_i for the case with ion interaction are an order of magnitude greater (Table 1). ΔE_i changes slowly with Z (ΔE_i=18 eV for Z=3 R_o and ΔE_i=24 eV for Z=100 R_o). The energy distributions of Au^+ and Si^{2+} ions for Z=100 R_o calculated with ion interaction are shown in Fig. 2b,d. They are quite similar in shape and have approximately the same ΔE_i. The same is valid for all reference planes.

CONCLUSION

We have calculated the virtual source size, the energy and angular distributions of Au^+ and Si^{2+} ions emitted in an AuSi LMIS. The distributions computed for both types of ions are similar in shape and have approximately the same FWHM. The obtained results indicate that Coulomb interaction between ions causes significant spreading of the energy and angular spectra and of the beam spot. The calculated virtual source diameters are smaller than the experimentally measured value due both to the small emission angle and small distance from the emitter considered in the simulations.

REFERENCES:

1. S.Georgieva, R.G.Vichev, N.Drandarov, Vacuum 45, No. 1 (1994).
2. J W Ward, R L Selinger, J Vac Sci Technol 19, 1062 (1981).
3. J C Wiesner, T E Everhart, J Appl Phys 44, 2140 (1973).
4. G L R Mair, J Phys D 21, 1654 (1988).

Vacuum Electron and Ion Technologies
D. S. Karpuzov (Ed.)

ISBN 1-56072-242-8

THE EFFECTS OF SOURCE TEMPERATURE AND EMISSION CURRENT ON THE MASS PECTRUM AND THE ANGULAR DISTRIBUTION OF THE IONS EMITTED BY AN AuSi FIELD ION SOURCE

S. Georgieva, N. Drandarov, B. Nikolov
Institute of Electronics, Bulgarian Academy of Sciences
72 blvd. Tzarigradsko shousee, 1784 Sofia, Bulgaria

An $Au_{82}Si_{18}$ field ion source has been investigated. The masspectrum and the angular distribution of the total emission current and of Au^{+}, Si^{++}, Au^{++} and Si^{+} ions have been studied at different temperatures, and a strong influence of temperature has been discovered. The angular distributions of Au^{+} and Si^{++} are composed of two peaks while the angular distributions of Au^{++} and Si^{+} have only one peak. When the temperature is increased, the two peaks of the angular distribution curves move closer one to another and the asymmetric outline of the angular distribution curves of the total emission current become smoother and acquire an ideally symmetric form. Increasing the emission current causes a broadening of the angular distribution curves.

Introduction

Liquid metal ion sources have been the subject of active investigation because of some potential advantages related to their application in focused ion beam (FIB) technologies, such as maskless ion implantation, lithography, ion etching, secondary ion masspectrometry and ion microscopy. Some of the metals necessary for circuit fabrication have high melting points and/or high vapor pressure. This requires that they should be used in the form of alloys. The alloy AuSi field ion source, emitting approximately 14% of its total current in the form of Si^{++} ions [1] is of special interest for GaAs doping and for ion lithography. The advantages and the disadvantages of the structures obtained with FIB employing this type of ion source have already been described and discussed [1,2,3,4].

Experimental

Investigations have been made with a ribbon field ion source, filled with the $Au_{82}Si_{18}$ alloy. The alloy was obtained as a result of melting in 10^{-3} Pa vacuum. The masspectrum of the source was measured by means of a Wien filter with an automatically controlled deflecting voltage. The output current was collected by means of a Faraday cup and was registered by a computer system. The Wien filter mass resolution has been estimated at $m\backslash\Delta m=13$ in the Au region and $m\backslash\Delta m=9$ in the Si region. Masspectra were obtained at two different temperatures, T of 500^{0}C and 600^{0}C, and at a total emission current, I of 20 and 40 μA respectively.

The angular distributions of I at different T between 400 and 1000^{0} C and different current values of 20 and 50μA have been investigated. The influence of T and I on the shape and the width of the angular distribution curves has been studied for Au^{+}, Si^{++}, Au^{++} and Si^{+}. A special device has been used for rotating the source apex around the axis of the Wien filter (fig.3), through a half-angle of up to 40°. Secondary electrons were suppressed by applying a negative voltage on an electrode situated in front of the entrance of the Faraday cup and also by smoothing the apertures. The angular distributions of each of the four mass - species of the ion beam have been measured at T between 400 and 900^{0} C and I equal to 20 and 30μA. All the tests were carried out at a vacuum of 10^{-4} Pa.

Results and Discussion

1. Masspectra

A typical masspectrum of the $Au_{82}Si_{18}$ alloy field ion source obtained by means of a Wien filter is shown in fig.1. The most intense peaks, corresponding to single atomic ions, that can be seen on the masspectrum, are of Au^{+}, Si^{++}, Au^{++} and Si^{+}. Also, intense emission of $AuSi^{++}$, $AuSi_{2}^{+}$ and Au_{2}^{+} cluster ions can be seen. $^{28}Si^{++}$ and $^{29}Si^{++}$ ions have been observed in the spectrum. Such a type of masspectrum, with cluster ions has been reported in references [1,2,5,6]. Wang et al. [1], Kubena et al. [2] have observed doubly charged ions of Si isotopes as well. When the temperature was increased an increase in the Au^{++}/Au^{+} ratio was observed. Komuro [7] attributed this to the field strength required for ion emission which depends on the composition of the molten alloy at the needle's tip. The observed decrease of the intensity of the peaks corresponding to Si with the increase of temperature, we associate with the decrease of Si concentration according to the liquidus line of the phase diagram of an AuSi ion source in which the Si concentration was lower than that at the eutectic point ($Au_{69}Si_{31}$). Si in AuSi alloy reacts with the W needle to form a WSi_2, at 700°C wetting temperature, and this should be taken into account when we explain the change of Si concentration with the temperature variation. Change in the Au^+, Si^+ and $AuSi^+$ peaks intensity with the variation of I has been observed. With the increase of the ion current the Si^+ and $AuSi^+$ peaks grow rapidly, while the Au^+ peak becomes smaller. The rest of the peaks observed remain almost without any change. The results obtained by us are similar to those of Machalett et al. [5]. The explanation of the changes observed in the different peaks intensities is possible only if the energy distribution

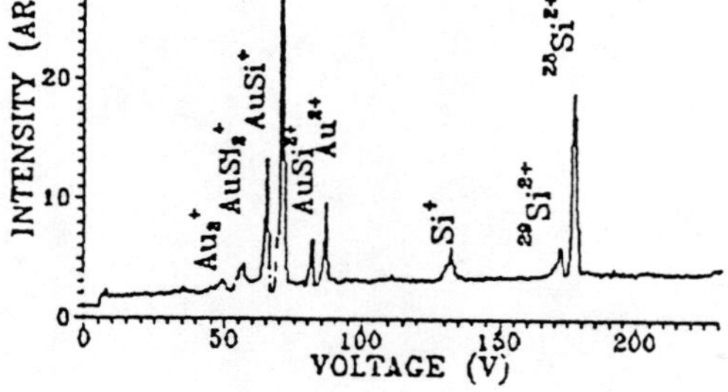

Fig.1.Masspectrum of the AuSi field ion source.

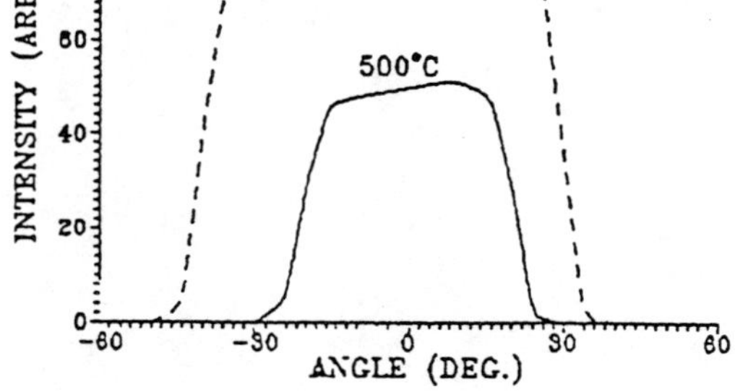

Fig.2.Angular distribution at 500°C, 20μA and 1000°C, 50μA

of each type of ions is taken into account. Change in intensity of the Au^+ peak with the ion current variation can be attributed to cluster ion decomposition due to collisions or to field induced breakup and to ionisation by collisions between ions or electrons and the evaporated atoms. Such additional mechanisms for formation of Au^+ ions could also explain the presence of low-energy tails on Au^+ energy distribution curves [8].

2. Angular distributions of the total ion current

The angular distribution curves of I obtained at T up to 700°C have asymmetric form (fig.2). With the increase of T these curves become smoother and their characteristic symmetric high temperature form is observed. We have also found broadening of the curves with the increase of T. Such asymmetric form and broadening of the curve with the increase of T has already been seen by Kulic et al. [9] for InGa field

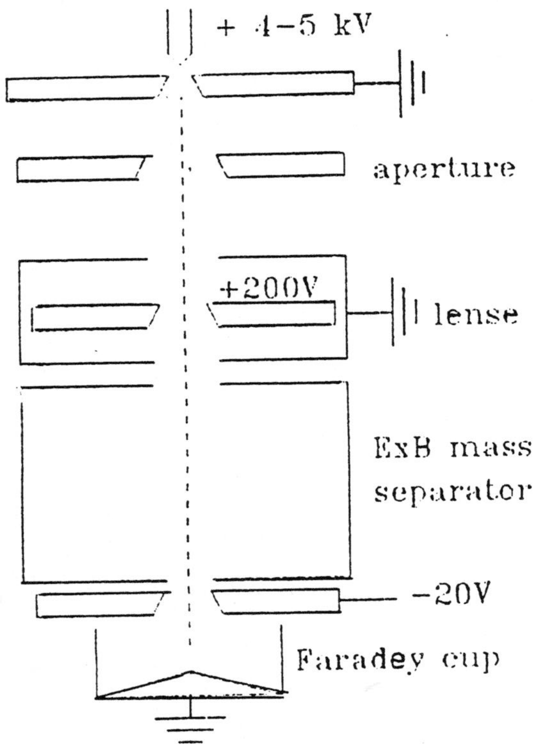

Fig.3.A special device has been used for rotating the source apex around the axis of the Wien filter.

alloy ion source. The asymmetric form can be associated with the geometric heterogeneity of the needle's tip. The continuous heating of the source at high T decreases the heterogeneity of the source and draws the angle distribution of the ions closer to the ideal form of the curve, with a maximum in the middle and steep slopes on both sides of the curve. The broadening of the angular distribution curve with the increase of the emission current we associate with the increased ion sputtering and with the possibility for the emerging of new emission tips. Such broadening of the angular distribution curve with the increase of the current has been observed by Kulic et al. [9], W. Ke-Li et al. [10], Wagner et al. [11] for Ga and Au LMIS.

3. Angular distributions of the different components of the ion beam.

The angular distributions of the ions of largest practical importance - Au^{+}, Si^{++}, Au^{++} and Si^{+} have been investigated at several different temperatures. All the angular distributions at T from $400^{0}C$ to $800^{0}C$ displayed the same character, while at T = $900^{0}C$ and $1000^{0}C$ they have another character. The typical angular

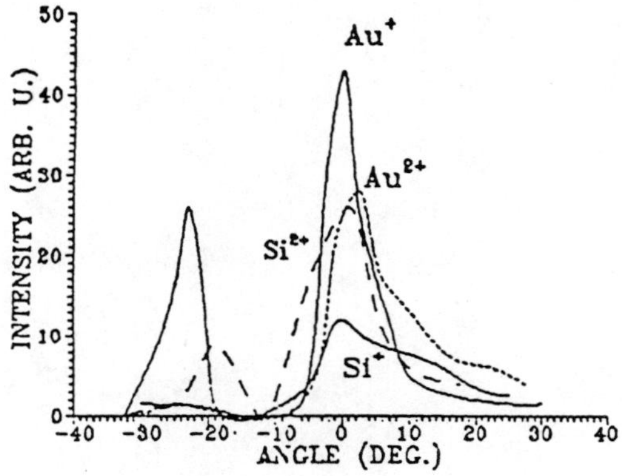

Fig.4.Angular distributions of the different components of the ion beam at $500^{0}C$ and 20μA.

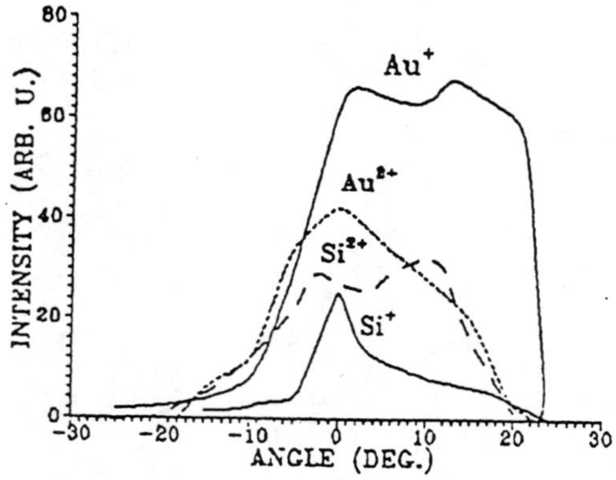

Fig.5.Angular distributions of the different components of the ion beam at $1000^{0}C$ and 30μA.

distributions at T up to 800°C are shown in fig.4, while the angular distributions at higher T are shown in fig.5. Au^+ and Si^{++} ions in all of the cases have angular distributions with two peaks, while Au^{++} and Si^+ ions display angular distributions with a single peak. The two Au^{++} and Si^+ peaks can be associated with the existence of two emission tips, which we observed when the AuSiBe emitting surface was frozen [14]. The evaporation field strength for Si^+ (4.43 V/A) and for Au^{++} (4.85 V/A) are higher than those for Si^{++} (3.14 V/A) and for Au^+ (4.02 V/A), respectively [12,13]. Single Si^+ and Au^{++} peaks are located at the same angle as one of Au^+ peaks. Therefore, there is reason to suppose that the field strength on one of emission tips is higher. Rao et al. [8] measured the angular distributions of ions emitted by an AuSi field ion source at 600°C and 4μA. They observed a dimple in the plots of Si^+, Si^{++},and Au^+ angular intensity at low angles. They related the presence of this dimple with the reduction of the electric field close to the axis due to the space charge. However, the very rapid fall in angular intensity of Au^{++} cannot be explained by the presence of such a space charge.

With the increase of T the two peaks of the angular distributions of Au^+ and Si^{++} broaden and come closely one to another, which is most probably associated with the transition to more unstable emission regime of the source and the related cluster emission.

Conclusion

The masspectra and the angular distributions of the totalemission current of each of the Au^+, Si^{++}, $Au^{++}$and Si^+ ions have been measured at different temperatures and values of the total emission current. It has been found that the angular distribution of Au^{++} and Si^+ have a single peak, while those of Au^+ and Si^{++} have two peaks, regardless of the variation of T. The increase of T makes the two peaks come closer one to another and make the form of the angular distribution of I more symmetric. The change of T induces a change of the intensity of some of the peaks in the masspectrum, too. The increase of the emission current broadens all angular distributions, and also changes the masspectrum.

References

1.V.Wang, J.Ward, and R.Seliger, J.Vac.Sci.Technol.19 (1981) 1158

2.R.L.Kubena, C.L.Anderson, R.L.Seliger, R.A.Jullens, and E.H.Stevens, I. Lagnado, J.Vac.Sci.Technol. 19 (1981) 916.

3.K.Gamo, T.Ukegava, and S.Namba, Jap. J. Appl. Phys. 19 (1980) L379.

4.K.Gamo, T.Ukegava, Y.Inomoto, K.K.Ka, and S.Namba, Microcircuit Eng. 80 ed R.P.Kramer (Delft: Delft University Press 1981) 283.

5.F.Machalett, R.Muhle, and I.Stiebriz, J.Phys.D: Appl.Phys. 20 (1987) 1417.

6.N.Drandarov, B.Nikolov and Tsv.Valchovska, Vacuum 42 (1991) 95.

7.M.Komuro, Jp.J.of Appl.Phys. 25 (1986) 1500.

8.S.Rao, A.E.Bell, G.A.Schwind, and L.W.Swanson, J. Vac. Sci. Technol. A8 (1990) 2258.

9.V.S.Kulic, V.K.Medved'ev, N.N.Popovich, V.I.Cherni, Ukr.Phys.J. 39 (1991) 1626.

10.W.Ke-Li, Chang Hua-Yi, D.Lei, Chen Tong, G.Jie, Qu Zhi-Can, Liu Xidoping, Proceeding of The Third Japan - China Joint Symposium on Accelerators for Nuclear Science and Their Applications.

11.A.Wagner, T.Venkatesan, P.M.Petroff, D.Barr, J. Vac. Sci. Technol. 19 (1981) 1186.

12.E.W.Mueller, T.T.Tsong: Field Ion Microscopy (Elsevier, New York 1969).

13.D.R.Kingham, and L.W.Swanson, Appl. Phys. A41 (1986) 157.

14.N.Drandarov, St.Georgieva, B.Nikolov, T.Donchev, Vacuum 45(1993)

Vacuum Electron and Ion Technologies
D. S. Karpuzov (Ed.)

ISBN 1-56072-242-8

SYSTEM FOR ENERGY DISTRIBUTION MEASUREMENTS OF IONS EMITTED FROM LIQUID METAL ION SOURCE.

T. Tcvetcov, N. Drandarov and B. Nikolov
Institute of Electronics, Bulgarian Academy of Sciences.
Blvd. Tcarigradsko shousee 74, Sofia 1784, Bulgaria

A five electrode ion optical system and cylindrical energy analyzer was designed and built. With this optical system and a quadropole mass spectrometer we can measure energy distributions of each kind of ions emitted from Liquid Metal Ion Sources. The five electrode emmersion lens works in deceleration mode and decelerates ions down to 80 eV. The energy analyzer was designed to have torroidal potential and can focus ions in two directions.

INTRODUCTION

In last decade a significant role in ion modification techniques is played by focused ion beams (FIB). The FIB are used widely as tools for material analysis in scanning ion microscopy and secondary ion mass spectrometry [1,2], ion beam assisted etching and deposition [3,4], for diagnostic and mask repair. Considerable rise of interest in use of FIB was observed after the invention of the liquid metal ion source (LMIS) [5,6].A liquid metal ion source has many attractive features such as simple structure, high brightness, and low power consumption. Despite of very large amount of works on LMIS a lot of questions on ion emission mechanisms have to be answered. These are how and where the various types of ions are formed, and how the composition of alloy LMIS influences the relative abundance of various type of ions and their energy distribution. The later is important for ion optic design to avoid chromatic aberration. To answer these questions one has to have a lot of experimental results on emission characteristics like mass-spectrum, energy and angle distribution of ions emitted from alloy LMIS. For these purposes we calculated and built an ion optical system for measurement of angular distribution of energy and mass separated ions,including two immersion lenses, a two electrode and a four electrode one, a cylindrical energy analyzer with toroidal field distribution. This system is coupled to a quadropole mass spectrometer. The ion optical system can retard the ion beam emitted from a liquid metal ion source down to 1% of initial ion energy and focus it on an entrance aperture of the energy analyzer.

The calculated energy resolution of the energy analyzer is about 1 eV at energy of 80eV.

EXPERIMENTAL SETUP AND CALCULATIONS

Our experimental setup consists of a commercial ultra-high vacuum chamber USU-4 and a commercial quadropole mass spectrometer. The mass spectrometer has mass resolution of about 1 AMU per unit charge over the hole mass range.

Fig.1. shows the optical column with the energy analyzer. The ion source together with the extractor can rotate around the axis passing through the emission point and we can obtain the angular distribution of the emitted ions. Ion optical system consists of two immersion lenses. The first one is two electrode type and reduces ion beam energy down to 18% of initial beam energy. Under the action of that lens the ion beam is focused at an intermediate aperture before second four electrode immersion lens. The later retards the ion beam at an energy down to 1% of the initial ion energy and gives a final focusing of the ion beam on the entrance aperture of the energy analyzer. For calculating the optical properties of this ion optical column we used a computer procedure for tracking the ion beam. This routine includes two main parts. The field-calculating part witch uses the method of integral equation to obtain the field distribution over the optical column. The second part calculates the trajectory of a given ion with initial energy, starting angle and X,Y,Z coordinates. The program can optimize one parameter to obtain best focusing. Fig.2 shows ion trajectories calculated at different starting angles. Corresponding potentials of each of electrodes are given too.

The electrostatic energy analyzer is designed as cylindrical one and possess a toroidal field distribution. The later we obtained when we put counter electrodes as shown in fig.3. This geometry ensures focusing in two directions, in contrast to conventional type cylindric energy analyzers, and then we enhance the transmittance of the analyzer. Our analyzer has the following physical parameters: The middle trajectory radius R is 51mm, the distance between the cylindrical electrodes is 10 mm, the width of analyzer is 20 mm, the distance between counter electrodes is 24 mm, the deflection angle is 90°. To minimize the effect of the influence of the energy analyzer ends on the electric field, we made a Herzog correction for the entrance and exit apertures of the analyzer [7]. Fig.4 shows the dependence of aperture diameter on the distance between the aperture and analyzer when zero potential is at the end of analyzer. The aperture diameter of our analyzer is 2mm.

Fig.5. shows a three dimensional view of the calculated potential distribution inside the energy analyzer. For these calculations the potentials are taken as follows: $U_a=-5.12V$, $U_b=4.64V$, in fig.5a $U_c=0$ V and in fig.5b $U_c=4$ V. U_a is the inner electrode potential, U_b is outer electrode potential and U_c is counter-electrode potential. It is seen clearly that when $U_c=0$, V near the middle radius the potential is similar to that of a conventional energy analyzer. When $U_c=4$ V the potential has toroidal symmetry, but the zero level potential is shifted towards the inner electrode. For a wide range of U_c the potential at R_m changes linearly when U_c changes. This dependence is shown on fig.6. In that case the potentials U_a and U_b remained constant.

CONCLUSIONS

An electrostatic system for energy distribution measurements was designed and built. This system is coupled to a quadropole mass-spectrometer, to obtain mass spectra of energy resolved ions. The system is designed to work with LMIS. With this setup we can measure angle resolved energy distributions of mass-separated ions. This will give us the possibility to investigate various type LMI sources for understanding better the ion emission processes.

ACKNOWLEDGEMENTS

This work was supported under the National Science Foundation.

Reference

1.R.Levi-Setti, Scanning Electron Microscopy, v.1 SEM Inc, Chicago (1983).

2.G.C.Allen and I.T.Brown, CEGB Res. 23, 3(1989).

3.Y.Ochiai, K.Shihoyama, T.Shiokawa, K.Toyoda, A.Masuyama, K.Gamo and S.Namba, Jap. J. Appl. Phys., v.25, No.7, July,1986,p.L526.

4.K.Gamo, D.Takehara, Y.Hamamura, M.Tomita and S.Namba, Microelectronic Engineering 5 (1986) 163 North-Holland.

5.Krohn VE, Ringo GR. (1975) Appl.Phys.Lett.27, 479-481

6.Clampitt R, Aitken K L and Jefferies DK (1975). J.Vac.Sci. Technol. 12, 1208.

7.R.Herzog. Elektronenoptische Zylinderlinsenwirkung der Streufelder eines Kondensators. Phys. Zs. 41, 18 (1940).

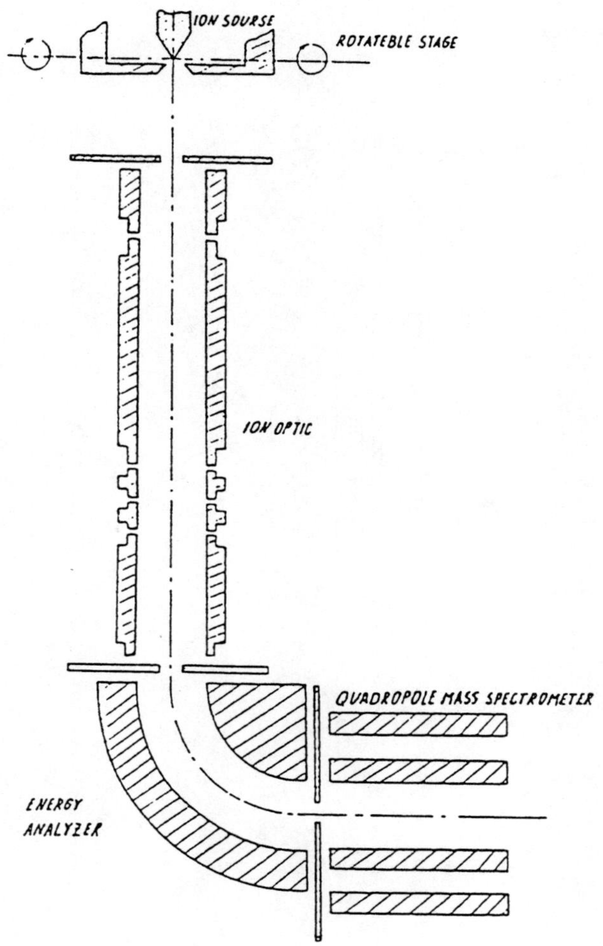

Fig.1 Optical column and energy analyzer.

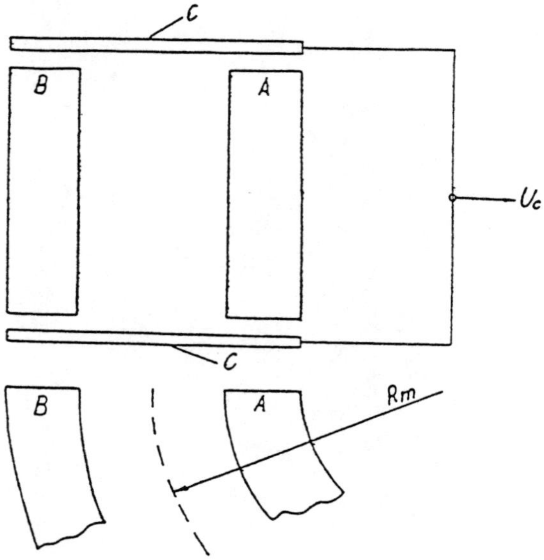

Fig.3 Schematic of the energy analyzer. "A" is inner electrode with radius R_a and potential U_a. "B" is outer electrode with radius R_b and potential U_b. "C" is counter electrodes with potential U_c. $R_m = \frac{R_a + R_b}{2}$

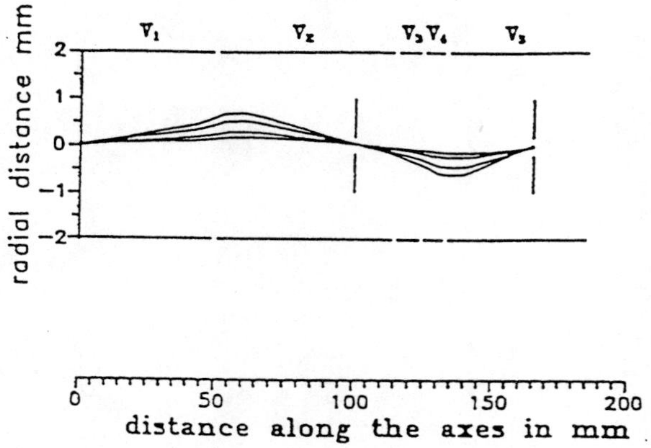

Fig.2 Calculated trajectories for different emission angles.

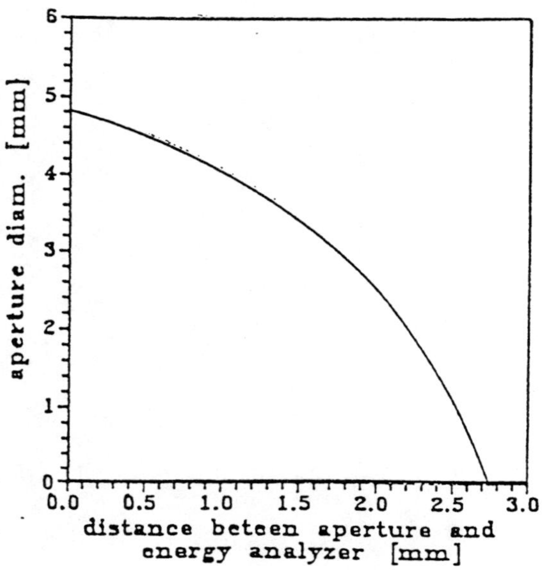

Fig.4 Dependence of aperture diameter plotted against the distance between aperture and energy analyzer.

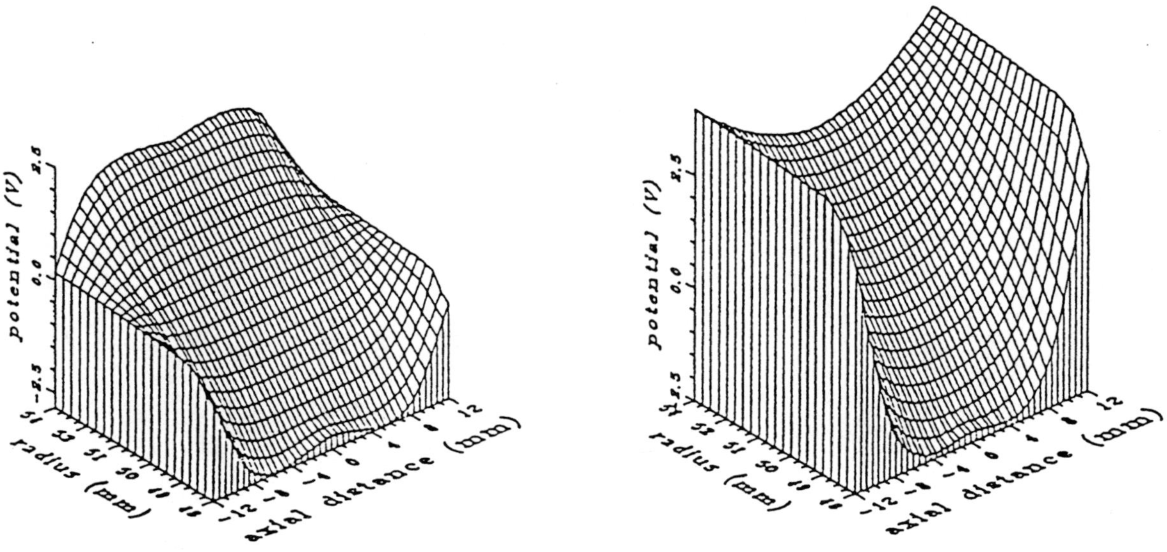

a) b)

Fig.5 Potential distribution inside the energy analyzer.

a) $U_a=-5.12V$, $U_b=4.64V$, $U_c=0V$

b) $U_a=-5.12V$, $U_b=4.64V$, $U_c=4V$

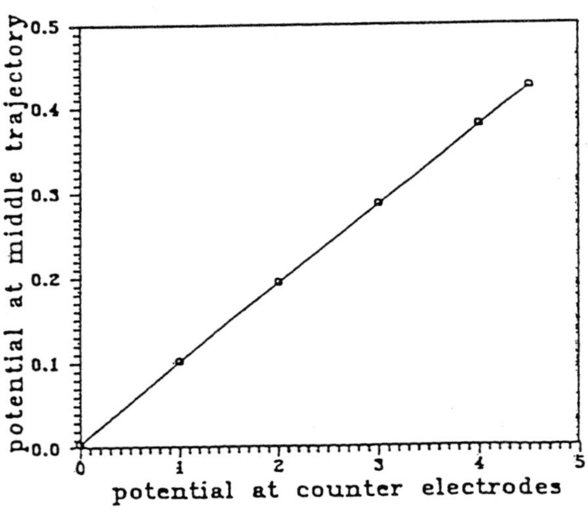

Fig.6 Dependence of middle trajectory potential against U_c. $U_a=-5.12V$ and $U_b=4.64$.

Vacuum Electron and Ion Technologies
D. S. Karpuzov (Ed.)

ISBN 1-56072-242-8

ON THE CONTROL OF ELECTRON BEAM AS A HIGH POWER HEAT SOURCE

Nikolai I Donkov and Dimitar L. Staikov
Institute of Electronics, Bulgarian Academy of Sciences,
Sofia 1784, Tzarigradsko shoussee 72, Bulgaria

1. INTRODUCTION

Electron beam (EB) as a high power heat source is widely used in vacuum coating systems, thin film production, tool coating and vacuum metallurgy. The lack of adequate models for the electron beam evaporating process makes it difficult to predict the optimum parameters for quality and low cost process control. The most widely used type of EB evaporators are shown in fig. 1.

The parameters characterizing the deposition process by the EB evaporation can be varied over a wide range by changing the EB heating power P_h. The power necessary to heat the EB evaporated material is a product of the anode current I_a and the anode voltage U_a, ($P_h = I_a \times U_a$). No problems would be faced during the EB heating process if I_a could be controlled adequately. I_a depends on the thermoemission from the filament and thus on the filament heating power P_f. In this case the manipulated variable is P_f and the controlled variable is I_a. The typically offset characteristic in fig. 2 is pertinent to the case of a tungsten filament (shown in fig. 3) of diameter d and length l, where l » d.

2. PROCESS MODEL

The power balance of the filament heating is:

$$P_f + P_0 = P_T + P_{em} + P_c + P_a + P_{conv}$$

where P_0 compensates the heat losses of the filament oxide film. In the case of $I_a \ll I_f$, $P_0 \approx 0$. The losses caused by the thermoionic emission can be neglected too, i.e. $P_a \approx 0$ (P_a does not exceed 2% of P_f). For working pressures below 10^{-4} Torr there are no convection losses and $P_{conv} \approx 0$. The thermoemissive losses are: $P_{em} = \varepsilon\sigma pl(T^4 - T_0^4)$ where ε is the total emissivity of the surface, σ is the Stefans-Boltzmann constant of radiation, p is the perimeter T is the filament temperature and T_0 is the room temperature. The total filament heating power P_f is given by: $P_f = I_f^2 R_f$, I_f is the filament heating current and R_f is filament resistance.

Neglecting the conductive losses, the equation of the stationary filament heat process takes the following form:

$$l\ p\ \varepsilon\ \sigma\ (T_m^4 - T_0^4) = I_f^2\ R_f$$

where T_m is the filament temperature at the wire's center.

In the case of contact heat losses ($P_c \neq 0$) the T-distribution is given by the differential equation:

$$-l\ k\ w\ \frac{\partial^2 T}{\partial x^2} = I_f\ R_f^2 - S\ \varepsilon\ \sigma\ (T^4 - T_0^4)$$

where k is the thermal conductivity, w is the cross sectional area of the filament, S is the surface area of the filament and x is the linear space coordinate.

The power necessary for the nonstationary filament heating process to occur is:

$$P_T = l\ w\ c\ \delta\ \frac{dT}{dt}$$

where c is the specific heat of tungsten and δ is the density.

On the basis of the power balance of filament heating, thermionic emission and electron bombardment heating processes, the transformation of P_f to I_a can be described by the equations (1)÷(3):

$$P_f = \varepsilon\ \sigma\ S\ (T^4 - T_0^4) \tag{1}$$

$$\frac{\partial T}{\partial t} = \frac{k}{c\ \delta}\ \frac{\partial^2 T}{\partial x^2} + \frac{p\ \varepsilon\ \sigma}{w\ c\ \delta}\ (T_m^4 - T^4) \tag{2}$$

$$I_a = S\ A\ T^2 e^{-b/T} \tag{3}$$

the coefficients A and b depend on the filament material.

The design of the I_a control system demands the synthesis of a linear mathematical model of the P_f to I_a transformation.

Bearing in mind the temperature intervals $T_{i+1} - T_i$, $i = 1 \div 4$ for the P_h ranges shown in fig. 2, equation (1) can be linearized as follows:

$$T = T_i + \Delta T$$

$$T^4 \approx T_i^4 + 4T_i^3\ \Delta T \tag{4}$$

$$P_{fi} = \varepsilon\ \sigma\ S\ T_i^4\ ;\quad P_f \approx P_{fi} + C_2 \Delta T\ ;\quad C_2 = 4\ \varepsilon\ \sigma\ S\ T_i^3$$

$$\Delta P_f = C_2\ (T_{i+1} - T_i)$$

The equation (3) can be approximated as:

$$I_a \approx S\ A_0\ \exp\left(-\ \frac{2.3b_0}{T}\right) \tag{5}$$

This relation is in good agreement [3] with experimental data. A_0 and b_0 depend on the type of filament material. Thus (5) can be linearized:

$$I_a = I_{ai}\left(1 + \frac{2.3b_0}{T_i^2}\Delta T\right) \; ; \quad I_{ai} = S\,A_0\,\exp\left(-\frac{2.3b_0}{T_i}\right)$$

$$C_4 = \frac{2.3b_0}{T_i^2} \; ; \quad \Delta I_a = C_4\,I_{ai}\,\Delta T$$

$$\Delta I_a = I_{ai}\,C_4\,\Delta T \tag{6}$$

$$\Delta I_{aimax} = I_{ai}\,C_4\,(T_{i+1} - T_i)$$

$$\Delta I_a = \frac{C_4}{C_2}\,I_{ai}\,\Delta P_f \tag{7}$$

If the magnetic field caused a space charge in the EB source, the I_a decrease is given by:

$$I_{ai} = S\,A_0\,K_i\,e^{-\frac{2.3\,b_0}{T_0}}$$

where K_i is the experimentally obtained coefficient.

With (4) in mind, and neglecting the T-distribution ($l \gg d$), the temperature increase per unit length is given by:

$$\frac{w\,c\,\delta}{4\,p\,\varepsilon\,T_i^3}\frac{d\Delta T}{dt} + \Delta T = \frac{p\,\varepsilon\,\sigma}{C_2}\,(T_{i+1}^4 - T_i^4)$$

with the boundary condition

$$\Delta T \longrightarrow T_{i+1} - T_i \qquad \text{when } T \longrightarrow \infty$$

The appropriate solution is:

$$\Delta T(t) = \frac{T_{i+1}^4 - T_i^4}{4T_i^3}\left(1 - e^{-\frac{t}{R}}\right) \tag{8}$$

$$R = \frac{c\,\delta\,w}{4\,p\,\varepsilon\,T_i^3}$$

From (6) and (7), for the transient response of the controlled member in the linearized intervals one can obtain:

$$\Delta I_{ai}(t) = \frac{I_{ai}\, C_4}{4T_i^3}\left(T_{i+1}^4 - T_i^4\right)\left(1-e^{-\frac{t}{R}}\right) \tag{9}$$

A simplified result is given by [3]

$$\Delta I_{ai}\ (t) \approx I_{ai}\ C_4\left(T_{i+1}-T_i\right)\left(1-e^{-\frac{t}{R}}\right) \tag{10}$$

The pulse response can be described by:

$$\Delta I'_{ai}\ (t) \approx \frac{I_{ai}\ C_4\left(T_{i+1}-T_i\right)\exp\left(-\frac{t}{R}\right)}{R}$$

The unit step responses for the 4 ranges from the offset characteristic are shown in fig. 4. To develop the control algorithm, it is necessary to create a discrete model of the evaporator as a controlled object.

The state - space model in the linearized range 4 is:

$$x(k+1) = A\ x(k) + B\ u(k)$$
$$y(k) = C\ x(k)$$

where x(k) is the five-dimensional state variables vector, y(k) and u(k) are the controlled members scalar output and input; y(k) corresponds to $I_a(k)$ and u(k) is proportional to $P_f(k)$. A and C are:

$$A = \begin{vmatrix} 0 & 1 & 0 & 0 & 0 \\ 0 & 0 & 1 & 0 & 0 \\ 0 & 0 & 0 & 1 & 0 \\ 0 & 0 & 0 & 0 & 1 \\ 0 & 0 & 0 & 0 & F \end{vmatrix} \qquad C = \begin{matrix} 1 & 0 & 0 & 0 & 0 \end{matrix}$$

$$B^T = 33.423\ \ 16.620\ \ 8.264\ \ 4.110\ \ 2.043$$

It's components are determined from the transient response (fig.5). The sampling time ΔT has been chosen to be 0.5s. F can be obtained from:

$$F = 1 - \frac{b_5}{\Delta P_{evimax} - \sum_{i=1}^{4} b_i}$$

In the same range 4, F is 0.503.

The transfer function of the evaporator in the "Z" domain is:

$$G_p(z^{-1}) = \frac{33.423z^{-1} + 0.08z^{-2} + 0.1z^{-3} + 0.01z^{-4} + 0.02z^{-5}}{1 - 0.503z^{-1}}$$

The pole z = 0.503 is less than unity and therefore the controlled object is asymptotically stable [5]

Effects as sputtering, oxidation etc. can influence R_f and d_f, which in turn can change the digital model's parameters and lower the control quality. Fig. 6 and fig. 7 shows the variation of the model's parameters with d. Calculations are made using equation 6 with ΔT = const and d-variable. The determination of d in real time can be performed easily using the available information for U_f and I_f.

$$d = 2\sqrt{\frac{I_f\,\rho_0(1 + \alpha\, T_i)}{\pi\, U_f}}$$

where ρ_0 is the specific resistance, α is the temperature coefficient of resistance. The block diagram of the control system is shown in fig. 8, where A-D converter (ADC) is optically isolated from the evaporator.

For the control law we have:

$$U_{DA} = H_I \sum_{i=0}^{k} R(i)-y(i) \quad - \sum_{j=2}^{6} H_j \hat{X}_j(k)$$

where H_j are the state vector feedback constants, R(k) is the reference signal, H_I is the integrator constant, $\hat{X}_j(k)$ are the estimated values (via finite time settling observer) of the state vector variables, that are not directly accessible [1]. The control law parameters can be calculate using the results [3].

Curve 1 in fig. 9 shows the control system response for unit step reference input with the algorithm for the model's parameter estimation, and curve 2 with a fixing parameters.

Curves 1 and 2 in fig. 10 shows the behavior of the I_a with control system and without control.

3. CONCLUSIONS

A mathematical description of the evaporator as a controlled object has been presented. The offset characteristic and transient responses are obtained.

An algorithm for model's parameters adaptation has been created. It takes into account the changes of the filament's diameter and the working ranges. Available adequate information for the model's parameter permits an "on line" tuning of the control law parameters.

REFERENCES

1. D. Auslender, M. Tomizuka & Y. Takahashi, *Proc. IEEE*, *66*, 113-24, 1978.
2. N. I. Donkov, *PhD Thesis*. Technical University - Sofia, Bulgaria, 1988.
3. N. I. Donkov, *J. Meas. Sci. Technol.*, *2*, 321-23, 1991.
4. N. I. Donkov, *VERHANDLUNGEN der D.P.G.*, Fruhyahrstagung-Regensburg, 3, p. 1013 Phisik verlag, Germany, 1992.
5. R. Iserman, *Digital Control Systems*, Springer, Berlin, 1981.
6. Y. Takahashi, M. Tomizuka & D. Auslender, *J. Dyn. Syst. Meas. Control*, *97*, 354-61, 1975.
7. Smirenin, *Radiotechnical Handbook*, Moscow, Gosenergoizdat, 1950

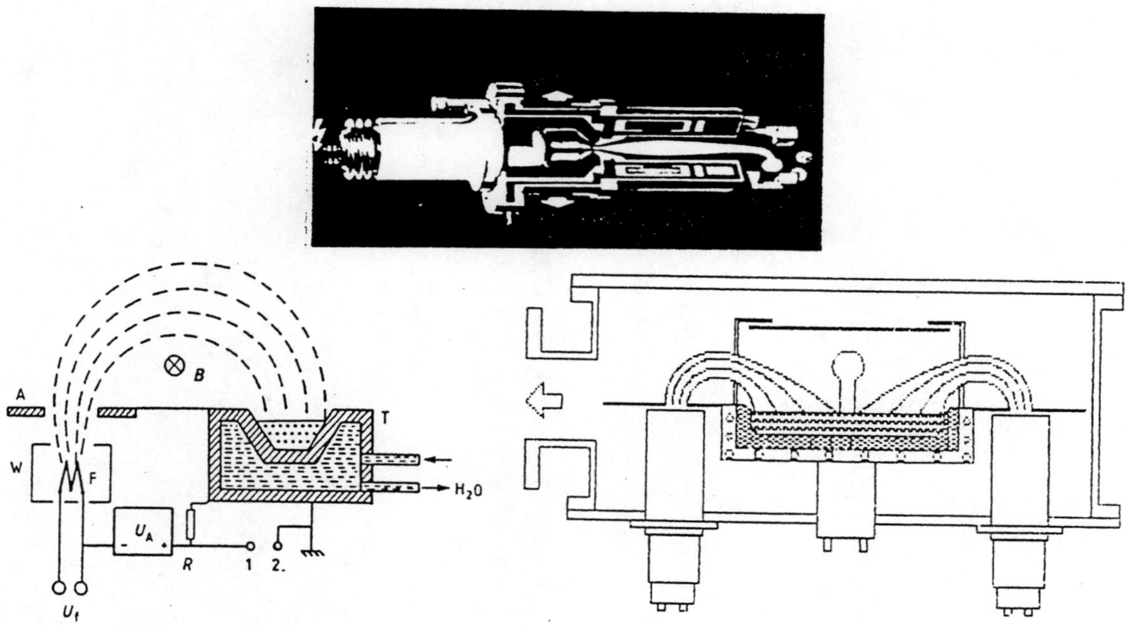

Fig. 1 Types of evaporators

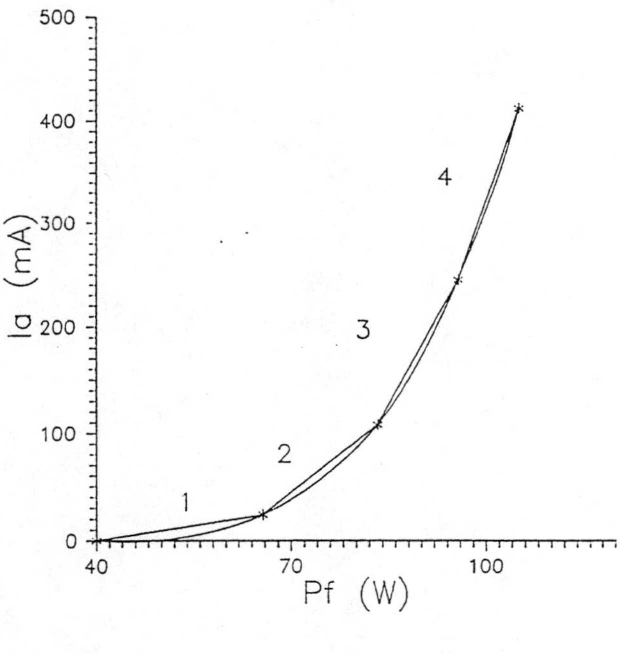

Fig. 2 The offset characteristic of the controlled object

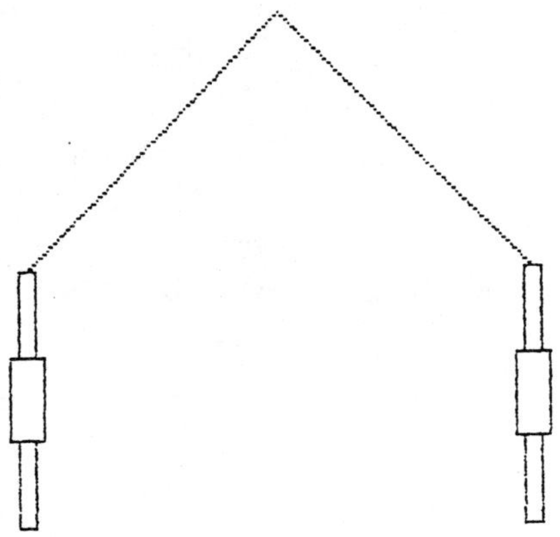

Fig. 3 Schematic drawing of the filament

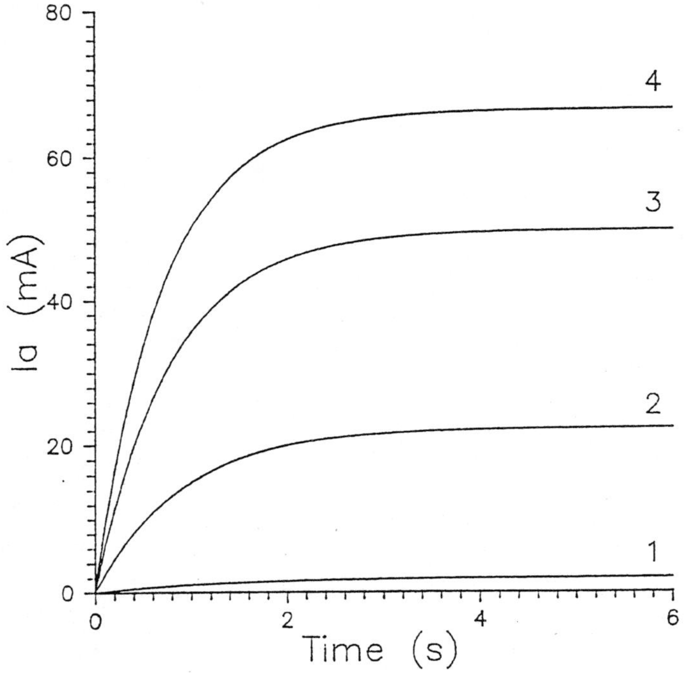

Fig. 4 The unit step responses of the object in the ranges 1-4

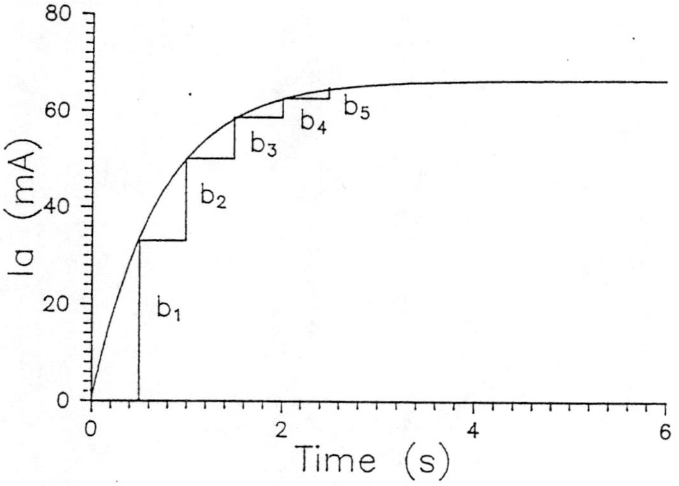

Fig. 5 The unit step response of the object and the digital model's parameter

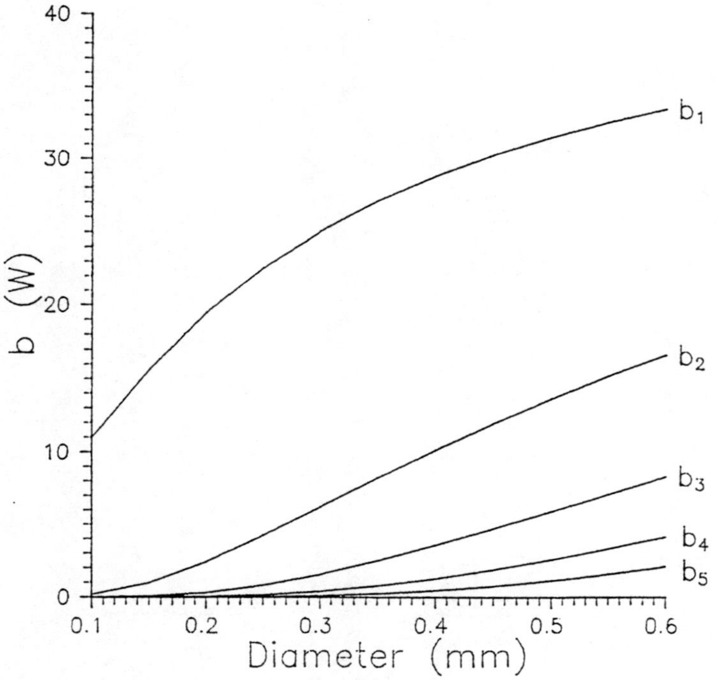

Fig. 6 The digital model parameter "b" as a function of "d"

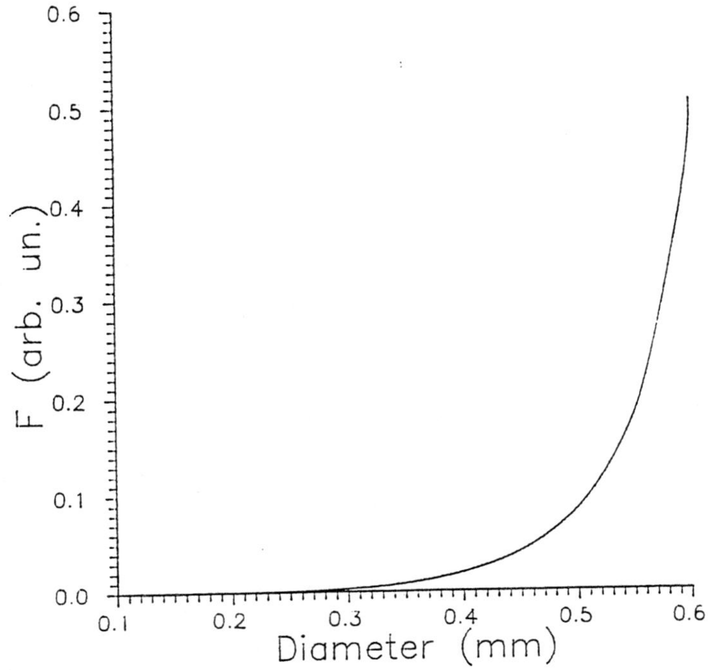

Fig. 7 The digital model parameter "F" as a function of "d"

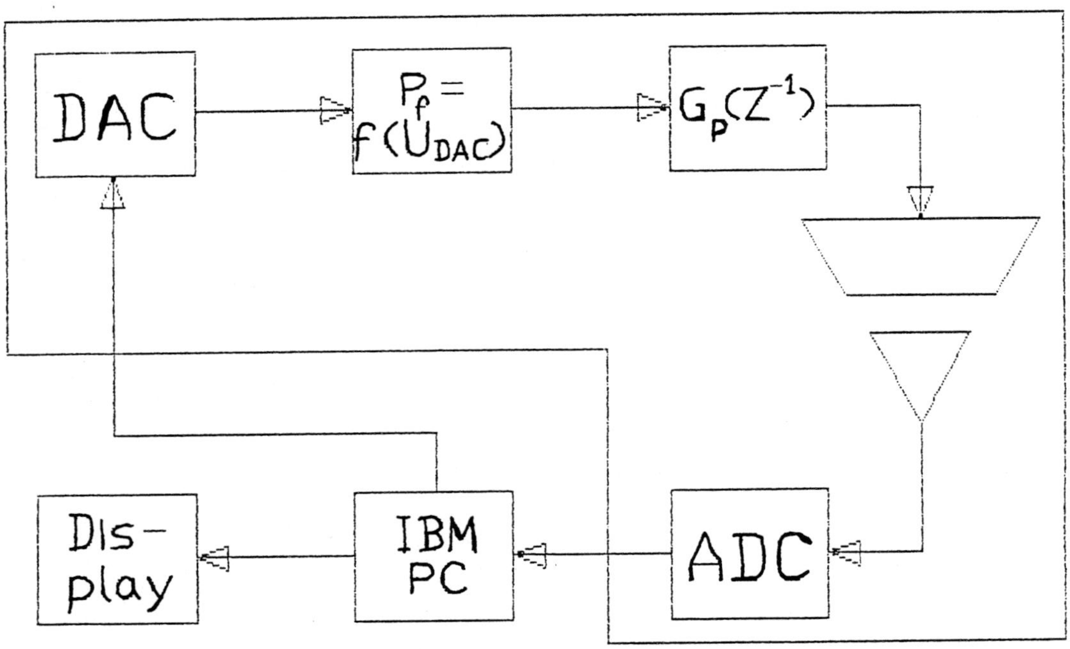

Fig. 8 The block diagram of the control system

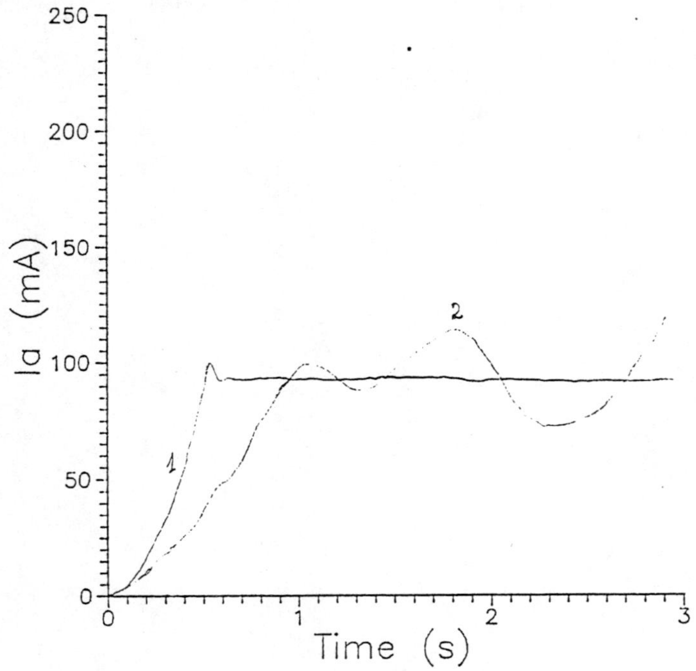

Fig. 9 The behavior of the controlled variables for unit step reference input with the model's parameter estimation and with a fixing parameter (curve 2)

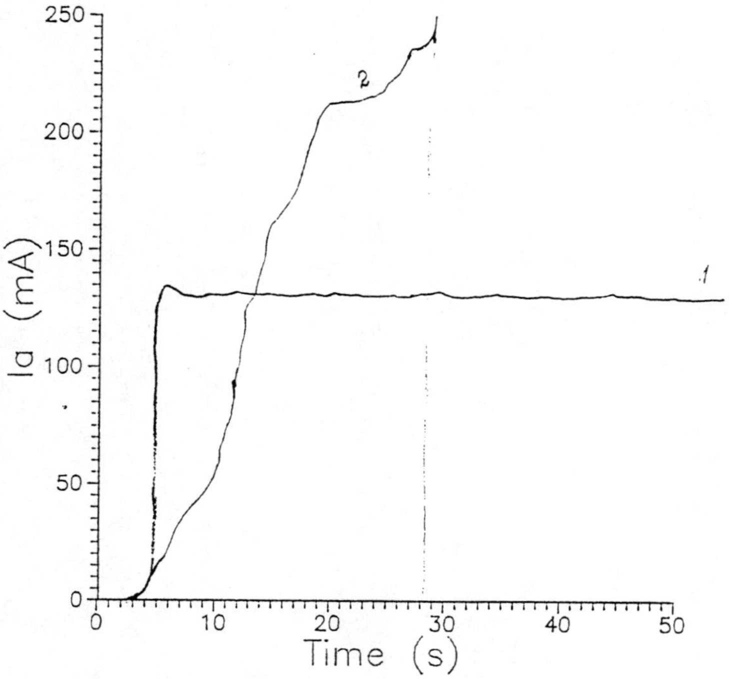

Fig. 10 The output variable for unit step reference input, with "I_a" control and without control (curve 2)

Vacuum Electron and Ion Technologies
D. S. Karpuzov (Ed.)

ISBN 1-56072-242-8

SEMICONDUCTOR QUANTUM WIRE STRUCTURES GROWN BY MOLECULAR BEAM EPITAXY: Growth Mechanisms and Application Perspectives

Marian A. Herman
Institute of Vacuum Technology, PL-00-241 Warsaw, and Institute of Physics, Polish Academy of Sciences, PL-02-668 Warsaw, Poland.

Recent developments in Molecular Beam Epitaxy, namely: (i) fractional monolayer growth on vicinal surfaces, and (ii) strained - layer MBE, have enabled crystallization of as - grown quantum wire (QWR) structures. The main advantage of as - grown QWRs is the real possibility of avoiding the charge carriers depletion effect inherent in post - growth processed structures.

Two MBE procedures, both resulting in as - grown QWR structures are described and shortly discussed in the paper. The first one is based on fractional monolayer growth, whereas the second uses strained layer epitaxy. The emphasis is put on the physical aspects of the growth mechanisms responsible for the formation of QWRs. The paper is concluded by a short outlook estimating the application perspectives of as - grown QWR structures in semiconductor devices.

1. Introduction

Crystal growth physics of low dimensional heterostructures (LDH) called also multidimensional quantum well (MD-QW) structures, belongs to the most fascinating problems in present day research on molecular beam epitaxy (MBE). Single or multiple quantum wells (1D-QW), quantum wires (2D-QW), and quantum boxes or quantum dots (3D-QW) create the MD-QW family.

LDH are the constituent elements of mesostructure electronics [1]. Growth of compound semiconductors by MBE makes possible fabrication of heterostructures with composition varying on a space scale of order ten to one hundred lattice constants. These has distinctive physical properties resulting from the occurrence of an intermediate size between two physical domains. The first is the microscale, or quantum effects domain of lattice constants, or the de Broglie wavelength λ_c of the charge carriers. The second is the macroscale domain associated with quasi-classical motion of the charge carriers. The motion of carriers in a direction where potential barriers created by heterointerfaces confine them to an area that has a size belonging to the microscale, is strongly quantized, while the carriers motion in directions where no such confining barriers exist, is quasi-classical. Fig.1 illustrates the HD-QW family and the variation of density-of-states of electrons with the increase of the quantization dimension in the QW structures [2].

As the dimension of QW increases from 1D to 2D or 3D, the degree of freedom in the free-electron motion decreases, leading to a change in $g_c(E)$. For the (3-i)-dimensional electron gas in the i-dimensional QW, $g_c^{(i)}(E)$ is expressed as follows [3]:

$$g_c^{(0)}(E) = \frac{(2m_c/\hbar^2)^{3/2}}{(2\pi^2)} \sqrt{E} \,, \tag{1}$$

$$g_c^{(1)}(E) = \sum_n \frac{m_c}{(\pi\hbar^2 L_z)} H[\, E - Ez(n) \,] \,, \tag{2}$$

$$g_c^{(2)}(E) = \sum_{n,l} \frac{(m_c/2\hbar^2)^{1/2}/(\pi L_y L_z)}{[\, E - Ey(l) - Ez(n) \,]^{1/2}} \,, \tag{3}$$

$$g_c^{(3)}(E) = \sum_{n,l,k} \frac{1}{(L_z L_y L_x)} \delta[\, E - Ex(k) - Ey(l) - Ez(n) \,] \,, \tag{4}$$

where m_c is the electron effective mass, E is the energy measured from the conduction-band edge Ec, $\hbar$ is Planck's constant, H(E) is a unit step function with $H(E \geq 0) = 1$ and $H(E < 0) = 0$, and $\delta(E)$ is the delta function. Ez(n), Ey(1), and Ex(k) denote the quantized energy levels with the quantum numbers n, 1, and k, respectively, over which summation should be carried out.

The area of application of 1D-QW structures in semiconductor devices is already considerable, extending from high electron mobility transistors through double barrier resonant tunneling microwave diodes to quantum well lasers and quantum well intraband photodetectors.

MD-QW lasers have been studied for the first time more than ten years ago [3], with the exciting conclusion that the threshold current of these lasers is much less temperature sensitive than that of a standard 1D-QW laser. Numerical examples of the calculated threshold currents versus operating temperature of MD-QW lasers are shown in Fig.2. A remarkable increase of optical gain and narrowing of the gain spectra of MD-QW lasers have also been demonstrated [2], what is shown in Fig.3. However, serious technological problems are still inherent in these lasing structures. This is the reason why until now only 2D-QW lasers (quantum wire lasers) have been grown with a reasonable success [4].

Two MBE growth procedures have been shown, so far, as resulting in as-grown quantum wire (QWR) structures. The first, realized on vicinal surfaces concerns serpentine superlattices (SSL) with quantum wires as parts of these structures [5]. The second, is realized on (100) on-axis substrates by the strain-induced lateral layer ordering (SILLO) process [6]. The main advantage of as-grown QWRs is the real possibility of avoiding the charge carriers depletion effect inherent in post-growth processed structures.

SSL structures are produced with a cyclic deposition and growth technique on vicinal substrates, taking the relative coverage between successive deposition cycles as a variable, precisely controlled parameter. During such a growth with MBE the shuttering of the molecular beams is under computer control and may be changed each cycle by a certain amount relative to the other cycles. Smoothly sweeping the per-cycle coverage back and forth through a range includinng monolayer coverage gives a structure with a continuosly varying tilt. 2D electronic confinement may be obtained at the wide places in the winding wells, where the structure turns a corner.

In the second MBE growth procedure it has been demonstrated that a strain-induced lateral periodic modulation of composition along the [110] direction with a periodicity as small as ~ 20 nm can be formed in vertical short-period superlattices (SPS) of $(GaAs)_2/(InAs)_2$ and $(GaP)_2/(InP)_2$ grown on nominally (100) GaAs and InP substrates, respectively. The strain induced by the deviation of superlattice periodicity from the substrate lattice constant in the SPS structure is the major driving force of the SILLO process.

2. Implementation of the MBE technology

The epitaxial growth process in which thin films made of semiconductors, metals or insulators, crystallize via reactions between thermal-energy molecular or atomic beams of the constituent elements and a substrate surface which is maintained at an elevated temperature in ultrahigh vacuum (UHV) ($p \leq 10^{-7}$ Pa) is called molecular beam epitaxy (MBE) [7].

The composition of the grown epilayer and its doping level depend on the relative arrival rates of the constituent elements and dopants, which in turn depend on the evaporation rates of the appropriate sources. The growth rate of typically 1 μm/h (1 monolayer/s) is low enough that surface migration of the impinging species on the growing surface is ensured. Consequently, the surface of the grown film is very smooth. Simple mechanical shutters in front of the beam sources are used to interrupt the beam fluxes, i.e., to start and to stop the deposition and doping. Changes in composition and doping can thus be abrupt on an atomic scale. The basic configuration of the MBE growth system is schematically illustrated in Fig.4.

In comparison to all other epitaxial growth techniques, MBE has an unique advantage. Being realized in an ultrahigh vacuum environment, it may be controlled in situ by surface sensitive diagnostic methods such as reflection high-energy electron diffraction (RHEED), Auger electron spectroscopy (AES) or ellipsometry [7]. These powerful facilities for control and analysis eliminate much of the guesswork in MBE, and enable the fabrication of sophisticated device structures by using this growth technique.

The majority of the UHV equipment used for MBE in the early stages of thie technique was bullt by commercial UHV manufacturers according to the instructions of the particular user. Since 1978, more-standardized MBE systems consisting of several basic UHV "building blocks", such as growth (reaction) chamber, substrate loading and preparation chamber, and surface analysis chamber, have become commercially available [7]. An example of such a commercially available MBE growth system is shown in Fig.5. This is the schematic drawing of the Perkin-Elmer MBE system PHI model 425 B. In this system the cassette with 10 substrate platelets fixed on their holders (mounting blocks) is loaded into the low volume entry chamber (a) which is then evacuated. The gate valve is opened, the elevator engages the cassette (b), and lowers it into the storage chamber (c). The elevator is raised, the fork (d) extracts the selected substrate holder (e) from the cassette, and transports it to the preparation chamber, where it can be transferred to the stage (f) or to the growth chamber. When moved to the growth chamber, it is inverted (g), transferred to the growth stage (h), and moved to the growth position (k). Before being transferred to the growth chamber, the substrate platelet is prepared for MBE growth process in the preparation chamber. Substrate preparation mainly means heat treatment in suitable conditions. This influences the quality of the substrate surface, and thus also to a large extent the quality of the epilayers grown by MBE. Usually the quality of the substrates prior to the MBE growth process is checked in the preparation-analysis chamber by means of surface analysis techniques (AES, LEED, or others). However, preparation and analysis are two different steps which are performed in a common vacuum chamber.

The enormous potential of MBE to produce structures of unequalled complexity can only be fully realized by utilizing effective automatic control. Stringent specifications for the accuracy and reproducibility of semiconductor structures such as superlattices [8], with layer thicknesses down to a few atomic layers, necessarily entail the use of a real-time computer control system. Real-time control ensures precise synchronization of events during the growth process. For example, shutter status and molecular beam fluxes must be assessed and adjusted simultaneously every one or two seconds in many growth procedures, especially when complicated device structures are to be grown with MBE.

The experimental arrangement of MBE enables one to study and control the growth process in situ in several ways. In particular, reflection high energy electron diffraction (RHEED) allows direct measurements of the surface structure of the substrate wafer and the already grown epilayer. It also allows observation of the dynamics of MBE growth [7]. Therefore, RHEED may be called an in-growth surface analytical technique.

The experimental geometry of RHEED is illustrated in Fig.6. Electrons having energy of typically 5-50 keV are incident on the substrate at a variable glancing angle ($0 \le \theta \le 5^0$). The diffraction of the incoming primary beam leads to the appearance of intensity-modulated streaks (or rods) normal to the shadow edge superposed on a fairly uniform background which is due to inelastically scattered electrons. It is obvious from the figure why this geometry is ideal for combination with MBE, where it is desirable to have the molecular beams impinging on the substrate surface at near-normal incidence.

In the conditions which are usually employed in MBE the de Broglie wavelength of the primary (incident beam) electrons is in the range of 0.17-0.06 Å and the penetration of the beam into the surface is low, being restricted to the outermost few atomic layers. Generally, the wavelength λ corresponding to an accelerating voltage V (expressed in volts) is given to a good approximation by [7]

$$\lambda \approx \frac{12.247}{[V\,(1 + 10^{-6}\,V)]^{1/2}} \quad [Å] \quad . \qquad (5)$$

Most geometrical aspects of the diffraction pattern can be interpreted on the basis of a limited penetration scattering model, i.e., a model which is kinematic in the diffraction sense. A detailed analysis of the diffracted beam intensity, particularly the way in which it changes with incident angle (θ in Fig.6), should in principle permit an exact determination of the complete structure of the surface unit cell. A set of data concerning the GaAs (001) surface has already been published, and a theoretical treatment also exists. However, it is very clear from many experiments that the incident beam penetrates deeper into the solid and, consequently, very strong multiple scattering effects (dynamical effects in the diffraction sense) dominate the diffraction process. This means that one has to be cautious when interpreting the intensity and apparent width of diffracted features [7].

In 1983 damped oscillations in the intensity of the specular and the difracted beams of the RHEED pattern were observed for the first time during MBE growth [9]. A typical example of the specular beam in the RHEED pattern from MBE-grown GaAs (001) surface is shown in Fig.6. The period of oscillation corresponds exactly to the growth of a single monolayer, i.e., a complete layer of Ga and As atoms (($a_0/2$) in the [001] direction) and is consequently independent of the azimuth of the incident beam and of the particular diffraction features being measured. The amplitude, however, is strongly dependent on both of these parameters. For evaluation of growth dynamics most of the information is contained in the specular beam. The observation of oscillatory effects in thin film growth studies is usually associated with a layer-by-layer growth process (i.e., two--dimensional growth).

If one equates changes in intensity of the specular beam in the RHEED pattern with changes in surface roughness, the equilibrium surface is smooth, corresponding to high reflectivity. Any roughening of this surface during growth can be related to local growth rate differences, although the overall growth rate does not vary and is determined only by the Ga flux. On application of the driving force (i.e., at commencement of growth) clusters are formed at random positions on the crystal surface, leading to a decrease in the reflectivity. This decrease can be predicted for purely optical reasons, since the de Broglie wavelength of the typically used electrons is ≈ 0.12 Å, while the bilayer step height is ≈ 2.8 Å, i.e., the wavelength is at least an order of magnitude less than the size of the scatterer, so diffuse reflectivity will result. Growth is not restricted to a single layer, but can recommence in a new layer before the preceding layer is complete. In the early stages, however, one layer is likely to be almost complete before the next layer starts, so the reflectivity will increase as the surface again becomes smooth on the atomic scale, with subsequent roughening as the next layer develops. This repetitive process will cause the oscillations in reflectivity to be gradually damped as the surface becomes statistically distributed over several incomplete atomic layers.

In Fig.6 a real space representation of the formation of the first two complete layers is shown [10]. This illustrates how the oscillations in the intensity of the specular beam occur. There is a maximum reflectivity for the initial and final smooth surfaces and a minimum (or maximum in diffuse scattering) for the intermediate stage when the growing layer is approximately half complete. It has, however, also been established that the amplitude of the intensity oscillations is dependent on the direction of the primary electron beam, being greater for the beam incident along [110] than [$\bar{1}$10] azimuths. This would suggest, if we take the optical model one stage further, that most of the steps which develop on the surface are along the [$\bar{1}$10] direction, as shown in Fig.6, i.e., they cause maximum diffuse scattering when their longer edges are normal to the incident beam. One may conclude, therefore, that growth occurs principally by a two-dimensional layer-by-layer process, but new layers are able to be started before preceding ones have been completed. The oscillation period provides a continuous and absolute growth rate monitor with atomic layer precision.

The phenomenon of RHEED pattern intensity oscillation has been routinely used by many MBE research groups to study the initial phase of the MBE growth mechanism of different materials. It has, however, also been applied for a considerable improvement of the growth procedure of very thin films and superlattice structures [7].

Neave et al. [12] examined RHEED oscillations occurring during growth of GaAs on vicinally cut (stepped) GaAs substrates at increasing growth temperatures. The principal feature of this work [Fig.7(a)] is the progressive reduction in the intensity of the RHEED oscillations until a critical temperature there are no oscillations. This effect may be interpretted as a transition from low-temperature growth by two-dimensional nucleation on the terraces to a high- temperature mode where increased mobility leads to direct incorporation of migrating adatoms into the growth front at the terrace edges. Thus, in the high temperature regime, growth proceeds by step propagation (step-flow mode), for which the mean surface step density remains approximately constant, in contrast to the two-dimensional nucleation growth mode, in which the step density varies as on a flat surface [Fig.7(b)]. Schematic illustration of the change in RHEED information as the growth mode changes from "step flow" to "two-dimensional nucleation" is shown in Fig.8.

3. MBE growth on vicinal surfaces. Tilted superlattices on GaAs substrates.

As already has been mentioned, RHEED intensity oscillations reflect the morphological change of the surface during growth by MBE. When the growth is dominated by two-dimensional nucleation, the surface morphology changes periodically and the RHEED intensity oscillates as a consequence. Under conditions where adatoms are predominantly incorporated into existing steps (step-flow growth mode) the surface morphology does not change significantly and RHEED intensity from the growing surface is approximately constant. Since the vicinal surface provides a staircase of steps, one can easily observe in situ the transition from two-dimensional nucleation to step-flow mode as the adatom mobility increases. It is clear that with RHEED intensity oscillation technique one may control the growth mode in MBE on vicinal surfaces [13].

During growth by step-flow mode, deposited atoms diffuse on an individual terrace of a substrate crystal until they become bound at a step. It is this concept that is central to the growth of planar superlattices and quantum wires. Depositing fractional monolayers of a $(GaAs)_m(AlAs)_n$ planar superlattice, with $m + n = p \approx 1$, one may grow on vicinal surfaces a so called tilted superlattice (TSL) [14].

In these superlattices, the interface planes are tilted with respect to the substrate surface plane, at an angle which can be controlled by adjusting the value of p or the orientation of the substrate surface. The tilt angle and period of the TSL provide a sensitive measure of the growth kinetics and the influence of variations in the growth parameters during MBE growth. Crystal growth proceeds by alternating depositions of between 1/2 and 1 monolayer of the column III atoms, with depositions of the column V atoms. This growth procedure may be referred to as alternating beam growth or ultrahigh vacuum atomic layer epitaxy (UHV ALE) [15].

An idealized representation of a TSL grown by UHV ALE is shown, together with the deposition algorithm, in Fig.9 [14]. The value of p, the superlattice period T, the step height d(= 2.83 Å), and the tilt α of the substrate plane with respect to (001) are related by

$$T = [\ pd/\sqrt{\tan^2\alpha + (1-p)^2}\ ] \ . \qquad (6)$$

A plot of T as a function of α, for selected values of p, is given in Fig.10(a). Equation (6) can be rewritten in terms of the angle between [001] and the TSL interface planes (β):

$$\tan\beta = (|p-1|/\tan\alpha) \ . \qquad (7)$$

Equations (6) and (7) are valid for $2/3 < p < 2$, $m < 1$, and $n < 1$. Plots of β as a function of α for selected values of p, and of β as a function of p for $\alpha = 2^{\circ}$ are given in Figs.10(a) and 10(b).

The TSL has been confirmed by TEM micrographs of all TSL layers grown in Ref.14. The results presented in Ref.14 demonstrate that the quality of the TSL, based on the criteria of long-range ordering and interface abruptness, is intimately tied to the crystal growth parameters. Uniformity of the substrate orientation over large areas is a critical prerequisite for successful growth of the TSL.

The last mentioned condition means that the steps array on the substrate surface should be perfectly periodic, i.e., uniform terrace lengths are indispensable for achieving a growth success in MBE of TSL structures [16].

In general, a virgin miscut sample will not have uniform terrace lengths. While the average spacing of steps is given by the ratio of the step height to miscut, it is clear from numerous surface science investigations that step spacings can vary significantly. For epitaxial applications one might attempt to overcome this situation through the growth of a buffer layer, a layer of the same material as the substrate. Assuming the buffer layer itself grows by step flow, and the atoms attach predominately at the up-step of each terrace, the surface terrace size distribution will sharpen. This is intuitively obvious when considering the mechanism of growth in this case. In a submonolayer deposition each terrace accumulates a fraction of a monolayer proportional to its length. The atoms diffuse to the terrace edge (the up-step) where they become part of the adjacent terrace. In general larger terraces tend to shrink and smaller terraces tend to grow towards the average value. (This is not so if the atoms have equal probability to attach at the up-step and down-step, or if they attach predominately at the down-step.

Consider a surface as schematically shown in Fig. 11. We assume that going from left to right only down-steps of uniform height occur. Each terrace is identified by its index n = 1,2,... and has a length L_n. Growth of deposit occurs at a rate r monolayer per unit time; the probability of an arriving atom to diffuse to the left up-step of a terrace and become attached there is 1 - P; the probability to attach at the right down-step, P. By introducing P we have essentially parametrized into a single parameter all the microscopic information related to the processes which lead to diffusion of an adsorbing atom over the terraces and steps and their eventual adsorption at a suitable step site. The probability P is related to the potential energy governing the movement of an adsorbed atom on the surface. Following Gossmann et al. [16] we can draw a potential-energy diagram of a stepped surface as schematically shown in Fig. 12(a). E_b and E_ν are the energy barrier, which may or may not be present, and the energy valley due to the step. Some limiting cases are immediately apparent; for example, if $E_b = 0$ and $E_\nu \rightarrow \infty$, then P = 0.5, or if $E_b \rightarrow \infty$ and $E_\nu \rightarrow \infty$, then P = 0. In general the absence of a barrier, $E_b = 0$, or for a perfectly reflecting barrier, $E_b \rightarrow \infty$, P is independent of the terrace length L. For intermediate values, P will in general be a function of L since the adatom concentration on the terrace depends on the length of the terrace. Fig. 12(b) gives schematically the values of P arising from a particular potential.

Calculations based on a model of step-mediated MBE growth, presented in Ref. 16 have demonstrated an analytical derivation of the realistic approach to uniform terrace lengths on the external surfaces of the buffer layer grown on stepped surfaces. The results show that growth interruption usually used in MBE to get smooth growing surfaces [7] is of no advantage and that in general the approach to uniform terrace lengths is quite slow. The width of the terrace length distribution varies approximately as the inverse 4th root of the deposited coverage. This will only occur if the atoms attach themselves predominately at the up-step of each terrace. Otherwise, the width of the distribution will grow without bounds. The model further predicts characteristic oscillations of terrace lengths in the vicinity of multiple height steps.

An interesting result concerning the MBE growth of TSLs has been demonstrated in Ref. 17. Periodic Al composition modulations have been observed to occur spontaneously during growth of AlGaAs on vicinal (100) substrates. The formation of the spontaneous Al modulation requires (a) the migration-enhanced epitaxy deposition and (b) one monolayer deposition of Al and Ga atoms per cycle. Coherent tilted superlattice (C-TSL) has been grown in the reported process. Cross-sectional transmission electron microscopy clearly shows that the C-TSL has the periodicity of the surface steps. It was also shown that the Al-rich regions of the C-TSL form at the bottom of steps before the As flux is established. These results indicate that growth kinetics dominates the C-TSL formation.

Migration-enhanced epitaxy (MEE) is a special modification of UHV ALE, realized with III-V compounds [15]. It has attracted much attention as one of the best methods to enhance the Ga atom diffusivity on surfaces during MBE growth. Although this mobility control has already been recognized as one of the key features in the TSL deposition, the growth kinetics during MEE deposition are still not well understood. Hence their understanding is of importance to improving the TSL perfection.

The following model, shown schematically in Fig. 13 has been proposed to account for the growth of C-TSLs. In the first phase of the C-MEE mode [see (a - 1) in Fig. 13], Al and Ga atoms come to the step ledges coherently. Both Ga and Al atoms on the surface have fairly large mobilities because the lack of As, and move around on the step surface. Even at step risers, Ga atoms are not incorporated. On the other hand, due to the stronger bonding, Al atoms are captured more easily by step risers. The picture corresponds to that of the low T_s growth of GaAs in the MEE mode [18]. Hence, Al atoms begin to accumulate at the bottom of the steps in this phase. Before As_4 molecules come to the surface, a phase separation into two as shown in (b-1) in Fig. 13 has occurred. Note that no C-TSL is formed when the As beam is incident on the surface in these phases (a-1) and (b-1). Hence, it is not likely that the phase separation occurs after incorporation of As. Then, As_4 molecules come to the surface (c-1), and epitaxial layers are formed (d-1), which are followed by the second cycle (a-2), (b-2),... . By repetition of these cycles, a C-TSL is formed. Note that this model relies on the difference in the attachment kinetics between movable Al and Ga, and does not depend on the phase separations in the underlayers. In the framework of this model, the higher T_s is, the less Al atoms are likely to be captured at step risers in the phase (a) and (b) in Fig. 13. This is one possible reason, in addition to the sublimation of GaAs, to the fact that no C-TSL is formed at the high growth temperature ($T_s = 650^{\circ}C$).

TSLs have not successfully been used to form two-dimensional confinement structures with good electrical and optical properties. One reason for poorly defined confinement is because of the thermodynamic difficulty of having fractional layers segregate fully on the vicinal terraces in the (Ga,Al)As system. Perhaps this may be avoided through a better understanding of the growth processes in this system or by turning to other systems such as the (Ga,Al)Sb system [5]. Another problem, intrinsic to the growth geometry, is that the tilt is not a well controlled parameter. This stems from the facts that the deposition and growth rates cannot be characterized to within less than about one percent and that growth rate variations across a substrate are typically greater than one percent. One percent errors in the growth rate variously give tilts of 14° and 30° on vicinal GaAs substrates misoriented 2° and 1°, respectively, from the (100) surface [14]. Electronic confinement energies, and resulting electrical and optical properties, will change with the tilt and are thus not well controlled in a TSL. Therefore, another cyclic deposition and epitaxial growth technique for vicinal surfaces has been presented [5]. The resulting structures, named serpentine superlattices (SSLs) for their purposefully meandering shapes, may be employed to obtain quantum wires. These structures inherently accommodate the geometric sensitivity to absolute growth rates.

As already mentioned in the Introduction, SSLs are produced with a cyclic deposition and growth technique on vicinal substrates, taking the relative coverage between successive deposition cycles as a variable, controlled parameter. During such a growth with MBE, for example, the shuttering of the molecular beams is under computer control and may be changed each cycle by a certain amount relative to the other cycles. Smoothly sweeping the per-cycle coverage back and forth through a range including monolayer coverage gives a structure with a continuously varying tilt, as is schematically represented in Fig. 14(a). Two-dimensional electronic confinement may be obtained at the wide places in the winding wells, where the structure turns a corner. The shape near these regions determines the energy spectrum of the confinement. Perhaps the most interesting case occurs for a linear ramping of the per-cycle coverage through a range including vertical growth. This instance gives a SSL with a cross section of periodically displaced parabolas having curvatures determined by the mis-orientation angle and the per-cycle coverage ramping rate $1/z_o$. The confinement energies are determined by z_o, the lateral well and barrier widths, and the barrier height. The confinement in this case is insensitive to errors in the absolute growth rates. An error shifts the vertices of the parabolas up or down in the growth direction but does not change their shape. Thus a confinement may be obtained that is quite insensitive to growth rate variations across an entire substrate.

4. Strained layer MBE. The SILLO process.

The possibility of growing high-quality epitaxial layers of different materials on lattice mismatched semiconductor substrates is a topic of considerable interest. The range of useful devices available with a given substrate is considerably enhanced by this method. During epitaxial growth, if the mismatch between a substrate and the growing epilayer is small, the first atomic layers which are deposited will be strained to match the substrate lattice and a coherent interface is formed. However, as the thickness of the growing epilayer increases, the homogeneous strain energy of the epilayer becomes large enough to make the nucleation of misfit dislocations energetically favorable. The thickness of the epilayer at which misfit dislocations are nucleated to relieve the homogeneous strain energy defines the critical thickness for the system. The existence of a critical thickness was first considered by Van der Merwe [19] and later by numerous authors [7].

Following Van der Merwe [20], we will use hereafter the term "misfit" to refer to the disregistry of the equilibrium interfacial atomic arrangements of the substrate and the overgrown epilayer. This disregistry results from differences in atomic spacings or lattice symmetries, which are characteristic of each of the two crystals in the absence of interfacial interaction between them.

In the important case of superlattice growth, where the thickness of the overlayer is comparable or equal to the thickness of the substrate, misfit is defined as

$$f_i = 2\frac{a_{si} - a_{oi}}{a_{si} + a_{oi}} \quad , \quad i = x,y \tag{8}$$

where a is the lattice constant (the natural lateral atomic spacing) and s and o designate the substrate (or the already grown epilayer in the layer sequence of the superlattice) and the overgrown layer, respectively.

Strained-layer superlattices (SLS) are structures comprising alternating layers of two or more materials with different equilibrium lattice constants, grown in a structure with a common in-plane lattice constant (see Fig.15). For sufficiently thin layers, the lattice mismatch is accommodated by uniform strain in the layers. Under these conditions, no misfit dislocations are generated at the interfaces, so high-quality crystalline materials are yielded.

SLSs can be grown with positive and negative strains in alternating layers, and thus no net forces, or with essentially all of the strains confined to one of the materials to yield a strain energy that increases with total SLS thickness. A typical structure designed to have no net strain consists of three principal regions: a bulk crystalline substrate, a buffer layer comprising an alloy with a graded composition (so that the lattice constant varies from that of the substrate to the average lattice constant of the SLS) and the SLS. This buffer layer provides a transition from the crystalline lattice to the SLS. The

SLS consists of many alternating thin (< 30 nm) single-crystal layers with equilibrium lattice constants that are alternately larger and smaller than the average constant. Each layer of the SLS is elastically strained and the alternating strains deflect dislocations that originate in the graded layer to the edge of the wafer. The dislocations are thus confined to the first few layers of the SLS to produce dislocation-free material in the upper layers.

The amount of strain that can be accommodated in equilibrium is determined by the energy balance between misfit dislocations and strain energy. The thinner the layer, the more strain it can accommodate. Very thin layers (of the order of tens of nanometers) can accommodate misfits of several atomic percent. In this case, the lattice spacings adjust in the plane with an accompanying change in the lattice spacing perpendicular to the layer. Thus crystals with cubic equilibrium structures undergo a tetragonal distortion, which can have major implications for the electronic and optical properties.

Strained layer MBE has been used recently for the realization of quantum wire structures [4]. It has been demonstrated [6] that a strain induced lateral layer ordering (SILLO) process occurs spontaneously when $(GaP)_n/(InP)_n$ or $(GaAs)_n/(InAs)_n$ short-period superlattices (SPS) are grown on (100)-oriented on-axis GaAs or InP substrates, respectively.

The epitaxial growth of $(GaAs)_n/(InAs)_n$ SPS on InP was carried out in a solid source molecular beam epitaxy system. A 3000-Å $Ga_xIn_{1-x}As$ buffer layer and a $(GaAs)_n/(InAs)_n$ SPS layer were successively grown on the on-axis (100)InP substrate. The SPS layer contains 100 to 170 pairs of $(GaAs)_n/(InAs)_n$ superlattices. The composition of the buffer layer was adjusted to closely match that of the InP substrate. Although there is a 7% lattice mismatch between GaAs and InAs, the average lattice constant of the $(GaAs)_n/(InAs)_n$ SPS is also nearly lattice matched to the InP substrate. The epitaxial growth of $(GaP)_n/(InP)_n$ SPS structure on the on-axis (100) GaAs substrate was performed in a gas source molecular-beam epitaxy system. In both cases, the layer thicknesses of both binary compounds within the SPS structure were calibrated through RHEED intensity oscillations. The SPS layers with $n \geq 3$, due to the critical thickness limitation, showed three-dimensional growth in relatively thin SPS layers and, therefore, were not studied. The substrate temperature was held at ~ 500°C throughout the growth of the SPS structures.

The microstructures of $(GaAs)_n/(InAs)_n$ and $(GaP)_n/(InP)_n$ SPS layers were first examined by cross-sectional TEM. From the positions of the satellite reflections of the TEM diffraction pattern, the periodicity of the SPS layer thickness has been determined. The percentage thickness deviation of the superlattice periodicity from na_0, i.e., (superlattice periodicity - na_0)/na_0, is denoted by ΔT, where the integer n is the number of monolayers of each binary compound within a period of SPS structure and a_0 is the lattice constant of the substrate material. Samples with n = 1 were studied first. When the superlattice periodicity is slightly deviated from a_0, i.e., $\Delta T > 4\%$, their cross-sectional TEM image contrast shows a marked difference in the [110] and $[\bar{1}10]$ directions for both $(GaP)_1/(InP)_1$ and $(GaAs)_1/(InAs)_1$ SPS structures. For samples with n = 2, a modulated image composed of dark and white fringes in the $[\bar{1}10]$ cross-section direction appears at a smaller ΔT than that of the sample with n = 1. Comparing to the uniform image for the sample with n = 1 and $\Delta T = 3.7\%$, a modulated image composed of very weak dark and light fringes parallel to the growth direction already appeared in the $[\bar{1}10]$ cross-section of the $(GaAs)_2/(InAs)_2$ SPS sample with $\Delta T = 4.2\%$. The image contrast of the dark and white fringes become stronger when the periodicity deviations are further increased.

Lateral modulation of the composition has been identified as the cause of the formation of the fringe contrast and alternating dark/light fringes. This is a lateral growth-induced [110] superlattice. The experimental results indicate that a lateral modulation of the composition exists only in SPS structures with a moderate to large deviation of the superlattice periodicity, ΔT, from na_0. Furthermore, this effect is enhanced when n increases from 1 to 2. Therefore, the lateral composition modulation process is strongly related to the magnitude of the induced strain in the SPS structure. Thermodynamically, the observation of both vertical long-range ordering and lateral compositional modulation (phase separation or lateral layer ordering) in these binary SPS structures is unfavorable. This is because the thermodynamic driving forces for their formation are opposite to each other. The existence of both long-range vertical ordering and lateral layer ordering suggests that the thermodynamic requirements are relaxed by other factors other than the formation enthalpies and entropies. One of these factors is the induced strain associated with the deviation of superlattice periodicity in the SPS structure from na_0. Taking into account elastic strain relaxation, Glas [21] showed that when the composition modulated in an epitaxial layer in directions parallel to the substrate surface, i.e., lateral layer ordering, it was more thermodynamically stable than its unmodulated counterpart. The implication of this model is that a strained epitaxial layer, if kinetically feasible, will undergo a lateral compositional modulation first, rather than a less favorable reaction of generating misfit dislocations to relieve the misfit strain. In addition, once a lateral modulation of the composition has started, it will continue with the same periodicity. The strain induced from the deviation of superlattice periodicity from na_0 in the SPS structure is the major driving force of the SILLO process. Schematic diagram (a) and bright-field TEM cross section (b) of the $Ga_xIn_{1-x}P$ multiple QWR heterostructure grown by MBE when applying the SILLO process is shown in Fig. 16 [22].

5. QWR structures grown by MBE

We will illustrate the MBE grown QWR structures on examples of quantum wire lasers, because for these devices QWRs are most frequently used at present.

Quantum wire semiconductor lasers offer a number of advantages over standard quantum well lasers, including lower threshold current densities, increased modulation bandwidths, narrower spectral linewidths, and reduced temperature sensitivities [3]. The realization of the benefits described above have been demonstrated by applying the SILLO process, described in Sect.4, to laser manufacturing [4].

The AlGaInP multiple QWR lasers were grown by a single-step gas source MBE [7], on (100)-oriented, on axis Si doped GaAs substrates. The schematic diagram of a typical laser diode is similar to the LED structure shown in Fig.16(a). With this structure the first visible $Ga_xIn_{1-x}P/Al_{0.15}Ga_{0.35}In_{0.5}P$ multiple QWR laser diodes have been demonstrated [4]. Threshold current densities were greatly reduced for lasers fabricated with contact stripes in the [110] direction, perpendicular to the QWR array. Electroluminescence of these lasers is highly polarized and corresponds directly to the dominant strain present in the device structure. These multiple QWR laser diodes show great promise due to the high density of QWRs formed and their ease of fabrication.

Another possibility of growing with MBE QWR laser structures is based on tilted and serpentine SL structures (see Sect.3). The schematic cross-sectional profile of the SSL geometry is shown in Fig.17. The curvature of the growth interface provides vertical confinement of the charge carriers in this structure. The degree of the confinement depends on the substrate tilt angle, α, the total thickness of the SSL, D_0, and the intentional variation of the growth rate, Δp. The shape of the vertical curved interface can be expressed as $y = \Delta p\, z^2/[(\tan \alpha)\, D_0]$. p, known as the tilt parameter, is defined as $p = m + n$, where m and n are the fractions of barrier and well material on a step.

If the deposition or growth rate is adjusted such that $p = 1$, the amount of material deposited per monolayer exactly covers a step and the interface between the well and barrier material will grow vertically. If, however, $p < 1$ or $p > 1$ interface will tilt one way or the other. $\Delta p = | p - 1|$ indicates the intentional deviation from the nominal p value of 1 and determines the slope of the interface.

It is obvious, that Δp, α, and D_0 plus the Al contents in the wires, barriers, and the claddings, respectively, are the parameters that are available to optimize the structure. The results of theoretical estimations concerning the optical gain characteristics of the SSL QWR laser are shown in Fig. 18. One may conclude that the considered laser structure, being still under laboratory investigation, is already promising and will probably become significantly improved in the nearest future.

6. Application perspectives of QWR structures

Concluding this paper let us follow the ideas and the suggestions published recently by Sakaki with coworkers [23] concerning the advantages of quantum microstructures.

Confined electrons in a quantum microstructure form quantized levels, subbands, or minibands in a well region, depending on the coupling strength through barriers. As the degree of confinement is increased, the density-of-state function becomes sharper and more discrete. Fig. 1 illustrates the schematic densities of states in a QW, QWR, and QB as a function of electron energy. These features could lead to very high efficiencies in quantum microstructure devices.

The first idea that a quantum microstructure could be applied to high-performance devices was proposed by Sakaki [3]. He found theoretically that the Coulomb scattering probability of electrons in a quantum wire is extremely strongly suppressed if most of the electrons are accommodated in a ground subband, namely, the quantum limit condition is fulfilled. This is because an electron traveling along the wire with initial state k_i can be scattered elastically only to the state $-k_i$ in momentum space, preserving a constant energy. Since this backward transition is accompanied by a large momentum exchange of $-2k_i$, its probability becomes extremely small. The estimated mobility in a GaAs QWR can be as high as 3×10^8 $cm^2V^{-1}s^{-1}$ at low temperatures for electron concentrations of (1-2) x 10^6 cm^{-1} and impurity-wire distances 20-10 nm.

Optical phonon scattering of electrons, which occurs at high temperatures and in high electric fields, has also been predicted to be suppressed in coupled QWR superlattices or periodic QB arrays if the miniband structure satisfies the following three conditions: (i) the minigap between the ground and second minibands is much wider than the average kinetic energy of electrons, so that the majority of electrons populate in the ground miniband (the quantum limit condition); (ii) the width of the ground miniband is narrower than the optical phonon energy E_{op} (≈ 36 meV in GaAs), so that both the phonon emission and its absorption are prohibited within the miniband; (iii) the first minigap is wider than E_{op}, so that no interminiband scattering is allowed.

High electron mobility as a result of the suppression of dominant scattering leads to increased acceleration rate and saturation velocity of electrons. This implies significant improvements in cut-off frequency and noise figure at ultra-high frequencies. In addition, the elimination of scattering results in an extremely long mean free path (≈ 40 μm if the mobility is 10^7 $cm^2V^{-1}s^{-1}$), which facilitates the construction of quantum interference devices.

Despite these advantages, it is difficult to fabricate quantum microstructures with a characteristic dimension of less than or equal to 25 nm at present. Elimination of size fluctuation, interface roughness, and contamination in the fabrication is required to obtain an efficient device. Quantum interference devices are likely to be among the promising applications. However, granularity noise arising from the probabilistic nature of quantum mechanics sets a fundamental limit on their performence.

One may, however, believe that the above-mentioned problems could be overcome by progress in fabrication technology and/or by the appearance of novel materials and device concepts in the near future.

REFERENCES

1. P.J.Price, IEEE Trans.Electron Devices, ED-28, 911 (1981).
2. M.Asada, Y.Miyamoto, and Y.Suematsu, IEEE J.Quantum Electron., QE-22, 1915 (1986).
3. Y.Arakawa and H.Sakaki, Appl.Phys.Lett., 40, 939 (1982).
4. E.M.Stellini, K.Y.Cheng, P.J.Pearah, A.C.Chen, A.M.Moy, and K.C.Hsieh, Appl.Phys.Lett., 62, 458 (1993).
5. M.S.Miller, C.E.Pryor, H.Weman, L.A.Samoska, H.Kroemer, and P.M.Petroff, J.Crystal Growth, 111, 323 (1991).
6. K.Y.Cheng, K.C.Hsieh, and J.N.Baillargeon, Appl.Phys.Lett., 60, 2892 (1992).
7. M.A.Herman, and H.Sitter, Molecular Beam Epitaxy: Fundamentals and Current Status, Springer, Berlin, Heidelberg, 1989.
8. M.A.Herman, Semiconductor Superlattices,Akademie Verlag, Berlin, 1986 (russian translation by Mir, Moscow, 1989).
9. B.A.Joyce, Rep.Prog.Phys., 48, 1637 (1985).
10. B.A.Joyce, P.J.Dobson, J.H.Neave, K.Woodbridge, J.Zhang, P.K.Larsen, and B.Boelger, Surface Sci., 168, 423 (1986).
11. J.P.Harbison, in: Concise Encyclopedia of Semiconducting Materials and Related Technologies, S.Mahajan, and L.C.Kimerling (eds.), Pergamon Press, Oxford, 1992, p.309.
12. J.H.Neave, P.J.Dobson, B.A.Joyce, and J.Zhang, Appl.Phys.Lett., 47, 100 (1985).
13. T.Shitara, J.Zhang, J.H.Neave, and B.A.Joyce, J.Crystal Growth, 127, 494 (1993).
14. J.M.Gaines, P.M.Petroff, H.Kroemer, R.J.Simes, R.S.Geels, and J.H.English, J.Vac.Sci.Technol., B6, 1378 (1988).
15. M.A.Herman, Vacuum, 42, 61 (1991).
16. H.J.Gossmann, F.W.Sinden, and L.C.Feldman, J.Appl.Phys., 67, 745 (1990).
17. M.Tsuchiya, P.M.Petroff, and L.A.Coldren, Appl.Phys.Lett., 54, 1690 (1989).
18. Y.Horikoshi, M.Kawashima, and H.Yamaguchi, Jpn.J.Appl.Phys., 27, 169 (1988).

19. J.H.Van der Merwe, J.Appl.Phys., 34, 123 (1962).

20. J.H.Van der Merwe, CRC Crit.Rev.Solid State Mater.Sci., 7, 209 (1978).

21. F.Glas, J.Appl.Phys., 62, 3201 (1987).

22. P.J.Pearah, E.M.Stellini, A.C.Chen, A.M.Moy, K.C.Hsieh, and K.Y.Cheng, Appl.Phys.Lett., 62, 729 (1993).

23. H.Noge, A.Shimizu, and H.Sakaki, Nanotechnology, 3, 180 (1992).

24. P.S.Peercy, in: Concise Encyclopedia of Semiconducting Materials and Related Technologies, S.Mahajan, and L.C.Kimerling (eds.), Pergamon Press, Oxford, 1992, p.521.

25. J.C.Yi, and N.Dagli, Appl.Phys.Lett., 61, 219 (1992).

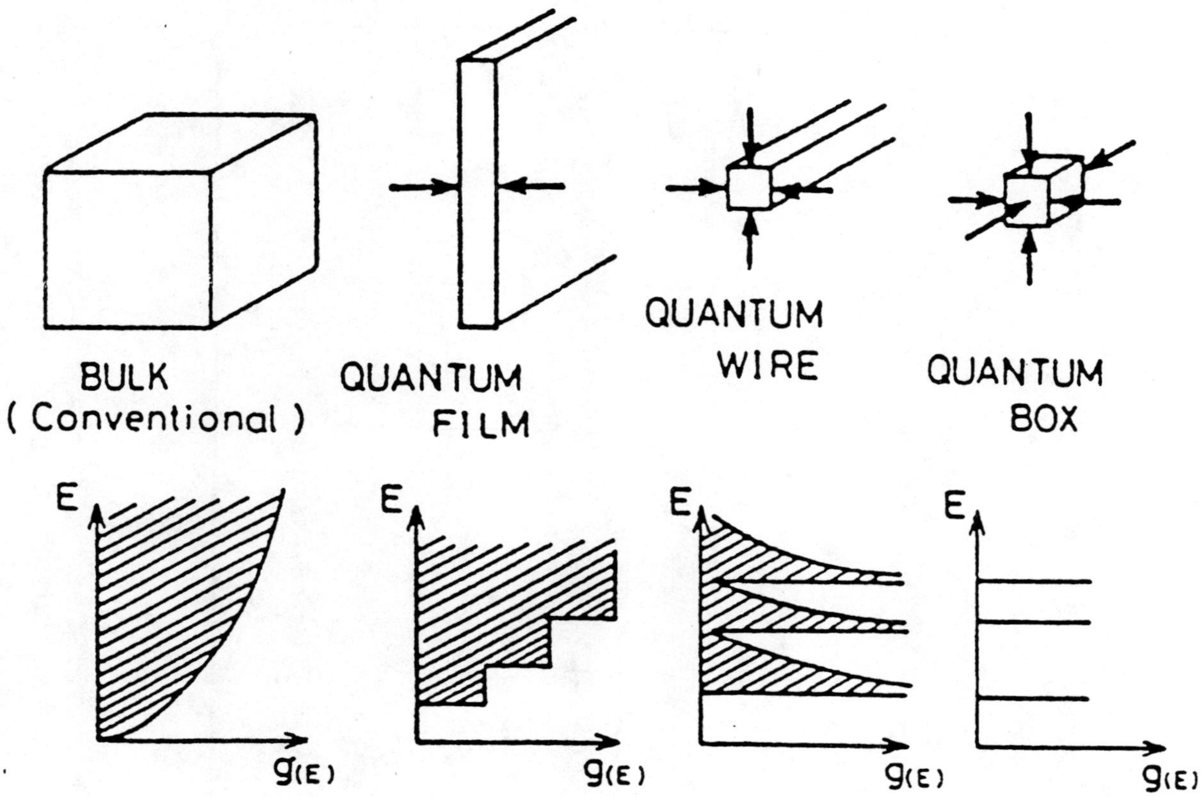

Fig. 1. Variation of density-of-states of electrons with the increase of the quantization dimension in quantum-well structures (form Ref. [2]).

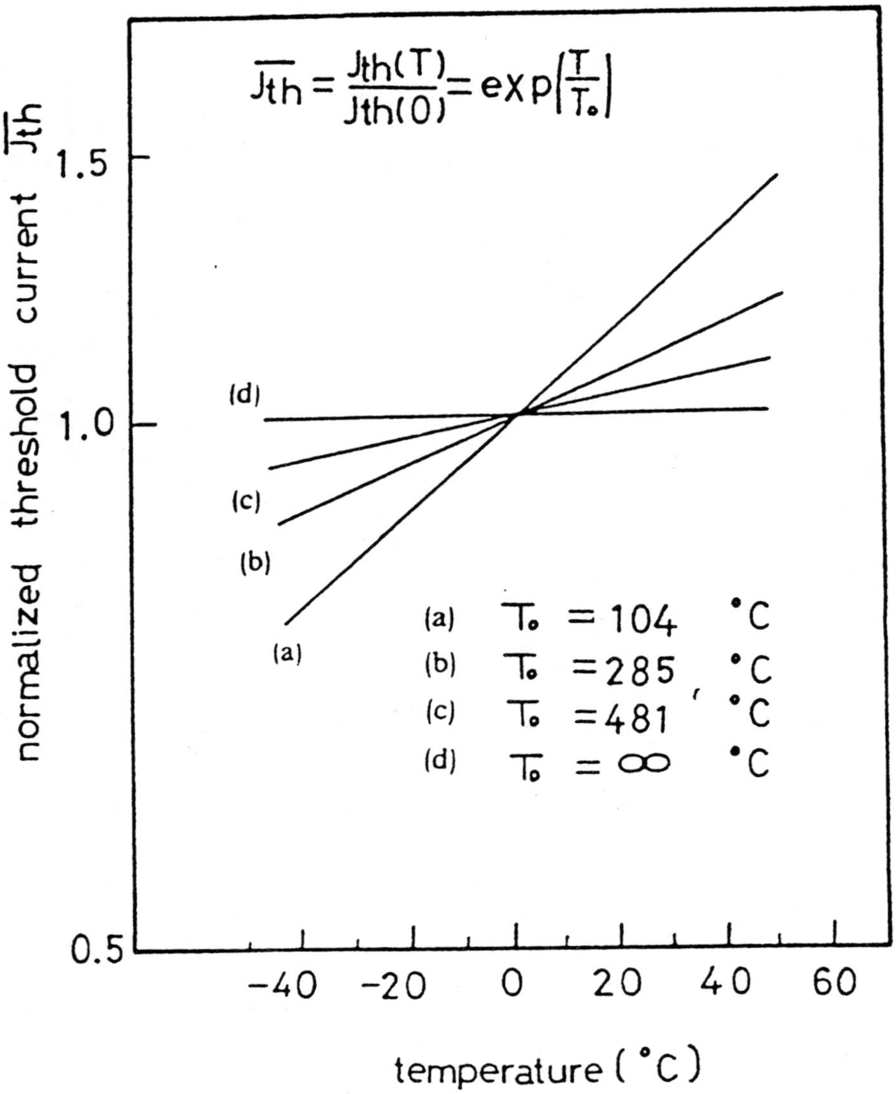

Fig.2. Numerical example of threshold current J_{th} calculated for conventional lasers (a) and quantum well lasers for (b) 1D-, (c) 2D-, and (d) 3D-QW structures. J_{th} is normalized by J_{th} at 0° C (from Ref.[3]).

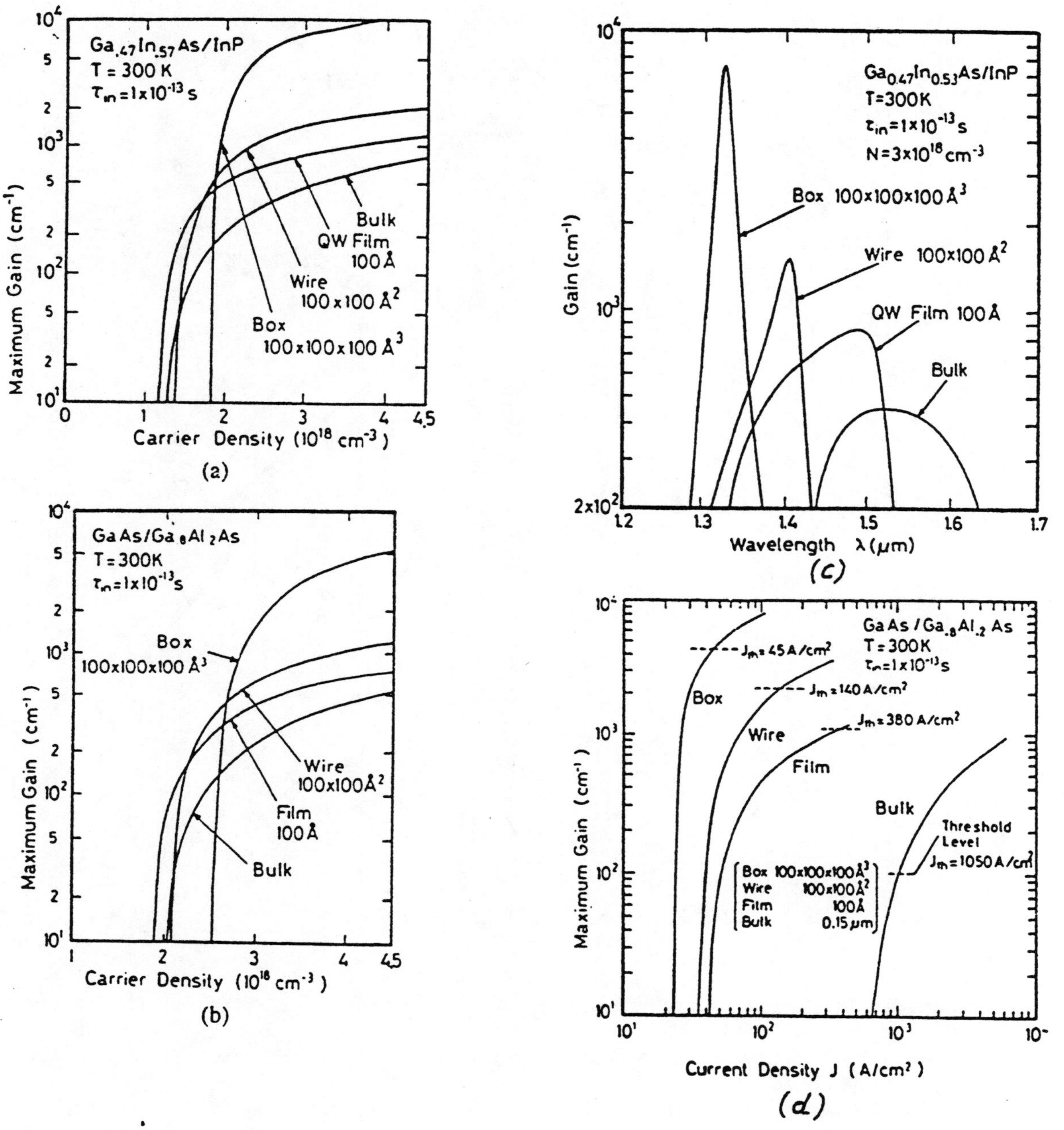

Fig. 3. Maximum gain as a function of carrier density calculated for (a) $Ga_{0.47}In_{0.53}As/InP$ and (b) $GaAs/Al_{0.2}Ga_{0.8}As$ quantum boxes, quantum wires, quantum films, and bulk crystals. Gain spectra calculated for $Ga_{0.47}In_{0.53}As/InP$ MD-QWs (c), and maximum gain (d) as a function of injection current density calculated for $GaAs/Al_{0.2}Ga_{0.8}As$ MD-QWs (from Ref. [2]).

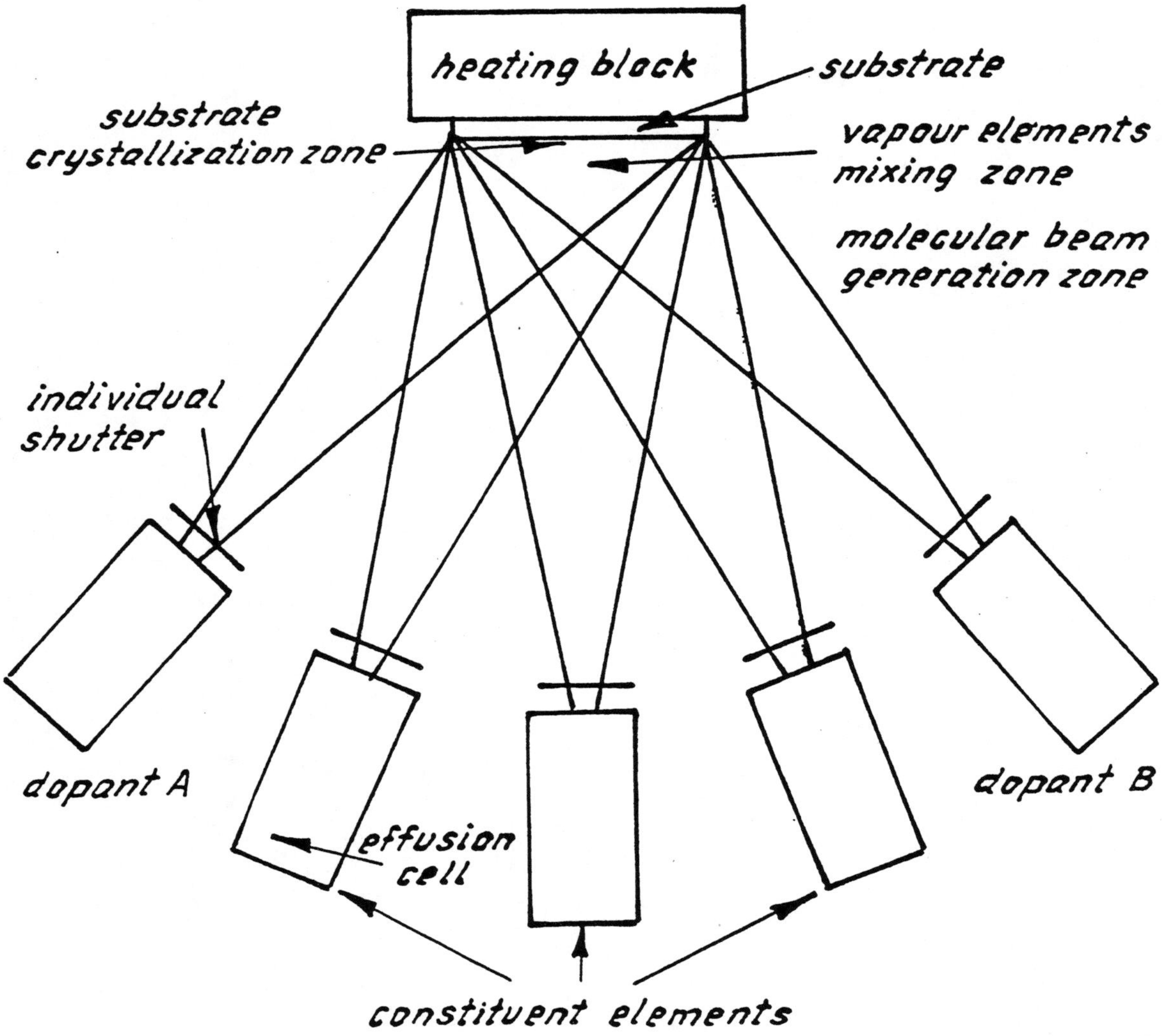

Fig. 4. Schematic illustration of the essential elements of a MBE system for fabrication of doped ternary compounds. The illustration borders represent the confines of the UHV system (from Ref. [7]).

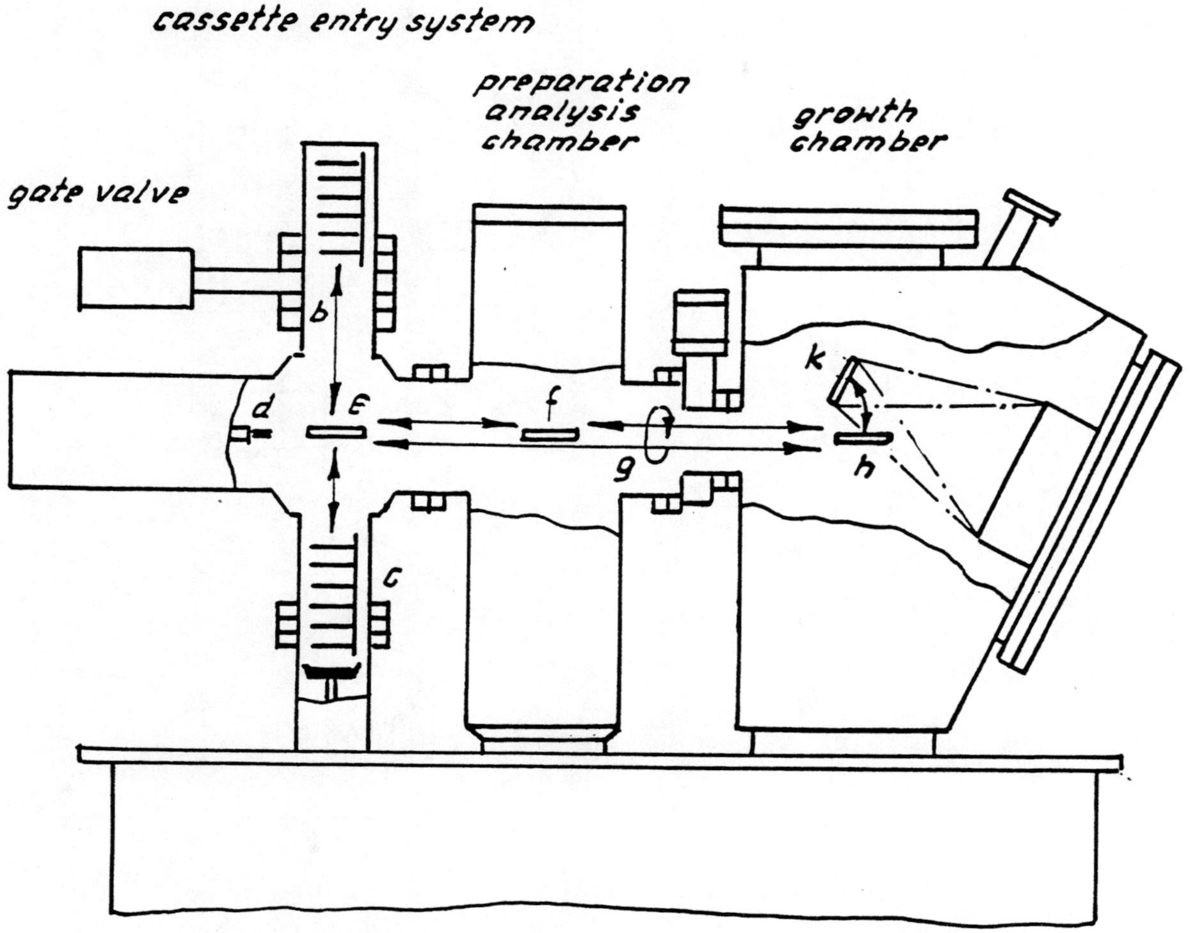

Fig. 5. Schematic diagram of the Perkin-Elmer MBE system PHI model 425B. After an advertising pamphlet of the Perkin--Elmer Company, USA.

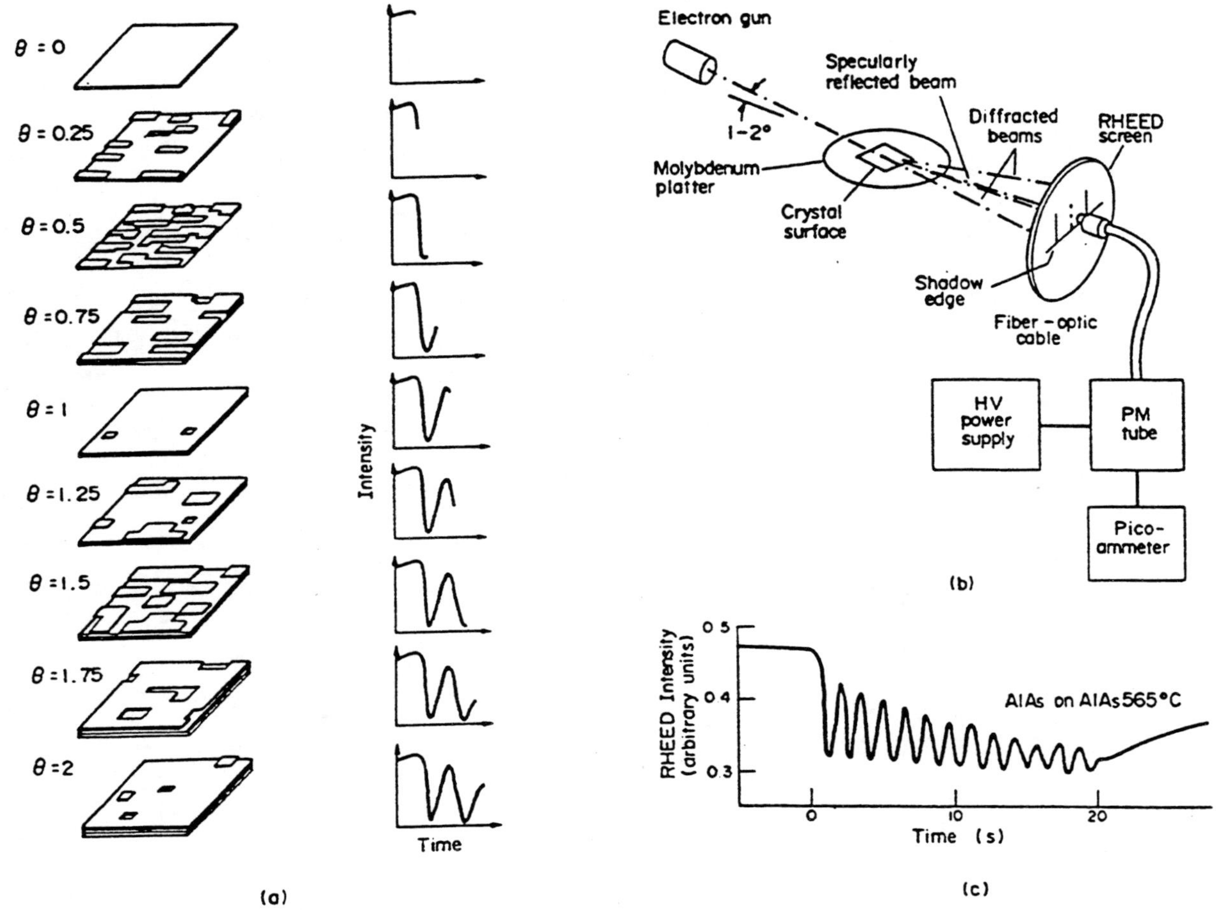

Fig.6. Schematic illustration of the experimental geometry of the RHEED system used in MBE growth chambers (b); typical example of the RHEED specular beam intensity oscillations from MBE-grown GaAs (001) surface (c), and (a) the real space representation of the formation of the first two complete layers of GaAs in the MBE process (from Ref.[11]).

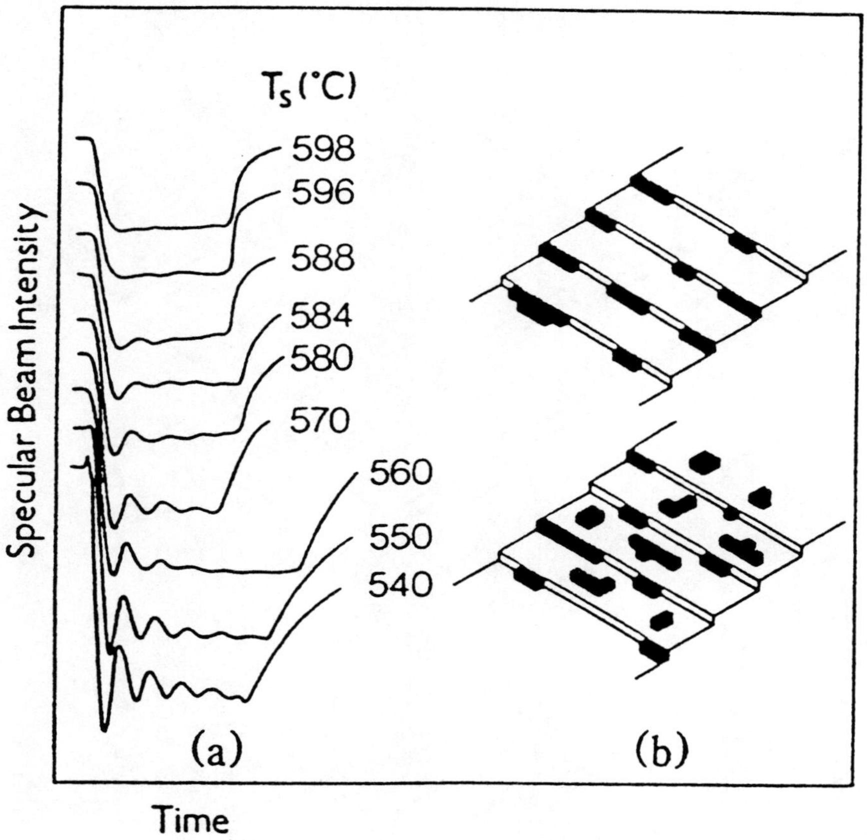

Fig. 7. a) RHEED measurements showing eventual disappearance of intensity oscillations with increasing substrate temperature T_S , b) schematic illustration of the transition from growth by 2D nucleation to step flow (from Ref. [12]).

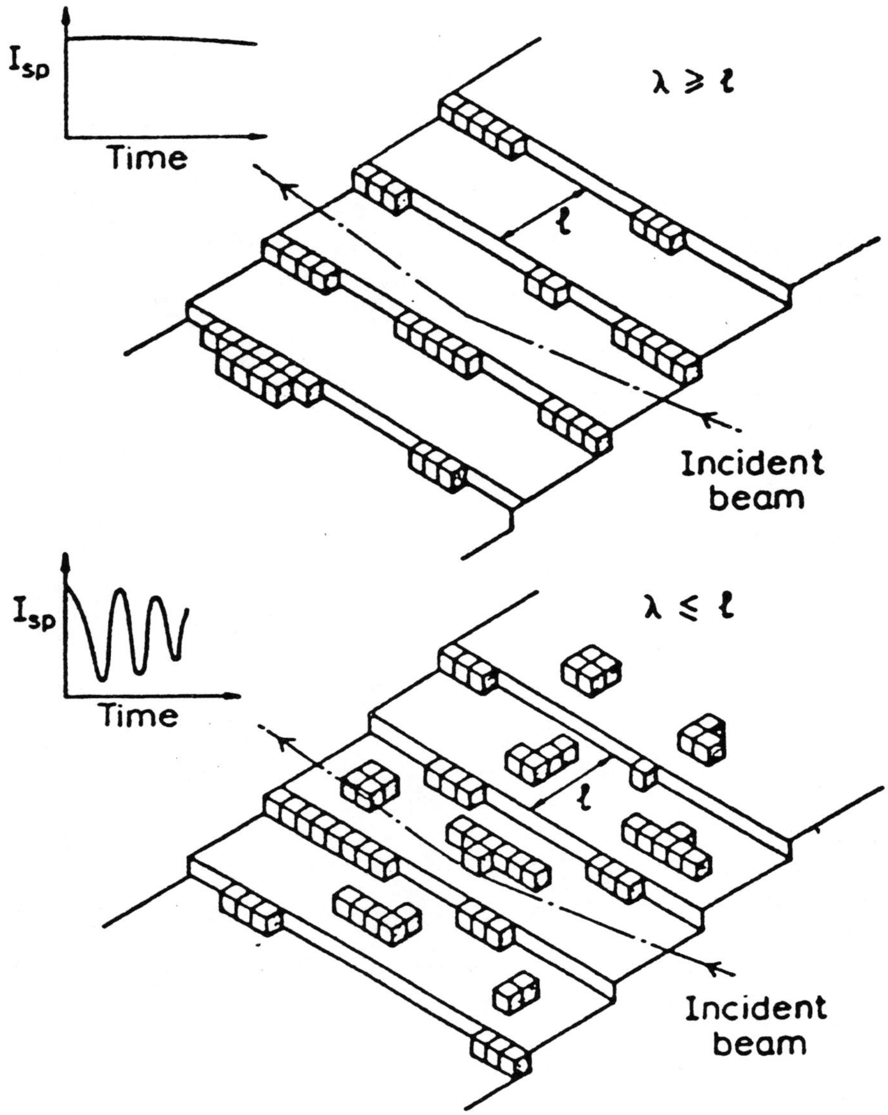

Fig. 8. Schematic illustration of the change in RHEED information as the growth mode changes from "step-flow" to "2D-nucleation" (from Ref. [12]).

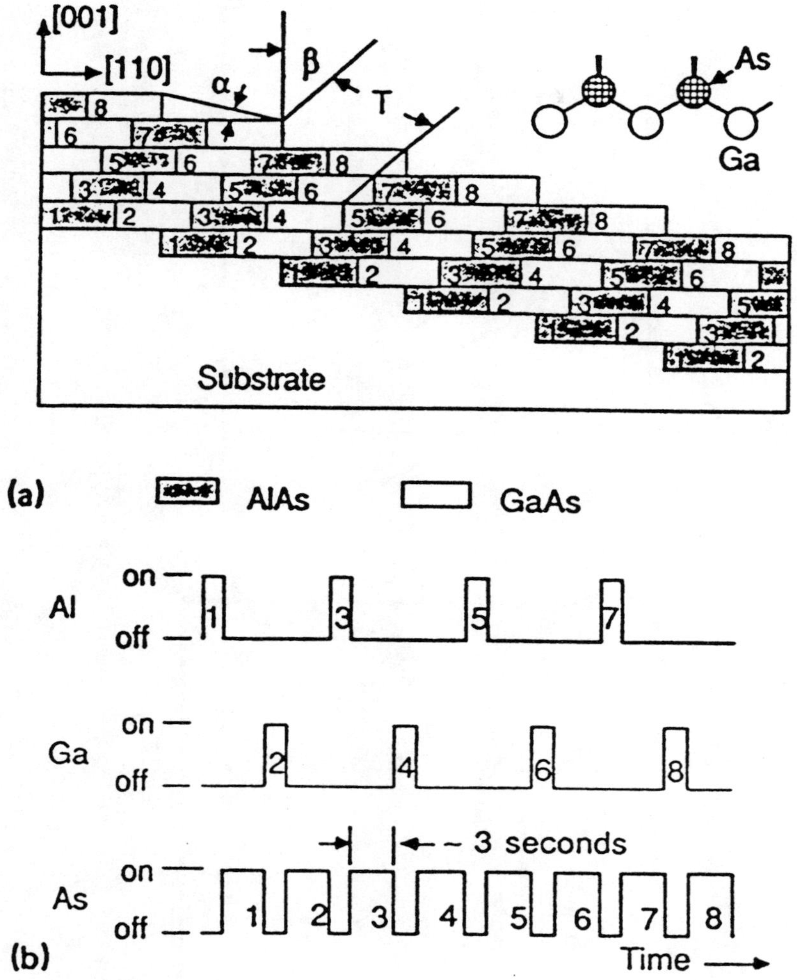

Fig.9. Idealized representation of a tilted superlattice grown by alternating-beam deposition. (a) Cross section, perpendicular to the step edges, of a superlattice with p = 1.25. The inset shows the arrangement of the As bonds with respect to the step edges. (b) Detail of growth, showing the order of atomic depositions. Numbers in (a) refer to partial monolayers deposited in (b) (from Ref.[14]).

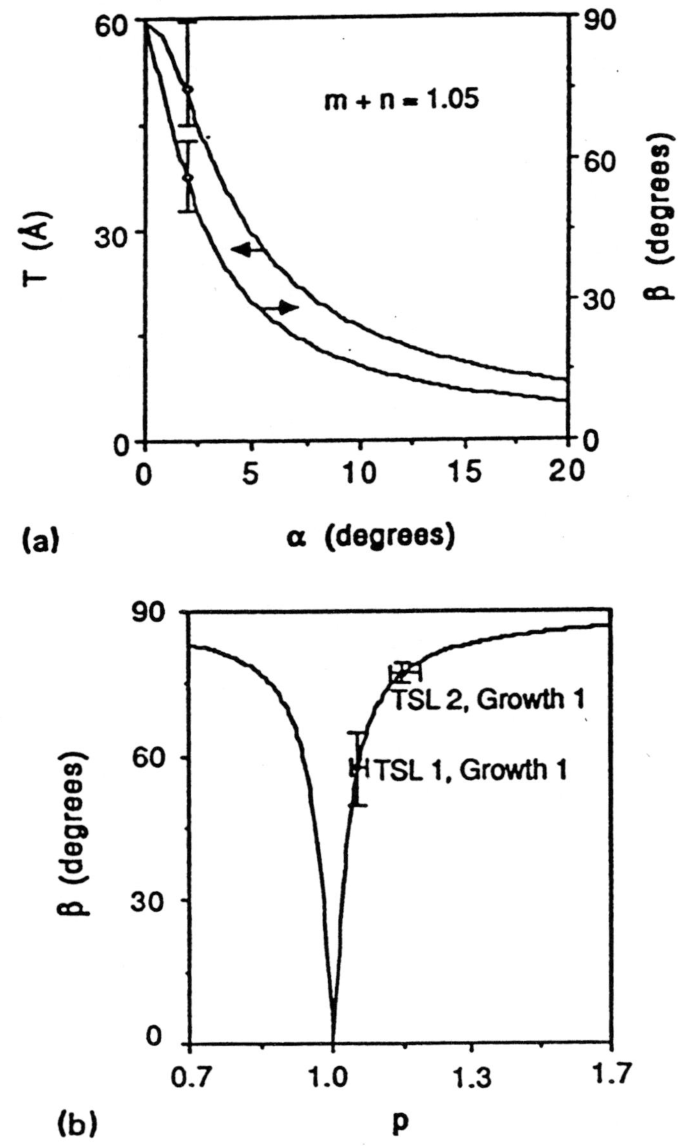

Fig. 10. (a) Plots of the TSL period T and tilt angle β as a function of the angle α between [001] and the surface normal for p = 1.05. (b) Plot of β as a function of p, for $\alpha = 2^{\circ}$. Experimental data for growth 1 are indicated (from Ref.[14]).

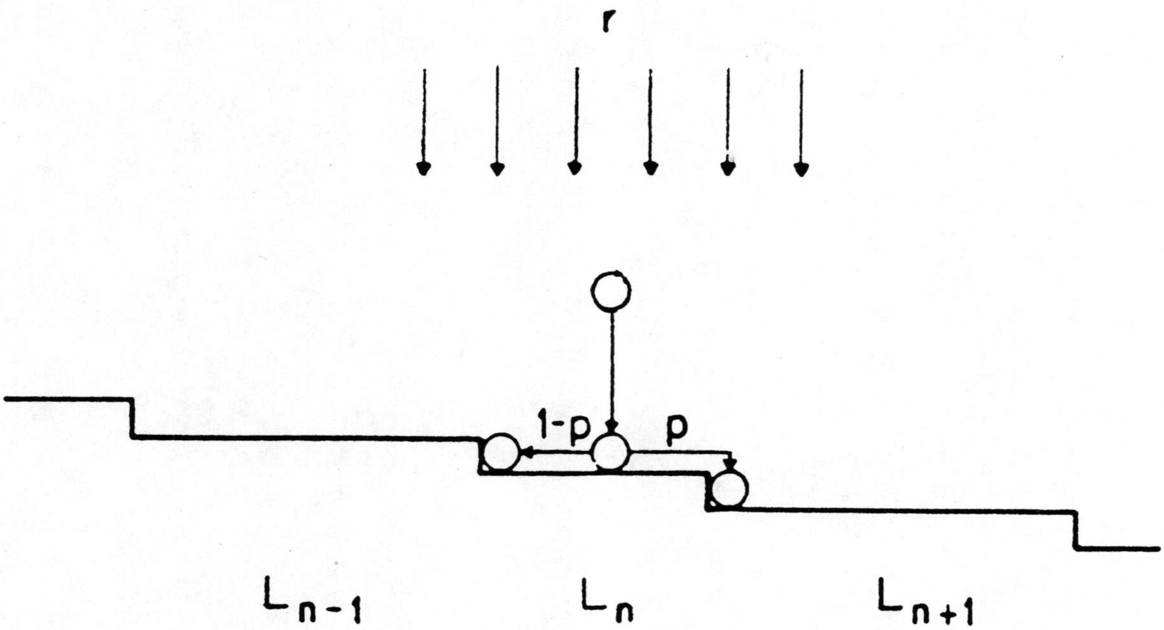

Fig. 11. Schematic sideview of the sample surface geometry used in the calculations presented in Ref. [16].

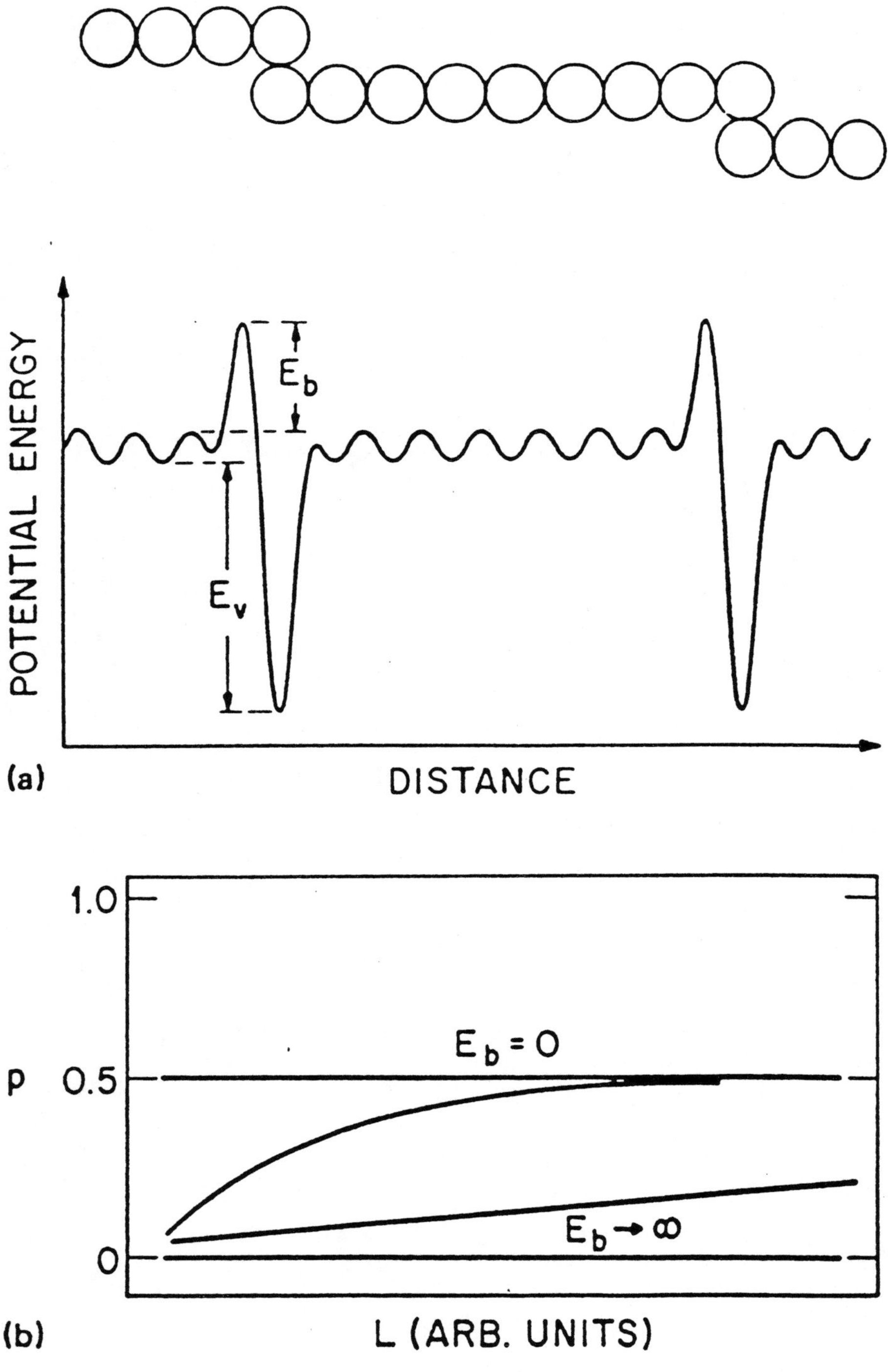

Fig. 12. (a) Schematic illustration of a stepped terrace (sideview) and the potential energy for an adsorbed atom. (b) Schematic illustrating the dependence of P on the shape of the potential and the terrace length L (from Ref. [16]).

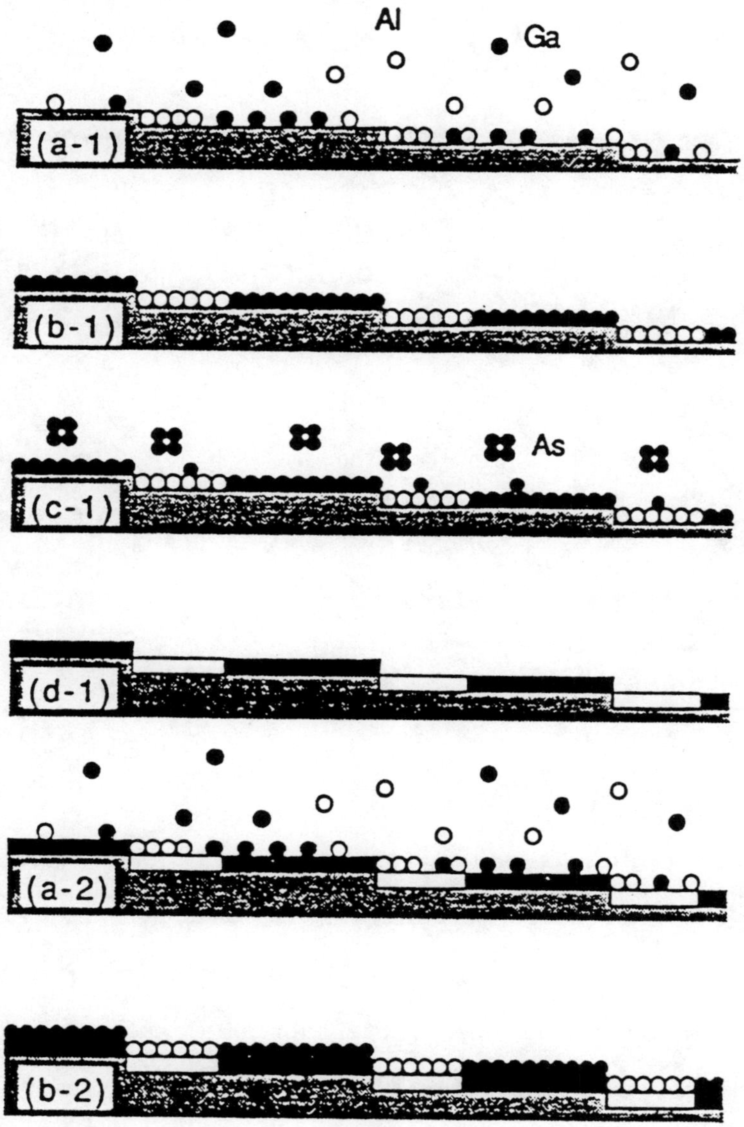

Fig. 13. Model describing the growth mechanism of the C-TSL. Figure (a-1) shows the coherent deposition of Ga (gray circles) and Al (open circles) atoms on a vicinal surface. At the end of the deposition (b-1), separated phases have been formed. Then, As (black circles) flux comes to the Ga and Al atoms (c-1) and completed epitaxial monolayers (rectangles) are formed on each step (d-1). Figures (a-2) and (b-2) show the second cycle (from Ref. [17]).

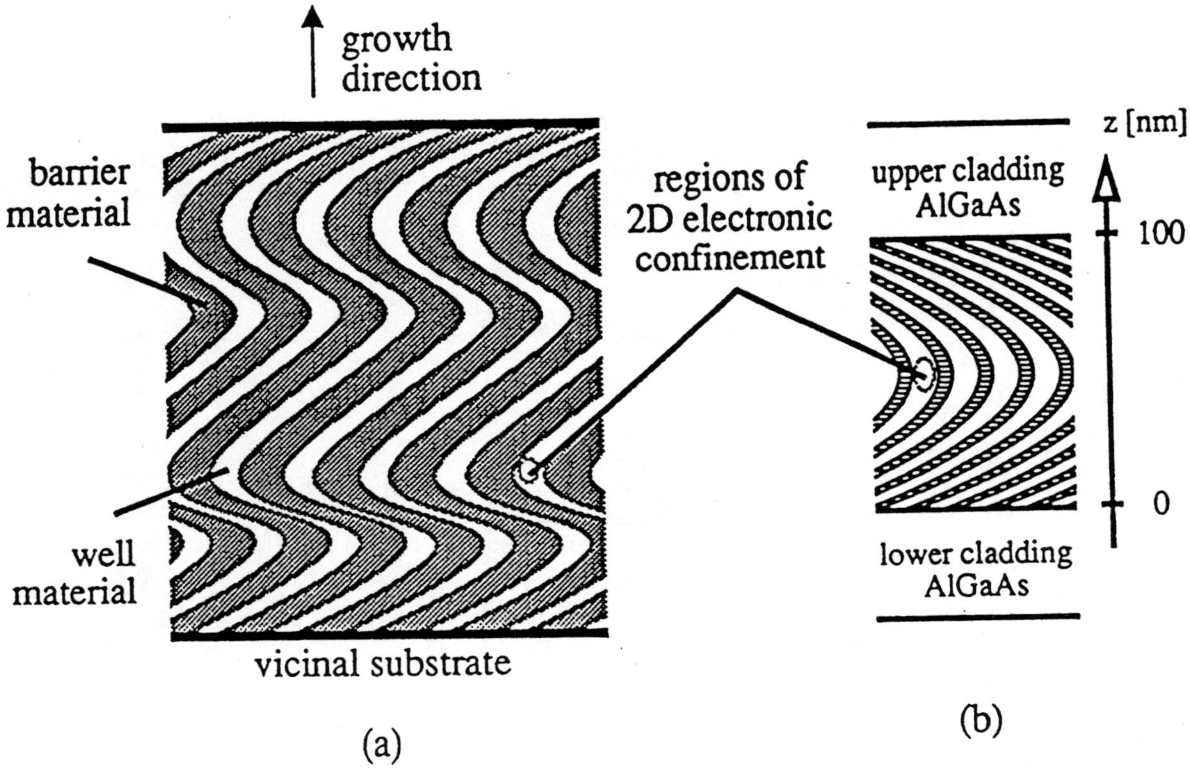

Fig. 14. A SSL cross section is shown (a) that would result from sweeping the per-cycle coverage back and forth through a range that includes exact monolayer coverage. At such places, where the tangent to the structure is vertical, electronic states are confined to two dimensions. The 2D potential energy minima near these regions may be characterized by the lateral barrier and well widths, lateral barrier height and the local curvature of the SSL. (b) Most of the simulated and grown structures have been single-crescent, parabolic geometries with barriers cladding the SSL region above and below (from Ref. [5]).

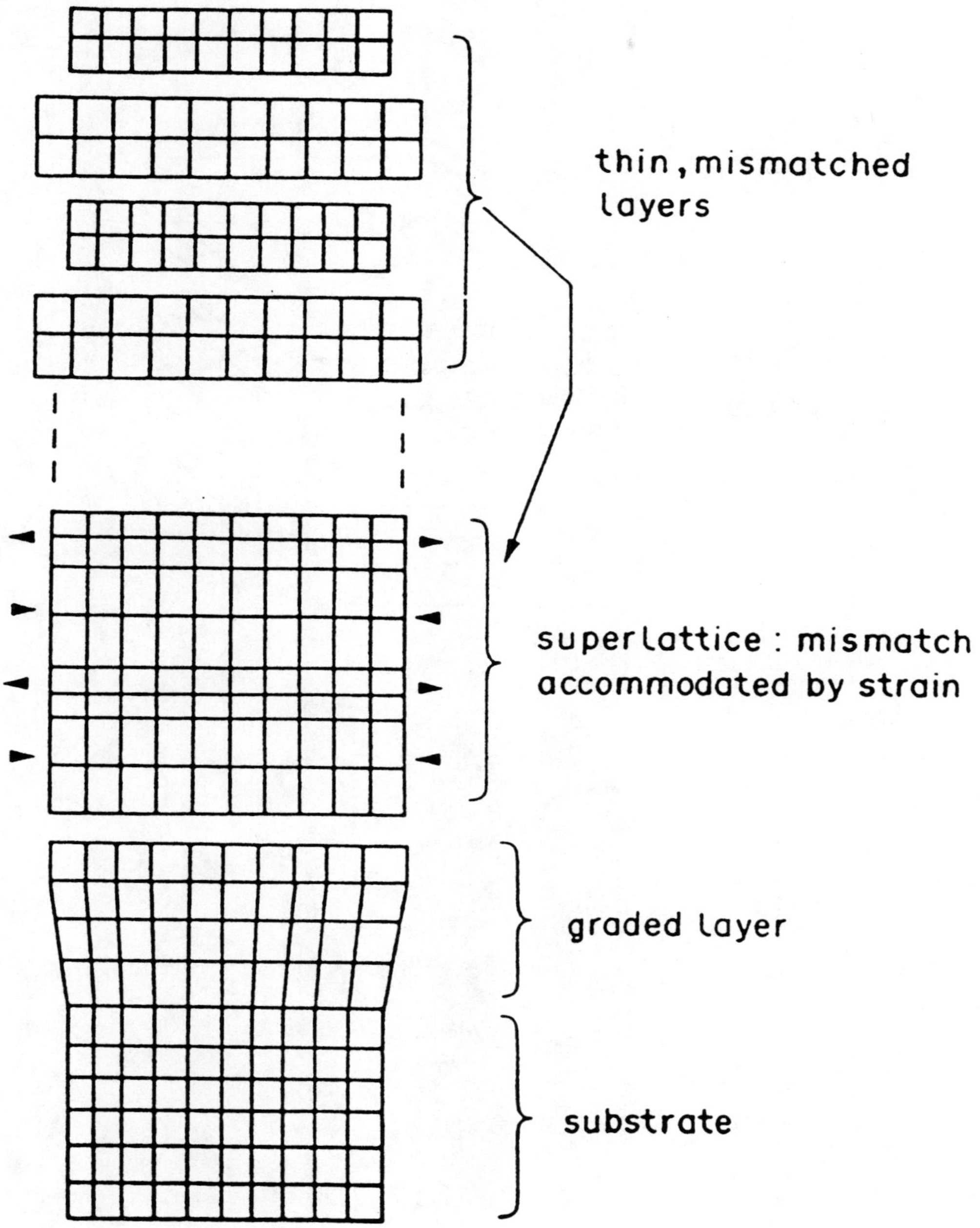

Fig. 15. Schematic illustration of the elements and formation of an SLS (from Ref. [24]).

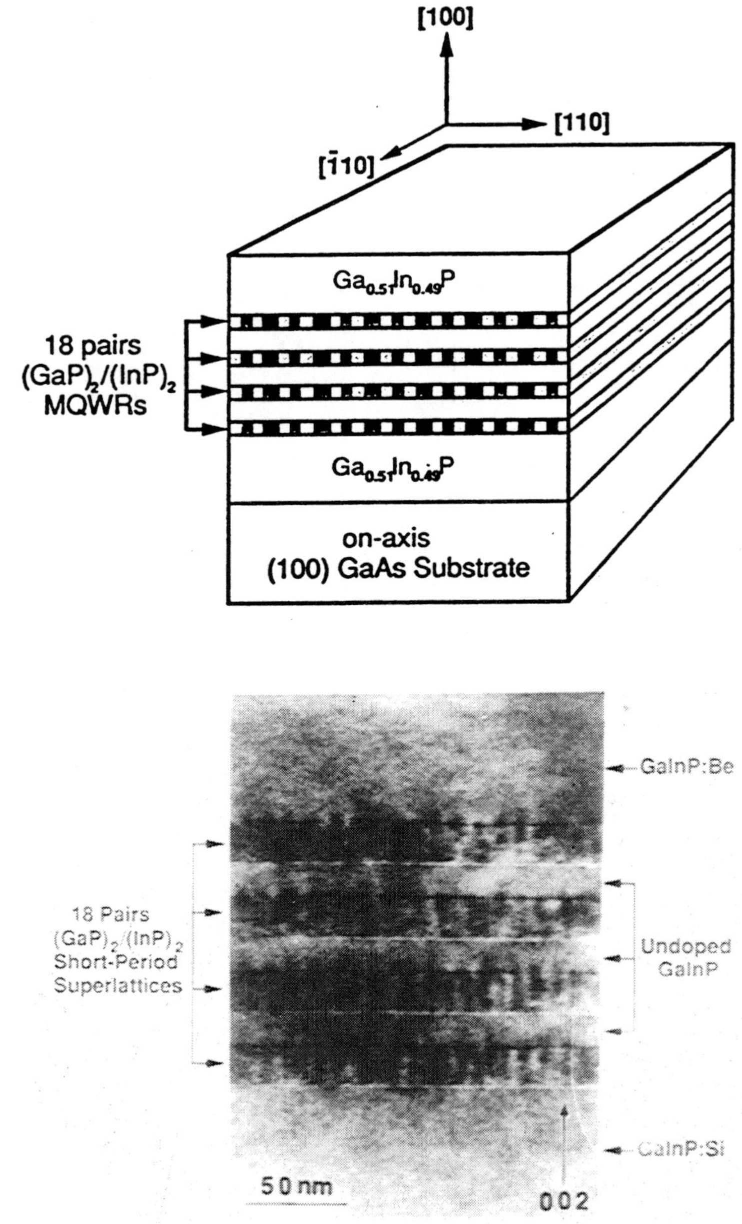

Fig. 16. Schematic diagram (a) and bright field TEM cross section (b) of the $Ga_xIn_{1-x}P$ multiple QWR heterostructure. A strained QWR array is formed parallel to the [110] direction. Lattice-matched $Ga_{0.53}In_{0.47}P$ p-type and n-type doped above and below, respectively form the QWR light emitting diode structure (from Ref.[22]).

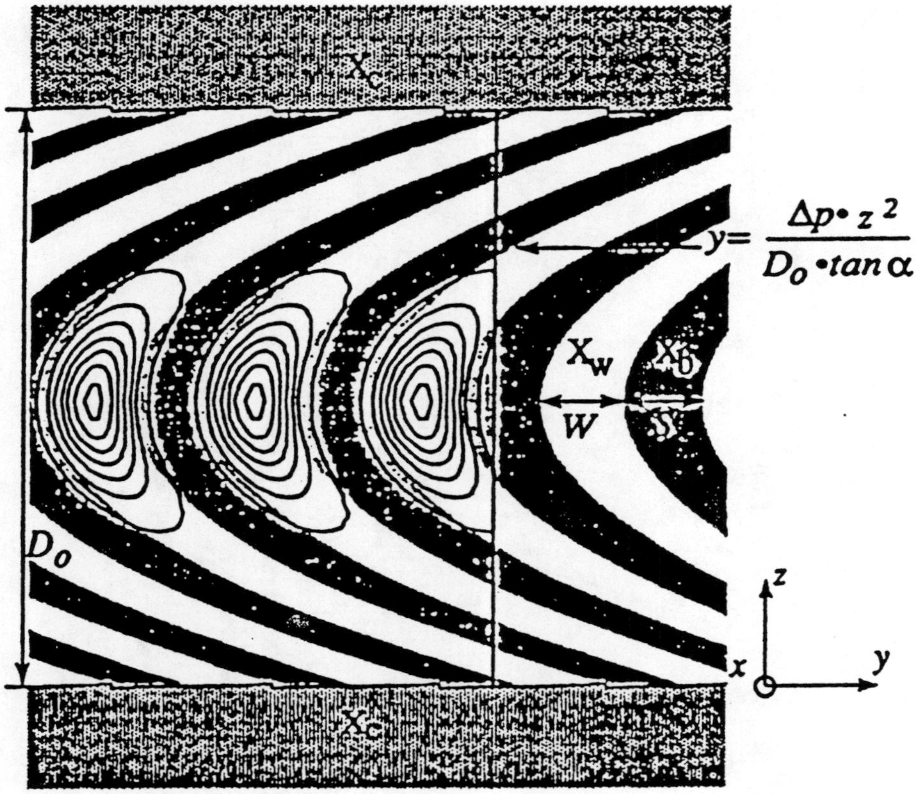

Fig. 17. Schematic cross-sectional profile of a SSL and its electron probability density for the lowest subband of the conduction band at Brillouin zone center. The gray and white areas represent $Al_{0.5}Ga_{0.5}As$ and GaAs regions, respectively (from Ref. [25]).

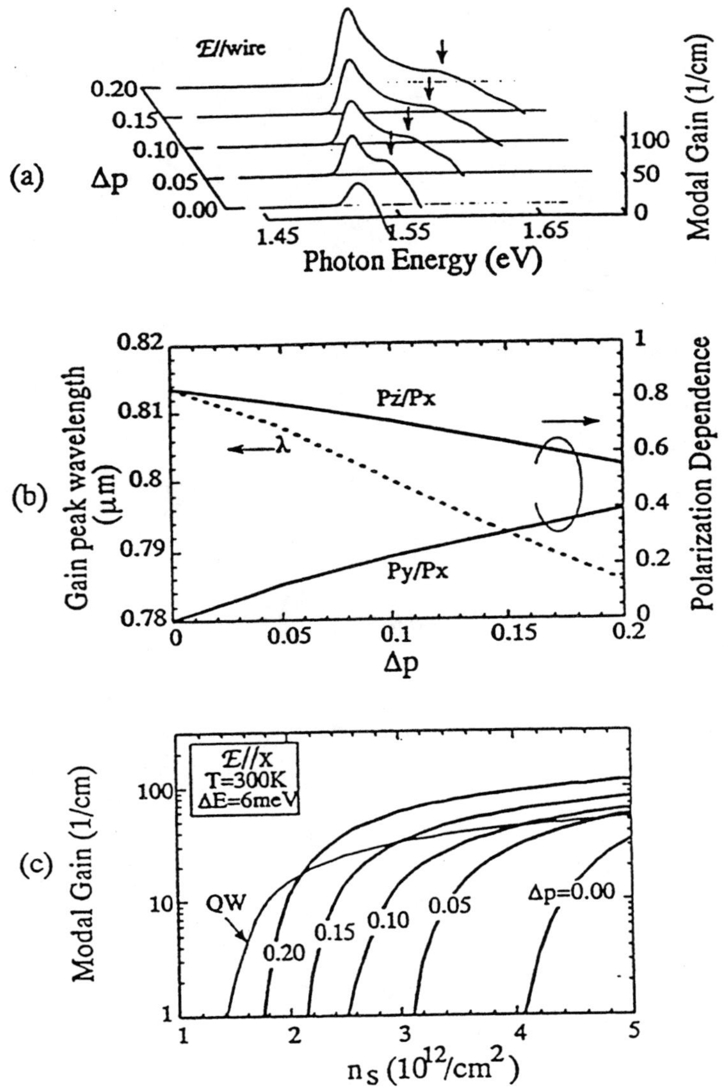

Fig. 18. Optical gain of a SSL with $x_b = x_c = 0.5$, $W = S = 5.4$ nm, $D_0 = 46$ nm, $T = 300$ K, and width $\Delta E = 6$ meV of the assumed Gaussian line shape function for modal gain: a) Modal gain spectrum at injected surface carrier density $n_s = 5 \times 10^{12}$ cm^{-2} with electric field E, polarized along the wire axis (arrows indicate the second gain peak); b) Gain peak wavelength and polarization dependence of the peak gain (P_x, P_y and P_z are the peak gains when the electric field is polarized in x, y, and z directions, respectively); c) Maximum modal gain as a function of injected surface carrier density for different Δp values. Electric field is polarized along the wire axis. QW is a 5.4 nm thick SQW of $GaAs/Al_{0.5}Ga_{0.5}As$ (from Ref.[25]).

Vacuum Electron and Ion Technologies
D. S. Karpuzov (Ed.)

ISBN 1-56072-242-8

ELECTRON CYCLOTRON RESONANCE (ECR) PLASMA DEPOSITION OF SILICON (oxy) NITRIDE THIN FILMS

J. Huran, L. Hrubcín and S. Hascík
Institute of Electrical Engineering. Slovak Academy of Sciences, Dúbravská cesta 9, 842 39 Bratislava, Slovakia

The properties of the SiN_x and SiN_xO_y films prepared by electron cyclotron resonance (ECR) plasma CVD were studied by laser ellipsometry, infrared absorption spectra and Auger electron spectroscopy. The effect of divergent magnetic field application on spatial distribution of N_2^+ ions for ECR plasma stream are disscused.

I. INTRODUCTION

For the formation of SiN_x and SiN_xO_y films the chemical vapor deposition (CVD) and particularly the plasma CVD, has been mainly used. The plasma CVD method has an advantage of low substrate temperature. However the method has a disadvantage connected with the degradation caused by hydrogen contamination from source materials such as silan (SiH_4) and ammonia (NH_3). In these deposition processes the hydrogen atoms incorporated with concetration of 10-20% in film deposited result in a decrease of the threshold voltage and in a degradation of the mutual conductivity in very large scale integration (VLSI) technology [1]. Therefore SiN_x films with extremely low concentration of hydrogen atoms are desirable. Pure nitrogen (N_2) has been used instead of NH_3 to eliminate the hydrogen contamination. However, in this case the film quality is insufficient comparing to the films prepared with NH_3. Recently a low temperature CVD method using ECR plasma has been developed to solve these problems as an attractive alternative for SiN_x formation. ECR microwave plasma exhibits the following characteristics: 1) generation of dense plasma (electron density of 10^{10}-10^{12} cm^{-3}), 2) generation of fast electrons, i.e. high electron temperature and 3) generation of highly ionized plasma (ionization of 10% and higher). Fast electrons play an important role not only in particle excitation and activation but also in direct ionization of the working gas [2].

II. EXPERIMENT

ECR microwave plasma was generated in the apparatus shown in Fig.1. Microwave power (600 W) was supplied through the rectangular waveguide from the magnetron generator, using magnetron in a continuous wave mode, to the resonance cavity with (TE 113) standing mode in vacuum at frequency of 2.45 GHz. Two magnetic coils are placed surrounding the microwave resonator to provide a magnetic field 87.5 mT, necessary for obtaining ECR condition. The currents through the upper and lower coils may be varied independently to shift the ECR position in the resonance chamber along the axial direction. The diameter and axial length of the

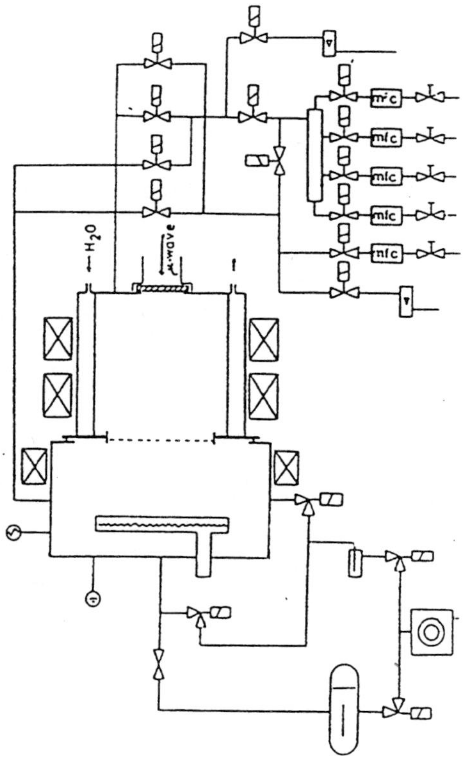

Fig.1 Schematic diagram of the microwave plasma stream apparatus

resonance chamber were 186 mm and 184 mm, respectively, while the deposition chamber have diameters of 360 and 160 mm, respectively. The plasma generated in the resonance chamber was accelerated through the apparatus hole toward the heated substrate holder of 210 mm in diameter, set at $z = 0$ in the center $r = 0$ of the deposition chamber. The magnetic field gradient was controlled by the current flow in the submagnetic coil located above the substrate plate. Nitrogen was introduced into the upper part of the reactor. Silane and nitrogen oxide were introduced directly into the vacuum chamber via a gas distribution ring. The system was pumped by a Roots blower with rotary forepump. The substrate holder was grounded and heated by a resistance heater, where its temperature was monitored by an imbeded thermocouple. Film thickness and refractive index (RI) were measured using a He-Ne laser ellipsometer (X = 632.8 nm). Fourier transformed infrared (FTIR) studies were performed with 300 nm silicon nitride deposited on bare silicon substrate. Optical emission spectroscopy with spatially resolved measurement of species was used to analyze and monitor the plasma in the visible and UV spectral emission.

III. RESULTS AND DISCUSSION

The emission of nitrogene molecules ($C^3Au - B^3A$) and nitrogene molecule ions (the first negative system $B^2\Sigma u^+ - X^2\Sigma g^+$) were observed. It is interesting that the emission intensity of $N_2^+$is remarkably stronger compared to that of N_2. Emisions of Si (288 nm) and hydrogen atoms Hβ (486 nm) were observed too. The emission species such as SiH, NH, Hα and H_2 were hardly observed. This fact may indicate that SiH_4 molecules in ECR plasma are decomposed sufficiently into Si and H atoms. Fig.2 shows the emission intensities of N_2^+ (391.4 nm) as a function of the distance from the center of the substrate holder. If we compare Fig. 2a with Fig. 2b we can see the influence of the operating pressure in the deposition chamber on the spatial distribution of N_2^+ ions. At a pressure of $P = 1$ Pa the homogeneity of N_2^+ distribution is very good across the whole measured area. But if the pressure decreases down to 5.10^{-2} Pa, the appearance of a discharge in the deposition area changes very drastically and the resulting N_2^+ distributions is shown in Fig.2b. At the surface of the substrate holder the N_2^+ concentration is constant across the area with diameter 70 mm. With increasing distance

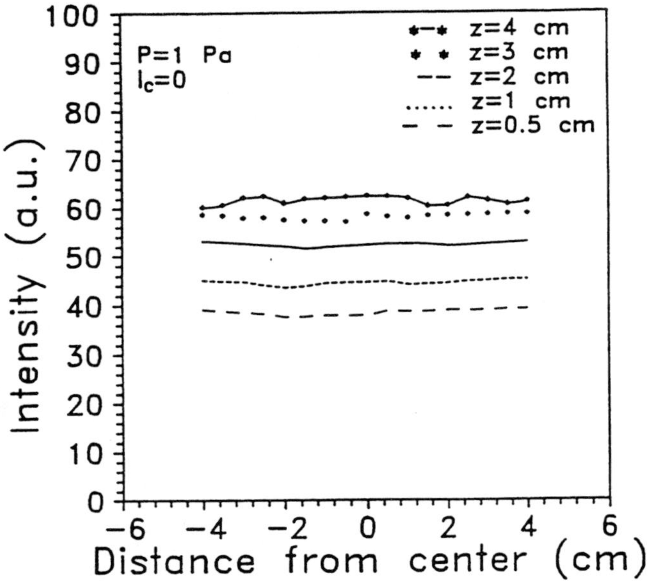

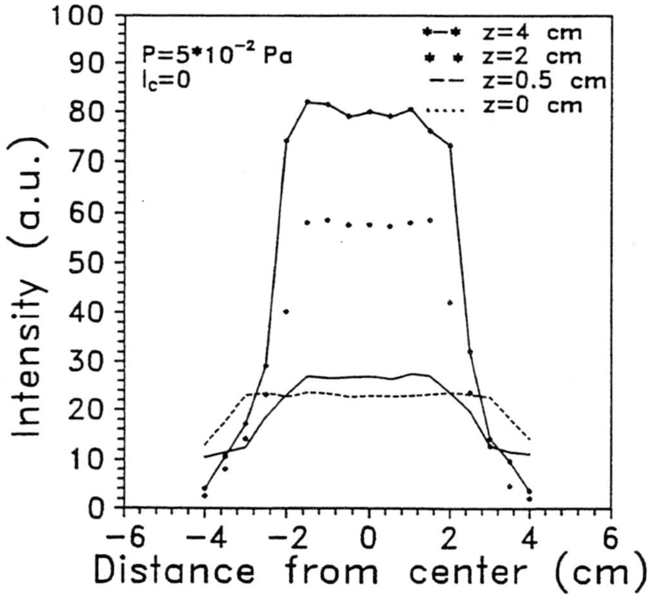

Fig.2 The spatial distribution of N_2^+
a) P = 1 Pa, b) P = $5x10^{-2}$ Pa

between the substrate and the holder the area of homogeneous distribution of N_2^+ decreases and at z = 4 cm the homogeneous circle diameter is only 35-40 mm.

Infrared analyses of SiN_x films show the presence of N-H (3360 cm^{-1}), Si-H (2160 cm^{-1}), Si-O (1170 cm^{-1}), Si-N (330 cm^{-1}) bonds (Fig.3). The hydrogen in SiN_x :H exists only in the form of N-H and Si-H bonds because no O-H bonds were observed. At microwave power of 200 W the absorbtion due to the Si-H bonds was found to be relatively large. But with increasing of the microwave power, the absorption at Si-H bond decreased. At a microwave power of 500 W the absorbtion due to Si-H bonds could be hardly detected.

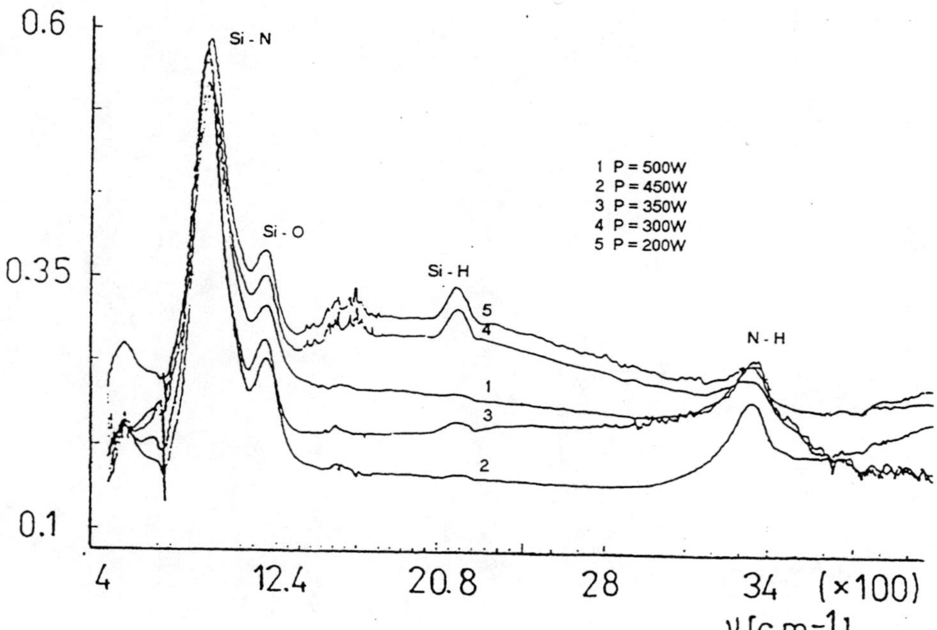

Fig.3 FTIR spectra of the deposited films

The refractive index has been used as a quality control index for the deposition process. Experimental results have shown that the RI of silicon nitride and/or silicon oxynitride films prepared by the ECR CVD process depend on the film composition and varies with the change in the Si/N/O ratio of the films. The refractive index of the presented films is plotted in Fig.4 as a function of the gas phase composition for the films deposited at a microwave power of 450 W and a chamber gas pressure of 0.1 Pa. From the figure we can see that with no nitrogen oxide in the deposition chamber (N_2O/SiH_4 = 0) the refractive index is nearly the same as for thermal CVD Si_3N_4 and varies from 1.95 to 2.1, depending on the SiH_4/N_2 ratio. Even small additions of N_2O into the gas mixture leads to the sharp decrease of the refractive index to the value of 1.55 at the gas flow ratio N_2O/SiH_4 = 5. If the N_2O/SiH_4 ratio increases above this value, the RI does not change any more. So, it is

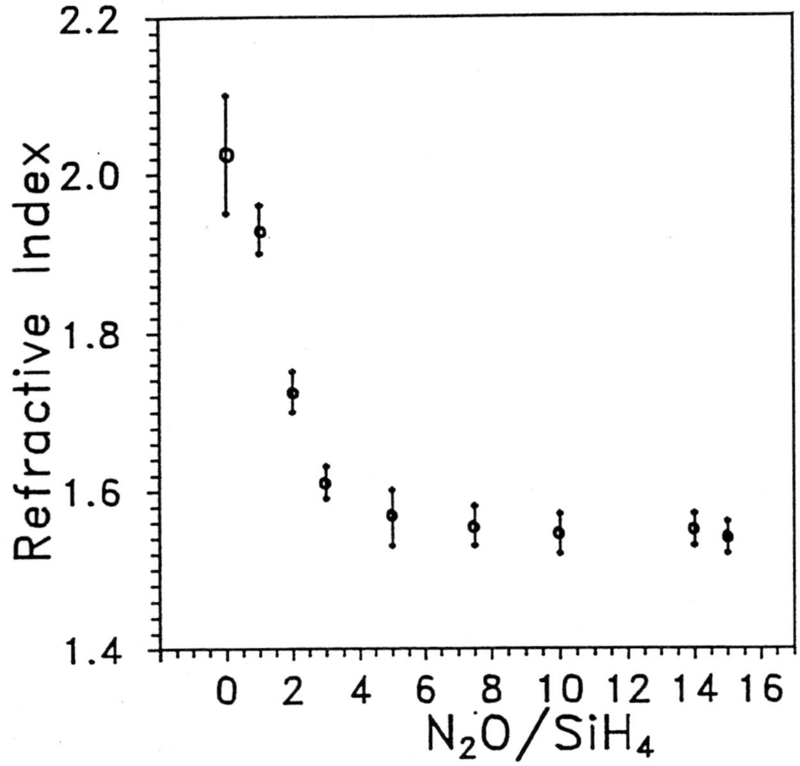

Fig.4 The RI as a function of gas flow ratio

possible to deposit either silicon nitride or silicon dioxide only by changing the N_2O/SiH_4 ratio from 0 to 5.

Figure 5 show typical compositional depth profile of silicon nitride and silicon oxynitride films determined by AES. It can be seen, that atomic rations Si/N, Si/N/O are approximately uniform.

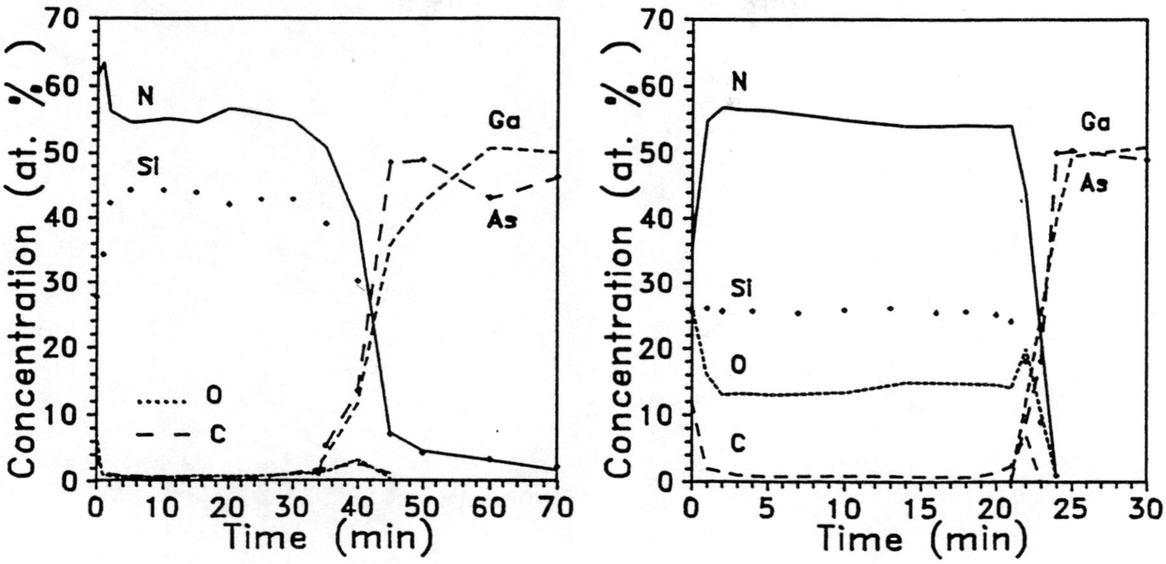

Fig.5 AES depth profile of a) SiN_x and b) SiN_xO_y

IV. CONCLUSIONS

In the ECR plasma, the optical emission intensities of nitrogen molecular ions, Si and H were much stronger than those in conventional rf plasma [3]. Therefore, nitrogen is effectively activated. SiH_4 is sufficiently decomposed into Si and H atoms. The working gas pressure strong influences the N_2^+ ions distribution. The physical characteristics of ECR plasma deposited silicon nitride and oxynitride films were analysed using various analytical techniques. FTIR analyses show that the films deposited under described conditions contain less amount of hydrogen bonded in Si-H bonds than those deposited by PE CVD method [4]. Summarizing, the results presented indicate that the microwave power density and the gas flow ratio has a strong influence on the resulting properties of the deposited films.

The authors are grateful to the Slovak Grant agency for science (grant No 2/999056/93) for support of this work.

REFERENCES

[1] Hirao,T. et al.: Jpn.Appl.Phys. **26** (1987), 2015

[2] Matsuoka,M. and Ono,K.: J.Vac.Sci.Technol.**A6** (1)(1988),25

[3] Manabe,Y. and Mitsuyu,T.: J.Appl.Phys. **66**(6) (1989), 2475

[4] Szulényi,F.. Huran,J.: Acta Phys.Slovaca **41** (1991), 368

Vacuum Electron and Ion Technologies
D. S. Karpuzov (Ed.)

ISBN 1-56072-242-8

SELECTIVE MULTILAYER MIRRORS FOR WHITE LIGHT BASED ON Ti AND Si OXIDES, PREPARED BY ION SPUTTERING IN VACUUM

Tamara Pencheva and Milen Nenkov,
Tu - Russe, Department of Physics,
Studentska str. No. 8, Russe-7017, Bulgaria

I. INTRODUCTION

One might face a number of problems in industry which are connected with the development and manufacturing of light sources with definite colours. For instance, in signal colour lamps used in transport, light colour might be corrected by using transmission colour filters. In spite of that, it is necessary in some cases to prepare new types of illuminators that match exact requirements in colour and spectral distribution of light intensity. Therefore in surgery and stomatology it is necessary the emitted light to be white, to have definite colour coordinates variations and minimal radiation in the infrared region of spectrum. In such cases colour correction of the light, emitted from a source, might be carried out by reflection from selective mirror which is transparent for the infrared radiation.

The present investigation is focussed on the optical characteristics and possibilities of multilayer selective systems for reflecting white light, based on Ti and Si oxides, prepared by ion-assisted sputtering in vacuum on glass substrates.

It is known that coatings with odd number of quarterwavelength layers with optical thickness $nd = \lambda_o/4$ posses good selective reflection in spectral region near λ_o [1,2]. If one decides to design multilayer selective reflector for white light which will use sun radiation, it is obvious that λ_o must be in the center of the visible region, which corresponds to human eye maximal sensitivity. Still if one uses any other light source and wants to obtain reflected light with predetermined colour, the structure of the reflection coating, number of layers and their optical thicknesses are not evident. The purpose of computer simulation, that is done in this work, was to investigate the capabilities of multilayer system based on Ti and Si oxides for selective mirror design with application to white light.

II. CALCULATION PARAMETERS

Computer investigation deals with some main parameters of reflected light from 3, 5, 7, 9-layer systems with structure, designed as air$/(TiO_2/SiO_2)_k$ $/TiO_2/$substrate, for standard light A, B and C-sources. Standard A-source corresponds to light from metal filament lamp at T = 2856 K, B-source - to direct sun light with characteristic temperature T = 4870 K, and standard C-sourse corresponds to day light with characteristic tempetrature T = 6774 K. Thus the results of our calculation can be used in analysis of illuminators supplied with real light sources from different types - metal filament lamps, luminescent lamps, metal-halogen lamps.

For determination of the reflected light colour some main parameters such as chromaticity coordinates (x, y), dominant wavelength λ_F and colour saturation m were used. The reflection efficiency was estimated with integral photometric reflection factor R_F.

Computing procedure consists of the following steps:

- Calculation of the reflection spectrum r_λ in the visible range $(0.38 - 0,77\mu m$, step $0.01\ \mu m)$ at normal incidence for multilayer system with determined number of layers N and characteristic optical thicknesses $4nd$. It was carried out in accordance with the characteristic Abeless' matrices, described in [1]. Parameter $4nd$ is equal to the wavelength, for which reflection of a multilayer system is maximal. The optical parameters of the layers are experimentally determined by the authors and are $n_{SiO_2} = 1.54, n_{TiO_2} = 2.35, n_o = 1.0$ (air).
- Chromaticity coordinates (x, y) and photometric reflection factor R_F calculations for light sources A, B and C.
- Determination of λ_F and m according to the position of the point with coordinates (x, y) on the chromaticity plot of the International Illumination Committee (IIC), 1931. Here λ_F and m are determined with respect to the achromatic stimul E, fig.1, with coordinates $x_o = 0.333, y_o = 0.333$ [3].

Illuminator quality for visible range depends on two independent groups of characteristics. Light efficiency is determined as a ratio between reflected light flow and consumed electrical power. This parameter is directly proportional to the photometric reflection factor R_F. It is possible to meet the colour requirements of reflected light, when the point with coordinates (x, y) is located in determined zone on the chromaticity plot of IIC. For instance, according to DIN-standard [4] for stomatologic illuminators, the white zone is determined from its edge boundary points $x_1 = 0.320, y_1 = 0.368; x_2 = 0.320, y_2 = 0.330; x_3 = 0.366, y_3 = 0.392; x_4 = 0.358, y_4 = 0.344$, fig. 1. The same information might be expressed by means of values of parameters λ_F and m .

Computing was done for multilayer reflection systems with equal and not equal optical thicknesses of separate groups of layers for λ_o in the region of $0.20 - 0.75\mu m$ with the main standard A, B and C-sources.

III. RESULTS AND COMMENTS

First, the above mentioned main parameters (x, y); R_F and m for multilayer reflection systems containing 3, 5, 7 and 9 layers with equal characteristic optical thicknesses, varied between 0.35 μm and 0.75 μm, were determined for A, B and C standard sources. In figures 2, 3 and 4, the curves of the chromaticity coordinates (x, y) connect points, which correspond to the colour of light reflected from systems with determined number of layers

for A, B and C sources respectively. Each point in any curve is calculated for different value of λ_o. It can be seen that, when the characteristic optical thickness of the layers is $4nd = 0.50 - 0.55 \mu m$, the colour of reflected light and the colour of the light source are close for A, B and C sources.

SOURCE B

	N	$4nd, \mu m$	x_o	y_o	$\lambda_F, \mu m$	$m, \%$	$R_F, \%$
	B		0.348	0.352	0.580	12.1	
	3*	0.550	0.364	0.386	0.570	24.2	60.0
max	5	0.550	0.377	0.414	0.570	37.1	78.3
R_F	7	0.550	0.389	0.438	0.570	48.4	88.1
	9	0.550	0.394	0.456	0.570	53.2	92.8
	3*	0.480	0.332	0.358	0.540	4.8	56.8
min	5*	0.500	0.321	0.362	0.520	5.0	72.0
m	7	0.500	0.308	0.366	0.520	6.4	79.0
	9	0.510	0.294	0.370	0.510	9.2	81.2

SOURCE C

	N	$4nd, \mu m$	x_o	y_o	$\lambda_F, \mu m$	$m, \%$	$R_F, \%$
	C		0.310	0.316	0.480	7.3	
	3*	0.550	0.327	0.357	0.550	6.9	60.0
max	5	0.550	0.342	0.393	0.560	19.0	78.2
R_F	7	0.550	0.357	0.424	0.565	34.0	87.9
	9	0.550	0.363	0.433	0.565	36.0	92.5
	3*	0.545	0.323	0.353	0.517	3.6	59.9
min	5	0.525	0.312	0.351	0.508	5.0	77.0
m	7	0.525	0.315	0.366	0.516	5.0	86.3
	9	0.520	0.307	0.367	0.518	7.1	89.7

SOURCE A

	N	$4nd, \mu m$	x_o	y_o	$\lambda_F, \mu m$	$m, \%$	$R_F, \%$
	A		0.448	0.407	0.584	58.0	
	3	0.550	0.454	0.426	0.580	64.0	59.9
max	5	0.550	0.460	0.440	0.580	72.0	78.2
R_F	7	0.550	0.466	0.449	0.580	78.0	88.0
	9	0.550	0.469	0.453	0.580	78.0	92.8
	3	0.375	0.312	0.375	0.515	8.0	21.0
min	5	0.440	0.296	0.411	0.520	15.0	39.3
m	7	0.460	0.280	0.430	0.520	19.0	47.3
	9	0.435	0.318	0.318	0.460	4.5	23.9

TABLE 1. Colour characteristics of standard A, B, C-sources and reflected from multilayer coatings radiation for the same sources. N is the number of layers, (x_o, y_o) - chromaticity coordinates of reflected light, λ_F - dominant wavelength, m - colour saturation, R_F - photometric reflection factor, $4nd$ - characteristic optical thickness

In fig. 2, 3 and 4 these points are located between corresponding dashed lines. Moreover, the reflected light colour depends on the layer optical thickness. One can see, that for B and C-sources the curves for 3 and 5 layers cross the white light DIN-standard zone. 7 and 9 layer curves for B and C, and 3 layer curve for A standard source are close to the zone.

As a result of our investigation it might be said, that photometric reflection factor R_F has maximum for all layer numbers and light sources, when characteristic optical thickness of the layers is $4nd = 0.55\mu m$. However, for this thickness, it is not possible to obtain minimal colour saturation of reflected light. It follows from the position of these points in the corresponding curves in the presented figures. In all cases the points for characteristic optical thicknesses $4nd < 0.55\mu m$ are more close to the achromatic point E. The data in table 1 provide opportunities to compare reflected light parameters for different values of layer optical thickness with maximal R_F and minimal m. All systems in DIN-zone for white light are denoted with *. In all tables the upper blok of data corresponds to maximal value of R_F and the lower blok - to minimal m.

It is possible to obtain reflected light that is more close to the white light varying optical thickness of layers, as it can be seen from the curves in fig. 2, 3, 4 and the data in table 1. Still one should note that the light source has considerable importance in gaining high value of R_F. It is possible to obtain reflected light from B and C-sources with colour saturation m, less than m for direct light from the same sources without significant decrease in R_F. Still the correction for A-source colour requires considerable decrease in R_F (by a factor of 2 - 3). These differences might be explained by the spectral distribution of emmited light energy. A-source radiation is most intensive in the red region of the spectrum and less intensive in the blue region. In order to correct the colour of this light source, the coating must reflect more intensively blue light and less intensively red light. As a result of this selectivity in reflection significant decrease in R_F is observed. The radiation of B and C light sources is more uniform in the visible region, and therefore, the necessary corrections are small, without significant decrease in R_F.

As it was mentioned above, the curves for A-source for all systems with equal optical thicknesses are out of DIN-zone (see fig. 4). It is possible to calculate some point coordinates in this zone, if the optical thicknesses of layers in the system are not equal to each other. During further investigation the layers in the system were dividing into two groups. The first one comprises the upper, the nearest to the air $N_1 = (N-1)/2+1$ layers, the second one - the lower $N_2 = (N-1)/2-1$ layers. For multilayer compositions with chromaticity coordinates close to white DIN-zone (fig. 4) variations in optical thickness for each group were made. For A-source and this type of multilayer system chromaticity coordinates of reflected light might be located in the white zone. The regions of optical thickness variations for both groups of layers and corresponding values of R_F are shown in table 2.

N	$(4nd)_1, \mu m$	$(4nd)_2, \mu m$	$R_{Fmin}, ^o/_o$	$R_{Fmax}, ^o/_o$
3	0.350 - 0.450	0.200 - 0.250	10.4	29.8
5	0.420 - 0.500	0.290 - 0.320	22.8	40.2
7	0.450 - 0.500	0.350 - 0.380	29.4	50.2
9	0.470 - 0.495	0.390 - 0.410	32.6	53.3

TABLE 2. Optical thicknesses of combined multilayer systems for A-source colour correction

Minimal values of R_F correspond to the points at the lower boundary of the white DIN-zone, near to the point E, while the maximum of this parameter is connected with the points, located close to the top of this zone with coordinates $x = 0.366, y = 0.392$ for all systems with A-source. However, the values of R_F in case of layers with different thicknesses are less than the maximal ones for the same number of layers, as it is shown in table 1.

For B and C-sources , combined with 7 and 9 layer reflection systems it is possible to obtain light with chromaticity coordinates in the white DIN-zone, when the layers are divided into two groups, as in the case of A-source. For B-source it might be reached, when the first group of layers is quarterwavelength for $0.47-0.55\mu m$, and the second one - for $0.20-0.35\mu m$. As a result, the maximal value of R_F for 7 layer system is $64.1^o/_o$ and for 9 layer system is $75.2^o/_o$. For C-source the first group of layer is quarterwavelength for $0.50 - 0.57\mu m$, and the second one - for $0.72 - 0.82\mu m$ with maxium of $R_F = 66.7^o/_o$ and $77.5^o/_o$ respectively. However, values of R_F in case of layers with different thicknesses are smaller again with respect to maximal ones for the same number of layers, as it is shown in table 1.

The role of both layer groups becomes clear from the data in tables 1 and 2. For A and B-sources with λ_F in the orange spectral region the first group of layers is quarterwavelength for blue and green light and reflects intensively in these spectral ranges. The second group of layers is quarter-wavelength for near ultraviolet region, which reflection decreases smoothly from blue to red. For C-source with λ_F in the blue spectral region the first group of layers is quarterwavelength for green light and the second group of layers - for red light. The combination of both groups of layers provides opportunities to correct the light sources colour. As a result the maximal value of the photometric reflection coefficient R_F has decrease, which depends on the spectrum of the light source.

IV. CONCLUSIONS

1. The photometric reflection coefficient has maximal values in case of multilayer reflection systems with equal thicknesses calculated for the central points in the visible region of spectrum. In this case the requirements for the reflected light colour are not matched for all considered light sources.

2. For B and C-sources which have chromaticity coordinates close to the

point of achromatic stimul E, the use of reflection systems with equal thicknesses and small number of layers (3, 5) provides opportunity to match the requirements of DIN-standard with respect to the white light.

3. Using multilayer systems with 7 and 9 layers and different optical thicknesses for the separate groups of layers for B and C-sources provides opportunity to obtain points in the white light region. The obtained values of R_F are less than the maximal ones for the same number of layers.

4. One is unable to obtain reflection light with chromaticity coordinates located in the white light region by using a standard A-source and multilayer reflection systems with equal layer thicknesses. It becomes possible if one uses reflection systems with diffrencies in thicknesses of separate groups of layers. As a result the photometric reflection coefficient R_F has strong decrease.

5. The colour correction of the source light by using selective reflection system might have less looses in reflected light intensity, when the chromaticity coordinates of the source are near to E-point. For this correction more effective proved to be the reflection systems with different thicknesses of separate groups of layers.

REFERENCES

1. Born M., and E. Wolf, 1970, Principles of Optics, Pergamon, New York.

2. Allen T., Properties of ion assisted deposited silica and titania films, SPIE Vol.325 Optical Thin Films, 1982, pp.93-100.

3. Judd D., G.Wyszecki, Color in business science and industry, 1975.

4. Standard DIN 67505.

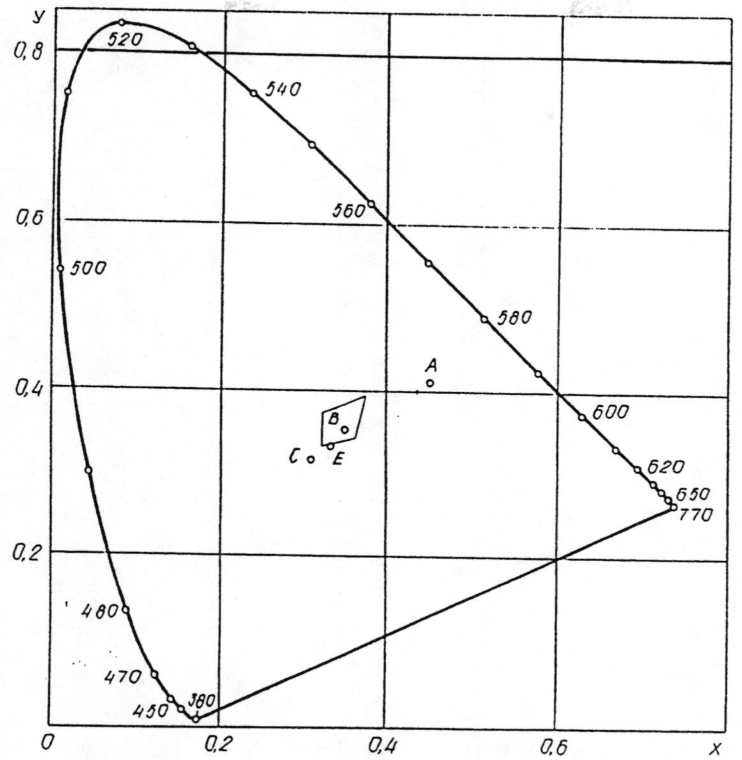

Figure 1. Chromaticity plot x, y (IIC,1931) with standard A, B, C-sources, achromatic point E and white DIN-zone

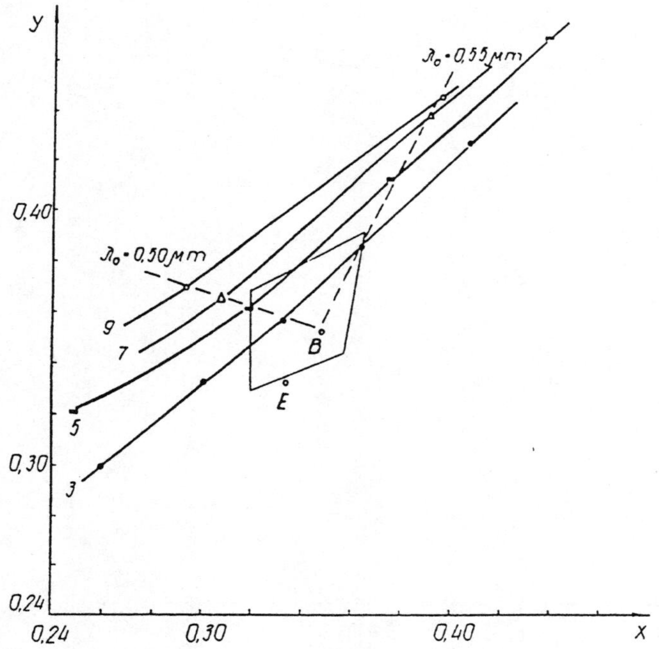

Figure 2. Chromaticity coordinates of light reflected from multilayer systems with determined number of layers for source B (equal optical thickness)

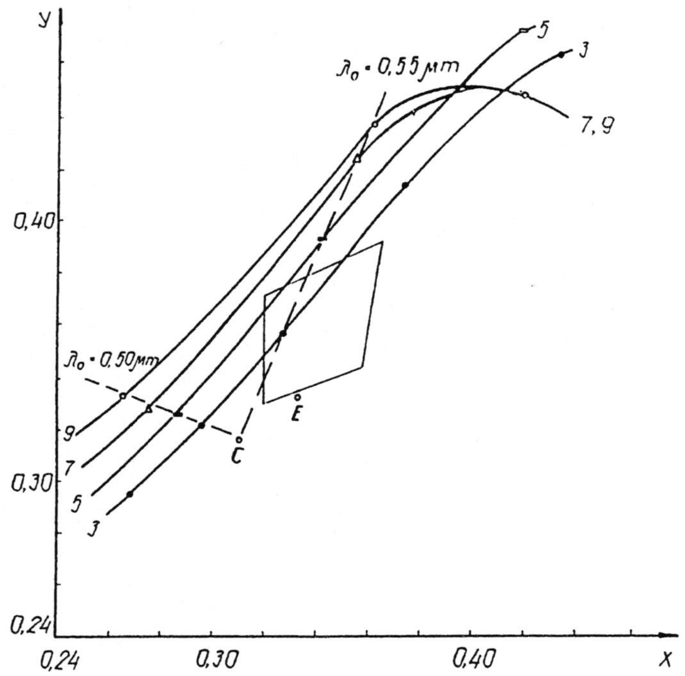

Figure 3. Chromaticity coordinates of light reflected from multilayer systems with determined number of layers for source C (equal optical thickness)

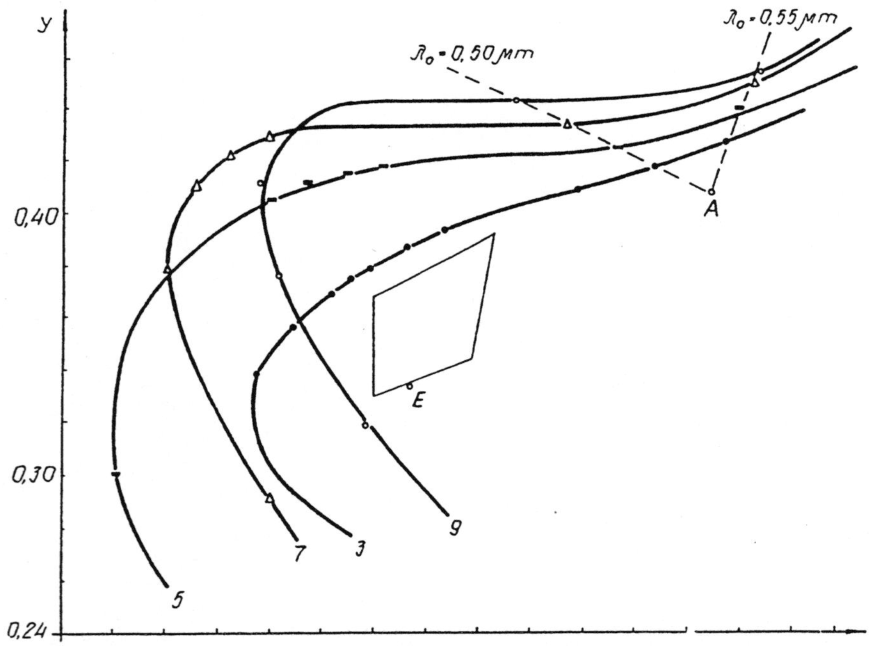

Figure 4. Chromaticity coordinates of light reflected from multilayer systems with determined number of layers for source A (equal optical thickness)

Vacuum Electron and Ion Technologies
D. S. Karpuzov (Ed.)

ISBN 1-56072-242-8

ANTIREFLECTION COATINGS ON Ge AND GaAs SUBSTRATES FOR 10.6 μm RADIATION

George Atanassov and Dimitar Popov
Department of Physics, Technical University - Rousse Bulgaria
Pei-Fu Gu
Department of Optical Engineering, Zhejiang University, Hangzhou 310027, China

Abstract

Antireflection coatings on Ge and GaAs substrates were designed for 10.6 μm on the basis of ZnS, ZnSe, PbF_2 and KCl thin films. Conventional resistive evaporation and resistive evaporation with Ar ion bombardment of the growing films with energies of 40 eV and 270 eV and current density of 120 $\mu A/cm^2$ were chosen as deposition techniques. The optical absorption loss at 10.6 μm was measured by photothermal deflection system arranged to measure local absorption loss in a focal point with power density of the pump laser beam about 5 kW/cm^2. A considerable decrease of optical absorption in single, double and three layer antireflection coatings on Ge substrates was observed in the samples deposited by resistive evaporation with Ar ion bombardment with energies of 40 eV. The samples bombarded with Ar ion energies of 270 eV showed about twice as big optical absorption as these bombarded with 40 eV. The optical coatings deposited on GaAs that contained KCl thin films were unable to withstand the pump beam power density and deteriorated during the test.

1. Introduction

Components of the transmissive optics for CO_2 lasers are generally manufactured from germanium, gallium arsenide, zinc selenide or alkaline halides single crystals such as sodium and potassium chlorides. The major assets of Ge and GaAs substrates are their high termoconductivity and mechanical hardness as well as their low dispersion of the refractive index in 8 - 12 μm spectral range. The main disadvantages of these substances are the temperature dependent free - carrier absorption and low figures of merit for optical distortion [1]. Still Ge and GaAs materials are widely used at laser powers lower than 0.3 - 0.5 kW especially as thin low - size optical components. For these components a considerable part of the absorption loss is predominantly located at the surfaces. Therefore, low loss optical coatings are needed in order to lower the total component absorption.

The design techniques for antireflection coatings (ARC) on high refractive index substrates such as Ge and GaAs were summarized by Macleod [2] and Aquilera et al.[3]. Special attention has been paid to three layer ARC as most expedient coatings for high-power laser application [4]. The most widely used design technique for three - layer ARC that contain two layers of the same material was elaborated by Mouchart and expressed in a useful diagram called Mouchart's diagram [5].

Ion bombardment of the growing films plays a vital role in condensing process and influences the substrate, substrate - film interface, film structure and film composition [6]. The ion - based techniques are mainly used in low loss thin film deposition. There is not full understanding of the ion - surface interaction phenomena, especially in the case of low-energy high - ion current bombardment. Still it was revealed that, in order to achieve atomic mobility on both substrate and coating and thus to ensure adhesion and homogenized coating, the condensation rate and ion - bombarding rate ratio is most important [7].

Quite similar deposition technique to those applied in the present work was used by Oh [8]. A small decrease in optical absorption of Ge thin films at 10.6 μm measured calorimetrically, was observed after ion bombardment with energy of 100 eV and current density of 20 $\mu A/cm^2$. It has also been demonstrated that Ar ion bombardment with energy of 300 eV and ion current density of 120 $\mu A/cm^2$ can considerably influence the optical absorption at 632.8 nm both in ZnS and MgF_2 thin films [9].

For characterization the weak absorption of thin films high sensitive measuring methods are needed. In addition, these methods should be able to give local absorptances in the films and substrate and thus to reveal the reasons causing the total absorption. Temple [10] showed that the total absorptance in a coating - substrate system depends on the relative power density distribution in depth of the system. The total absorption is expressed as a sum of the local absorptances at the boundaries "air - coating", "film - film", "coating - substrate" and the volume absorptances in the layers and substrate. The method of photothermal deflection seems most expedient for local absorption measurements [11]. Extinction coefficients that do not exceed some 10^{-7} can be measured by this method and separation of surface and volume absorptances are possible [12,13].

In the present work design of single, double and three layer ARC on Ge and GaAs substrates for 10.6 μm was carried out. Some of these ARC were deposited by conventional resistive evaporation and by resistive evaporation with Ar ion bombardment of the growing films with two different ion energies and relatively high ion current density of 120 $\mu A/cm^2$. The total absorptances were measured by photothermal deflection system at 10.6 μm. Some assumptions regarding the ion - surface interaction phenomena are discussed.

2. Design of ARC and Experimental Details

The design of single, double and three layer ARC on Ge and GaAs substrates for 10.6 μm was carried out on the basis of Shuster's diagram [2], Mouchart's diagram [5] and exact synthesys methods [4]. As high refractive index materials ZnS and ZnSe were used and as low refractive index materials PbF_2 and KCl were used. In the case of three layer ARC the thickness of the upper layer was chosen to be big enough to protect the second PbF_2 or KCl layer against atmospheric moisture. The thicknesses of the other two layers were evaluated from near-to-zero reflection conditions and after that the distribution of the normalized electric field and the total absorptance were calculated. Only solutions with minimal absorptance for the chosen materials were preferred.

The following systems with ARC were chosen for calculation the optical spectra, for practical deposition and absorption measurement:

Ge//1.00 ZnS//Air; Ge//0.712 ZnSe/0.581 PbF_2//Air;
Ge //0.245 ZnSe/0.258 PbF_2/ 0.428 ZnSe // Air; (1)
GaAs//0.516 ZnSe/0.348 KCl/0.231 ZnSe//Air;

as the numbers express the film optical thicknesses in $\lambda/4$ units (λ = 10.6 μm).

ARC were deposited by conventional resistive evaporation or by ion - assisted resistive evaporation. The optical thickness of the layers was monitored by photometric system with monochromator in visible light using glass witnesses. The monitoring wavelength was different for each layer and was estimated by simple derived relation

$$\lambda_c = \frac{\lambda . h . n_c}{k . n} \qquad (2)$$

where h is the optical thickness in $\lambda/4$ units (λ = 10.6 μm), k is the monitored extrema number, n and n_c are the film refractive indices at λ and λ_c respectively.

For ion - assisted deposition a Kaufman - type ion source previously described by Gu et al. was used [9].

The transmission spectra of the samples in 2.5 - 25 μm spectral range were measured by NICOLET DX spectrophotometer.

For measurement the absorptance of the samples a photothermal deflection system was used. The schematic diagram of this system is given in Fig. 1. The pump laser 1 is a cw CO_2 single mode laser with output power of 6 Watt. A beam splitter 2 splits about 25 % of the pump power for power monitoring 3. The mechanical chopper 4 with a linear frequency of 24 Hz cuts off another 50 % of the rest power. Through a lens with focal length 6 cm the pump beam is focused on the sample 5. In the focal point we assumed to have Gaussian distribution of the intensity with Gaussian radius of about 0.012 cm and power density of about 5 kW/cm^2. After passing through the sample the laser beam gets to absorber 6. The probe laser 7 is He-Ne laser with output power of 2 mW. Through a lens with a focal length of 6 cm that provides focal point radius much less than that of the pump beam, the probe beam scans the heated sample until the maximal slope of the bulged region is found. The probe beam detector 8 is a quadrant position detector that might be settled either in transmission mode 8A or in reflection mode 8B, when the samples are respectively transparent or impervious with respect to probe beam. The modulated deflection signal is measured by lock-in amplifier 10.

All substrates were optically polished and had diameter of 1.8 cm and thickness 0.04 cm for Ge and 0.15 cm for GaAs. For a modulation frequency of 24 Hz (in the case of Ge substrates) the typical thermal length is 0.07 cm which is bigger than the sample thickness. Therefore, at these conditions, mainly the total absorptance of the sample has been measured. The total absorptance of an uncoated Ge sample is 0.1 % and comprises the volume absorptance (α.l = 0.08 %) and the surface absorptance, which is about 15 % of the volume absorptance for each surface [1]. These data were used for system calibration. The system constant was verified by comparison of the measured absorptance of a single $\lambda/4$ ZnS layer with those measured by another independent methods and proved to have an error less that 10 %. GaAs samples had bigger thickness and, in order to use similar calibration procedure, the modulation frequency was reduced to 3 Hz.

3. Experimental Results and Discussion

The absorptance of the samples was obtained by substituting the total absorptance of bare substrate from the total absorptance of the coated sample. The results for Ge substrates are given in Table 1 and Fig. 2.

Table 1. Experimental results for absorptance of different coatings deposited on Ge substrates

Deposition technique	AR coating	Absorptance, %
Conventional resistive evaporation	Bare Ge substrate	0.10
	Ge // ZnS // Air	0.171
	Ge // ZnSe / PbF_2 // Air	0.113
	Ge // ZnSe / PbF_2 / ZnSe // Air	0.123
RE - IAD Ar-ion energy of 40 eV	Bare Ge substrate	0.10
	Ge // ZnS // Air	0.104
	Ge // ZnSe / PbF_2 // Air	0.082
	Ge // ZnSe / PbF_2 / ZnSe // Air	0.099
RE - IAD Ar-ion energy of 270 eV	Bare Ge substrate	0.10
	Ge // ZnS // Air	0.413
	Ge // ZnSe / PbF_2 // Air	0.140
	Ge // ZnSe / PbF_2 / ZnSe // Air	0.224

Before discussion one should point out that the chosen ion current density of 120 $\mu A/cm^2$ provides about 10^{18} bombarding Ar ions on the substrate during the typical deposition time of 5 minutes. This number is of about the same order as the number of ZnS, ZnSe, PbF_2 or KCl molecules condensed on the substrate for the same time. The penetration depths of the bombarding ions in the coatings were negligible even for the more energetic bombardment with 270 eV ion energies [6]. Sputtering effects, although presented, might not be taken into consideration because of their smallness. Thus one could expect that Ar ions would be delivering energy mainly to influence the molecule mobility on the surface causing film smoothness and homogeneity enhansement and therefore lowering absorptance.

It could be seen in Fig. 2 that the ion bombardment had significantly influenced the absorption of single ZnS coating. The lower energy bombardment caused a coating absorptance decrease and confirmed the preliminar expectation. The more energetic ion bombardment with ion energy of 270 eV leads to considerable raise in the absorptance. The same effect was observed for the absorption of ZnS films at 632.8 nm after bombardment with Ar ions with energy of 300 eV [9]. One could propose that this higher absorptance might due to a small layer decomposition at higher ion energy bombardment.

Slightly different situation is observed in double and three layer ARC that contain both ZnSe and PbF_2 layers. ZnSe alike ZnS shows decreased absorptance when bombarded with 40 eV and increased absorptance when bombarded with 270 eV. Still in double and three layer ARC this effect is more vague due to presence of both PbF_2 layer and additional "film - film" interfaces. Unlike ZnS and ZnSe, PbF_2 layers showed reduced absorption at 270 eV more than at 40 eV ion bombardment. Such a behaviour is easy to find out taking into account that PbF_2 layers generally show slightly lower than unity relative density and contain small amount of capillarly absorbed water. This water absorption effect takes place especially when PbF_2 layer contacts air as it takes place in double layer ARC. More energetic ion bombardment can contribute to PbF_2 layer densification. As a result, the total absorptance in double layer ARC deposited with 270 eV bombardment increases just a little due to these two competive effects in ZnSe and PbF_2 films. In three layer ARC PbF_2 is not in contact with air and, therefore, the observed changes in the total absorptance might due mainly to the changes in ZnSe absorption.

With respect to these different influences of the ion bombardment on ZnSe and PbF_2 layers it is expedient to use different ion bombardment regimes - lower energy ion bombardment for ZnS and ZnSe layers and higher energy ion bombardment for PbF_2 layers. Such a selective bombardment will be a purpose for author's further investigations.

ARC on GaAs substrates were unable to withstand the pump power density of 5 kW/cm^2 regardless of the deposition technique. During heating the upper ZnSe layer had pilled off and after its removal many cracks were found in the lower KCl layer. It was assumed that KCl thin films, despite their softness, have high thermal extention which makes them not compatible to ZnSe films.

4. Conclusions

In order to deposit low loss ARC on Ge and GaAs substrates three different deposition techniques were preferred - conventional resistive evaporation and resistive evaporation with Ar ion bombardment of the growing films with ion energy of 40 eV and 270 eV. An ion current density of 120 $\mu A/cm^2$ was chosen to provide ionized particle rate of the same order as the rate of the condensing particles.

The local optical absorptance was measured by photothermal deflection technique at 10.6 μm arranged to provide pump beam power density of about 5 kW/cm^2.

Single, layer ARC comprising a $\lambda/4$ ZnS thin films showed decreased absorptance when bombarded with ion energy of 40 eV. This decrease is supposed to be due to increased condensing particle mobility on the substrate during deposition. The absorptance of $\lambda/4$ ZnS layer when bombarded with 270 eV increased, supposedly due to a small material decomposition. The same behaviour as ZnS showed ZnSe layers in double and three layer ARC.

PbF_2 layers in double and three layer ARC showed decreased absorption when bombarded with energy of 270 eV, supposedly due to their densification and reducing their water contain. The absorption performance of double and three layer ARC is a combination of the absorptances in all layers.

KCl layers in three layer ARC on GaAs substrates are unable to withstand the pump beam power density of 5 kW/cm^2 , supposedly due to their higher extention compared with that of ZnSe layers.

From practical point of view low energy (40 eV) high current Ar ion bombardment for ZnS and ZnSe layers and 270 eV for PbF_2 layers might be recommended for reducing the total absorptance in ARC on Ge substrates.

Reference

1. J. A. Savage, Infrared Optical Materials Their Antireflection Coatings, Adam Hilger Ltd., Bristol, 1985;
2. H. A. Macleod, Thin Film Optical Filters, Second Edition, Adam Hilger Ltd., Bristol, 1986;
3. J. A. Aquilera et al., Antireflection coatings for germanium IR optics: a comparison of numerical design methods, Appl. Opt., 27, 2832, 1988;
4. C. L. Nagendra and G. K. M. Thutupally, Design of three layer antireflection coatings: a generalized approach, Appl. Opt., 27, 2320, 1988;
5. Jacques Mouchart, Thin film optical coatings. 1. Optical coating stabilities, Appl. Opt., 16, 2486, 1977; 2. Three layer antireflection coating theory, Appl. Opt., 16, 2722, 1977;
6. P. J. Martin and R. P. Netterfield, Optical Films Produced by Ion - Based Techniques, Ed. by E. Wolf, Progress in Optics, vol.23, Elsevier Science Publishers B. V., 1986;
7. G. Carter and D. G. Armour, The interaction of low energy ion beams with surfaces, Thin Solid Films, 80, 13 - 29, 1981;
8. Thomas I. Oh, Broadband AR coatings on germanium substrates using ion - assisted deposition, Appl. Opt.,27, 4255, 1988;
9. Pei - Fu Gu et al., Properties of dielectric coatings produced by ion assisted deposition, Appl. Opt., 28, 3318, 1989;
10. P. A. Temple, The measurement of absorption in thin films by laser calorimetry, SPIE, vol. 652, 272, 1986;
11. M. Commandre, I. Bertrand, G. Albrand and E. Pelletier, Measurement of absorption losses of optical thin film components by Photothermal Deflection Spectroscopy, SPIE, vol. 805, 1987;
12. Y. - X. Nie, I. Bertrand, Separation of surface and volume absorptions by photothermal deflection, J. Appl. Phys., 65 (2), 438, 1989;

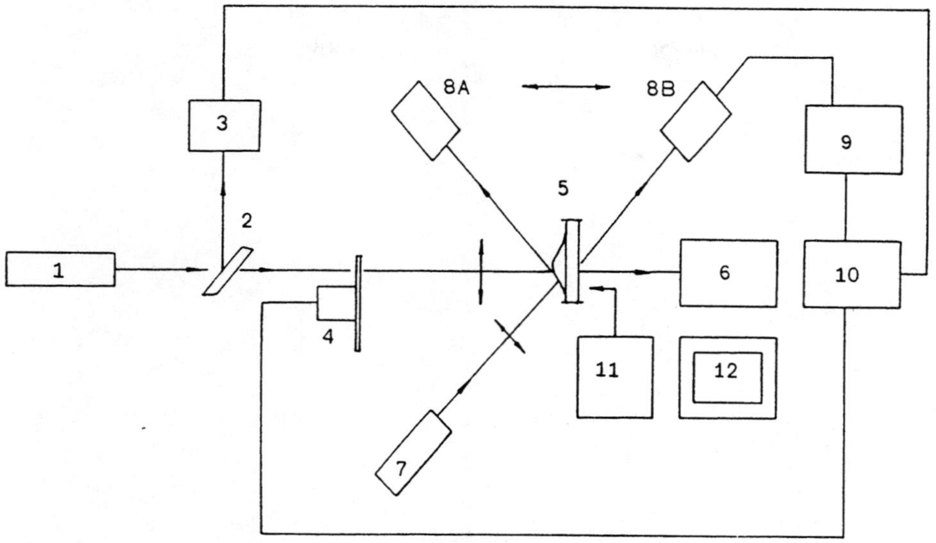

Fig. 1. Scheme of the experimental set-up: 1 - pump CO2 laser; 2 - beam splitter; 3 - power monitor; 4 - mechanical chopper; 5 - measured sample; 6 - absorber; 7 - probe He-Ne laser; 8 - detector; 9 - preamplifier; 10 - lock-in amplifier; 11 - motor driver; 12 - computer;

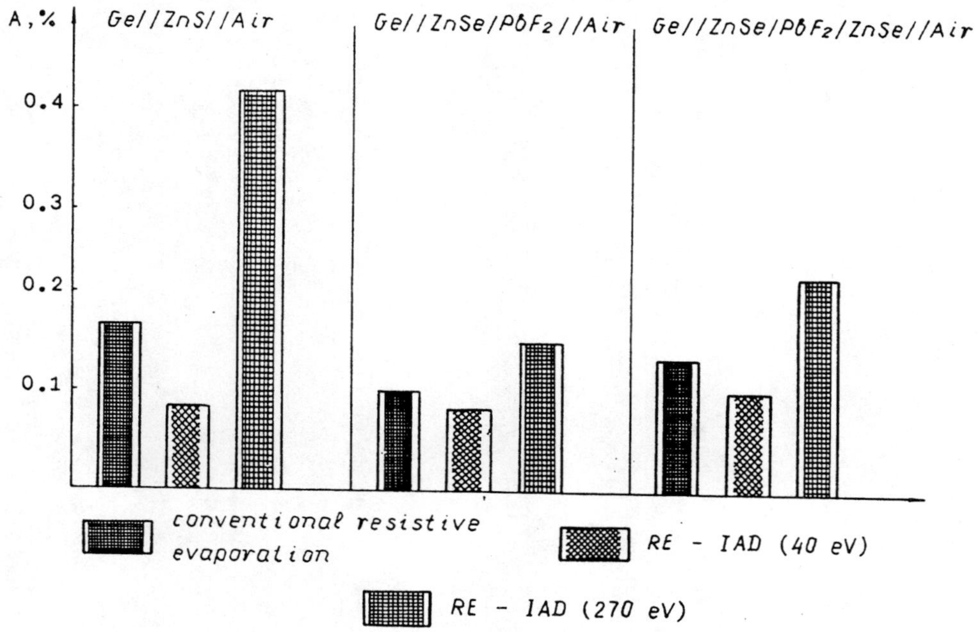

Fig. 2. Comparative diagram for absorptances of single, double and three layer ARC on Ge substrates deposited by different techniques;

Vacuum Electron and Ion Technologies
D. S. Karpuzov (Ed.)

ISBN 1-56072-242-8

SOLID- PHASE INTERACTIONS IN THIN-FILM Cu/Ti SYSTEM

Tz. Uzunov, St. Stojanov, and S. Lambov
Faculty of Engineerings and Pedagogy, Sliven Technical University, Sofia
Sliven 8800, Bulgaria

The solid-phase interactions at isothermal annealing of multi-layer Cu/Ti systems deposited on different substrates have been investigated. The sequence and annealing temperature and time for formation of intermetallic compounds of systems in an excess of copper and titan have been studied. The obtained results have been compared with the well-known structural diagram of Cu/Ti system.

I. Introduction

The "massive" Cu/Ti system has been an object of many investigations, but a definitive structural diagram is still missing. The different authors[1-3] show different number of intermetallic compounds. There is no common opinion about the stoichiometry, structure, areas of homogenity and temperature ranges of existing of some compounds[4-7]. In ASTM card-index[8] eleven intermetallic compounds with different stoichiometry and structure are shown (Table 1).

The thin-film Cu/Ti system has been less investigated. The authors of paper[9] have established that no solid-phase interactions occur at room temperature. The epitaxial growth of Ti-layer, obtained by isothermal deposition /vaporization/ on monocrystalline copper <111> substrate, has been studied in paper[10]. During the annealing of the system the Ti/Cu (400°C, 10 min) and $TiCu_3$-phases (550°C, 10 min) were registrated. The crystallization of amorphous layers from the alloys $Cu_{48}Ti_{52}$ and $Cu_{60}Ti_{40}$ deposited by magnetron sputttering has been investigated by Marshall et al[11,12]. In the first case at 433°C the equilibrium Cu_3Ti_2 phase, which initially occurs together with the metastable Cu_4Ti and Cu_4Ti_3 phases, is formed. In the second case at 385°C an intermediate longperiod orderly phase, which can be scrutinized as a γ-CuTi structure enriched with titan atoms, is observed.

TABLE 1

Stoichiometry and Structure of the Intermetallic Compounds in Cu/Ti System

Compound	Structure	Lattice parameters, A		
		a	b	c
Cu_4Ti	orthorhombic	4.530	4.342	12.930
β- Cu_3Ti	"	2.585	4.527	4.351
β- Cu_3Ti	"	5.170	4.351	4.527
Cu_3Ti	"	5.45	4.42	4.30
Cu_2Ti	"	4.363	7.977	4.478
Cu_3Ti_2	tetragonal	3.13	-	13.95
Cu_4Ti_3	"	3.13	-	19.94
γ- CuTi	"	3.108	-	5.887
δ- CuTi	"	4.436	-	2.813
$CuTi_2$	"	2.9458	-	10.7861
$CuTi_3$	"	4.158	-	3.594

At 405°C this phase disintegrates in the equilibrium γ-CuTi and $CuTi_2$ phases.

The growth kinefics of CuTi and Cu_7Ti_2 (Cu_3Ti) phases in the thin-film Cu/Ti system in an excess of copper has been investigated by Gershinskii et.al.[13] in the temperature range 400-500 C and by Liotard et. al.[14] in the range 350-450°C.

2. Experimental technique.

Two types of multi-layer Cu/Ti systems in an excess of Cu(specimens type A and B) and Ti(specimens type C and D) have been examined.

The thin copper and titan layers were obtained by magnetron sputtering in an Ar atmosphere at a pressure $5x10^{-2}$ Pa. The layers were deposited at room temperature and their thickness was measured by means of resonance-frequency method with silica plate.

Specimens type A contained 10 bi-layers Cu/Ti as every layer had a thickness of 100 nm (total thickness of 2 μm). The layers were deposited on polycrystalline Cu substrate. The atomic ratio Cu : Ti = 1,5 : 1. Specimens type B were prepared upon preliminarily thermally oxidized <111> Si-plates (the thickness of the oxide layer was about 0,4 μm). The seperate layers were deposited in the following sequence : at first a copper layer with a thickness 100 nm, after that 10 bi-layers Ti/Cu with a thickness of the titan layers 100 nm and 200 nm for the copper ones and finally a titan layer with a thickness 50 nm. The obtained system contained 21 sub-systems, each with a thickness 150 nm (100 nm Cu and 50 nm Ti). The atomic ratio Cu : Ti = 3 : 1.

Specimens type C were prepared upon polycrystalline Ti substrates. They contained a system of 6 bi-layers Cu/Ti as the thickness of Cu layers was 100 nm and of Ti-ones-200 nm. The atomic ratio Cu : Ti = 1 : 1,33. Specimens type D were prepared on thermally oxidized <111> Si-plates in the following sequence : at first a Cu layer with 100 nm thickness deposited after that 5 bi-layers Ti/Cu with a thickness of the Ti layers 600 nm and of the Cu ones 200 nm and finally a Ti-layer with 300 nm thickness. The system contained 11 sub-system each with 400 nm thickness (Cu - 100 nm and Ti - 300 nm). The atomic ratio Cu : Ti = 1 : 2.

The specimens were isothermally annealed in silica ampules at a residual gas pressure 10^{-2} Pa.

The annealing regimes for specimens types A and B are given in Table 2, and those for types C and D - in Table 3.

The phasic composition of specimens before and after annealing was determined by X-ray analysis with X-ray difractometer DRON UM-1 with Co-K_α emission. The identification of the formed phases was done by the interplane distances according to ASTM card-index[8] and the results are given in Table 1. Some specimens of type B and D after unsticking from the substrate and after chemical and electrochemical thinning[5] were examined by means of transmission electron microscope Tesla BS-500 at beam voltage 90 kV.

TABLE 2

Influence of the Annealing Regime Upon the Phasic Composition of Specimens Type A and B

Table 2

Annealing regime		Specimens A		Specimens B	
Temperature, °C	Time, min	Registrated Phases	$\frac{C_{\gamma}\text{-CuTi}}{C_{\beta}\text{-Cu}_3\text{Ti}}$	Registrated Phases	$\frac{C_{\gamma}\text{-CuTi}}{C_{\beta}\text{-Cu}_3\text{Ti}}$
25	-				
220	1080				
260	1080				
300	60		2.3		3.0
370	30		2.0		2.8
420	30		1.9		0.7
470	30		1.5		0.14
520	30		1.1		0.11
600	30		0.4		0.10
740	30				
920	30				

TABLE 3

Influence of the Annealing Regime Upon The Phasic Composition of Specimens Type C and D

Annealing regime		Specimens C			Specimens D		
Tempera- ture, ^{o}C	Time, min	Registrated Phases	$\frac{C_{\gamma-CuTi}}{C_{\beta-Cu_3Ti}}$	$\frac{C_{\gamma-CuTi}}{C_{CuTi_2}}$	Registrated Phases	$\frac{C_{\gamma-CuTi}}{C_{\beta-Cu_3Ti}}$	$\frac{C_{\gamma-CuTi}}{C_{CuTi_2}}$
25	-						
320	30		2.6			2.9	
370	30		2.5			2.5	
400	30		1.7			2.1	
440	30			2.9		2.7	
470	30			1.2		5.6	7.5
540	30			0.9			0.9
640	30			0.5			0.05
740	30						

3. Experimental results and discussions

3.1. Thin-film Cu/Ti system in an excess of Cu.

The schematic diffractograms of systems in an excess of copper (specimens type B) after different annealing regimes are shown in Figure 1. Similar diffraction picture show the specimens type A, but since the copper amount for them is not unlimited at high annealing temperatures, other quantitative phase correlations were observed.

Figure 1-a shows that the copper and titan layers do not interact during the deposition process at room temperature. In both types of specimens γ- CuTi phase was firstly registrated after annealing at 260°C for 18 hours.

β- Cu Ti phase was registrated after annealing at 300°C for 1 hour (Fig. 1-b).

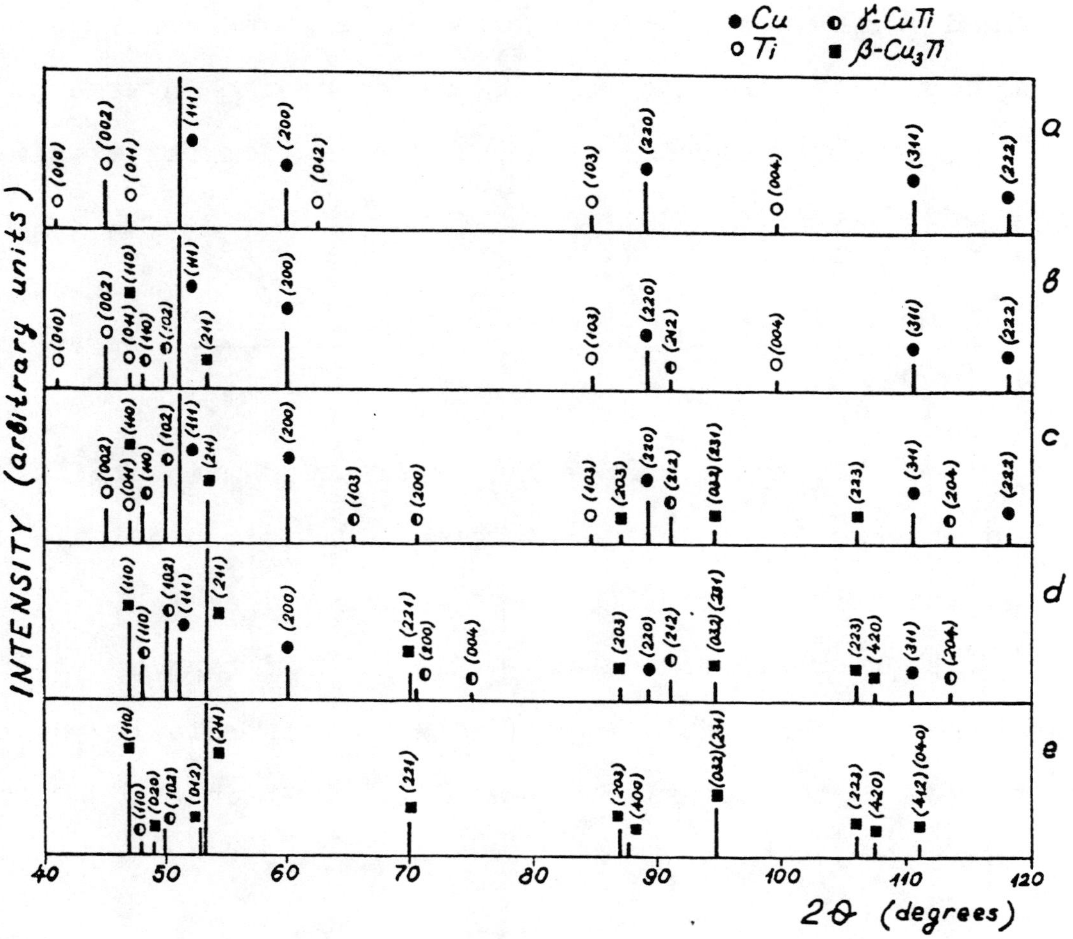

Fig.1. Schematic diagrams of the most characteristic X-ray diffractograms of multi-layer Cu/Ti specimens (type B): without annealing (a) after annealing at 300°C for 60 min (b), at 370°C for 30 min (c), at 420 C for 30 min (d) and at 520°C for 30 min (e).

The most intensive peak <211> of β - Cu Ti phase at Breg's angle $2\theta = 53,5^{o}$ is also present on the diffractograms made after annealing at $260^{o}C$ for 18 hours but with a very slight intensity and large width.

The relation of the formed quantities expressed in vol. % of the intermetallic $C_{\gamma\text{-}CuTi}$ /$C_{\beta\text{-}Cu_3Ti}$ phases was determined from the diffractograms. The phase analysis was carried on with reading the textures by the inverse pole method[15]. The total density of all n diffraction reflexes was used according to the equation :

$$\frac{C_{\gamma\text{-}CuTi}}{C_{\beta\text{-}Cu_3Ti}} = \frac{\frac{1}{n_\gamma}\sum_{i=1}^{n_\gamma}(S_\gamma/I_\gamma)(hkl)_i}{\frac{1}{n_\beta}\sum_{j=1}^{n_\beta}(S_\beta/I_\beta)(hkl)_j} \qquad (1)$$

were S_{hkl} is the determined by the diffractogram intensity of the respective diffraction peak $\{hkl\}$. I_{hkl} is the calculated integral intensity of refflection from the planes $\{hkl\}$[16].

The results of relation (1) at different annealing regimes are given in Table 2 and Table 3. This relation gives also an idea of the relative speed of γ - CuTi and β - Cu Ti phase formation.

With increase of the annealing temperature (up to $420^{o}C$) the diffraction maxima intensity of the non-interacted Cu and Ti phases progressively dicreases (Figure 1"a-e"). Above this temperature (Table 2) for specimens type B Cu and Ti phases were not registrated. For specimens type A only Cu lines remain as their intensity begin slowly to increase because of the element diffusion from the substrate to the layers. When in the system (specimens type A and B) there is still non-reacted Ti, the quantity of the intermetallic γ - CuTi and β - Cu Ti phases grows up almost simultaneously with temperature increase. This is confirmed by the insignificant changing of $C_{\gamma\text{-}CuTi}/C_{\beta\text{-}Cu_3Ti}$ relation in the temperature range 300-370°C (Table 2).

After spending of the free titan, the stable phase continues to growth according to the solidphase reaction

$$Cu + \gamma - CuTi \longrightarrow \beta - Cu_3Ti \qquad (2)$$

with different rate for the two types of specimens defermined by the quantity of the free copper.

For specimens type A the necessary copper for the superior reaction (2) in the first minutes of the annealing is at the expense of the copper layers, but after that it comes from the substrate by a diffusion way. Its quantity is less than that in specimens type B. In these specimens the copper quantity in the layers is double bigger and the diffusion way is minimal. That's the reason why for specimens type A the relation (1) changes lightly in the wide temperature range 420-600°C, whereas for specimens type B only in the range 420-470°C it changes sharply (Table 2).

The lattice constants of β - Cu_3Ti phase determined by the diffractograms after annealing at 500°C and 600°C (a = 5,163 Å, b = 4,388 Å and c = 4,520 Å) are smaller than those for ideal stoichiometry (Table 1). Probably the difference is due to Ti shortage in the phase. In confirmation of our assumption is the defined by Liotard et. al.[14] stoichiometry of the same phase: $Cu_{3,12}Ti_{0,88}$.

After annealing at 740°C for 30 min in specimens type A the predominant phase is copper. Upon the diffractograms except the lines of Cu and β-Cu_3Ti phases other four slight diffraction lines are observed. These lines belong to a phase which neither can be identified as a compound between copper and titan nor as a compound of copper and titan with oxygen or nitrogen (from the residual atmosphere in the ampules). The diffraction lines of the nonidentified phase (in Table 2 it is marked with X) correspond to the interplane distances a_1 = 2,303 Å, a_2 = 1,880 Å, a_3 = 1,702 Å, a_4 = 1,330 Å as their intensities are as 5 : 1 : 1 : 2.

The diffractograms of the same type specimens registrated after annealing at 320°C for 30 min contain two very slight and wide lines corresponding to interplane distances 1,990 Å and 2,170 Å. It can be supposed that the annealing at this temperature leads to a contact melting of the Cu/Ti layers. (According to papers [1,2] the system has two eutectic points at temperatures 880°C and 960°C). The formed liquid phase interacts with the copper from the substrate forming solid solution on the basis of Cu with slight inclusions of β- Cu_3Ti or high-temperature β'- Cu_3Ti and Cu_4Ti phases (Table 1). The observed diffraction lines can be refered to every one of the mentioned three phases [8].

The micro-structure of the multi-layer system in an excess of copper after different annealing temperatures is shown in Figure 2. On the first photoes ("a" and "b") γ-CuTi and β-Cu_3Ti phases at an opening stage of growth can be seen, while on the other one ("c") the β-Cu_3Ti phase is predominant.

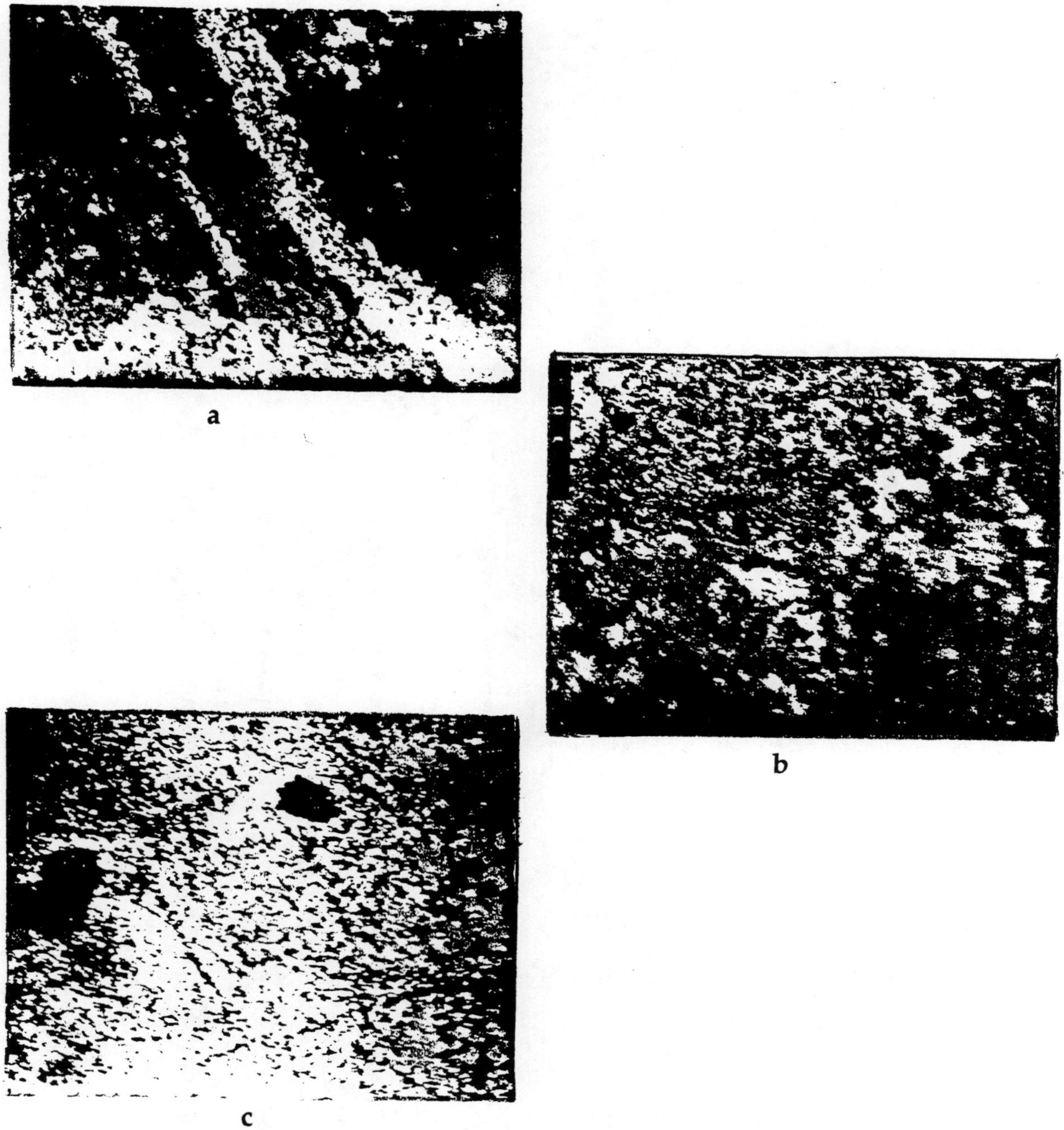

Fig.2. Structure of multi-layer Cu/Ti system (specimens type B) after annealing at the following regimes: a - 300°C, 60 min (18000); b - 370°C, 30 min (10000); c - 600°C, 30 min (10000).

3.2. Thin-film Cu/Ti system in an excess of Ti.

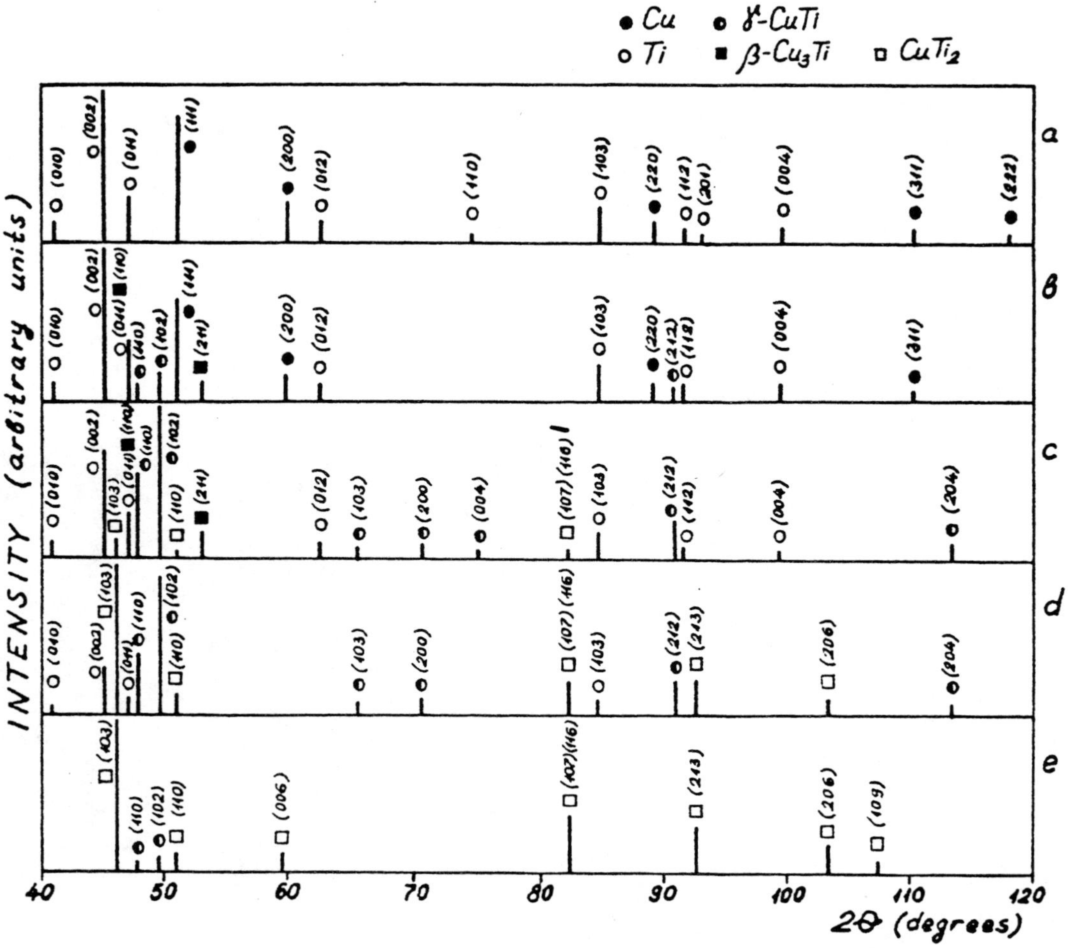

Fig.3. Schematic diagrams of the most characteristic X-ray diffractograms of multi-layer Cu/Ti specimens type D: as deposited (a) and after annealing for 30 min at: 370°C (b); 470°C (c); 540°C (d) and 640°C (e).

The obtained diffractograms of multi-layer Cu/Ti specimens type D in an excess of Ti are shown schematicaly in Figure 3. During the deposition the layers of copper and titan do not interact (Figure 3-a). After annealing γ- CuTi and β- Cu_3Ti were formed at first (Figure 3-b) as the starting stoichiometry did not influence the sequence. While in the system there was a residium of non-reacted copper, the quantity of both phases grew simultaneonsly and the relation (1) decreased slowly with temperature increase (Table 3).

When the free Cu is over, the stable γ-CuTi phase continues to grow at the expence of β- Cu_3Ti compound. Immediately before its spending on the interphasic boundary γ- CuTi/Ti a new rich of titan $CuTi_2$-phase comes into existence (Figure 3-c).

For specimens type C the CuTi phase is registrated after annealing at 440°C for 30 min together with Ti and γ-CuTi phases as the last is predominant: the β-Cu_3Ti phase is absent in the system (Table 3).

The CuTi compound grows according to the solidphase reaction

$$Ti + \gamma\text{- CuTi} \text{ ------> } CuTi_2 \quad (3)$$

The increase of annealing temperature (up to 640°C) leads to gradual increase of CuTi quantity at the expence of γ -CuTi.

The quantitative relation (in vol. %) $C_{\gamma\text{-CuTi}} / C_{CuTi_2}$ varies from 2,9 to 0,5 (Table 3).

The necessary for the reaction (3) titan after it has been spent from the layers begins to come from the substrate. After annealing at 740°C for 30 min only Ti and $CuTi_2$ are present in the system.

In specimens type D $CuTi_2$ phase is registrated at higher temperature (470°C, 30 min) because of the bigger Cu quantity in the starting layers. At this annealing temperature the single case when three intermetallic γ-Cu Ti, β-Cu_3Ti and CuTi are together registrated (Figure 3-c) occurs as γ-CuTi phase is predominant while the rest two are in very small quantities (Figure 3-c and Table 3). For this type of specimens the relation (4) at very narnow range of annealing temperature sharply changes - from 7,5 at 470°C to 0,9 at 540°C (Table 3). Probably the reason is the bigger width of the titan layer in comparison with those in specimens type C, where after spending the titan from the layers the last by diffusion enters the reactionary substants zones. After annealing at 640°C for 30 min on the diffractograms except the CuTi lines, two slight lines of γ-CuTi are present as the quantity of this phase does not exceed five vol. % (Figure 3-e and Table 3).

3.3. Generalization

In Figure 4 the established sequence of the registrated intermetallic compounds in the thin-film Cu/Ti system after isothermal annealing is shown.

The solid-phase interaction in the system begins at annealing temperature 260°C as the first registrated intermetallic phase is γ-CuTi. At higher temperature on the interphase Cu/γ-CuTi boundary of β-Cu_3Ti phase comes into existence. Probably it is posible both phases spring up together, but the lower growth rate of β-Cu_3Ti does not allow its unique registration at 260°C.

The joint growth of both phases according to the solid-phase reaction

$$Cu + Ti \text{ ------> } \gamma\text{-}CuTi + \beta\text{-}Cu_3Ti \quad (5)$$

stops after one of the reagents is fully spent.

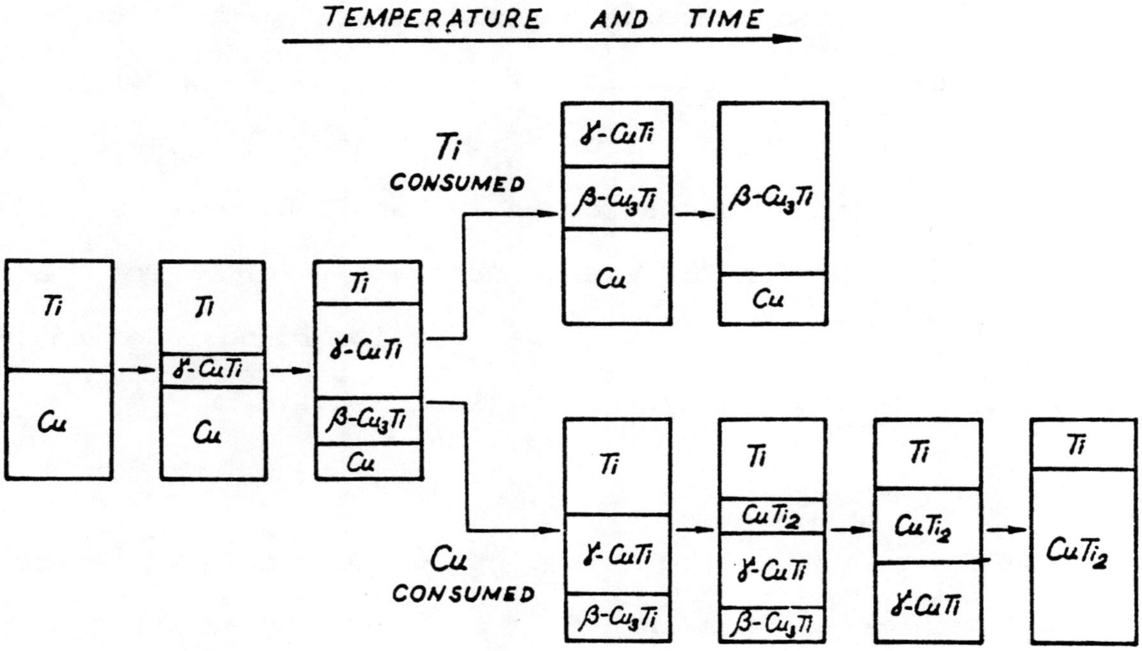

Fig.4. Scheme of formation of the intermetallic compounds in thin-film Cu/Ti system.

If Ti is consumed, the growth of the equilibrium phase β-Cu_3Ti countinues according to the reaction (2) at the expense of the γ-CuTi compound. The final phases in the system are Cu and β-Cu Ti.

If Cu is spending γ-CuTi phase continues to growth according to the reaction:

$$Ti + \beta\text{- } Cu_3Ti \text{ ------> } \gamma\text{- } CuTi \quad (6)$$

From the so obtained experimental data, an unambiguons conclusion, if the $CuTi_2$- phase comes into existence after the full spending of β-Cu_3Ti compound or still during the reation (6), can not be done. On the grounds of the mere case

(Figure 3-c) when the three intermetallic phases have been observed together, we can suppose that $CuTi_2$ phase comes into existence still when there is a residium of β-Cu Ti compound in the system. The initiation of the second phase during the growth process of the first one, i.e. before the spending of one of the reagents, is considered in detail by other authors[17]. After β-Cu_3Ti compound is spent, the growth of γ-CuTi phase stops while $CuTi_2$-phase growth goes off according reaction (3). The final phases in the system are Ti and $CuTi_2$.

4. Conclusion

In contrast to the investigations of "massive" systems of other anthors, in the thin-film Cu/Ti system as a result of solidphase interactions at isothermal annealing the formation of only three intermetallic compounds has been registrated.

The temperatures and sequence of the intermetallic compounds formation for four types of multi-layer systems have been established. In the systems in an excess of copper the intermetallic γ-CuTi and β-Cu_3Ti phases have been registrated, whereas in the systems in an excess of Ti - a $CuTi_2$ phase more. The three phases are together observed at 470°C in systems deposited on neutral substrate. The final intermetallic phase in systems in an excess of copper is β-Cu_3Ti, whereas in those in an excess of titan - $CuTi_2$.

In the systems in an excess of copper deposited on Cu substrate except Cu and β-Cu_3Ti phases an unidentified phase has been registrated after annealing at 740°C for 30 min.

Literature

1. Ф.Шанк. Структуры двойных сплавов. Москва, 1973.
2. М.Хансен, К.Андерко. Структуры двойных сплавов. Москва, 1962.
3. Диаграммы состояния металлических систем. Выпуск XII, Москва, 1966, Выпуск XXVIII, Москва, 1985.
4. W.B.Pearson. The crystal chemistry and physics of metals and alloys. New York, London, Sydney, Toronto, 1972.
5. W.Schatt, H.-Y.Ullrich u.a., Kristall und Technik, 13, 185, (1978).
6. B.Grezeta, M.Stubicar, N.Cowlam and R.Trojko. Philosophical Magazine A., 55, 227 (1987).
7. M.J.Kaufman, R.D. Shull. Metallurgical Transactions A, 17, 575 (1986).
8. American Society for Testing Materials, X-ray Department, Philadelphia.
9. V.Simic, Z.Marinkovic, Naucno-tehnicki pregled. God. XXXV, 1985, N 10, S.25.
10. J.Vahakangas, E.Wiliams, K.Park, Physikal Rewiew B, 1986, v.33, N 4, p.2281.
11. A.F.Marshall, Y.S.Lee, D.A.Stevenson, Acta Metall., 1983, v.31, p.1225.
12. A.F.Marshall, Y.S.Lee, D.A.Stevenson, Acta Metall., 1987, v.35, p.61.
13. A.E.Gershinskii, A.A.Khoromenko, E.I.Cherepov, Phis. Stat. Sol. (A), 1975, v.31, p.61.

14. J.L.Liotard, D.Gupta, P.A.Psaras and P.S.Ho J.Appl. Phis., 1985, v.57, p.1895.

15. Практические методы в электронной микроскопии. Под ред. Одри М. Глоэрам, Ленинград, 1980.

16. Структура и физические свойства твердого тела. Под ред. проф. Палатника, Киев, 1983.

17. U.Gosele and K.N.Tu, J.Appl.Phys., 53, 3252 (1982).

Vacuum Electron and Ion Technologies
D. S. Karpuzov (Ed.)

ISBN 1-56072-242-8

COMPOSITION OF THIN OXIDE FILMS DEPOSITED BY REACTIVE MAGNETRON SPUTTERING OF Sn : Sb and Cd : Sn TARGETS

D. Popov, P. Mashkov and P. Docheva
Department of Physics, Technical University - Rousse,
7017 Rousse, Bulgaria

Oxide thin films have been deposited by reactive magnetron sputtering of Sn : 8 wt.°/$_o$$Sb$ and $Cd : Sn(2:1$ at.°/$_o$) target in $Ar + O_2$ atmosphere at different partial pressures of O_2 and different discharge currents. On the base of a simple phenomenological model of the deposition process a relation between the film composition and process parameters has been obtained. The dependence of film optical transmittance, refractive index and specific resistance on the process parameters support the proposed model.

1. INTRODUCTION

Ion sputtering of metal targets using gas discharge in reactive $Ar+O_2$ atmosphere is widely used technique for deposition of thin oxide films [1 - 3]. The film composition, film structure and stoichiometry depend mainly on the discharge current and oxygen partial pressure. The target and substrate characteristics (sputtering yield, secondary electron emission coefficient, sticking coefficients to the surface of the growing film, substrate temperature etc), geometry in the vacuum chamber and target poisoning (oxidation) during the sputtering in oxygen atmosphere can also influence the film properties.

The deposition parameters, geometry in the vacuum chamber and the processes which takes place on the target and substrate surfaces also determine the stability of the deposition process. In the case when the system is unstable a hysteresis effect can be observed. The influence of the process parameters and geometric factors on the stability of the deposition process during the reactive diode and magnetron sputtering has been intensively studied in the recent years [4,5], but the problems concerning the film composition obtained at different deposition conditions are not solved in details up to now.

In this paper, on the base of a phenomenological model of the sputtering process in reactive oxygen atmosphere, a relation between the film composition and the parameters of the deposition process has been obtained. The experiments with deposition of oxide films by magnetron sputtering of $Sn : Sb$ and $Cd : Sn$ targets support the proposed model.

2. MODELLING OF THE DEPOSITION PROCESS

We suppose that the deposition is realized by sputtering in gas discharge of reactive $Ar+O_2$ atmosphere. The modelling of the deposition process has been made under the following assumptions: 1. the target is sputtered using a planar circular magnetron with a uniform distribution of the ion current density on the erosion region; 2. the substrate is situated in a plane parallel to the target surface just above the target center, as target- substrate distance is h (fig.1); 3. the sputtering and secondary electron emission due to the oxygen ion bombardment are neglected; 4. scattering of the sputtered particle has also been neglected and the spatial distribution of atoms in sputtered fluxes is supposed to be cosine.

In general case the sputtering takes place under such conditions that the surface of the metal target is partially covered with oxides (poisoned). Thus, during the sputtering of partially oxidized target the following fluxes of atoms participate in establishing the film composition: flux of metal atoms sputtered from the whole erosion region of the target; flux of oxygen atoms sputtered from the oxidized part of the erosion region and flux of oxygen molecules comming from the reactive atmosphere.

The number dN_M of metal atoms sputtered from an arbitrary element dS on the target surface and deposited on a unit area around the point M on the substrate (fig. 1) per unit time is [6,7]

$$dN_M = \frac{\alpha_M S_M j cos^2\theta \rho}{\pi e(1+\gamma)r^2} d\rho d\varphi. \quad (1)$$

On deriving the relation (1) a cosine distribution of sputtered atoms has been assumed. In this relation j is the discharge current density; S_M- sputtering yield of metal atoms sputtered from the target surface; γ- secondary electron emission coefficient of the target surface, α_M- the sticking coefficient of metal atoms to the surface of the growing film; r - the distance between the element dS and point M; θ- angle between the substrate (or target) plane and the direction of r; ρ and φ- the polar coordinates of the element dS; e - elementary electric charge.

Using the relations $r^2 = h^2 + \rho^2$ and $cos\theta = h/r$ (see fig.1) and integrating (1) over ρ (from R_1up to R_2) and φ(from 0 to 2π) one can obtain the number N_M of atoms sputtered from the whole sputtering region of the target and deposited on an unit area around the point M of the substrate per unit time

$$N_M = \frac{\alpha_M S_M h^2 I}{\pi e(1+\gamma)(R_1^2+h^2)(R_2^2+h^2)}, \quad (2)$$

where R_1 and R_2 are respectively the inner and the outer radia of the sputtered region and $I = \pi j(R_2^2 - R_1^2)$ is the discharge current. In general case the sputtering yield and secondary emission coefficient γ depend on the degree of the target poisoning θ_t as

$$\begin{aligned} S_M &= S_{MM}(1-\theta_t) + S_{M_O}\theta_t, \\ \gamma &= \gamma_M(1-\theta_t) + \gamma_O\theta_t. \end{aligned} \quad (3)$$

Here S_{MM} and γ_M are respectively the sputtering yield of metal atoms and secondary electron emission coefficient of the nonoxidized part of the target surface and S_{M_O} and γ_O are respectively the same quantities but for an oxidized part of the target surface. The degree of poisoning (oxidation) θ_t is defined by the ratio between the number of oxygen atoms chemisorbed to the target surface and the total number of places where such atom can be chemically captured.

In a similar way one can obtain the expression

$$N_O = \frac{\alpha_O S_O \theta_t h^2 I}{\pi e(1+\gamma)(R_1^2+h^2)(R_2^2+h^2)}, \tag{4}$$

where N_O is the number of oxygen atoms sputtered from the oxidized part of the target surface and deposited on unit area around point M per unit time; S_O and α_O are respectively the sputtering yield of oxygen atoms and their sticking coefficient to the surface of the growing film.

There is another flux of oxygen atoms which are coming from the reactive atmosphere. The flux density $N_O^{'}$ of such atoms deposited around the arbitrary point of the film surface is derived using the kinetic theory of gases and is given by the expression

$$N_o^{'} = \frac{2\alpha_o P_{O_2}}{(2\pi mkT)^{\frac{1}{2}}}, \tag{5}$$

where P_{O_2} s the oxygen partial pressure, m- mass of O_2 molecule, T absolute temperature and k Boltzmann's constant.

In a steady state the film composition is defined by the ratio

$$\theta_s = \frac{N_O + N_O^{'}}{N_M} \tag{6}$$

Thus using (1), (4), (5) and (6) one obtains

$$\theta_s = \frac{\alpha_O S_O \theta_t}{\alpha_M S_M} + \frac{2\pi\alpha_O e(1+\gamma)P_{O_2}(R_1^2+h^2)(R_2^2+h^2)}{(2\pi mkT)^{\frac{1}{2}}\alpha_M S_M I h^2}. \tag{7}$$

If the metal target is an alloy of two metals (A and B) the ratio

$$\theta_s^{'} = \frac{N_O + N_O^{'}}{N_{MA} + N_{MB}} \tag{8}$$

could be used as a measure of the film composition, N_{MA} and N_{MB} being the flux densities of sputtered A and B atoms. N_{MA} and N_{MB} are defined by the relations similar to relation (2).

Thus using the above mentioned procedure one obtains that in this case

$$\theta_s^{'} = \frac{\alpha_O S_O \theta_t}{\alpha_{MA} S_{MA} + \alpha_{MB} S_{MB}} +$$

$$+ \frac{2\pi\alpha_O e(1+\gamma)}{(\alpha_{MA} S_{MA} + \alpha_{MB} S_{MB})(2\pi mkT)^{\frac{1}{2}}} \cdot \frac{P_{O_2}}{I} \cdot \frac{(R_1^2+h^2)(R_2^2+h^2)}{h^2} \tag{9}$$

3. EXPERIMENTAL TECHNIQUE

Oxide thin films have been deposited by reactive magnetron sputtering of $Sn:8$ wt.°/$_o$$Sb$ and $Cd:Sn(2:1$ at.°/$_o)$ target in $Ar+O_2$ atmosphere at constant total pressure $P_t = 0,8Pa$ and different oxygen partial pressures P_{O_2}. The ratio P_{O_2}/P_t has been varied in the range 2 - 8 °/$_o$ and discharge current in the range 0 - 250 mA. A circular planar target has been sputtered as the sputtering region is ring like with inner and outer radia respectively $R_1 = 1,2$cm and $R_2 = 3,8$cm. Glass substrates have been used situated above the center of the target, the target - substrate distance has been varied in the range 5 -15 cm. All depositions have been carried out in conditions avoiding the target surface poisoning (sputtering occurs in an elemental mode).

The light transmition spectra in the visible region have been measured by spectrophotometer "Specord UV VIS". Using the transmition spectrum the refractive index is calculated by the method discribed by Manifacier et all [8]. For measuring the film conductivity a four point method has been used.

4. EXPERIMENTAL RESULTS AND DISCUSSION

The optical and electrical properties of oxide thin films deposited by reactive magnetron sputtering of $Sn : Sb$ and $Cd : Sn$ targets in $Ar + O_2$ atmosphere are strongly dependent on the film composition. We shall note here again that as a measure of the composition the ratio between the concentrations of oxygen and metal atoms has been used. In (7) and (9) this ratio is marked as θ_s and θ_s' for sputtering of a single - component or two - component metal targets respectively. One can see from relations (7) and (9) that the film composition is essentially dependent on the process parameters P_{O_2} and I and the geometric factor $g = (R_1^2 + h^2)(R_2^2 + h^2)/h^2$. If the sputtering occurs in an elemental mode (unpoisoned target) $\theta_t = 0$ and θ_s (or θ_s') becames a linear function of the ratio P_{O_2}/I and geometric factor g .

The dependence of the refractive index n on the geometric factor g for thin oxide films deposited by reactive magnetron sputtering of Sn : 0,8 wt.°/$_o Sb$ target at $P_{O_2} = 0,055Pa$ and $I = O,25$ A is shown in fig. 2. The curves 1, 2 and 3 correspond to refractive indices obtained at 530 nm, 600 nm and 700 nm respectively.

Comparing the linear dependences shown in fig. 2 with the relation (7) and (8) at $\theta_t = 0$ one can propose a linear relation between n and θ_s when the sputtering takes place in an elemental mode

$$n = n_M + a\theta_s, \qquad (10)$$

where n_M is the refractive index of a metal thin film (at $\theta_s = 0$, $n = n_M$). If we neglect the influence of Sb on the refractive index then at $\theta_s = 2$, $n = n_O$ is the refractive index of SnO_2 and $a = (n_O - n_M)/2$. Thus the film characteristics "refractive index" depends linearly on the film composition if oxide thin films are deposited by reactive magnetron sputtering of Sn target. One could expect similar relation for other film characteristics as well.

In fig. 3 one can see the measured values of the specific resistance ρ and transmittance T of oxide thin films deposited by reactive magnetron sputtering of $Cd : Sn(2 : 1$at.°/$_o)$ target at different discharge current I and different percentage of oxygen in the reactive atmosphere. The oxygen percentage is difined by the ratio P_{O_2}/P_t. From these results one can obtain the values of $(P_{O_2}/P_t)_m$ and I_m for which ρ and T reach minimum (ρ_{min}) and maximum(T_{max}) values respectively. Every couple of I_m and $(P_{O_2}/P_t)_{max}$ defines a point in I_m - $(P_{O_2}/P_t)_m$ plane. The results are shown in fig.4.

Straight lines with a same slope have been obtained for ρ_{min} and T_{max}. Thus for all experiments maximum transmittance and minimum resistance have been reached at the same ratio of $I_m/(P_{O_2}/P_t)_m$ while the values of ρ_{min} have been obtained at larger values of discharge current than the values of T_{max}. These results again confirm the supposition that the film optical transmittance and specific resistance are dirrectly related to the film composition. Definite values of these characteristics (ρ_{min} and T_{max}) have been obtained at the same $I/(P_{O_2}/P_t)$and (according to (9)) at the same film composition.

5. CONCLUSSIONS

On the base of a simple phenomenological model of the deposition process a relation between the film composition and process parameters (such as discharge current, partial pressure of O_2 etc.) have been obtained. This relation also shows the dependence of the film composition on the geometry in the vacuum vessel.

The results from the measurements of the optical transmittance, refractive index and specific resistance of the deposited thin oxide films show that in the case of tin oxide films the film refractive index is proportional to the film composition while the other two characteristics (transmittance and specific resistance) have a maximum and minimum value at the same ratio P_{O_2}/I when $Cd : Sn$ sputtering target was used. These results support the proposed model of the film deposition.

LITERATURE

1. R. P. Howson, M. I. Ridge, Thin Solid Films, 77 (1981) 119
2. W. D. Munz, I. Heimbach, S. R. Reineck, Thin Solid Films, 86 (1981) 175
3. M. J. Brett, R. W. McMahan, J. Affinito , R. R. Parson, J. Vac. Sci. Technol., A 1(2) (1983) 352
4. S. Kadlec, J. Vyskocil, Vacuum, 37 (1987) 729
5. S. Berg, H.-O. Blom, T. Larson and C. Nender, J.Vac.Sci.Technol. A5, 202(1987)
6. S. Berg, T. Larson , C. Nender and H.-O. Blom, J.Appl.Phys. 63, 887(1988)
7. D. Popov, P. Docheva, S. Navassardian, Bulg. J. Phys., 15 (1988) 453
8. J. C. Manifacier, J. Gasiot, J. P. Fillard, J. Phys. E, 9 (1976) 1002

FIGURES

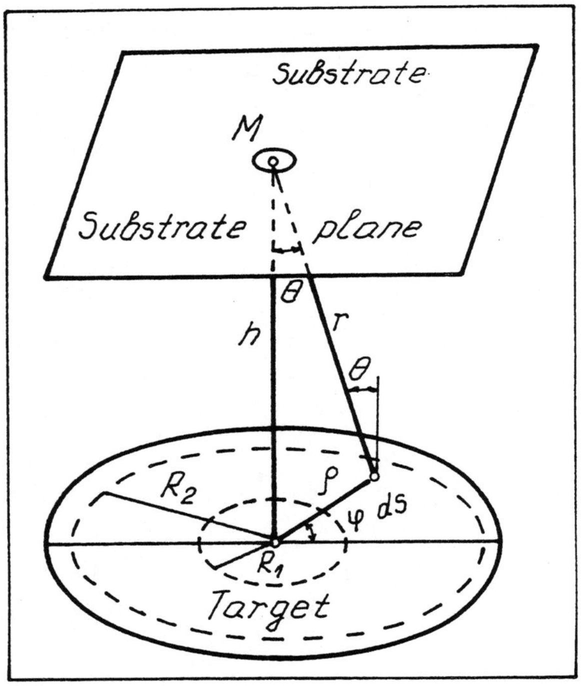

Fig.1. Geometry of vacuum vessel used for reactive magnetron sputtering and deposition of oxide thin films.

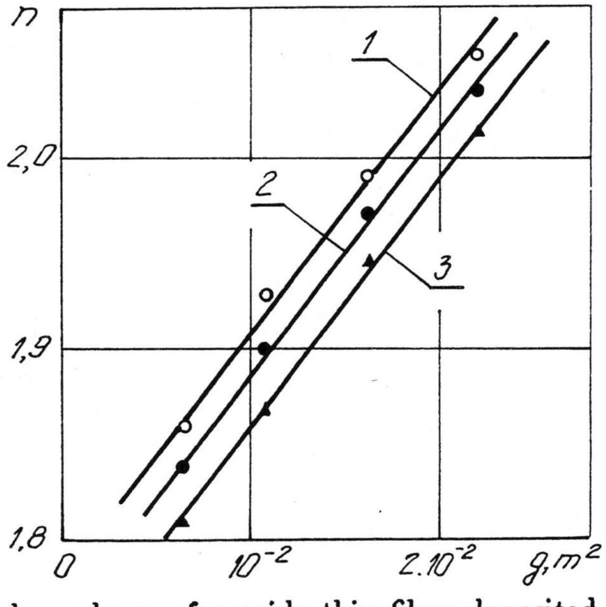

Fig.2. n - g dependences for oxide thin films deposited by reactive magnetron sputtering of Sn : 8 wt.$^o/_o Sb$ in reactive $Ar + O_2$ atmosphere at $P_{o_2} = 0,055 Pa$ and discharge current $I = 0,250$ A. n - is the film refractive index, $g = (R_1^2 + h^2)(R_2^2 + h^2)/h^2$ - geometric factor as R_1 and R_2 are the radii of the sputtering region and h - target to substrate distance. The lines 1, 2 and 3 correspond to refractive indices measured at 530, 600 and 700 nm respectively.

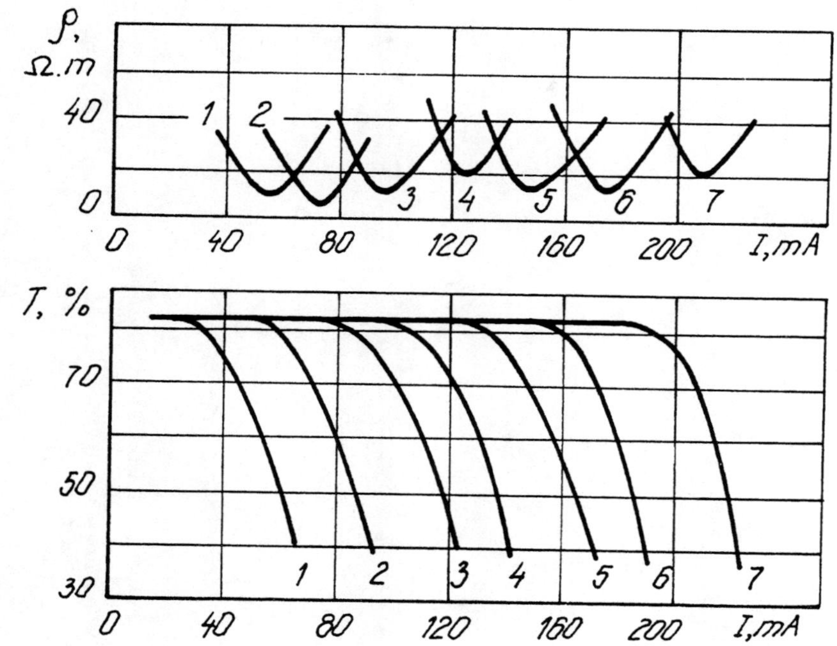

Fig.3. ρ - I and T - I dependences for oxide thin films deposited by reactive magnetron sputtering of Cd : $Sn(2:1$ at.$^o/_o)$ in $Ar + O_2$ atmosphere at different percentage P_{O_2}/P_t of oxygen. ρ and T are respectively the film specific resistance and film transmittance at 632 nm, I is the discharge current. Curves 1, 2, 3, 4, 5, 6 and 7 correspond to $P_{O_2}/P_t = 2^o/_o, 3^o/_o, 4^o/_o, 5^o/_o, 6^o/_o, 7^o/_o$ and $8^o/_o$ respectively.

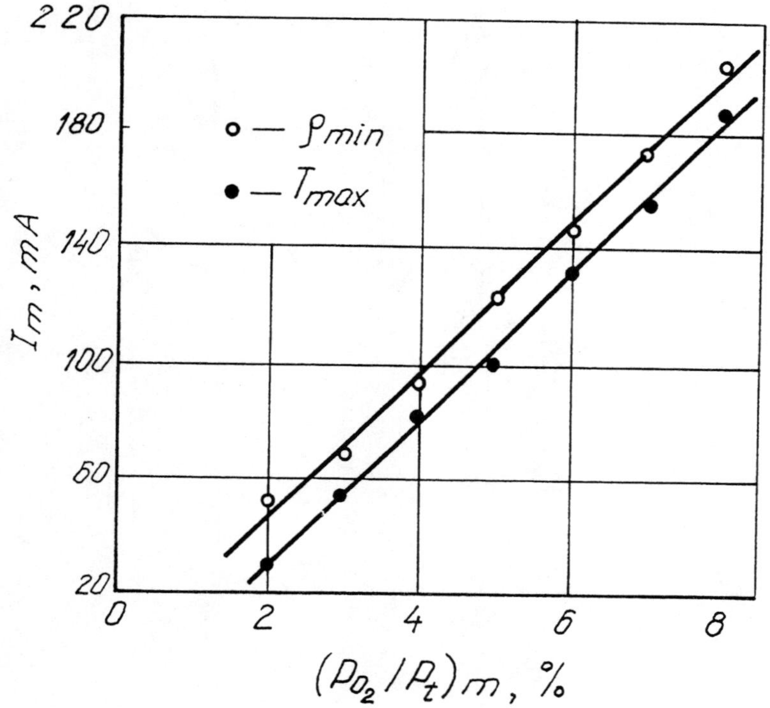

Fig.4. Discharge current I_m and oxygen percentage (P_{O_2}/P_t) at which specific resistance has a minimum value and transmittance at 632 nm has a maximum value. The empty and full circles correspond to ρ_{min} and T_{max} respectively.

Vacuum Electron and Ion Technologies
D. S. Karpuzov (Ed.)

ISBN 1-56072-242-8

RIE OF VACUUM DEPOSITED POLYIMIDE THIN FILMS

G. Danev, E. Spasova, K. Popova* and D. Dimitrov
Central Lab. of Photoprocesses, Bulg. Acad. Sci. G. Bontchev str., bl. 109, Sofia 1040, Bulgaria
*Institute of Electronics, Bulg. Acad. Sci., 72 Tzarigradsko Chaussee, Sofia 1784, Bulgaria

Abstract

The resistance of vacuum deposited **PI** films **(VDP)** to **RIE** in oxygen and CF_4 has been investigated. The films have been etched in reactive radio frequency plasma and the resulting surface observed by **SEM**. Strong influence of the deposition conditions on the film structure and etched surface morphology has been found. A comparison with **PI** films deposited from solution shows different film structure and surface morphology.

1. Introduction

Polyimides **(PI)** are finding various applications in microelectronics due to their excellent physical and chemical properties. **PI** films are used as dielectrics in multilevel electronic structures and as planarizing bottom layers in multilayer resist systems **(MRS)**. For this application, deposition of **PI** by spin coating is not compatible with the trend for **"all-dry"** patterning technologies. As an alternative, the solventless technique of vacuum deposition has received considerable attention[1-3], allowing coating of wider range of substrates and greater control on film thickness. Oxygen reactive ion etching **(RIE)** was reported recently to be an effective tool for "sectioning" of polymer films[4]. We investigate in this paper the influence of the deposition parameters and post deposition treatment on vacuum deposited polyimide **(VDP)** films. The **VDP** surface morphology and microstructure have been observed in scanning electron microscopy **(SEM)** after **RIE** in oxygen to various film depths.

2. Experimental

VDP films were deposited in vacuum from thermally heated *Knudsen* vessel sources. **PI** films from 0.5 to 2 μm have been prepared by simultaneous evaporation of two monomers: oxydianiline **(ODA)** and pyromellitic dianhydride **(PMDA)** from two separate sources. A uniform film thickness was obtained by rotation of the substrates at 120 rpm during evaporation. After deposition of **VDP** films additional thermal treatment in air was carried out to achieve imidization. Conventional **PI** layers have been prepared by spinning polyamide-carbon acid[5] or photoresist Selectilux HTR 3-200. Two types of substrates have been used: **n**-type **Si** wafers (100) and soda lime glass substrates coated with a 200 A thick chromium layer[6].

A modified commercial **rf** parallel plate system was used for the **RIE** experiments. The samples were placed on the water-cooled cathode. During **RIE** it was covered with 2 mm thick carbon plate of the same diameter to avoid redeposition from the cathode and contamination of the etched surface. Most of **RIE** parameters have been fixed to allow for comparison between different samples. The etched surfaces morphology was observed directly in a **SEM** model JSM-T 200.

3. Results and discussion

PI films produced by different techniques strongly differ in their surface morphology as illustrated in fig.1. Films produced from solution (fig.1 a,b) have well

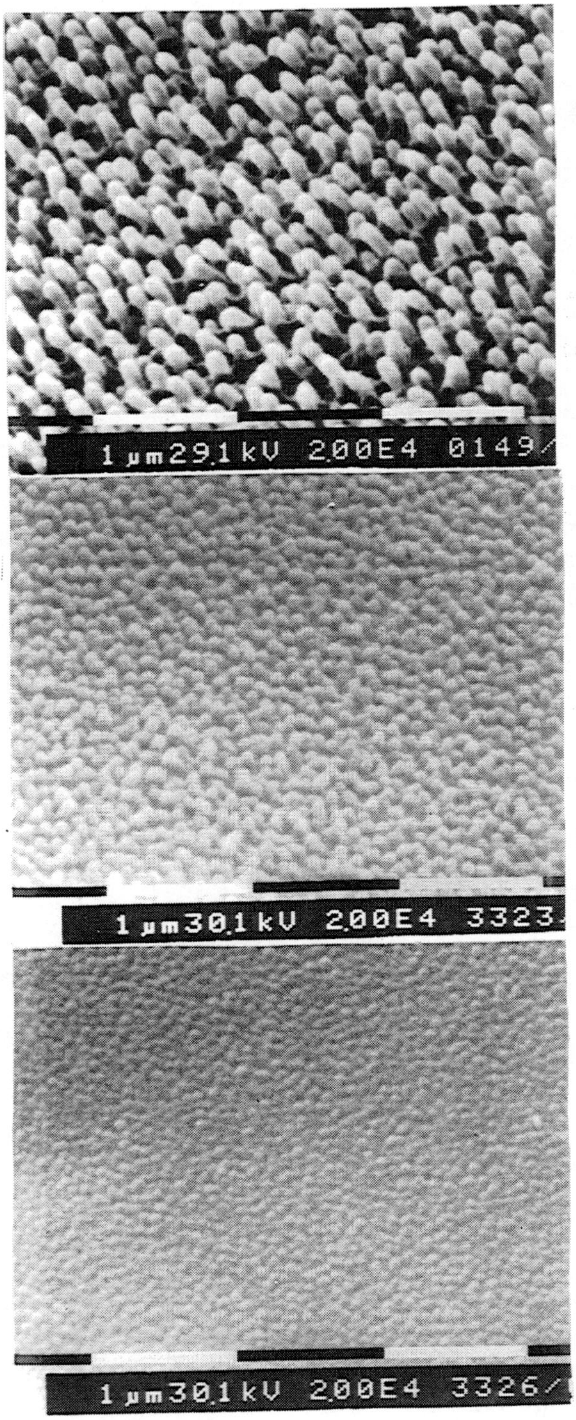

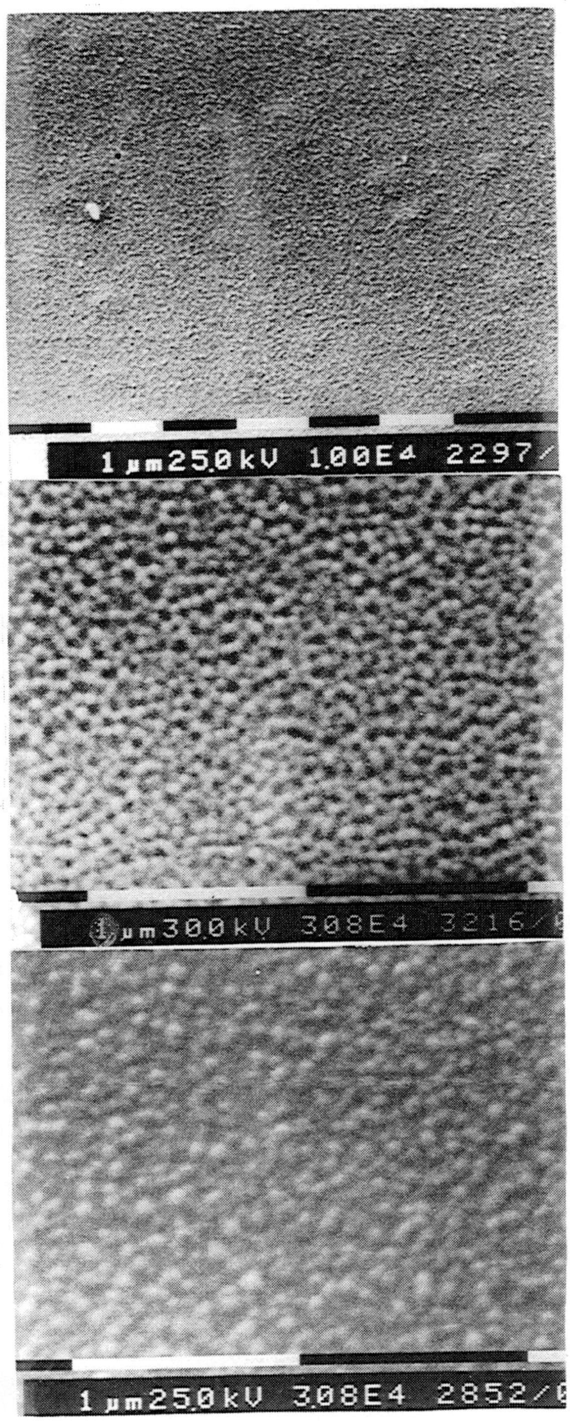

a

b

c

Fig.1 Etched surface of PI films after 2,5 min RIE-O_2, 50 mTorr; 0,1 W/cm^2; a/ PI film from solution; b/ Sellux HTR 3-200; c/ VDP film

Fig.2 Surface morphology of VDP films (2 A/s; ODA:PMDA 1:1) a/ before RIE; b/ after 2,5 min RIE c/ after 5 min RIE

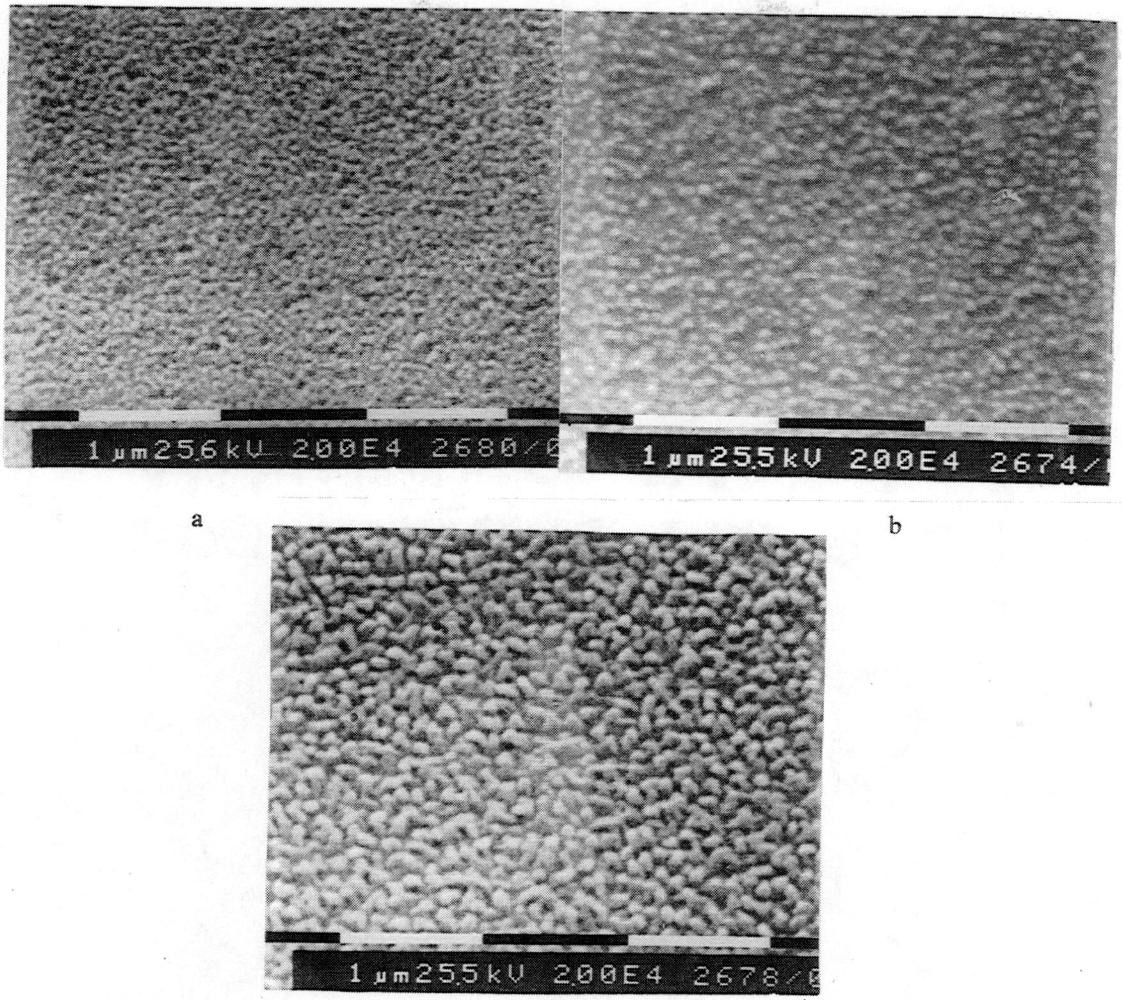

a b c

Fig.3 Influence of deposition parameters on VDP film structure observed after 5 min RIE-O_2; 50 mTorr; 0,1 W/cm^2
a/ deposition rate 2 A/sec; ODA:PMDA 2:1
b/ deposition rate 2 A/sec; ODA:PMDA 1:1
c/ deposition rate 5 A/sec; ODA:PMDA 1:1

Table 1

Polyimide etch rates at 50 mTorr oxygen RIE, 0.1 W/cm^2

VDP deposition parameters		Thermal treatment	Etch rate nm/min
Deposition rate	Source monomers rel.cont		
2 A/s	2:1	standard	102
2 A/s	1:1	standard	120
5 A/s	1:1	standard	130
2 A/s	2:1	no	112
5 A/s	1:1	no	140

PI film produced from solution

a

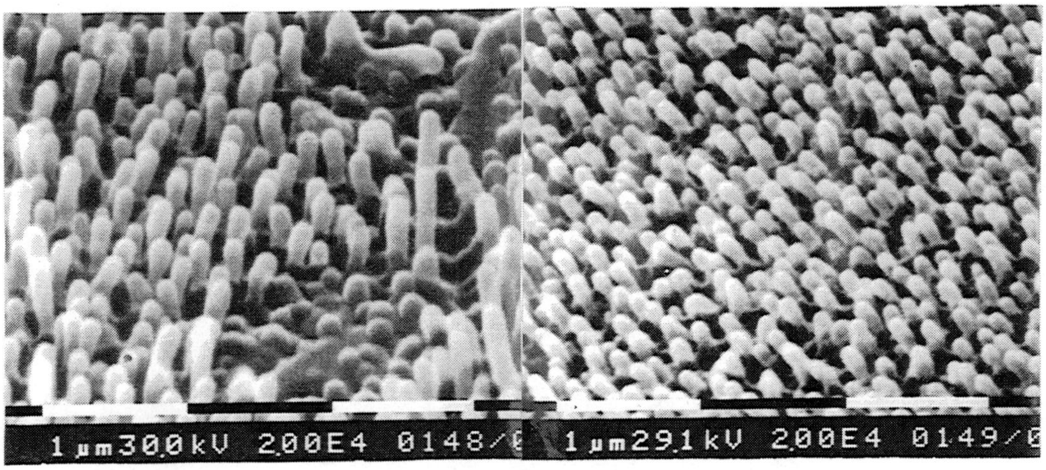

b

VDP films 2 A/sek ODA:PMDA 2:1 -c/ and d/ ODA:PMDA 1:1 -e/ and f/

c

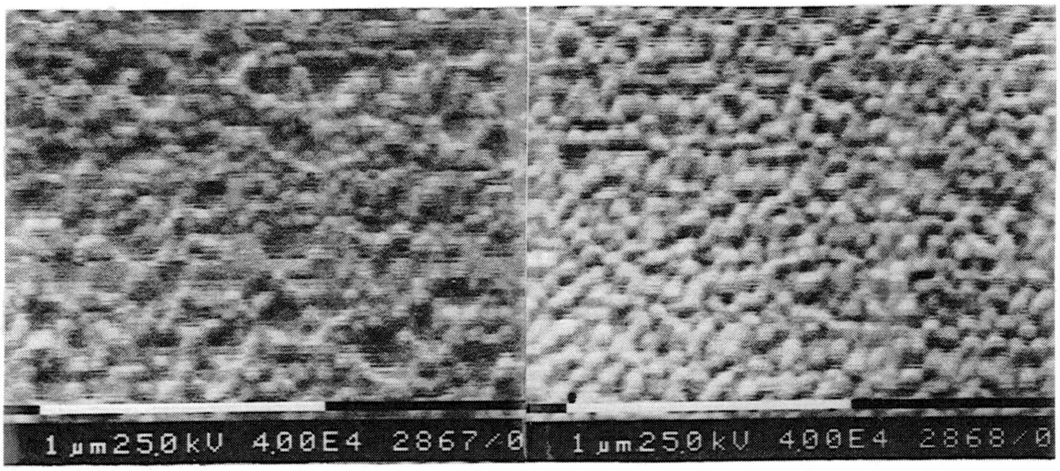

d

e

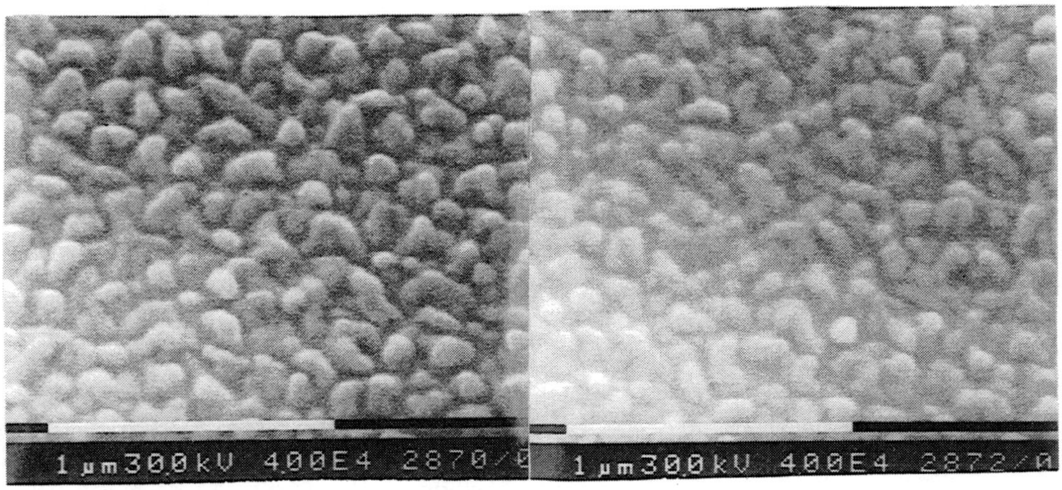

f

Fig.4 Film structure observed after 5 min RIE-O_2; 50 mTorr, 0,1 W/cm^2; a/ c/ and e/ without thermal treatment; b and d/ 1 hour at 150° C and 1 hour at 250° C; f/ 1 hour at 150° C defined grain or oriented columnar structure while evaporated films (fig.1c) are more compact with randomly distributed finer grains. The highly oriented columnar structure of **PI** films produced from solution can be attributed to the influence of the solvent leading to stronger adhesion between the film and the substrate. In contrast, **VDP** films do not exhibit any oriented structure. This particular film morphology probably explains the factor of two smaller optical anisotropy of **VDP** films compared to **PI** films prepared from solution[2,7-8]. The **VDP** film surface before etching is reasonably smooth and uniform (fig.2a). Film removal by **RIE** process, based on the combined effect of ion bombardment and surface chemical reactions, shows a gradual change of structure with depth. This is illustrated in fig.2b,c, which shows the **VDP** surface after etching for 2.5 and 5 min respectively. The film structure is strongly influenced by changes in deposition parameters. Changing of the relative content of source monomers from 1:2 **(ODA:PMDA)** to 1:1 changes the film structure from compact to granular (fig.3a,b). The case of **VDP** film deposition from unstoichiometric ratio of reactants (1:2) was investigated by Mack et al. and Saini et al.[9,10]. They explain the compact structure with crosslinking in the bulk polymer due to excess of the diamine component and imine bond formation. A higher deposition rate (5 A/s) leads to bigger grains with better defined grain boundaries (fig.3c). The effect of deposition parameters on the film etching resistance was evaluated by measuring the etch rates. Etch rates of the investigated **VDP** films at standard **RIE** conditions in oxygen are summarized in table 1. The etch rates are the same order of magnitude as the etch rates of **PI** films produced from solution (100 nm/min) and can successfully be used for **MRS** bottom layer patterning. The **VDP** etch rate increases from 102 nm/min in the case of unstoichiometric film (fig.3a) to 120 nm/min for the stoichiometric one (fig.3b) and to 130 nm/min in the case of increased deposition rate (fig.3c). We conclude that the **VDP** etch rate is correlated with film structure which in turn is dependent on the deposition parameters. The effect of thermal treatment on **VDP** films was also investigated. Fig.4 shows the influence of the thermal treatment on the **PI** layer morphology. The thermal treatment effect is more strongly expressed for **PI** layers morphology obtained from solution (fig.4a,b). By **VDP** layers the thermal treatment leads to homogenization of the structure (fig.4c,d). When the thermal treatment of the **VDP** is very mild (up to 150°C), the influence on the structure is hardly noticeable (fig.4e,f). The absence of a solvent in this case leads to a decrease of the interlayer stresses, as well. Additional baking for 1 hour at 250°C was sufficient for film densification and this sequence (1 h 150°C + 1 h 250°C) was adopted as a standard. The effect of postdeposition thermal treatment on film performance was confirmed by measurements of the etch rate of films deposited under the same conditions. Thermally treated **VDP** films exhibited in general 10% lower etch rates in oxygen-**RIE** compared to thermally untreated films (table 1). When CF_4 is used as the reactive gas the **VDP** etch rate decreases by approx. a factor of two. This is an acceptable etching resistance for **VDP** films for use in **RIE** of working layers in CF_4 with **MRS** containing a bottom layer of **VDP**.

4. Conclusions

RIE was used to observe changes in **VDP** film structure and to estimate film resistance as a component of resist systems. The influence of various deposition parameters and postdeposition thermal treatments was investigated. It was found that dense **VDP** films with a fine structure can be deposited at low deposition rates and relative content of the source components **ODA** and **PMDA** 1:1. Additional thermal treatment improves film resistance to **RIE**. Etching in oxygen is recommended for polyimide patterning.

This paper is supported by Ministry of Education and Science - grant **HT-121**.

5. References

1. Y.Takahashi, M.Iijima, K.Inagava, A.Itoh, J.Vac.Sci.Technol.,A5(4), p.2253 (1987)
2. J.Salem, F.Sequeda, J.Duran, W.Lee, J.Vac.Sci.Technol.,A4, p.369 (1986)
3. M.Grunze, R.Lamb, J.Vac.Sci.Technol.,A5, p.1685 (1987)
4. D.McClure, C.Muller, R.Tabar, J.Vac.Sci.Technol.,A6 (3), p.960 (1988)
5. J.Shopov, P.Kirova, K.Batev, E.Venkova, Bulg. Patent 55053 (1982)
6. R.Stoycheva, K.Petkov, A.Buroff, J.Signal AM, S5, p.347 (1980)
7. G.Danev, E,Spassova, K.Popova, Thin solid Films 228, p.302 (1993)
8. K.Nakagawa, J.Appl.Polym.Sci., 41, p.2051 (1990)
9. R.Mack, H.Patterson, M.Cook, C.Carlin, J.Polym.Sci., Polymer Letters Edition,27, p.25 (1989)
10. A.Saini, C.Carlin, H.Patterson, J.Polym.Sci.,Polymer Chem.Edition,30, p.419 (1992)

Vacuum Electron and Ion Technologies
D. S. Karpuzov (Ed.)

ISBN 1-56072-242-8

ELECTRON-BEAM DEPOSITED TiC FILMS STRUCTURE AND OPTICAL PROPERTIES

J. Eneva, S. Kitova, K. Petkov, V. Dimov*
Central Laboratory of Photoprocesses, Bulgarian Academy of Sciences,
G. Bonchev str., bl. 109, 1113 Sofia, Bulgaria
*Institute of Applied Mineralogy, Bulgarian Academy of Sciences, Rakovski str. 92, 1000 Sofia, Bulgaria

Abstract

The influence of the parameters of the electron-beam evaporation of thin TiC films on their composition, structure, optical characteristics and applicability to digital laser recording has been studied. TEM, SEM and electron diffraction have been used for the examination of the films and their laser recording properties.

1. Introduction

Thin films of refractory substances as carbides, nitrides, oxides and silicides have a wide-ranging application in microelectronics, medicine, etc. Recently such refractory substances in combination with polymers are of interest as media for digital laser recording[1-3] as well as for submicron lithography with excimer lasers[4]. The carbide of Cr, Fe, Ti, etc. are very attractive because of their high chemical stability and light absorption in wide spectral region (UV to IR)[1-3].

As it is known the film properties can be strongly affected by the preparation conditions. Due to the high melting point of the carbides, only e-beam units can be applied for their preparation.

In the present paper the e-beam evaporation is used for the deposition of thin TiC films on the polymer sublayer. The aim is to study the influence of the preparation conditions on the film composition, structure and optical properties. The possibilities for digital laser recording in thin TiC films are also investigated.

2. Experimental

TiC films, thickness range 5-140 nm, were obtained by e-beam evaporation with ESV-6 (Leybold) electron gun at vacuum $>6.10^{-4}$Pa and beam voltage of 12 kV. By varying the beam current from 35 to 80 mA a deposition rate - 0.015 - 1.2 nm/s was achieved.

TiC films were deposited on glass plates preliminarily spin coated with 300 nm thick layer of polyvinyl alcohol.

The optical transmittance and reflectivity of TiC films were measured by Specord (Zeiss) double beam spectrophotometer UV-VIS having an additional, specially designed attachment for the reflectivity.

The films were examined in Philips EM 420T transmission electron microscope with bright-field technique and electron diffraction.

The laser recording was realized in dynamic regime at rotation rate of 540 rpm with pulse Ar laser (λ= 488 nm), pulse duration of 200 ns and diameter of the focused laser beam on the film surface of 0.8 μm (FWHM)[5]. The recorded spots were analyzed by Amplival pol u (Zeiss) optical microscope and JEOL T-200 scanning electron microscope. Calibrated SEM micrographs were used the size of the recorded spots to be measured.

3. Results and discussion

3.1. Composition and structure of TiC films

TEM micrographs and electron diffraction patterns of the TiC films prepared at different deposition rate (V_d) and respectively with different film thickness (d) are given in Figs. 1 and 2. As seen the films have an extremely fine grained structure altered insignificantly by V_d. The thicker films are obtained with a larger grain size, which, however, remains < 4 nm within the thickness range studied. This is an important prerequisite for achieving high resolution both of the digital laser recording and submicron lithography.

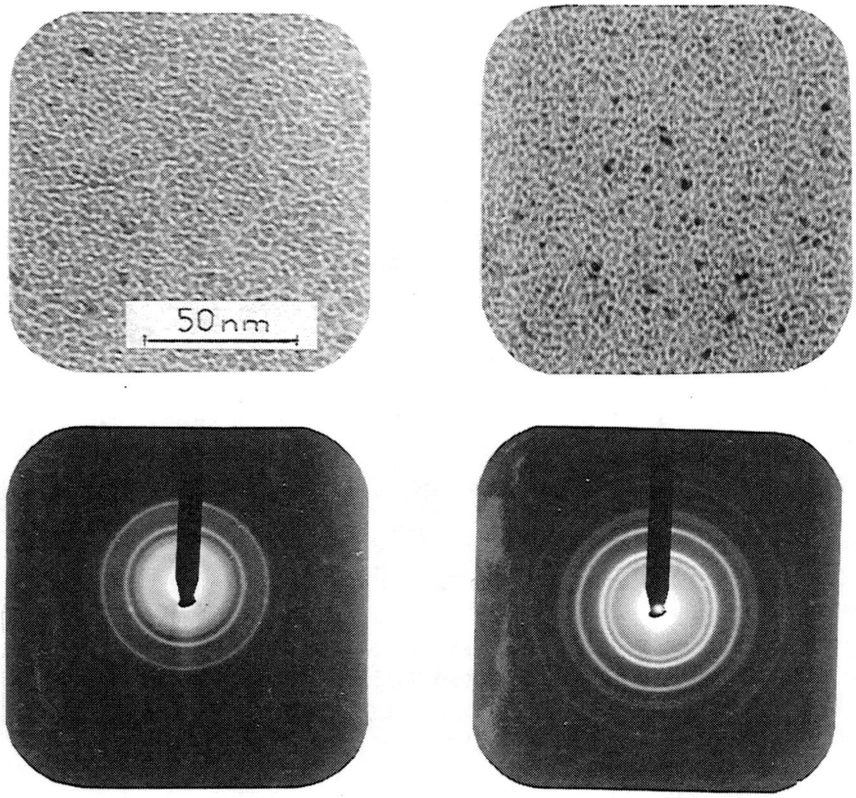

a) b)

Fig. 1 TiC films obtained at the deposition rate V_d:a) 0.017 nm/s; b) 0.74 nm/s (constant d = 26 nm).

The electron diffraction patterns show that the increase of V_d (Fig. 1) as well as of d (Fig. 2) leads to the deposition of films with higher degree of crystallinity. In all cases, however, the TiC films have a stoichiometric composition which is a indication for their non-critical and reproducible preparation.

3.2. Optical properties of the TiC films

Among the optical characteristics, the absorptance (A) and reflectivity (R) of the films are most important for their application to the digital laser recording or submicron laser lithography.

It was found that the spectral dependences of A and R have identical character for films prepared at different V_d and d. Fig. 3 illustrates the change of R and A with

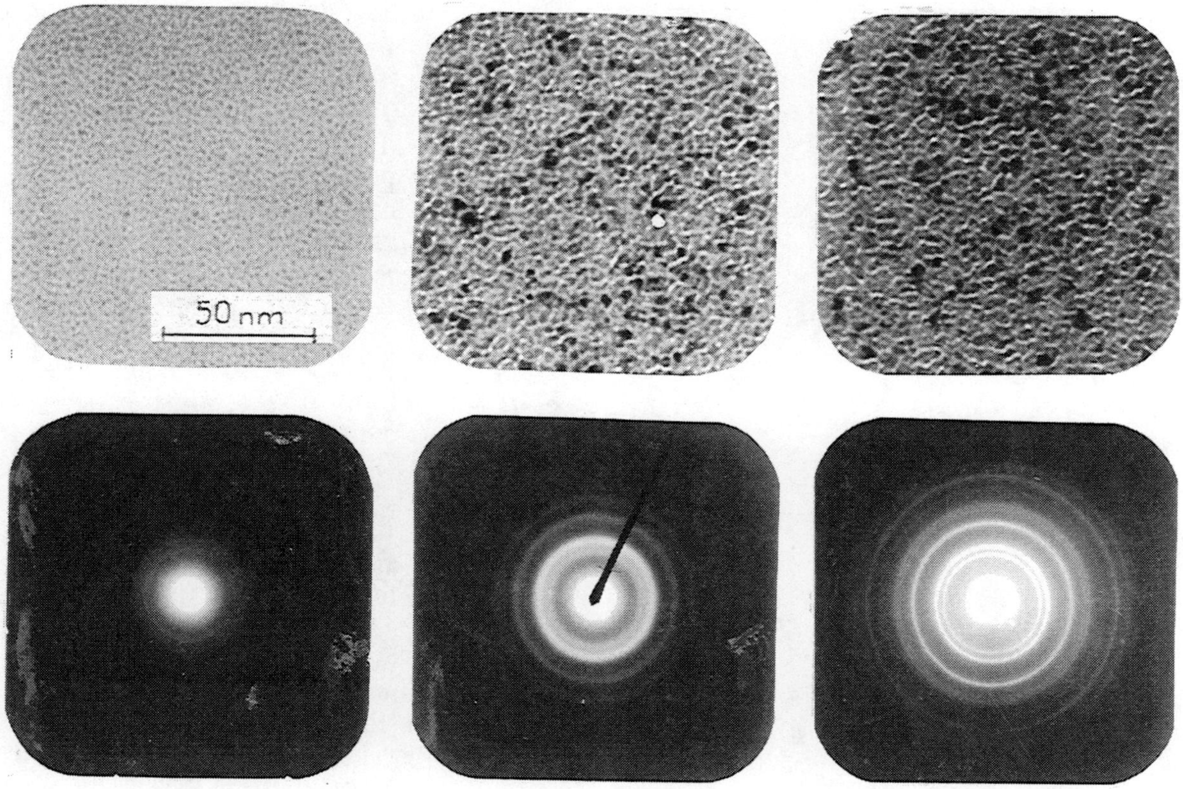

a) b) c)
Fig. 2 TiC film with thickness d: a)5 nm; b) 40 nm; c)140 nm (constant V_d = 0.2 nm/s).

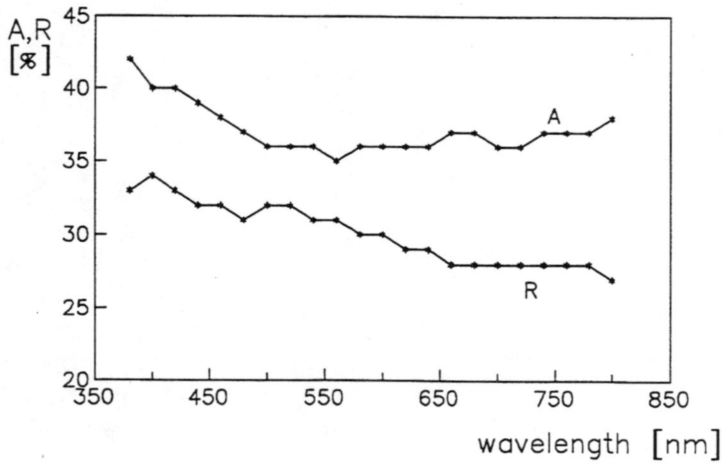

Fig. 3. Dependance of the absorptance (A) and reflectivity (R) with wavelength for TiC film (40 nm thick, V_d=0.2 nm/s).

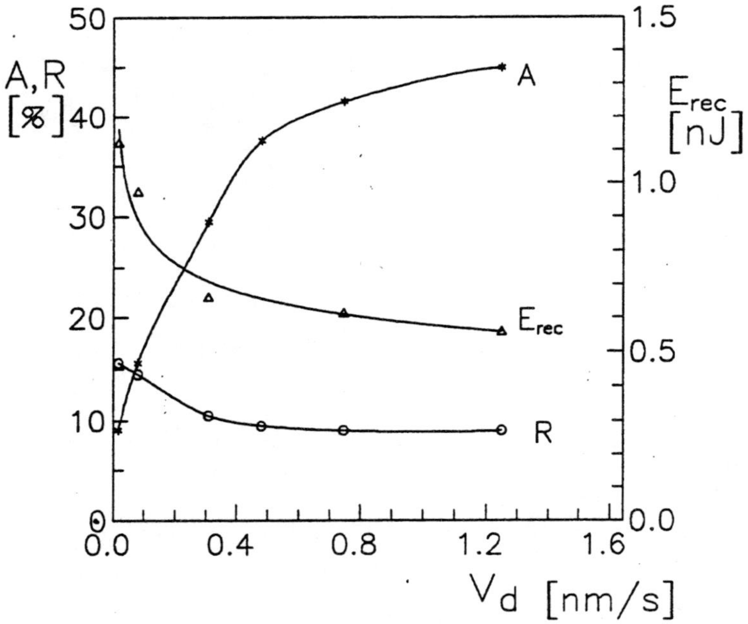

Fig. 4. Dependence of the absorptance (A) reflectivity (R) and laser recording energy (E_{rec}) on the deposition rate V_d for 26 nm thick TiC films at $\lambda = 488$ nm.

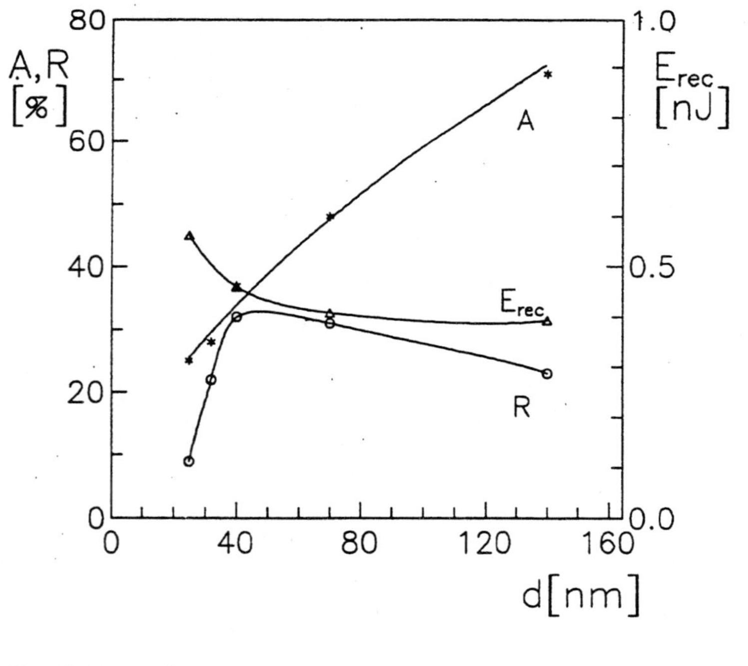

Fig. 5. Dependence of R, A and E_{rec} on the film thickness d (constant $V_d = 0.2$ nm/s, $\lambda = 488$ nm).

wavelength for 40 nm thick TiC film ($V_d = 0.2$ nm/s). From the very small variation of A and R ($\Delta A, \Delta R < 10\%$) within the 380 - 800 nm region, it can be expected that the TiC films can be used for laser recording with laser devices in wide spectral region (from UV to near IR).

Figs. 4 and 5 present the absorptance and reflectivity of TiC films as dependent on the deposition rate V_d and film thickness d at $\lambda = 488$ nm.

The reflectivity is almost the same within the range of V_d studied (Fig. 4), as can be assumed from the negligible change of the mean grain size with V_d (Fig. 1). It should be noted that at thickness of 26 nm R is lower than 20% and does not meet the requirements for laser recording (20% < R < 45%). As seen in Fig. 5 the reflectivity rises with the growth of the thickness and films with d > 30 nm have R high enough to be suitable for laser recording.

The observed dependence of the absorptance on the film thickness (Fig. 5) is predictable and it is due to the higher quantity of the absorbing material per unity area and the larger grain size of the thicker films. The abrupt rise of A with increasing of V_d is, however, not expected since the mean grain size of the films changes insignificantly with V_d. Most probably this dependence is due to the increase of the degree of film crystallinity with V_d.

3.3. Digital laser recording

The most important criteria for the selection of a suitable medium for laser recording with well focused laser beam are the sensitivity, resolution and reflection contrast.

The laser energy E_{rec} at which the recorded spot size is equal to the laser beam diameter (0.8 μm) is usually accepted as a measure of the film sensitivity. For easier comparison the variations of E_{rec} with V_d and d are also given in Figs. 4 and 5. As seen the recording energies E_{rec} meeting the requirements ($<$ 1 nJ) are achieved with TiC films deposited at $V_d > 0.1$ nm/s. Simultaneously only films thicker than 30 nm have appropriate reflectivity value ($R > 20\%$) for providing a high contrast of recording.

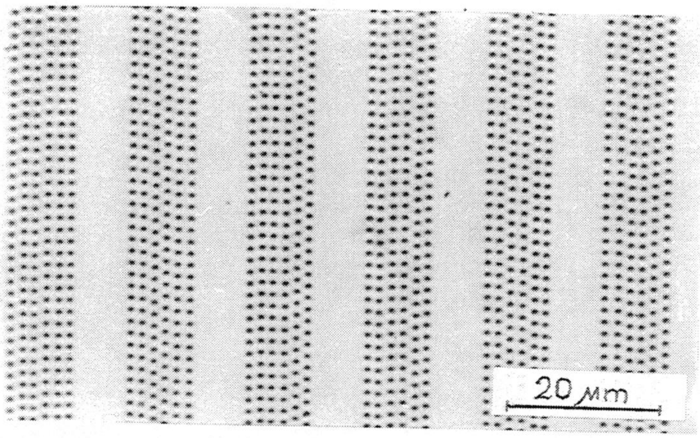

Fig. 6 Optical micrograph (reflected light) of spots recorded in 40 nm thick TiC film (V_d = 0.2 nm/s).

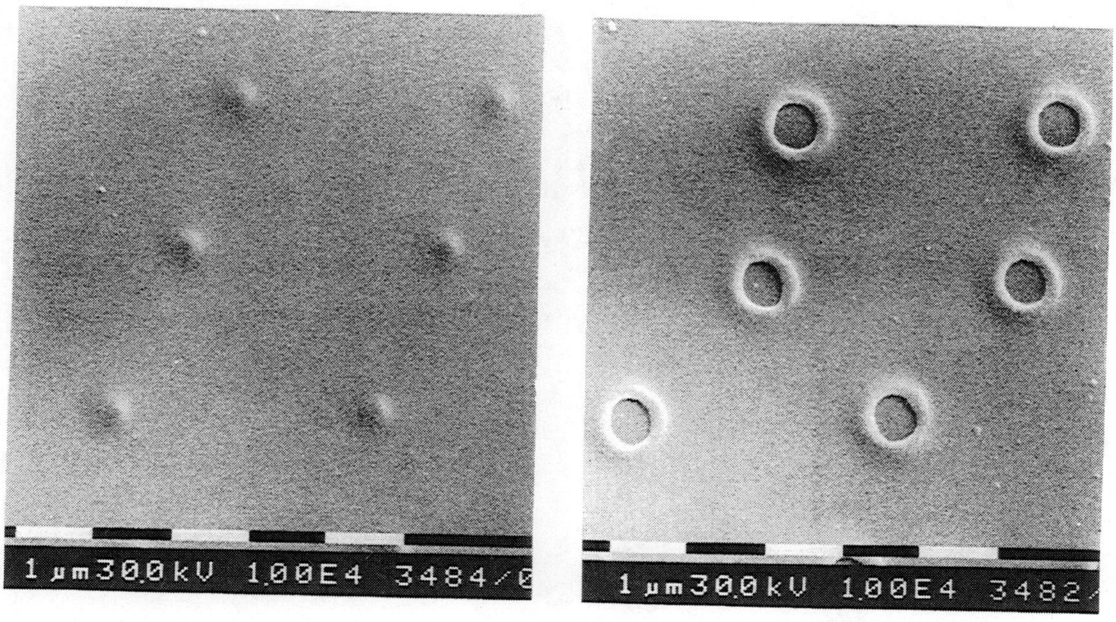

a) b)

Fig. 7 SEM micrographs of spots recorded at different laser energy a)bubble recording b) ablative recording.

Fig. 6 presents an optical micrograph of spots, recorded in TiC film. The reflectivity of the film changes abruptly in the irradiated spots, which means that high contrast of recording can be realized.

SEM investigations show that at energy close to E_{rec}, the observed changes of R are caused by a bubble recording (Fig. 7a) which transforms into an ablative (Fig. 7b) with the increase of the laser energy. As seen in the Fig. 7a the minimal spot size, which is a measure for the resolution power of the recording film is in the submicron region ($< 0.6\ \mu m$).

It should be noted that formation of bubbles is hindered and critical for films with $d > 100$ nm regardless of their high sensitivity ($E_{rec} < 1$ nJ), most probably due to the reduced film flexibility.

4. Conclusions

The results show that TiC films obtained by e-beam evaporation have stoichiometric composition, extremely fine grained structure (< 4 nm) and a degree of crystallinity depending on V_d and d. Their optical properties can be varied also by the film thickness and deposition rate.

The obtained laser recording parameters - Erec, resolution and contrast indicate that 30 - 100 nm thick TiC films prepared at $V_d > 0.1$ nm/s can be used as optical recording media at $\lambda = 488$ nm. From the negligible change of A and R with the wavelength, it can be expected that TiC films are also suitable for laser recording with semiconductor laser ($\lambda = 820$ nm).

Acknowledgements.

The authors would like to thank Dr. I.Konstantinov for the design of the original attachment allowing the measurement of the absolute reflectivity by Specord spectrophotometer.

This work has been accomplished with the financial support of Bulgarian Ministry of Education and Sciences - contract No X - 106/1991.

5. References

1. Maffitt at al., *US Patent* **4 470 053** (1984).
2. T.Rinchart at al., *US Patent* **4 556 893** (1985)6
3. J.Eneva, S.Kitova, J.Assa, K.Petkov, Proceedings of the 7th International School on Condensed Matter Physics, Varna'92, Ed. J.Marshall, N.Kirov, A.Vavrek, World Scientific, p.377 (1992).
4. I.Spasova, G.Danev - to be published
5. A.Poleshchuk, E.Churin, Y.Yurlov, B.Mednikarov, *J. Imaging Sci.*, **30**, 132, 1986.

Vacuum Electron and Ion Technologies
D. S. Karpuzov (Ed.)

ISBN 1-56072-242-8

RAMAN AND TEM STUDY OF SiO_2:Ge SPUTTERED THIN FILMS SHOWING VISIBLE LUMINESCENCE AT ROOM TEMPERATURE

M. Sendova-Vassileva, N. Tzenov, M. Tzolov and D. Dimova-Malinovska
Central Laboratory for Solar Energy & New Energy Sources, Bulgarian Academy of Sciences, 72 Tzarigradsko Chaussee blvd.., 1784 Sofia, Bulgaria
V. Dimov
Institute of Applied Mineralogy, Bulgarian Academy of Sciences, 92 Rakovski str., 1000 Sofia, Bulgaria

Samples obtained by rf magnetron co-sputtering in Ar gas of Ge chips placed on a SiO_2 target and annealed after deposition in Ar gas atmosphere were studied. The appearance of nanocrystals after annealing is demonstrated by Raman spectroscopy – a sharp band appears at around 300 cm^{-1} on a broad structure, similar to that of a-Ge. There is no strong dependence of the Raman FWHM on the Ge concentration.
The changes and crystallization taking place in the samples after annealing are observed by TEM. Crystallites of sizes between 5 and 30 nm are observed depending on the Ge concentration.
Visible photoluminescence is measured after annealing. Its shape, position and intensity depend on the deposition conditions and temperature and duration of annealing. Its origin is discussed.

1. Introduction

Recent reports of visible photoluminescence (PL) from porous silicon have stimulated considerable interest in the scientific community and suggest that the material will have uses in optoelectronic applications. However the physical mechanism resposible for the PL is not well understood yet. One of the proposed mechanisms is quantum confinement [1]. A similar explanation has been put forward for the room temperature visible photoluminescence of Ge microcrystals in SiO_2 thin films observed by some authors [2]. However recently another explanation namely quantum confinement in the surface layer of the crystallites has been given by Kanemitsu [3]. The physical quantum confinement model has been challanged by the chemical quantum confinement model which emphasizes the similarity of the luminescent and vibrational properties of porous Si with those of long known siloxene derivatives [4].

In this paper we report some results concerning the luminescence in SiO_2:Ge sputtered thin films, containing Ge microcrystals.

2. Experimental

The samples were prepared by rf magnetron cosputtering of SiO_2 and Ge. The target was a 10 cm in diameter suprasil (SiO_2) plate on which 4 to 16 small pieces of Ge wafer (5x8 mm^2) were placed. Some pure SiO_2 samples were produced under the same conditions. The sputtering was performed in either $Ar+H_2$ or only Ar gas of 0.5 Pa with rf power of 100 or 200 W. The substrates were Si wafers, fused silica plates and NaCl plates for the TEM measurements and they were either not heated or heated to 350 or 450^{0}C. Samples of two main thicknesses were produced - thick ones ($\sim 1\mu m$) for optical and Raman measurements and thin ones (less than $0.1\mu m$) for TEM measurements. After deposition the samples were annealed in Ar gas atmosphere at temperatures up to 800^{0}C for different lengths of time (from half an hour to three hours). A group of samples was annealed using a 40 W CO_2 laser in vacuum of the order of 10^{-2} Pa at temperatures between 800^0C and 1200^0C.

Raman spectra were taken using a SPEX 1403 double monochromator and a photon counting system. Luminescence was excited with the 488.0 nm Ar^+ laser line and measured at room temperature on the same equipment but this time a GaAs based PMT (RCA C31034) was used.

The TEM study was carried out on a Philips EM420T electron microscope operated at 120 keV, equipped with an EDAX X-ray microanalyser with Si (Li) detector which can detect elements from Na to U. The samples were wet stripped from the NaCl substrates and mounted on electron microscopic grids prior to annealing.

PDS spectra of the annealed and not annealed spectra were measured and compared.

3. Results

The deposited samples containing Ge have the Raman spectrum of amorphous germanium. After annealing a sharp band appears at about 300 cm^{-1} similar to the one characteristic of c-Ge but broader which is explained [5,6] by the small size of the Ge crystallites embedded in the SiO_2 matrix. Fig.1 shows Raman spectra of three samples deposited and annealed in identical conditions differing only in the number of Ge pieces put on the target. As can be seen there is a slight decrease in the FWHM of the Raman band when the concentration of Ge increases, but this effect is very weak.

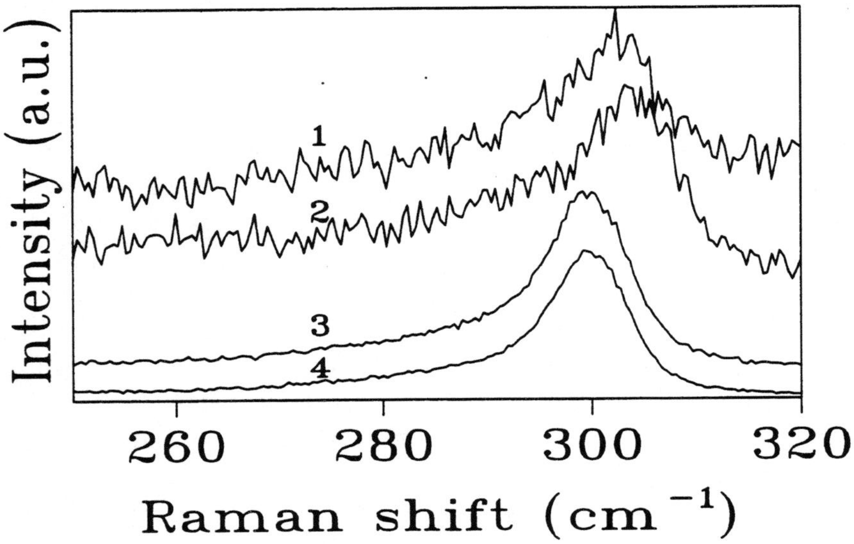

Figure 1. Raman spectra of four samples annealed at 800^0C for 1 hour differing in the number of Ge pieces on the target (1) 4 pieces, (2) 8 pieces, (3) 12 pieces, (4) 16 pieces.

There are some interesting tendencies that we find in our data.

In all samples the Raman band is asymmetric with a greater width on the low energy side (Fig.1). In samples grown and annealed in identical conditions differing only in the substrate temperature during the deposition the width of the Raman line increases with substrate temperature. All being equal the width of the Raman line in the thin samples was substantially smaller than in the thick ones (Fig.2). One explanation could be that the width is connected to a distribution of crystallites differing in size and for a thicker sample this distribution is wider. This could be corroborated by the fact that too long annealing leads to wider Raman lines - e.i. a wider distribution of crystallites in differing surroundings, strain, etc. As can be seen on Fig.2 the 300 cm^{-1} band appears on a broader structure which proves that under these annealing conditions not all of the Ge in the samples is crystallized.

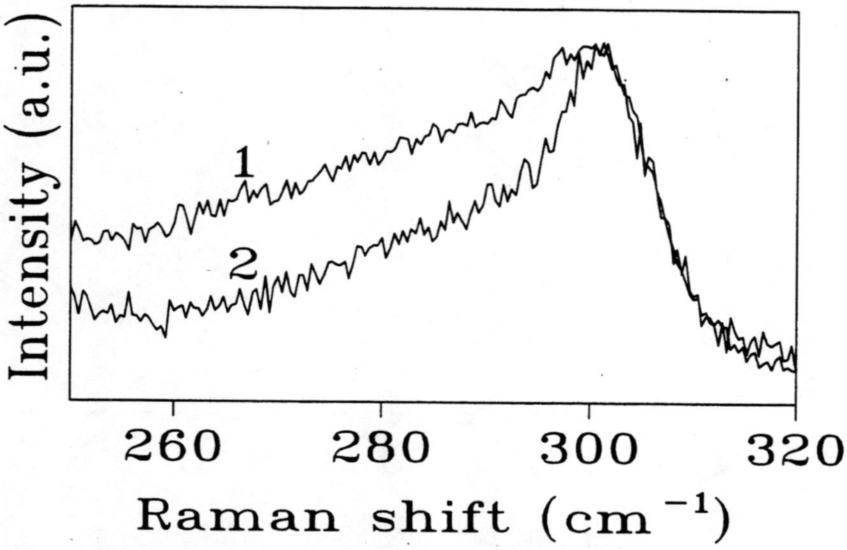

Figure 2. Raman spectra of two annealed samples differing only in thickness (1) 0.075μm, (2) 1.7μm.

Fig.3 shows the dependence of the Raman phonon frequency shift on the FWHM of the Raman bands in our samples and the theoretical curve for the case of spherical quantum dots [6]. It can be seen that only for a few samples there is an agreement between the experimental and theoretical data. There is a displacement towards negative values of the Raman shift for most of the samples. This can be a result of strains in the samples [6].

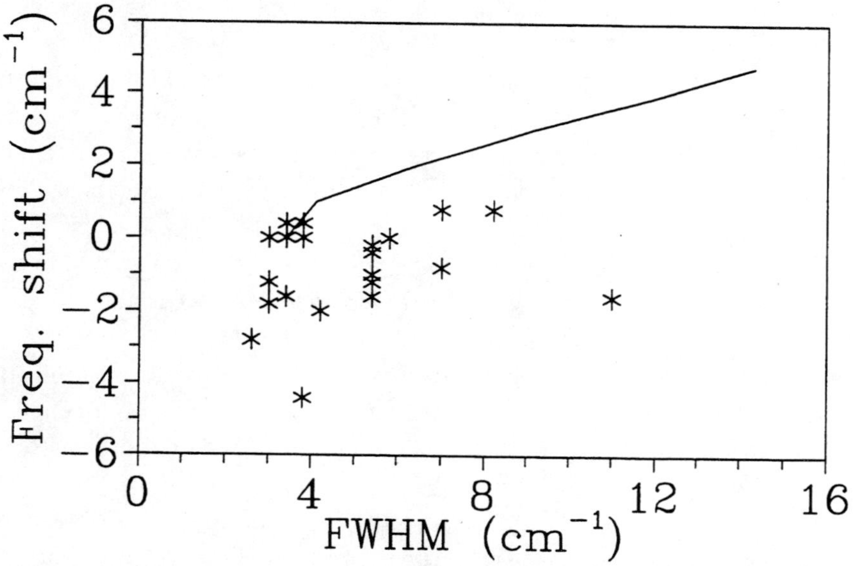

Figure 3. Dependence of the Raman phonon frequency shift on the FWHM of the Raman line for our samples and the theoretical curve for spherical quantum dots.

Fig. 4 displays the luminescence spectra of several samples grown and annealed in different conditions to show the variation in position and width of the spectra we have obtained. Samples containing Ge do not luminesce before annealing. The spectrum of a sample of pure SiO_2 deposited and annealed under similar conditions is included, too (curve 1). The films were deposited on polished Si surfaces and the spectra show interference fringes. The luminescence maximum is between 560 and 700 nm and its width varies from 110 to 300 nm.

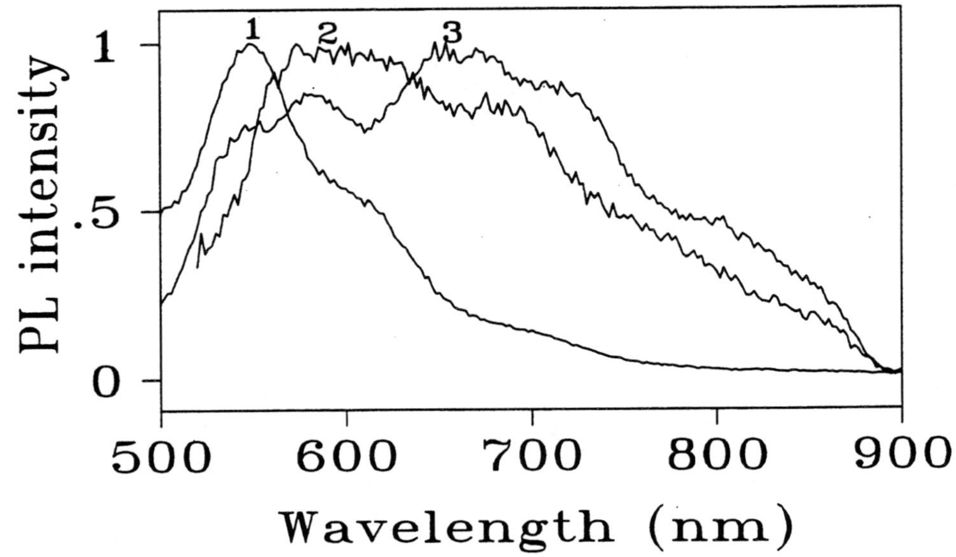

Figure 4. Normalized luminescence spectra of annealed samples (1) SiO_2 sputtered without Ge, (2) sputtered with 12 pieces Ge, (3) sputtered with 8 pieces Ge.

The peak position of the luminescence shows a tendency to shift to lower wavelength as the concentration of Ge in the sample increases all else being equal. The shift is of the order of 100 nm for the whole range of concentrations we have used. The intensity of luminescence decreases with Ge concentration. This decrease is more pronounced with increasing Ge pieces put on the SiO_2 target from 4 to 12 (this behaviour is not demonstrated in fig. 4, where normalized spectra are presented). At the same time the FWHM of the luminescence band decreases.

For samples deposited at higher substrate temperature the width of the luminescence band grows without any measurable change in the position of the maximum of the luminescence band. Thicker samples show broader and less intensive bands than thin samples without any change in position. Longer times of annealing lead to a decrease in the intensity of luminescence.

The PDS results show that increased Ge concentration in the annealed samples leads to a decrease in the density of states in the gap and a decrease in the Urbach energy.

All this is true about samples containing Ge.

Samples grown without Ge show a maximum of luminescence at about 560 nm and the width of the band is 110 - 120 nm. Unannealed SiO_2 films show luminescence, too. The width, position and intensity of the band are not influenced by the conditions of deposition and annealing.

Fig. 5 shows the electron microscopic pictures and diffraction patterns obtained by TEM. Two samples differing in Ge concentration have been studied. The first one was obtained by placing 12 pieces of Ge wafer on the target and the second one by placing 16 pieces. The other conditions of preparation of the two samples were identical. They have been annealed in two regimes - at 300^0C and 700^0C for 30 min. The samples annealed at 700^0C crystalline dots in amorphous matrix can be seen by TEM (Fig.5b,c). The size of the crystallites is 5-10 nm for the first sample and 18-30nm for the second one. The polycrystalline selected-area electron diffraction (SAED) patterns for the two samples annealed at 700^0C show sharp Debye rings

(Fig.5b,c - insets). Laue spots on the Debye rings appear for the sample sputtered with more germanium (Fig.5c - inset). Detailed analysis reveals that for the first sample these rings correspond to c-Ge, but in the second sample they correspond to tetragonal crystal GeO_2. In the latter case a monocrystalline conversion beam electron diffraction (CBED) pattern was obtained in which a monocrystal of GeO_2 in [211] orientation parallel to the electron beam is identified (Fig.5d). Formation of GeO_2 crystals in the films deposited with higher Ge concentration is a puzzling result which has to be studied more carefully. It will be noticed that Kanemitsu [7] has reported about the presence of a-GeO_2 in his samples prepared by the same method we have used. The identification was performed using software developped at the electron microscopy laboratory of the Institute of Applied Mineralogy.

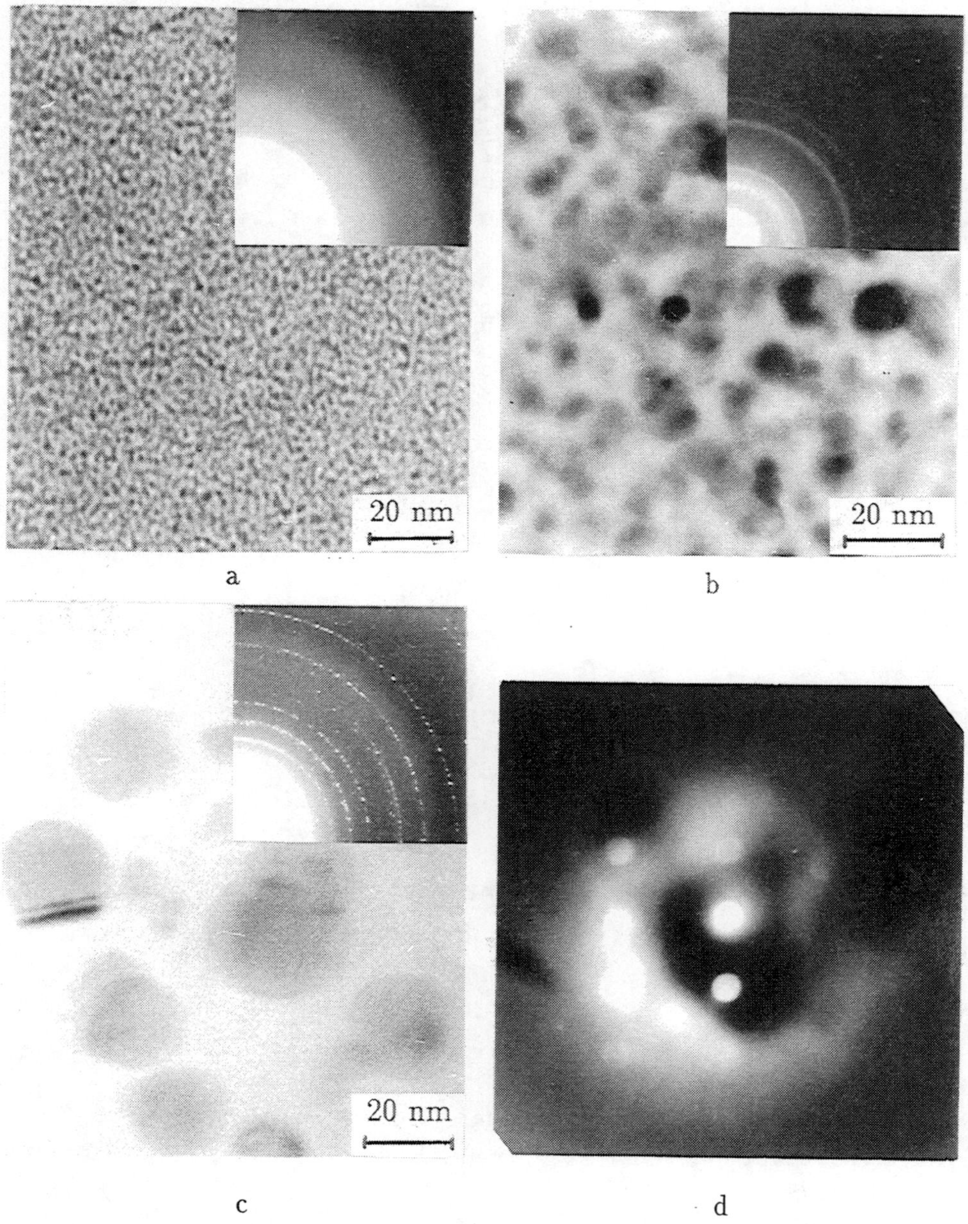

Figure 5. TEM results: (a) sample annealed at 300^{0}C for 30 min – amorphous, (b) sample sputtered with 12 pieces Ge and annealed at 700^{0}C for 30 min – Ge crystallites, (c) sample sputtered with 16 pieces Ge and annealed at 700^{0}C for 30 min – GeO_2 crystallites, (d) monocrystalline CBED pattern from the latter sample.

4. Discussion

Our results show that the luminescence peak energy increases with Ge concentration while if we assume that quantum confinement is responsible for the luminescence the peak energy should decrease with Ge concentration because then the size of the nanocrystals is expected to get bigger as our TEM results show. This contradiction could be explained by presuming that as the size of the crystallites grows a smaller proportion of them contribute to luminescence and the luminescence band is increasingly dominated by the SiO_2 matrix luminescence centred at 560nm. In samples sputtered from a target with higher Ge concentration in which our TEM results demonstrate the appearance of GeO_2 crystallites these crystallites might contribute to the luminescence as well as our latest results show [8].

Higher Ge concentration leads to samples with less defects, less disordered samples - with less Raman and luminescence band widths, lower density of states in the gap and lower Urbach energy.

Longer annealing above 1 hour leads to a wider distribution of luminescence sites, a broader Raman band, a decrease in the intensity of luminescence but not necessarily broader luminescence bands. The fact that longer annealing which should lead to bigger Ge crystallites in fact results in broader Raman bands leads to the conclusion that in this case the Raman band width is not determined by crystallite size but by other factors - strain, defects, etc.

Our proposition to explain our results is as follows. The width of the Raman line and the width of the luminescence spectrum are both connected with a distribution of sites. The luminescence is partly due to quantum size effects in Ge crystallites and partly due to luminescent centres in the SiO_2 matrix and GeO_2 crystallites. Probably germanium creates centres of non-radiative recombination in the films and that is why unannealed samples containing Ge have no measurable luminescence.

Our results on stain-etched germanium [8] lead us to put forward the hypothesis that the luminescence there is due to GeO_2 in the presence of -OH and H. Further study will ellucidate whether the luminescence in both cases has similar origin.

Acknowledgements

This work has been supported by the Bulgarian Ministry of Science and Education (contract F234) and by the Commission of the European Communities (project 7839).

References

1. L.T. Canham, *Appl. Phys. Lett.* **57** (1990) 1046.
2. Y. Maeda, N. Tsukamoto, Y. Yazawa, Y. Kanemitsu and Y. Masumoto, *Appl. Phys. Lett.* **59** (1991) 3168.
3. Y. Kanemitsu, H. Uto and Y. Masumoto, ICAS 15, Cambridge (UK), 6-10 Sept. 1993.
4. M.S. Brandt, H.D. Fuchs, M. Stutzmann, J. Weber and M. Cardona, *Solid State Commun.* **81** (1992) 307.
5. M. Fujii, S. Hayashi and K. Yamamoto, *Jap. J. Appl. Phys.* **30** (1991) 687.
6. H. Munder, C. Andrzejak, M.G. Berger, U. Klemradt, H. Luth, R. Herino and M. Ligeon, *Thin Solid Films* **221** (1992) 27.
7. Y. Kanemitsu, ICAS 15, Cambridge (UK), 6-10 Sept. 1993.
8. M. Dimova-Malinovska, N. Tzenov and M. Tzolov, ICAS 15, Cambridge (UK), 6-10 Sept. 1993 and submitted for publ. to *Solid State Commun.*.

Vacuum Electron and Ion Technologies
D. S. Karpuzov (Ed.)

ISBN 1-56072-242-8

OPTIMAL POSITIONING OF THE TWO ELECTRON BEAM SOURCES FOR THIN FILM DEPOSITION OF ALLOYS

Al. Stanchev, D. Simeonova, Ch. Tenchov
Institute of Electronics, blvd. "Tzarigradsko chaussee" 72, Sofia 1784, Bulgaria
A. Shopov
Institute of Microelectronics, Sofia 1184, Bulgaria

ABSTRACT

Tantalum and boron have been deposited simultaneously from two separate electron-beam sources. The layer obtained was studied using X-ray Photoelectron Spectroscopy (XPS) analysis. The presence of TaB_x compound is shown. In this work we report the optimal geometrical parameters of the substrate positioning with respect to the two EB sources. Factors, influencing the [Ta]/[B] ratio are discussed .

1. INTRODUCTION

Thin films of borides, obtained by sputtering, e. g. TaB_2, have high corrosion resistance, temperature coefficient of resistance about zero and high stability under heavy mechanical loads [1].

Our initial attempt to evaporate a TaB_2 alloy by using a single electron-beam source failed to produce an alloy thin film. The reason was found to be the dissociation of TaB_2 at high temperatures (even below the melting point). Since the pressures of B and Ta were quite different, the deposited thin films contained only B. This has been confirmed by Auger Spectroscopy Analysis (AES). Therefore, we tried an evaporation from two different electron beam sources (one evaporating Ta and the other - B). It is well known, that during the deposition process of the multicomponent layers of two or more atomic sources, stoichiometric problems occur. In such a case the ratio of the fluxes, arriving at the substrate must be taken into account.

2. EXPERIMENTAL SETUP

The sources are situated in a vacuum chamber kept at a pressure of about 10^{-4} Pa. The sources are the well-known electrostatic evaporators (*Statron*), operating with two independent power supplies (Figure 1). The screens Sc_1 and Sc_2 were carefully put into their proper positions, since the necessary power for evaporation depends heavily on the distances d_1 and d_2. The currents of the evaporators were chosen so that the molten material pools were both observed with a diameter of about 6 mm. A previous investigation of the temperature distribution over the surface of a molten metal heated by an electron-beam [2], allows us to consider that the temperature differences are rather small with respect to the melting temperature. The corresponding rates of evaporation were calculated following [3], thus enabling us to estimate the ratio of the two vapour fluxes. The ratio of the molar fluxes is given by the following relation:

$$\frac{\nu_{d_2}(Ta)}{\nu_{d_1}(B)} = \frac{\nu_{e_2} S_{a_2} \cos\varphi_2 \cos\Theta_2 (r_1)^2}{\nu_{e_1} S_{a_1} \cos\varphi_1 \cos\Theta_1 (r_2)^2}, \quad (1)$$

where ν_d and ν_e being vapour fluxes falling on the substrate and leaving the evaporating material surface respectively, S_a is the accomodation coefficient, φ, Θ and r are corresponding to the parameters on the figure 1. In order to obtain the stoihiometry of TaB_2 the value of this ratio has to be 1/2. In figure 2 the parameters x, α, Θ_1 and Θ_2 for $h = 100$ mm, $h = 150$ mm and $d = 80$ mm are presented. The curves are obtained under the assumption that the coefficients of accomodation are equal. The figure shows extremently great angles we must keep out. Undesirable residual gas effect may take place in the angles range above 45°. Hence, on deposition process arrival vapour flux of boron should be reduced, that means the surface temperature of boron source has to be retained below 2350 K, i. e. $f_e < 7.64$.

The arrival ratio $R_g = \frac{\nu_g}{\nu_d}$, shown in figure 3 for Ta is nearly one order of magnitude greater than that for B. At residual gas pressures about 10^{-4} Pa to 10^{-3} Pa the parameter R_g can significantly reduce the optimal range especially on boron sublimation. Figure 4 corresponds to a range of the stoichiometry ratio $1.9 \le x \le 2.1$ for different $f_e = \frac{\nu_{e2}}{\nu_{e1}}$. It is seen (figure 4) that in the substrate area above Ta vapor source deviation of the stoihiometric range is smaller than that in the area above B vapor source. The stoichiometric changes are calculated in tantalum evaporation — boron sublimation case. Taking ito account the latter particularity the deposition process requires accurate control of falling fluxes.

In our initial experiments carried out with parameters $D = 80$ mm, $h = 150$ mm, $x = D$, $\alpha = 0$ and $f_e = 7.64$ the assumption $S_{a1} = S_{a2}$, i. e. the residual gas effect, was tested.

3. PRELIMINARY RESULTS AND DISCUSSIONS

In the preliminary tests of the experimental set-up we used a substrate of Si. The deposited film sample was studied using X-ray Photoelectron Spectroscopy (XPS).

The depth from which we can acquire information of XPS analysis is 3λ, where λ is the mean free path of the photoemitted electrons. In our case $\lambda \approx 25$Å. In order to obtain information at greater depths, we removed the upper layer of the film by bombarding the sample with Ar ions at an energy of 3 kV, the total sputtering time for the whole Ta_yB_{1-y} alloy layer being 140 min.

The deconvolution of the spectrum shows pure Ta, oxidized tantalum (Ta_2O_5), and TaB_x. In the same layer the $1s$ spectrum of B has two peaks: one corresponds to oxidized boron - B_2O_3; the other is shifted with respect to $1s$ line of the pure B, which is clear evidence that we have obtained a compound of TaB_x. The presence of Ta_2O_5 and B_2O_3 is probably due to oxidation in the upper layer at room temperature. Below the upper layer (after 1 min sputtering time) pure Ta and/or TaB_x is detected in the $4f$ spectra of Ta. The presense of tantalum compound is evidenced by the different ratio of intensities of $4f_{7/2}Ta$ and $4f_{5/2}Ta$ peaks at 1 min and 81 min compared to that for pure Ta. The intensity of $4f_{7/2}$ is always higher than that of $4f_{5/2}$ in pure Ta. Further proof for the existence of TaB_x was discerned from $1s$ spectra of B recorded for the same times. Namely, both lines are located at the same energy position shifted from that for pure B. In the region of sputtering times between 1 min and 81 min no changes of $4f$ of Ta and $1s$ of B shapes were observed. The $[Ta]/[B]$ ratio decreases but it is still higher than 1.

The $[Ta]/[B]$ ratio of the obtained film is which is plotted on figure 5 as a function of the sputtering time, curve 1. The curve 2 shows the appearance and the increase of the Si signal, revealing the film-substrate interface. First, let's note that this ratio firmly establishes the existence of Ta_yB_{1-y} alloy. Ratio values neigher than 1/2 mean a considerable presence of pure Ta together with TaB_x. The significantly lower boron concentration in the film than calculated above may be explained by two effects. On one hand, the interaction between Ta and B fluxes, as well as between the absorbed atoms decreases the accomodation coefficient drastically. On the other hand B atoms can reevaporate in the form of B_2O_3 or BO_x which are volatile. This hypothesis will be tested in a further study.

4. CONCLUDING REMARKS

As a conclusion we have experimentally established that Ta_yB_{1-y} alloy films can be obtained by deposition technique using two separate electron beams. Further work is required to measure precisely the vapor fluxes at given positioning of the substrate and some other kinetic parameters. We intend to use two microbalances simultaneously for these purposes.

Another scheme to optimize the quality of the deposited alloy is to find an optimal substrate temperature which would provide the best conditions for generating TaB_x structures. This work is also under way.

5. ACKNOLEGEMENTS

This work has been sponsored by Bulgarian Ministry of Education and Science, National Founadation "Scientific Research", according to the contract $SR - P202$.

5. REFERENCES

1. S. D. Donchev, A. V. Shopov, P. S. Dishkov, *Bulg. J. Phys. 15 (1988) 1.*

2. Ch. Tenchov, D.S. Simeonova, V.A. Mitev , submitted to *Vacuum* (1993).

3. V P Glushko, *Termodinamicheskie Svoistva Individualnih Veshestv, vol. 4,* (Moscow, 1982), p.72, (in Russian).

4. D. Brigs, M. P. Seach, *Practical Surface analysis by Auger and X-ray photoelectron spectroscopy,* Jonh Wiley and Sons, New York, 1983.

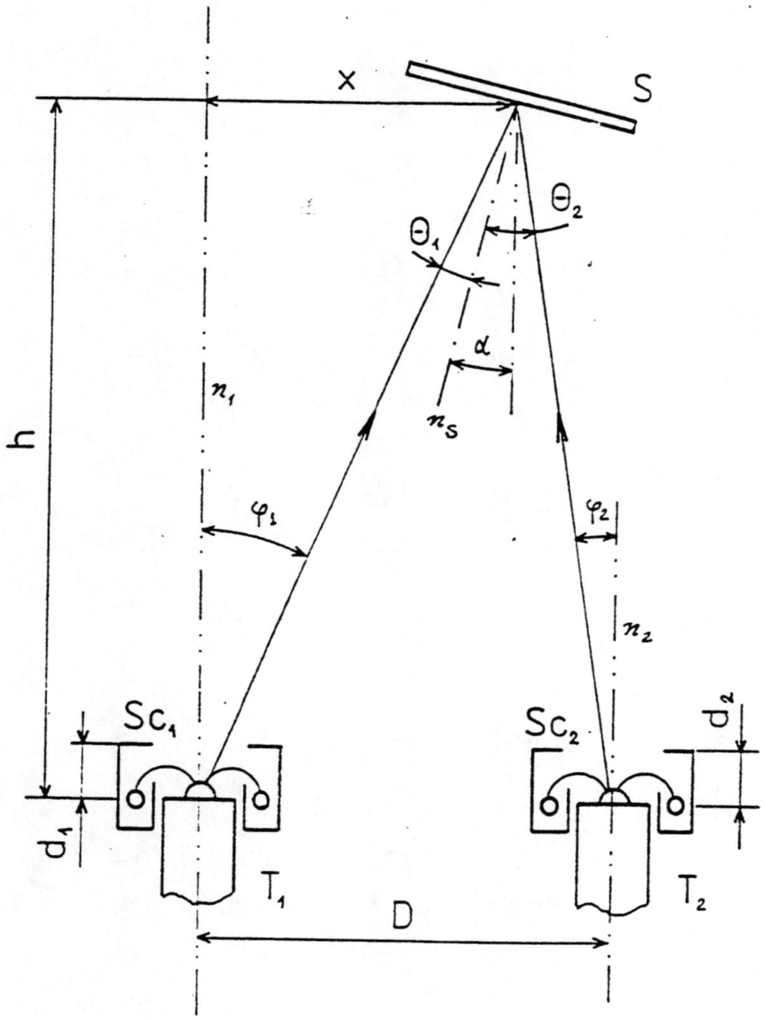

Figure 1. Experimental Set-up: T_1, T_2 — water - cooled Cu crucibles, S- substrate, n_1, n_2, — normals with respect to crucibles. n_s — normal with respect to substrate, Sc_1, Sc_2 — screens. , φ_1, φ_1 — angles for B and Ta resp. Θ_1, Θ_2 — angles with respect to the substrate for B and Ta resp. α — angle between $n_{1,2}$ and n_s, D — distance between the two crucibles, d_1 and d_2 — distances between the crucibles and screens resp. x. h — horizontal and vertical distance with respect to the T_1

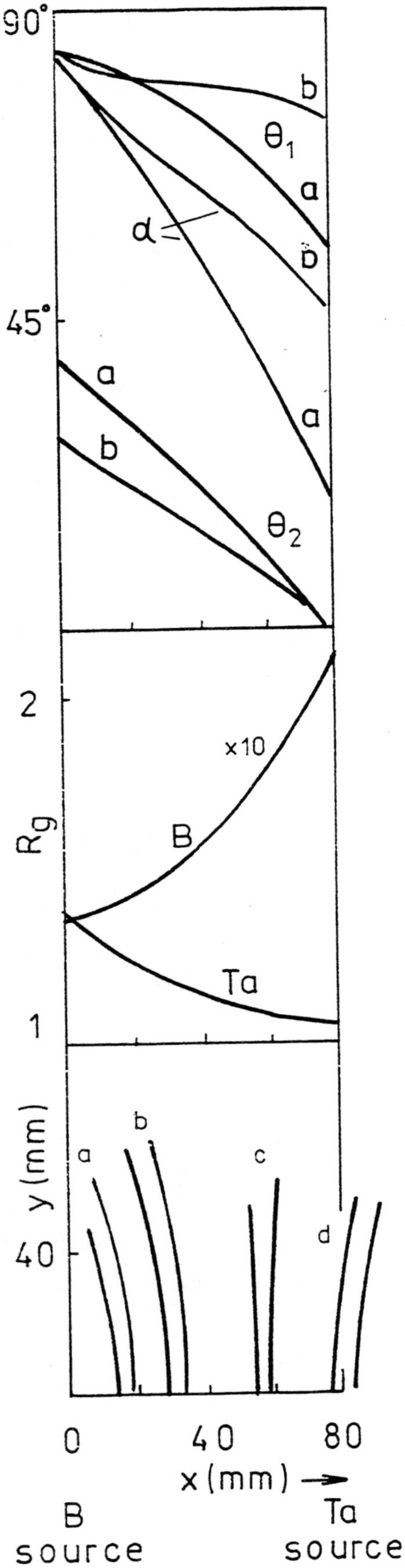

Figure 2. The dependences of α, Θ_1, Θ_2 versus x position at the condition $\frac{v_{3_2}}{v_{3_1}} = 1/2$, $f_e = 7.64$, $h = 100$ mm (a), 150 mm (b); B, Ta — molten

Figure 3. The arrival ratio R_g for Ta and B respectively. $h = 100$ mm, 10^{-3} Pa, $f_e = 7.64$, $\alpha = 0$ Ta, B — molten

Figure 4. Calculated stoichiometric ranges in the films on the substrate plane, $h = 100$ mm, $\alpha = 0$, $f_e = 1.5$(a), 2 (b), 3.5 (c), 5.7 (d); Ta — molten, B — sublimated

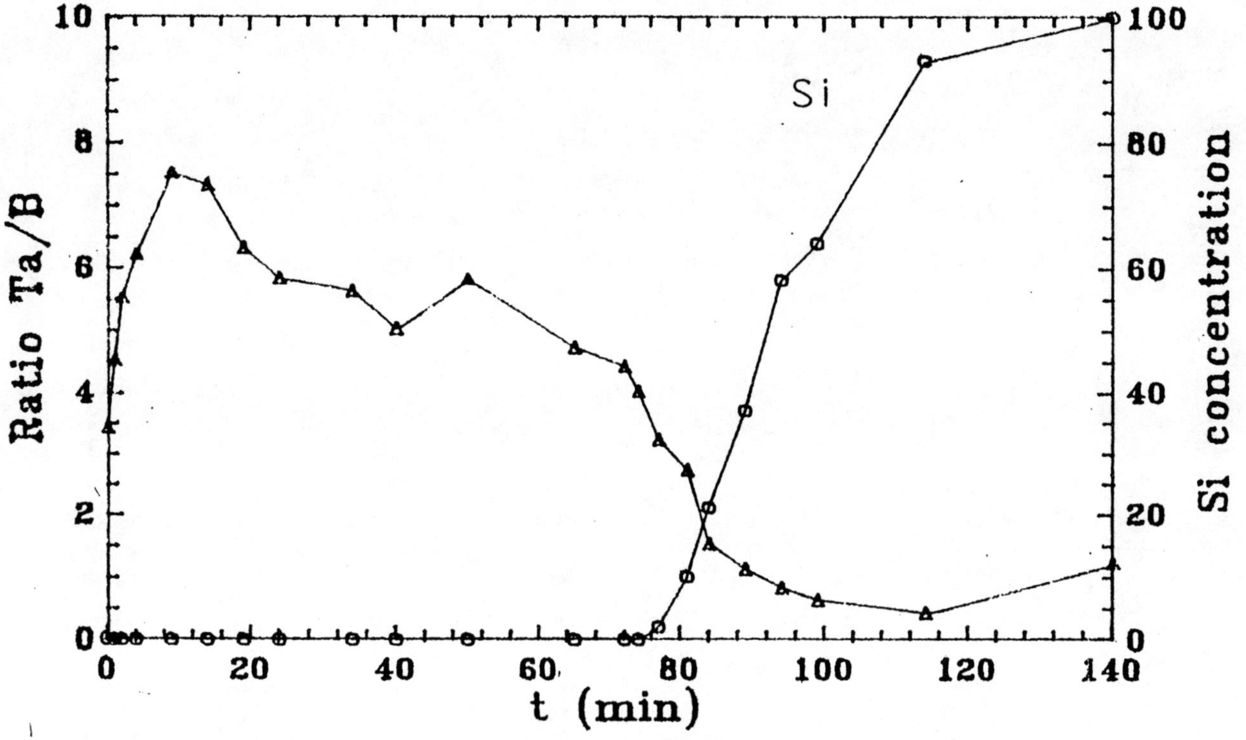

Figure 5. Dependence of Ratio [Ta]:[B] on sputtering time

Vacuum Electron and Ion Technologies
D. S. Karpuzov (Ed.)

ISBN 1-56072-242-8

IN SITU DEPOSITION AND PASSIVATION OF HTSC-FILMS BY SPUTTERING METHODS

E. Steinbeiss, K. Steenbeek, T. Schüler, T. Eick and W. Michalke
Institut für Physikaliche Hochtechnologie, Helmholtzweg 4, PF 100239,Jena, Germany

1. Introduction

High quality HTSC-films of the REBaCuO- system can be prepared now by all established thin film technologies, especially by laser ablation, sputtering, coevaporation and MOCVD. Therefore the international activities are now concentrating on the development of different devices by means of HTSC-films like bolometers, SQUIDs and passive microwave devices. In such devices the REBaCuO films usually must be combined with further functional films for fitting special substrates, for the separation of different HTSC-planes, for the induction of grainboundaries, as well as for passivation and contacting the film system [1]. Therefore the development of HTSC devices is connected with several additional requirements on the deposition technology. The main goal is the conservation of noncontaminated smooth interfaces with high crystallographic perfection by an in situ deposition of the whole film system with compatible deposition steps. These problems are solved for each deposition method in a specific manner. In the following we will focus on the specific problems of the magnetron sputtering technique.

2. In situ deposition of epitaxial REBaCuO-films by magnetron sputtering

High quality REBaCuO-films can be prepared only by epitaxial growth of single phase films with the exact stoichiometric composition with respect to the cation ratios and the oxygen content, requiring:

- the application of monocrystalline substrates with a low misfit of the lattice parameter and high surface quality
- high substrate temperatures between 700°C and 800°C
- the exact control of the film composition and
- the choise of a suitable oxygen partial pressure during film growth in agreement with the conditions of the phase diagram.

Fig1 SEM-picture of a $Gd_1Ba_2Cu_3O_{7-x}$ film sputtered at 52 Pa

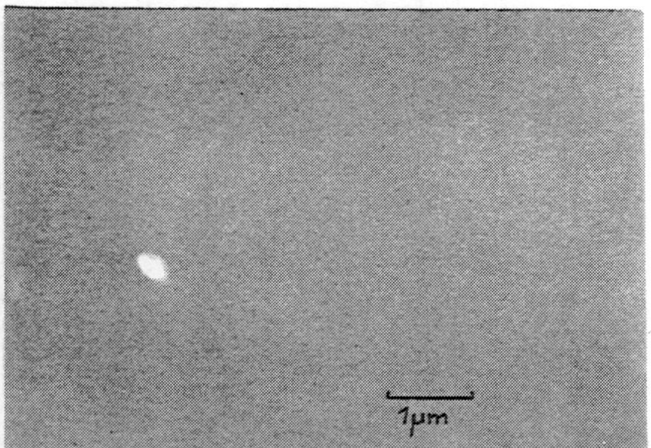

Fig. 2 SEM-picture of a $Gd_1Ba_2Cu_3O_{7-x}$ film sputtered at 18 Pa

The high O_2-partial pressure necessary for phase stability at 800°C according to the phase diagram of Bormann et al. [2] and the disturbing influence of energetic particles usually involved with sputtering processes [3] are the main and specific problems for the deposition of high quality HTSC-films by sputtering methods. These have been solved by the application of high pressure sputtering with special off axis substrate arrangements to prevent the bombardment of energetic particles on the film surface.

Basic disadvantages of this method are

- the noncompatability of the process parameters with the optimal deposition conditions for other materials
- the sensitive influence of convection processes on the deposition profile.

Therefore the further development of sputtering methods for HTSC-film deposition was strongly influenced by experimental results showing that high quality single phase REBaCuO-films could be prepared even at much lower oxygen partial pressures below 10 Pa [4] which is clearly outside the phase boundary of the phase diagram of Bormann et al.

This result can be explained by the assumption, that the growing film reacts mainly with the activated oxygen component from the magnetron discharge, for which a modified phase diagram is valid, similar to that of ozone determined recently by Sawa et al. [5]. Besides dissociated oxygen ionized O^{2+} may be one of the most important components, which can be easily detected by emission spectroscopy of the magnetron discharge [6]. The strongly increased reactivity of the activated oxygen enables the sputtering of single phase ReBaCuO-films even at total pressures of 10 Pa and below, which has the advantage, that the surface morphology can be controlled additionally by optimizing the total pressure (Fig. 1 and Fif. 2).

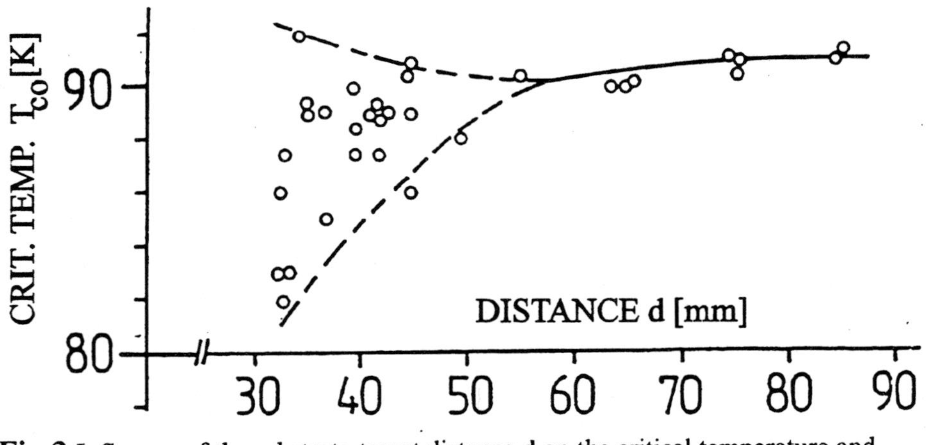

Fig.3 Influence of the substrate-target distance **d** on the critical temperature and the deposition rate for YBCO films sputtered with a 4"-planar magnetron

On the other hand for low pressure sputtering additional effort is necessary to prevent the increased influence of nonthermalized energetic particles on the composition of the growing films. Besides the application of quasi off axis substrate arrangements further steps like e.g. the increase of the substrate-target distance (Fig. 3) are necessary to restrict the negative influence of the plasma [4]. The observed plasma effects can be explained by the bombardment of charged particles accelerated by potential differences between plasma and film surface. These cause increasing deviations of film stoichiometry with rising plasma density and electron temperature.

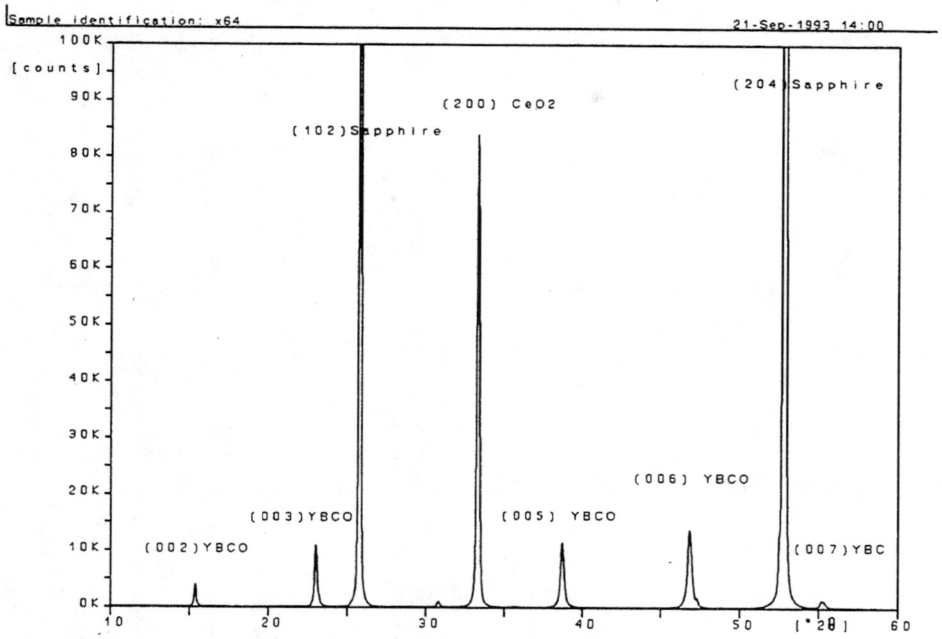

Fig. 4 BXRD spectrum for YBCO on CeO_2 buffered sapphire

Therefore these disturbing plasma effects can be suppressed effectively by reducing the plasma density and electron temperature near the substrates by application of suitable anodes and special screens. Unfortunately the reduction of plasma density also lowers the concentration of the activated oxygen, so that from kinetical reasons fully oxidized films can be grown only at relatively low deposition rates around 100 nm/h.

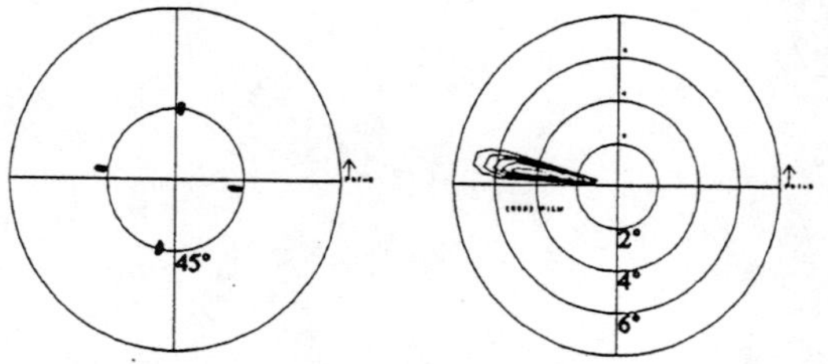

(103) film (005) film

Fig. 5 Pole figures of the (103) and the (005) reflex of a GdBaCuO film on a $SrTiO_3$ substrate surface inclined 5° to (100)

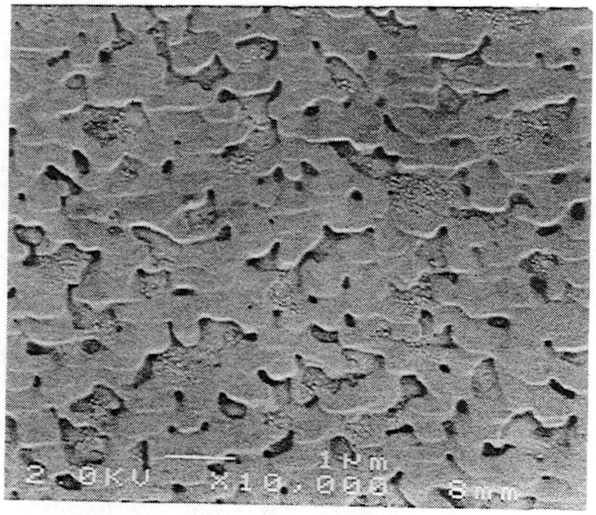

Fig. 6 SEM-picture of the same film used in Fig.5

Nevertheless the low pressure sputtering technique described above allows now the reproducible deposition not only of single REBaCuO-films on $SrTiO_3$ but also the compatible deposition of high quality HTSC-films together with different further functional films in film systems suitable for many device applications. This can be illustrated by several

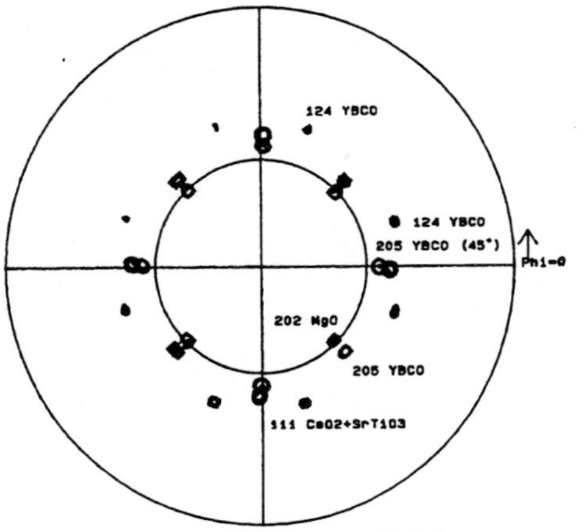

Fig. 7 Pole figure of an YBaCuO-film with a 45° grain boundary induced by a MgO seed layer

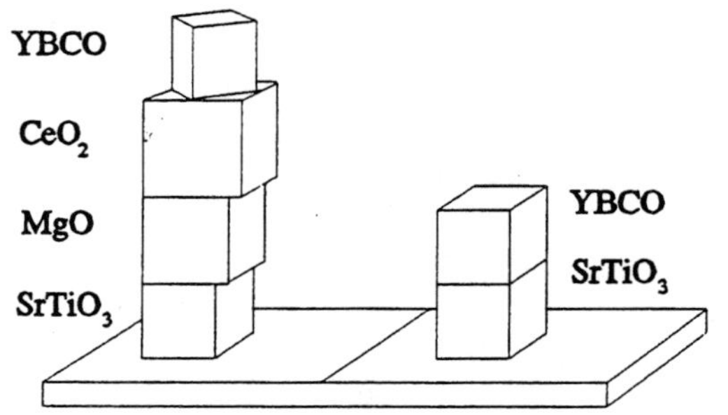

Fig. 8 Crystallographic orientation of the film system resulting from fig. 7

examples showing the epitaxial growth of REBaCuO-films on epitaxial buffer layers (Fig. 4), on tilted (100) substrate surfaces for the preparation of contacts with low resistance (Fig. 5, Fig. 6) and the turn of the epitaxial orientation of REBaCuO-films on epitaxial seed layers (Fig. 7, Fig. 8). The critical current densities of these HTSC-films are throughout above 10^6 A/cm^2 with top values up to 7.10^6 A/cm^2 at 77 K.

3. Passivation and contacting of HTSC-film system

Usually the epitaxial HTSC-films on substrates and buffer layers with an increased lattice misfit have a more or less increased defect density compared with REBaCuO on $SrTiO_3$. This causes only a little deterioration of the superconducting properties but an increased instability of the films with respect to ageing and thermal stresses under reduced oxygen pressure (for example during ion etching for patterning) or H_2O- and CO_2-containing atmospheres because of an enhanced outdiffusion of oxygen and a decomposition to hydroxides and carbonates along crystal defects [7].

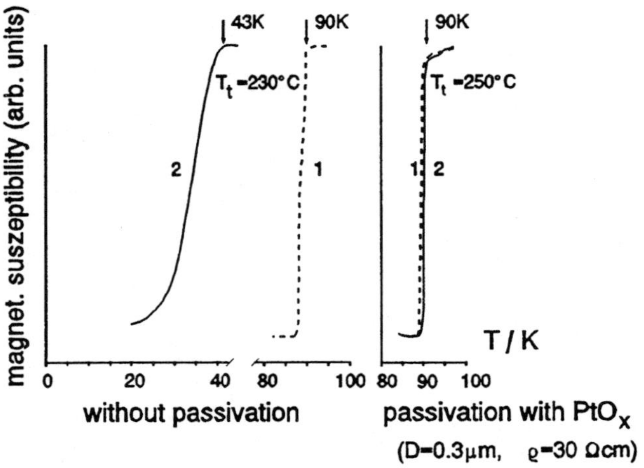

Fig. 9 Temperature dependence of the magnetic susceptibility of GdBaCuO films (1: after preparation, 2: after annealing at T_t, 1 hour in air)

The application of such film systems therefore requires an effective sealing by suitable passivation layers. In order to prevent any contamination the passivating films must be deposited immediately (in situ) after the REBaCuO deposition and after oxidation, which means, at low substrate temperatures. Therefore plasma-activated deposition methods like sputtering are the most suitable methods for the deposition of high quality passivating films, especially of oxide and nitride films known as effective diffusion barriers with high resistivities.

Unfortunately the widely used Si- and Al-compound films disturb the superconducting properties of REBaCuO-films by interdiffusion processes and have the additional disadvantage, that the contacting of the HTSC-films is complicated by the critical steps of opening and cleaning the contact windows. Therefore we have investigated a new class of passivation layers, the noble metal oxides which can be deposited by reactive sputtering of noble metal targets at substrate temperatures between 100°C and 200°C and total pressures around 1 Pa [8]. As shown in Fig. 10, the resistivity of the films can be easily varied by the composition change of the O_2/Ar-sputtering gas. The passivated REBaCuO-films show an increased stability against temperature stress in different environments compared with nonpassivated films (Fig. 9).

The noble metal oxide films can be reduced to the metallic state by annealing in air. The transformation temperatures are 150°C for AuO_x and 500°C for PtO_x.

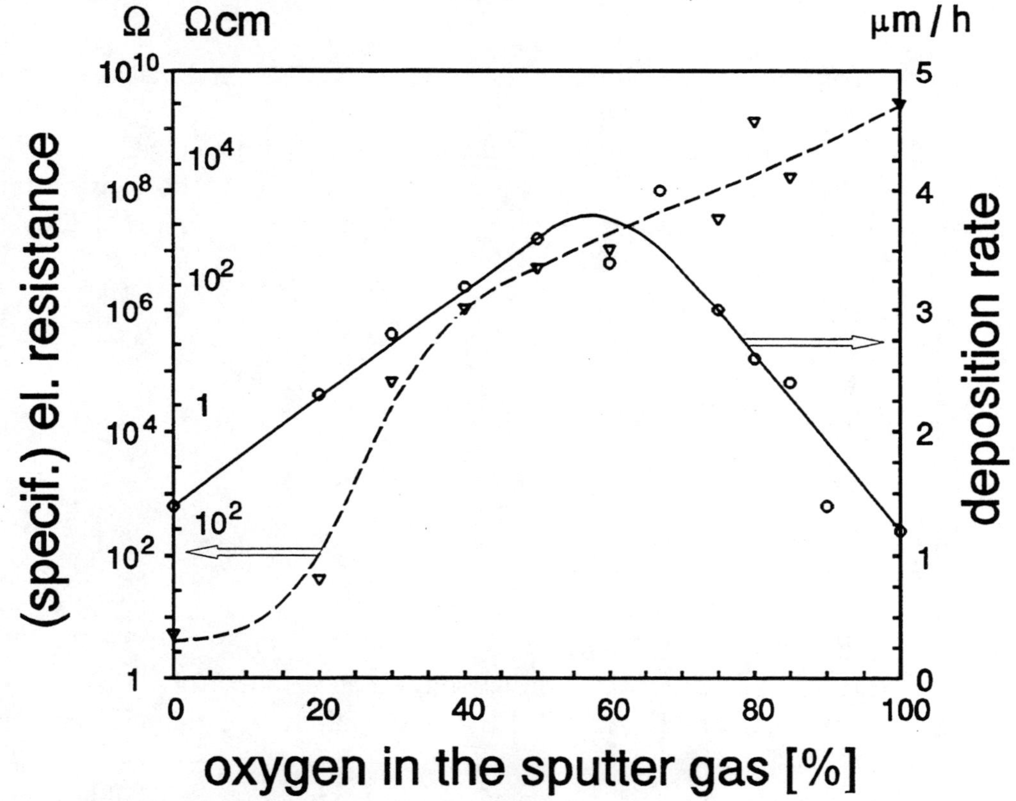

Fig. 10 Resistance and deposition rate of PtO_x-films as a function of the oxygen content in sputter gas

Local heating by a laser beam with suitable power provides low resistive contacts to the REBaCuO-film, which can be bonded for example by ultrasonic bonding methods (Fig. 11 and Fig. 12).

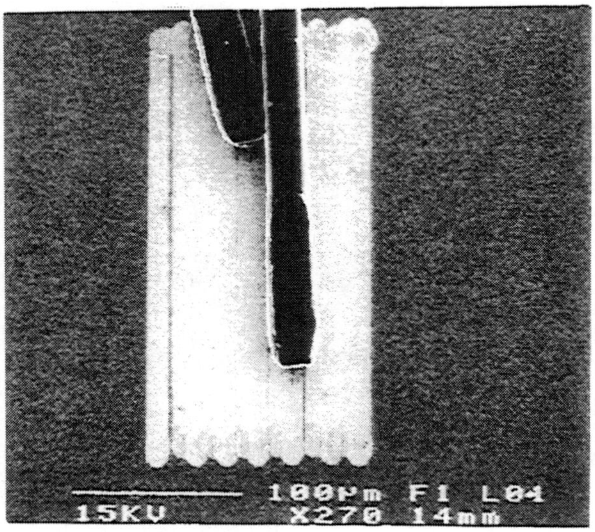

Fig. 11 SEM-picture of a Pt-contact prepared by laser annealing of the PtO_x-film and subsequent Al wire bonding

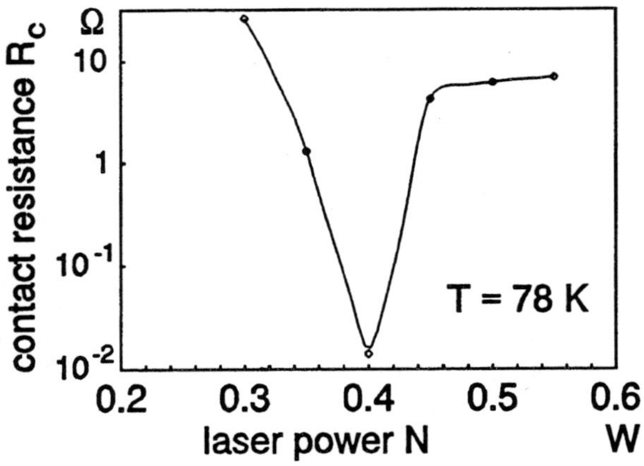

Fig. 12 Contact resistance of wire bonded contacts scanned with different laser energy densities

4. Conclusions

We can conclude, that low pressure magnetron sputtering with special off axis substrate arrangements is a useful deposition method for the preparation of high quality REBaCuO-films together with epitaxial buffer layers for different purposes in HTSC-devices in a compatible manner. The in situ deposition is a necessary condition for the conservation of pure interfaces of high crystallographic perfection, which enables an epitaxial growth of high quality films throughout the whole film system.

5. Acknowledgement

This work was supported by BMFT, Grant Nr. FKZ13N5926.

6. References

[1] R. H. Ono: MRS Bulletin 17 (1992), 34
[2] R. Bormann, J. Nolting: Appl. Phys. Letters 54 (1989), 2148
[3] J. J. Hanak, J. P. Pellicane: J. Vac. Sci. Technol. Vol. 13, No1 (1976), 406
[4] T. Schüler, E. Steinbeiβ, G. Bruchlos, Th. Eick: phys. stat. sol. (a) 134 (1992), K25
[5] A. Sawa, S. Kosaka, H. Obara, K. Aoki: IEEE Trans Appl. Superconductivity 3 (1993), 1088
[6] X.X.Xi, X. D. Wu, A. Inam, Q. Li, D. Hemmick, A. Findikoglu: Appl. Phys. Lett. 57 (1990), 96
[7] C. A. Copetti, J. Scubert, W. Zander, H. Sollner, U. Poppe, Ch. Buchal: J. Appl. Phys. 73 (1993), 1339
[8] K. Steenbeck, E. Steinbeiβ, Th. Eick, T. Köhler, L. Redlich, H. Schmidt: Proceed. of the European Conference on Applied Superconductivity EUCAS'93 Okt.4-8, 1993 in Göttingen, Germany

Vacuum Electron and Ion Technologies
D. S. Karpuzov (Ed.)

ISBN 1-56072-242-8

PLASMA TREATMENT EFFECT ON HIGH-TEMPERATURE SUPERCONDUCTORS

Zhong Quan Yang, A.G. Ulyashin, NV.Shlopak, Y.A. Bumai, V.A. Esepkin, and A.V. Alifanov
Belarussian State Polytechnic Academy, Skaryna Ave. 65 Minsk 220027 Belarus

1. Abstract

The influence of atomic hydrogen, oxygen or argon exposure on the superconductivity of YBaCuO(123) and BiSrCaCuO(2212 and 2223) ceramics has been examined by the critical current density J_c, critical temperature T_c and X-ray diffraction measurements. It is shown that exposure of high-T_c superconductors to a low power density oxygen, hydrogen or argon plasma at 80 K, 300 K, 473 K and 673 K leads to an increasing of T_c and J_c in some case. The improvement of superconductivity can be explained by the passivation of dangling bonds and defects on grain boundaries of these materials.

2. Introduction

Superconductivity of the bulk high-T_c materials is highly sensitive to amount of oxygen vacancies and properties of grain boundaries. The vacancy concentration as well as properties of grain boundaries is thought to be influenced considerably by low power density oxygen or hydrogen plasma treatment. An increase of T_c by oxygen or argon etching of $YBa_2Cu_3O_{7-x}$ has been observed earlier /1,2/. However, the low power density plasma treatment and related variation of J_c have not been investigeted in details. It should be to note that the interest in atomic hydrogen effects on high-T_c superconductors derives from its well known ability to passivate the electrically active defects and impurities as well as surface dangling bonds in semiconductors /3/. In this work we present the results of atomic hydrogen, oxygen as well as argon effects on T_c and J_c of YBaCuO (123) and BiSrCaCuO (2212 and 2223) high-T_c superconductors.

3. Experimental procedure

Samples with nominal composition $YBa_2Cu_3O_{7-x}$ were prepared by high-purity powders of Y_2O_3, CuO and $BaCO_3$ with small excess of $BaCO_3$ in respect to stoichiometric (123) composition. The heat treatment consisted in firing the mixture of the initial material in the alumina crucible at a temperature 930°C for the time of 15-20 hours in air with intermediate grindings. Finally, the pressed pellets or rings from the powders were sintered at 940°C in air for 10-20 hours and slowly cooled with furnace. Single 123 phase samples were obtained. The resistivity of the samples becomes zero at 85-93 K and the critical current density is 50 - 100 A/cm^2.

Samples with nominal composition $Bi_2Sr_2CaCu_2O_z$ were prepared from high purity powder reagents of Bi_2O_3, $SrCO_3$, $CaCO_3$ and CuO according to /4/. The appropriate amount of powders was mixed, calcined at 800-870°C, ground, cold pressed into disk shaped pellets and heated in an alumina crucible at 840°C for 17 hours and heated again at 850°C for 24 hours. Then these pellets were ground well and pressed into rings or pellets and heated at 840°C for 9 hours. The samples were taken out from furnace and cooled to room temperature. Single 2212 phase samples were obtained. Some of the obtained samples were heated again at 845°C for 5 hours and quenched to room teperature. This procedure was repeated for 6 times. By this way 2212 phase quenched samples were obtained and marked by $Bi_2Sr_2CaCu_2O_z$ (Q). The resistivity of the samples becomes zero at 78-80 K and the critical current density is about 10 A/cm^2.

Samples with nominal composition $Bi_{1.85}Pb_{0.35}Sr_{1.9}Ca_{2.1}Cu_{3.1}O_y$ were prepared from metal nitrate aqueous solution according to /5/. After evaporation of solution the received powder was heated at 840°C for 10 hours with following grinding. Then powder was pressed into pellets and then heated at 850°C for 75-100 hours. Finally, these pellets were ground and cold pressed into rings and pellets. The pressed rings and pellets were heated at 850°C for 10 hours and furnace-cooled. Samples close to single 2223 phase with small content of 2212 phase were obtained and marked by $Bi_2Sr_2Ca_2Cu_3O_y$ (c) and $Bi_2Sr_2Ca_2Cu_3O_y$(n) separately. The difference between these two kinds of samples is: the superconducting phase volume is more than 40% for the $Bi_2Sr_2Ca_2Cu_3O_y$ (c) samples and less than 30% for the $Bi_2Sr_2Ca_2Cu_3O_y$ (n). The resistivity of these samples becomes zero at 105-115 K and the critical current density is 200 - 400 A/cm^2 for $Bi_2Sr_2Ca_2Cu_3O_y$(c) and 100 - 200 A/cm^2 for $Bi_2Sr_2Ca_2Cu_3O_y$ (n).

The critical current densities and transition temperatures of high-Tc superconductors were measured by contactless technique. A toroidal superconductor specimen has been used as tertiary winding of small ferrite transforme as described in /6/. In some control measurements the conventional four-probe method has been used to measure the resistance of the samples. The temperature of the samples is controlled by calibrated semiconductor detector. The structure of the samples was examined by means of X-ray diffraction.

The samples were exposed to hydrogen, oxygen or argon plasma of dc glow discharge with power density <0.1 W/cm^2 for 0.5-2 hours at 80 K, 300 K, 473 K and 673 K. It should be noted that such atomic hydrogen treatment is often used for hydrogenation of semiconductors /3/.

4. Results and discussion

Figures 1-**2** show the average changes in Tc difference (ΔT) and in Jc difference (ΔJ) versus oxygen, hydrogen and argon plasma treatment time (t) separatly. It can be seen that an increasing of Tc for $YBa_2Cu_3O_{7-x}$, $Bi_2Sr_2CaCu_2O_z$(Q) and $Bi_2Sr_2Ca_2Cu_3O_y$ (c) systems which are more sensitive to amount of oxygen vacancies than $Bi_2Sr_2Ca_2Cu_3O_y$(n) and $Bi_2Sr_2Ca_1Cu_2O_z$ takes place under hydrogen and oxygen plasma exposure at 300 K and 80 K. Since the diffusion of hydrogen and oxygen at 80 K is negligible this increasing can be explained by the passivation effects on grain boundaries of these materials.

The critical density increasing in common is consistent with ΔTc increasing.

After argon plasma treatment the critical current density Jc of $YBa_2Cu_3O_{7-x}$ and $Bi_2Sr_2Ca_2Cu_3O_y$. (n) was changed slightly but Tc of $Bi_2Sr_2Ca_2Cu_3O_y$ (n) was increased in 1.5 - 2 K and this corresponds to the results presented in /1/.

The results were depended on the preparing technique of the samples. The properties of samples $Bi_2Sr_2CaCu_2O_z$ were not changed after oxygen plasma treatment at 473 K and 673 K, but the critical current density J_c of $Bi_2Sr_2CaCu_2O_z$ (Q) phase quenched samples was increased in 3 - 5 times and T_c on 12 K.

X-ray diffraction patterns of the samples after plasma treatment show stable angle value of peaks for all samples without appearing of impurity phases. According these results we conclude that the observed increasing of T_c and J_c isn't caused by a compressive stress as in /2/ but by passivation effects at grain boundaries. The improved parameters J_c and T_c of examined superconductors after low power density plasma treatment are stable to heating up to 850°C. The improvement of superconductivity of high-T materials is unexpected because most of earlier invesigations show that hydrogen incorparated in high-T superconductors from gas phase /7/ or ion implanted one /8/ lowers both T_c and J_c.

5. Conclusion

According to these results we conclude that in case of argon plasma exposure the observed increasing of T_c is not caused by a compressive stress but due to plasma treatment stimulation of unintentional passivation of grain boundaries. Thus the improvement of the electrical properties of high-T_c superconductors after any plasma treatment can be explained by the hydrogen or oxygen passivation of dangling bonds and defects on grain boundaries of these materials.

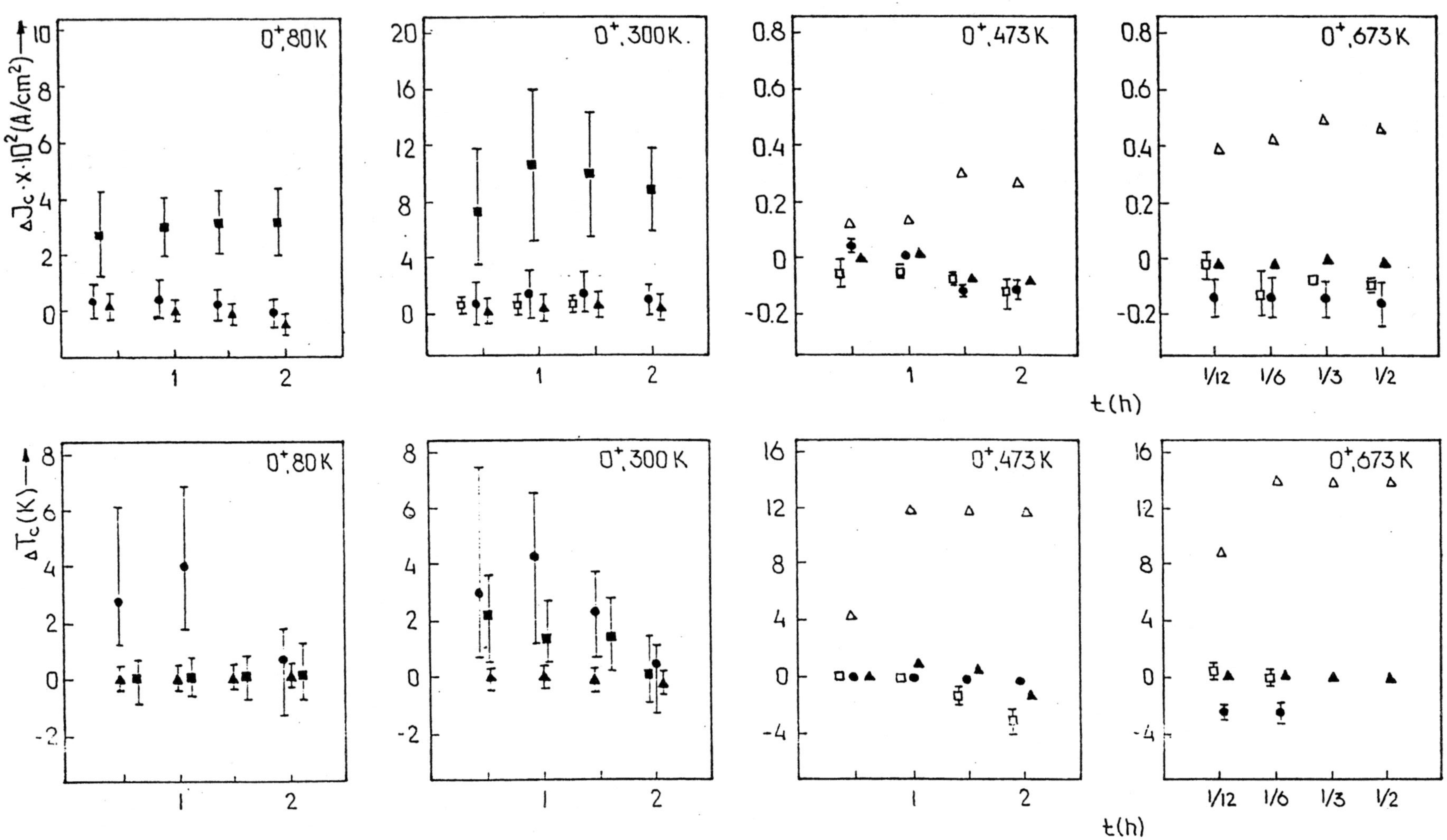

Fig. 1: Average changes in Tc and Jc differences versus oxygen (O+) plasma treatment time (t) at 80K, 300K, 473K or 673K: (■ - $Bi_2Sr_2Ca_2Cu_3O_y$ (c), □ - $Bi_2Sr_2Ca_2Cu_3O_y$ (n), ▲ - $Bi_2Sr_2Ca_1Cu_2O_z$, △ - $Bi_2Sr_2Ca_1Cu_2O_z$ (Q), ● - $Y_1Ba_2Cu_3O_{7-x}$)

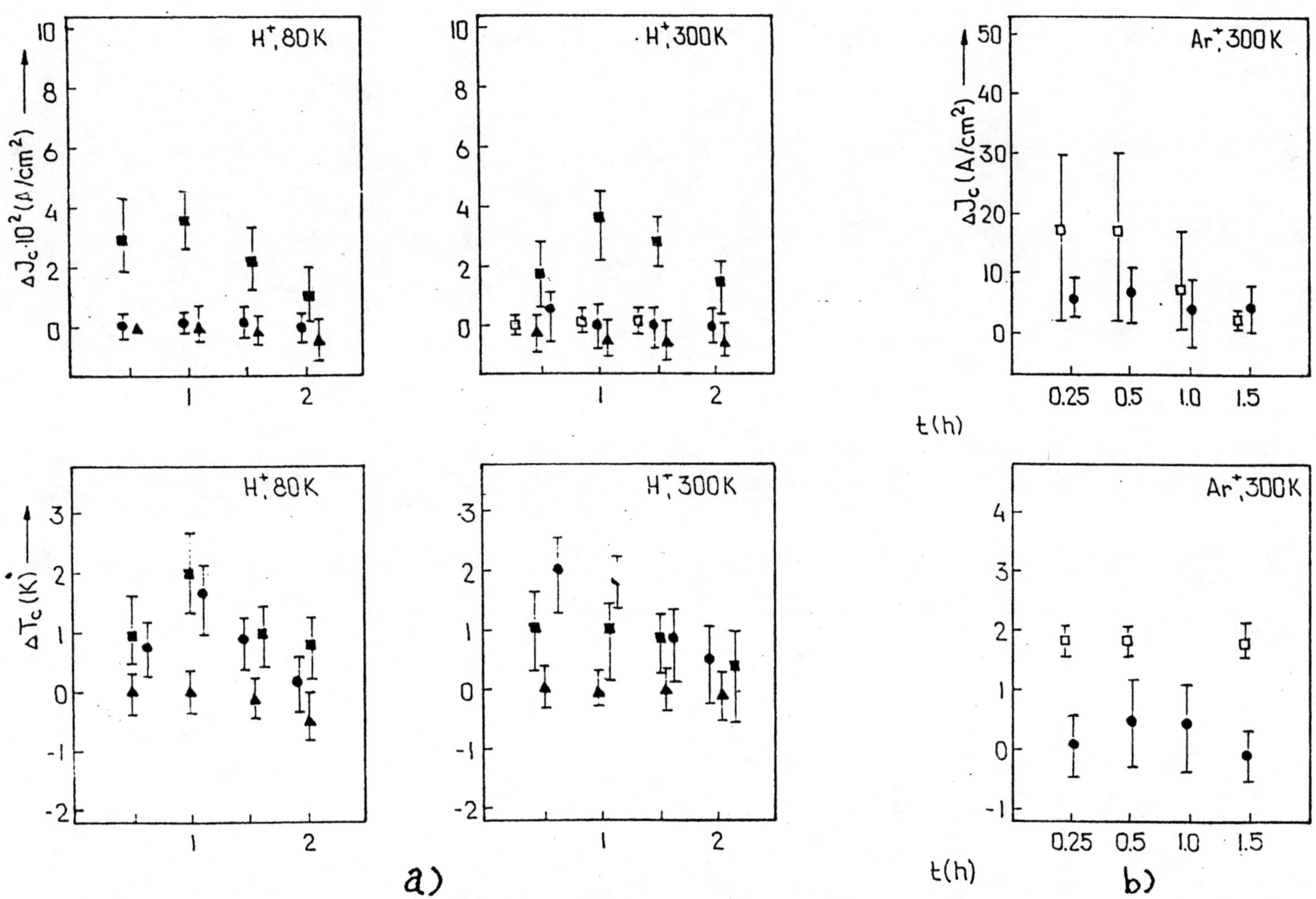

Fig. 2: Average changes in Tc and Jc differences versus a) hydrogen(H+), b) argon (Ar+) plasma treatment time (t) at a) 80K or 300K, b) 300K: (■ - $Bi_2Sr_2Ca_2Cu_3O_y$(c), □ - $Bi_2Sr_2Ca_2Cu_3O_y$(n), ▲ - $Bi_2Sr_2CaCu_2O_z$, Δ - $Bi_2Sr_2Ca_1Cu_2O_z$ (Q), ● - $Y_1Ba_2Cu_3O_{7-x}$)

References

/1/ Y. Nishi, S. Moriya, N. Inoue, S. Tokunaga and T. Shima, J. Mater. Sci. Lett. 7, 281 (1988).

/2/ Y. Nishi, Y. Takahashi, S. Takagi, T. Shima and N. Inoue, Nucl. Instr. and Meth. B59/60, 1415 (1991).

/3/ S. Pearton, J. Corbett and S. Shi, Appl. Phys. A43, 153 (1987).

/4/ T. Ishida, Japan J. Appl. Phys. 28, L573 (1989).

/5/ A. V. Dubovitskii, E. F. Makarov, M. K. Makova, V. A. Merganov and B. N. Topnikov, Superconductivity: physics, chemistry, technique. 4, 1024 (1991).

/6/ E. Harris, J. Bishop, R. Havvil and P. Ward, J. Phys. C: Solid State Phys. 21, L673 (1988).

/7/ J. Erxmeyer, G. Dortmann, J. Steiger, O. Boebel and Weindinger, J. Less-Com. Met. 172-174, 419 (1991).

/8/ T. Kato, K. Usami, J. Kuniya, and S. Matsuda, Jap. J. Appl. Phys. 27, L1104 (1988).

Vacuum Electron and Ion Technologies
D. S. Karpuzov (Ed.)

ISBN 1-56072-242-8

INFLUENCE OF THE STOICHIOMETRY ON THE SUPERCONDUCTING PROPERTIES OF YBCO THIN FILMS PREPARED BY SPRAY PYROLYSIS

V. Lovchinov and G. Mihova
Institute of Solid-State Physics, Bulgarian Academy of Sciences, 72, Tsarigradsko Shose Blvd, 1784 Sofia
I. Nedkov, I. Razkazov, R. Todorovska
Institute of Electronics, Bulgarian Academy of Sciences 72, Tsarigradsko Shose Blvd, 1784 Sofia

ABSTRACT

Superconducting YBCO films (T_c =88 K, J_c = 4000 A/cm^2, (77 K, 0 T) were prepared using the spray deposition technique and melt-textured growth processing. The influence was studied of the metal atoms ratio Y:Ba:Cu in the initial sol,utions on the structure and properties of the films obtained. The effect is discussed of the different substrates and of the buffer layer between the substrate and the YBCO film on the film's properties. The influence is also followed of the film thickness on its superconducting characteristics. The existence is demonstrated of an optimal thickness corresponding to a maximal value of J_c.

Introduction

Several methods, such as screen printing [1], spin coating [2], and chemical spray pyrolysis [3] were applied for the preparation of HTSC materials. The main disadvantage of the films prepared by chemical methods is the low critical current density. Recently, some interesting experimental results, considering the so-called melt-textured growth (MTG) were published [4, 5]. YBCO superconducting films with J_c = 4800 A/cm^2 (77 K, 0 *T*) were prepared by this process [5].

The main feature of the MTG technology proposed by Ban et al. [4, 5] for the preparation of thick superconducting YBCO films is their fast heating (2 min) up to 1020 - 1030 ^{0}C in argon. We consider that further studies of the MTG process applied to the preparation of thick HTSC films will lead to the improvement of the film properties.

The first aim of the present work was to reproduce and, possibly, exceed the values of T_c and J_c reported by Ban et al. [4, 5]. Secondly, we tried to clarify the influence of the various preparation parameters on the phase formation and film microstructure.

Experimental Procedure

The YBCO films were prepared following the technology described in detail elsewhere [4, 5]. We selected the temperature regime, proposed by Ban et al., that ensures the highest J_c values. Furthermore, some additional procedures and studies were performed differing from those reported in [5]. The raw materials ($Y(NO_3)_3.5H_2O$, $Ba(NO_3)_2$, and $Cu(NO_3)_2.3H_2O$, minimum purity 99.5%) were used to prepare three separate solutions with the Y:Ba:Cu ratios 1:2:3, 1:3:3, and 1:4:3. Distilled water was used as a solvent. In order to avoid the barium nitrate precipitation, 20 vol.% of ethyl alcohol were added to all solutions. The solutions prepared were deposited on single-crystal substrates of YSZ, $SrTiO_3$, MgO, sapphire, GGG, and YIG. The preliminary treatment of the substrates involved mechanical and chemical polishing, followed by heating at 400 ^{0}C for 10 min in oxygen performed immediately before the film deposition. Some of the YSZ, MgO and $SrTiO_3$ substrates were (100) oriented. The different solutions were sprayed on the substrates using helium as a carrier gas. The substrates temperature was kept between 240 and 280 ^{0}C and was controlled by means of a thermocouple. The next step in the films production consisted in a 30-min heating in air at 500 ^{o}C in order to pyrolyze the nitrates. The rapid heat-treatment procedure was carried out as described in [5].

Following the procedure just described, three series of samples were prepared using the three different initial solutions (Y:Ba:Cu ratio of 1:2:3, 1:3:3, and 1:4:3). Layers with four different thicknesses (approximately 1, 2, 3 and 15 μm) were made with each of the solutions. The thickness was determined using a surface profilemeter. Their values were employed for evaluation of the critical current density J_c. The critical current I_c and the critical temperature T_c were measured by means of the standard d.c. four-probe technique at zero magnetic field.

The phase and structural analysis of the samples were performed using X-ray diffraction (XRD), electron probe microanalysis (EPMA) and scanning electron microscopy (SEM).

Experimental Results and Discussion

The investigation of the effect of the different Y:Ba:Cu ratios in the initial solution on the superconducting characteristics of the films prepared was prompted by our earlier SEM studies of the substrate-sprayed film interface. The high temperature (1020 ^{0}C) used in the MTG technology for a short time causes interaction between the substrate and the film deposited.

Table 1 Superconducting properties of the melt textured YBCO films (* denotes 100-oriented substrates).

Substrate	Film thickness [μm]	T_c^0 [K]	J_c(77 K, 0 T) [A/cm^2]
YSZ*	0.85	87.6	1500
YSZ*	1.5	87.6	4000
YSZ*	3	88.5	2500
YSZ*	15	86	100
YSZ	0.8	86	200
YSZ	2	86	300
YSZ	3	86.5	450
YSZ	15	86	10
MgO*	3	79.6	10
$SrTiO_3$	3	---	---
Sapphire	3	---	---
GGG	3	---	---
YIG	3	---	---

The XRD analysis demonstrated that the compounds formed at the interface with the different substrates contain barium, which was observed also in other works [6-10]. This means that a deviation from the stoichiometric 1:2:3 ratio takes place in this part of the film due to barium depletion. Thus, besides the 1:2:3 ratio, we chose the solutions with ratio 1:3:3 and 1:4:3. Unfortunately, the YBCO films formed by deposition of 1:3:3 and 1:4:3 on every type of the substrates were not superconducting (Table 1). In the case of 1:2:3 films, the buffer layer formed at the interface between film and substrate probably stimulates the growth of grains oriented predominantly along the c axis. This result is illustrated in Fig.1, where the SEM photograph shows the surface of the MTG layer with 1:2:3 composition sprayed on a (100)-oriented YSZ substrate.

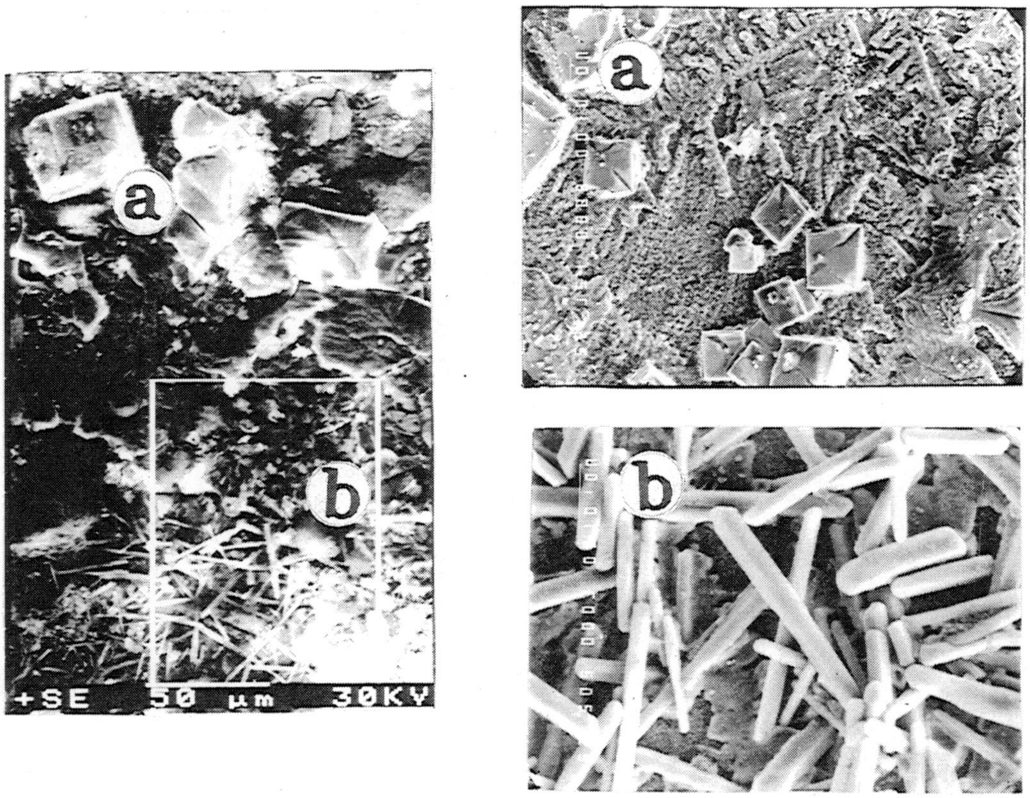

Fig.1. SEM photograph of a YBCO (1:2:3) layer on a YSZ (100) substrate, prepared by MTG spray process.

Two types of structures - plate-like and needle-like - can be observed. Fig.1.a and Fig.1.b show these structures with greater magnification.

The EPMA analysis revealed the elements ratio in the plates as 1:2:3, while in the needles as 2:1:1. Chemical etching of the two structures uncovered a $BaZrO_3$ layer. The samples of this type exhibited the maximal critical current density, which in our case was J_c = 4000 A/cm^2 (77 K, 0 T). The superconducting transition temperature for these films was T_c = 87.6 - 90.3 K (ΔT_c = 2.7 K). We consider that the formation of a plate-like structure, as well the presence of the 2:1:1 phase in the layer, are the main causes for the high J_c values. Perhaps, other effects is connected with properties of the grain boundaries, or some improvement of junctions with Josephson-like properties, also play a certain role.

This assumption was confirmed by the data obtained for MTG films sprayed on unoriented YSZ substrates. Fig.2 shows a SEM photograph of the surface of such a film.

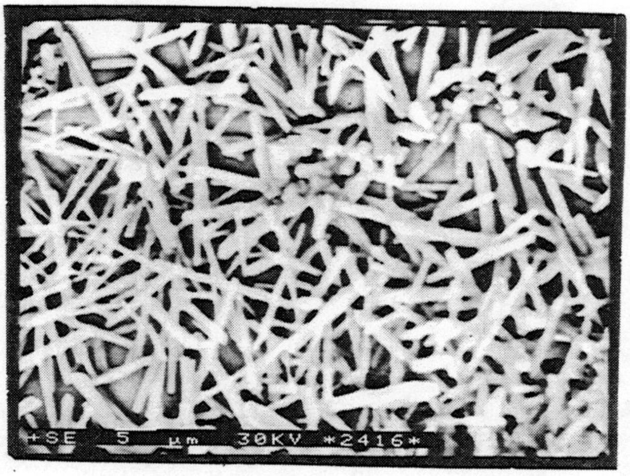

Fig.2. SEM photograph of a [1:2:3] MTG film sprayed on an unoriented YSZ substrate.

This film's thickness was twice as large (3 μm) as that of the film on an oriented substrate presented in Fig.1. One can easily distinguish c-axis oriented needle-like grains, while no other structures are present. The EPMA analysis showed a 1:2:3 ratio of the elements in these grains. We should emphasize here that the similar shape and the different content of the grains in the films in Fig.1.b and Fig.2 are due to the identical orthorhombic structure of the grown 1:2:3 and 2:1:1 phases. However, the absence of a plate-like structure and of a 2:1:1 phase in the thicker films sprayed on unoriented YSZ substrates leads to values of the critical current density lower by almost one order of magnitude: the maximal value was in this case J_c = 450 A/cm^2, while the transition temperature was T_c =86.5 - 89.5 K, in spite of the identical conditions for preparation of all films.

The films produced following the same MTG-technology on the other types of substrates ($SrTiO_3$, sapphire, GGG, and YIG) did not exhibit superconductivity for any of the 1:2:3, 1:3:3, or 1:4:3 solutions. Superconducting films (T_c = 79.6 - 82.7 K, J_c = 10 A/cm^2) were only obtained when a 1:2:3 solution was sprayed on (100)-oriented MgO substrates.

Chemical etching of the films on MgO substrates did not reveal the presence of a buffer layer. This fact, in our opinion, is the reason for its worse superconducting characteristics (T_c and J_c). As asserted by several authors - [11,12], the Mg ions diffuse into the YBCO, while, on the other hand, Cu diffuses into the substrate, which disrupts the film's stoichiometry. When a buffer layer is present, these processes can be eliminated, or at least brought to a minimum.

The necessity of the presence of a buffer layer suitable for the growth of oriented grains of orthorhombic 1:2:3 YBCO phase has been recently widely discussed [10,13]. The $BaZrO_3$ compound, formed as a buffer layer on YSZ substrates, has a lattice constant (a = 4.185 Å) close to that of the YBCO phase; the absence of such a layer (the case of MgO substrates), or the presence of one with unsuitable properties ($SrTiO_3$ substrates), result in worse superconducting characteristics, or in a complete lack of superconductivity. This assumption is in good agreement with our results presented in Table 1, as well as with those of other authors, who did not observe superconductivity in MTG films sprayed on $SrTiO_3$ [4,5].

Fig.3 presents (in log-log scale) the results of our further study concerning the influence of the thickness of the sprayed MTG films on their superconducting characteristics T_c and J_c.

The experiments were carried out on films deposited on YSZ substrates only, since, as shown above, only they possessed relatively high T_c and J_c. Our data demonstrated that the increase of the film thickness affects slightly the critical temperature T_c, while, in contrast, an optimal thickness was observed for the J_c above which its value dropped sharply. The SEM analysis disclosed that above this thickness both the number of intergranullar defects and of microcracks increase, which strongly influences J_c. Obviously, the behavior of J_c should not be only considered as being the results of the film's texture, but other factors should also be taken into account.

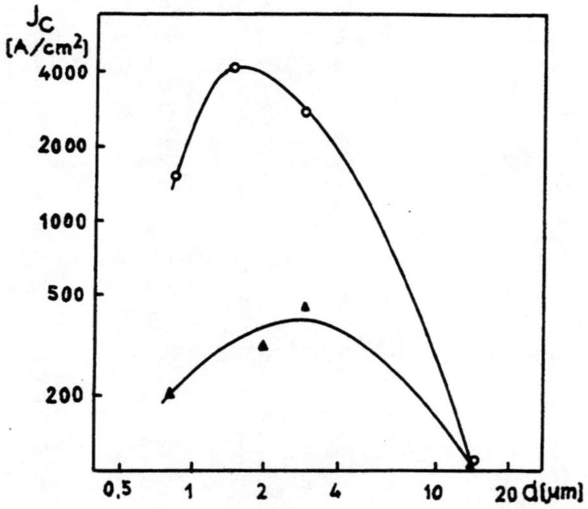

Fig.3. Critical current density J_c (77 K, 0 T) of films formed on YSZ (100)-oriented (o) and unoriented (Δ) substrates as a function of the film thickness.

Conclusion

The increase of the Ba content in the initial solution did not improve the YBCO thick films superconducting properties. Departure of the initial solution composition from the 1:2:3 ratio led to to disappearance of superconductivity in 1:3:3 and 1:4:3 MTG YBCO films deposited by spray pyrolysis.

The (100)-oriented YSZ was chosen as the best substrate for melt-textured film growth. For this substrate we obtained the maximal J_c (about 4000 A/cm^2) confirmed by the resultsreported elsewhere [4, 5].

We also established the existence of an optimal thickness for spray deposited MTG films, above which a decreasing of J_c is seen; this is related to the increased number of microcracks and other defects in the film after this thickness is exceeded. Determining its precise value thickness requires a more detailed study of the parameters influencing it.

References

1. H. Koinuma, T. Hashimoto, T. Nakamura, K. Kishi, K. Kitazawa and K. Fueki, *Jpn. J. Appl. Phys.*, **26** (1987), L 761
2. P. May, D. Jedamzik, W. Boyle and P. Miller, *Supercond. Sci. Technol.*, **1** (1988) 1
3. M. Kawai, T. Kawai, H. Masuhira and M. Takahashi, *Jpn. J. Appl. Phys.*, **26** (1987), L 1740
4. E. Ban, Y Matsuoka and H. Ogawa, *J. Appl. Phys.*, **67** (1990), 4367
5. E. Ban, Y. Matsuoka, H. Ogawa and K. Kurosawa, *Journal of Alloys and Compounds*, **187** (1992), 193-205 JALCOM 302
6. W.P.T. Derks, H.A.M. van Hal and C. Langereis, *Physica C*, **156** (1988), 62
7. M.J. Cima, J.S. Schneider, S.C. Peterson and W. Coblenz, *Appl. Phys. Lett.*, **53** (1989), 710
8. J. Tabuchi and K. Utsumi, *Appl. Phys. Lett.*, **53** (1988) 606
9. A. Bailey, G. Alvarez, G.J. Russel and K.N.R. Taylor, *Cryogenics*, **30** (1990) 599
10. Junji Kabuchi and Kazuaki Utsumi, *Appl. Phys. Lett.* **53**, No7 (1988)
11. S.J. Golden, H. Isotalo, M. Lanham, J. Mayer, F.F. Lauge and M. Ruhle, *J. Mater. Res.*, **5** (1990) 1605
12. T. Teruyama and T. Itoh, *J. Electrochem. Soc.*, **137** (1990) 336
13. D.K. Fork, S.M.Garrison, Marilyn Hawley and T.H. Geballe, *J. Matter. Res.*, **7**, No 7 (1990)

Vacuum Electron and Ion Technologies
D. S. Karpuzov (Ed.)

ISBN 1-56072-242-8

LOW DAMAGE DRY ETCHING OF $YBa_2Cu_3O_{7-\delta}$ SUPERCONDUCTING NANOMETER STRUCTURES AT CRYOGENIC TEMPERATURES

R. Barth, B. Spangenberg and H. Kurz
Institut für Halbleitertechnik II, Rheinisch-Westfälische Technische Hochschule (RWTH) Aachen, Sommerfeldstr. 24, D-52074 Aachen, Germany

We report a systematic study about the influence of substrate temperature during dry etching on the superconducting properties in $YBa_2Cu_3O_{7-\delta}$ nanobridge structures. The substrate temperature has been varied between 300 K and 20 K. High energy electron-beam lithography has been used to fabricate submicron bridges between large area contact pads. The linewidth has been varied from 1.5 µm to 200 nm whereas the length has been kept constant at 10 µm. It is shown that LN_2 cooling of the substrate is an efficient method to minimize ion beam induced degradation in $YBa_2Cu_3O_{7-\delta}$ nanostructures. Compared to 300 K substrate temperature, an increase in the critical current density j_c at 77 K by an order of magnitude is reported for 400 nm wide structures. As a result of further reduced substrate temperature (20 K) during the dry etching process, 200 nm wide $YBa_2Cu_3O_{7-\delta}$ structures have been fabricated with 64.5 K zero resistance temperature.

1. INTRODUCTION

There is much interest in the application of high-temperature-superconductor (HTSC) thin films in active and passive devices. For many devices such as flux-flow-transistors, SQUIDs or Dayem bridges, it is necessary to fabricate submicron structures. Several different approaches have been reported for submicron patterning of $YBa_2Cu_3O_{7-\delta}$ (YBCO) thin films. Inhibit patterning [1, 2], wet etching [3] and dry etching methods in combination with electron-beam lithography (EBL) [4, 5, 7] are reported. However, it is a general problem that properties of superconducting HTSC submicron structures degrade drastically when their width is reduced [2, 4, 5].

Especially for submicron patterning of HTSC thin films, we developed a modified three-level resist system which ensures minimal degradation of superconducting properties during the lithographic process steps [4]. The topic of this work is to investigate the influence of the dry etching parameters on the superconducting properties in the submicron range.

One reason for deterioration of superconductivity might be damage induced in the films during dry etching. Ion bombardment locally elevates the film temperature, resulting in an increase in defects and dislocations. Superconductivity appears to be extremely sensitive to defects such as oxygen deficiencies [6]. In a previous work, we identified oxygen effusion as the main reason for degradation of superconductivity [7]. Oxygen effusion is a temperature-dependent process and thus can be suppressed by lowering the substrate temperature. As a matter of fact we demonstrate that efficient substrate cooling during the dry etching process is a successful method to prevent the degradation of superconductivity in YBCO submicron structures.

2. EXPERIMENTAL

The samples used in this study are 70 -100 nm thick highly epitaxial c-axis-oriented YBCO films deposited on [100] $SrTiO_3$ substrates, either by off-axis laser ablation [8] or DC magnetron sputtering [9]. The transition temperatures are typically $T_c > 88$ K and critical current densities at 77 K $j_c(77K) > 10^6$ A/cm^2.

In order to reduce the contact resistance and the resulting noise during electrical characterization, large ohmic contact regions were defined by shadow mask evaporation technique of 100 nm Au onto the as grown YBCO film.

Subsequently, a 25 nm Pt protection layer is deposited on the YBCO film followed by a 50 nm Ti intermediate layer. A photoresist image layer is used for the fabrication of the submicron structures.

First, optical lithography is applied to define coarse contacts to the bridge regions with four-terminal layout. The photoresist structure is transferred into the intermediate layer by BCl_3

reactive ion etching (RIE). After ultrasonic assisted resist stripping in acetone, a 90 nm thick polymethylmethacrylate (PMMA 950k) electron-beam resist is spin coated onto the substrate. High energy electron-beam-lithography is used to define the submicron bridge structures. The resist pattern is transferred into the Ti-layer via RIE again. All mentioned lithographic process steps are described in more details in [10]. In a subsequent Ar/O_2 ion beam etching (IBE) step the Ti-pattern is transferred into the Pt-layer and the HTSC thin film. The Ti-mask is used as an extremely resistant mask due to the formation of a thin TiO_x layer.

The IBE process is performed in a 3" Veeco microetch sputter system. The system consists of a main chamber fitted with a 3" Kaufman-type ion source and a diffusion pump (3000 l/sec) backed with a mechanical pump. Typical base pressures are in the range of 3×10^{-7} Torr. The Kaufman source delivers an highly homogeneous ion beam with variable energies between 0.4 and 2.0 keV. Beam neutralization is performed by a hot filament. The typical ion beam energy and current density in this study are adjusted to 500 eV and 0.35 mA/cm^2, respectively. This results in the relative low sputter rate of approximately 5-7 nm/min. Ion beam energy has been reduced to minimum to minimize ion beam induced defects in the HTSC submicron structures.

To perform the dry etching study of the HTSC thin films in the wide temperature range from 300 K down to 20 K, the substrate holder of the sputter system has been modified. The original water-cooled substrate holder with rotation and tilting options has been exchanged by a continuos flow cryostat, mounted in the centre of the ion beam. The distance between substrate and ion source remains unchanged.

3. RESULTS AND DISCUSSION

A typical submicron bridge structure (w = 200 nm, l = 10 µm) fabricated by the above mentioned technology steps is shown in Fig.1. The constant width of the bridge and the steepness of the sidewalls clearly demonstrate the maturity of the lithographic process.

For efficient substrate cooling, low thermal contact resistance is essential. Best thermal contact between sample and cryostat has been achieved by substrate mounting with silver paint. Ion bombardment and thermal radiation of the hot filament results in a local elevation of the HTSC film temperature. Therefore, the efficiency of substrate cooling has been checked by *in situ* measurements of the HTSC film temperature during the IBE process. For this purpose, a $SrTiO_3$ substrate with 10 µm x 1 µm YBCO bridge structures has been mounted on the cryostat with silver paint. One bridge has been electrically connected and its resistance versus temperature has been measured *in situ*. The sample has been cooled down and stabilized at 80 K, well below the resistive transition of the bridge (T_c = 86 K). After exposing the substrate to the ion beam, the substrate temperature starts to rise. When the critical temperature of the bridge is reached, the resistance increases. After 5 min ion bombardment, thermal equilibrium is reached and the resistance stops rising. By comparing this equilibrium value with the measured R(T) curve one can determine the actual film temperature and thus the IBE-induced temperature elevation. With the above mentioned ion beam parameters the measured local temperature did not rise more than 10 K.

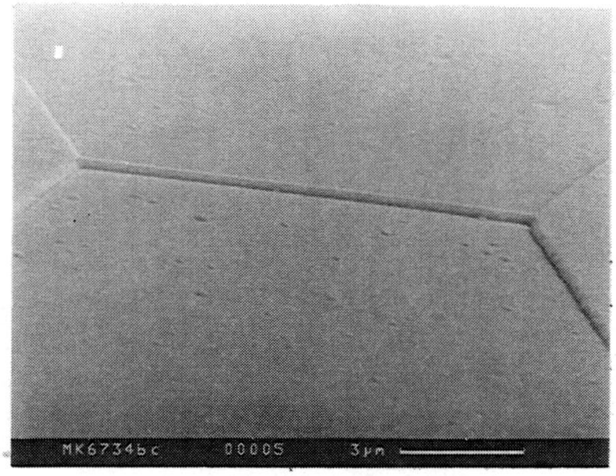

Figure 1. Typical $YBa_2Cu_3O_{7-\delta}$ bridge fabricated with Ar/O_2 ion beam etching at 20 K substrate temperature (length l = 10 µm, width w = 200 nm).

In order to avoid numerous tests to evaluate the temperature dependence of the sputter rate for the calculation of the individual sputter time a simple method for endpoint detection is essential. Additionally the film thickness is often uncertain within a few percents. Endpoint detection with mass spectrometry is too involved. Alternatively it can be realized quite simple during IBE of HTC films on isolating substrates by an *in situ* measurement of the film resistance.

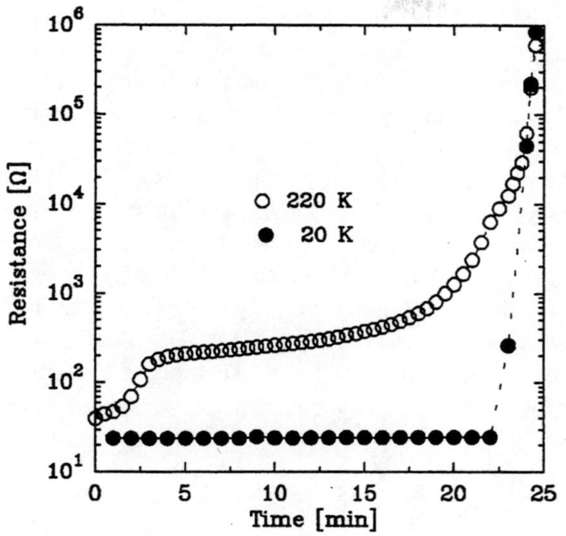

Figure 2. Endpoint detection during ion beam etching by *in situ* measurement of the film resistance. The abrupt increase of the resistance permits precise endpoint detection.

To achieve stable electrical contacts within the full temperature range (300 K - 20 K), Au wire is bonded onto the sample. For resistance measurements the Au wire is connected to a Keithley 196 System via electrical feedthroughs.

Figure 2 shows the film resistance versus sputter time for two different substrate temperatures. At 20 K the HTSC film is in the superconducting state during the IBE process. The resistance remains constant at the value of the contact resistance with increasing time. After 22 min sputter time the remaining HTSC material turns into the normal state and the following abrupt increase of the resistance determines the endpoint. For temperatures $T > T_c$ the parallel resistance of the 25 nm Pt-layer and the HTSC film is measured. As an example the R(t) curve at 220 K is shown. A first increase of the resistance can be observed when the Pt/HTSC interface is reached. In the following, the resistance is determined by the constant decrease of the film thickness due to IBE. The endpoint is detected by the abrupt increase of the resistance.

The influence of substrate temperature during dry etching is studied by comparing the resistive transition and critical current density of 10 µm long YBCO bridges for widths (w) ranging from 0.4 to 1.5 µm. Figure 3 shows the temperature dependence of the resistance of bridges etched at different temperatures. The curves are normalized to the resistance at T = 100 K

The results in the upper part of Fig. 3 are obtained for an effective substrate temperature of 300 K, the lower part of Fig. 3 refers to LN_2-temperature. In the case of 300 K substrate temperature, Fig. 3 shows the typical broadening of the resistive transition for narrow bridges. The zero-resistance temperature $T_c(0\ \Omega)$ is reduced to 80, 73 and 70 K for widths of 1.5, 0.6 and 0.4 µm respectively. The midpoint of the resistive transition (T_c^{mid}) decreases from 86 K for 1.5 µm wide bridges to 83.8 K for 0.4 µm wide lines, T_c^{mid} hence varies by 2.2 K for different values of w. With LN_2 substrate cooling, the results are quite different. The resistive transitions of bridges with different w overlap very well. There is a negligible difference in the midpoint of the resistive transition (0.07 K). On the other hand the zero resistance temperature is influenced by a residual pedestal. Therefore $T_c(0\ \Omega)$ decreases form 81 to 78,5 and finally 77 K for widths of 1.5, 0.6 and 0.4 µm. Smaller structures are nonsuperconducting. A further reduction of sputter temperature to 20 K results in superconducting 10 µm long and 200 nm wide YBCO structures with a zero resistance temperature of $T_c(0\ \Omega)$ = 64.5 K.

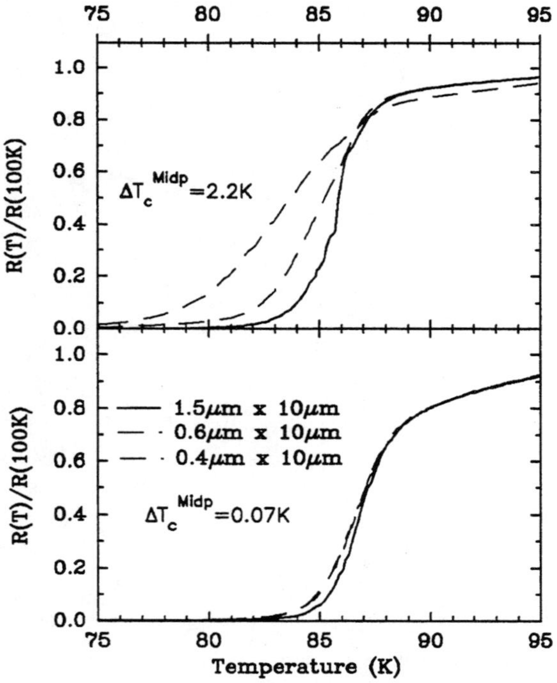

Figure 3. Influence of substrate temperature during IBE on the resistive transition of 10 µm long YBCO bridges. The curves are normalized to the resistance at 100 K. Upper part: water cooled substrate (300 K), lower part: LN_2 cooled substrate (77 K)

We suppose that the monotonously reduced zero resistance temperature, which is displayed in the resistive pedestals, results form local fluctuations of the superconducting properties. T_c as well as j_c are determined by that part of the bridge with the lowest transition temperature. Local variations of superconductivity may originate in two different ways. First, sidewall damage induced by IBE may create defects like oxygen effusion and dislocations, migrating across the bridges. In the case of 20 K sputter temperature they are reduced to a minimum, resulting in highest transition temperature in 200 nm wide, 10 μm long YBCO bridges. Second, growth related defects on submicrometer scale may dominate superconductivity in long (10 μm) submicron bridge structures. The later defects are independent on the substrate temperature during IBE, so they are responsible for the pedestals to a certain degree.

For technical application of YBCO submicron structures critical current density at 77 K is of interest. Thus Fig. 4 displays the dependence of j_c(77 K) of 10 μm long YBCO bridges on line width, achieved at different substrate temperatures during dry etching.

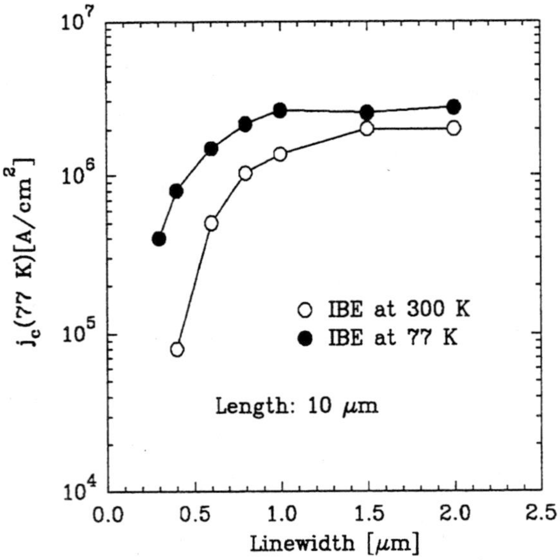

Figure 4. Influence of the substrate temperature during dry etching on the critical current density at 77 K of 10 μm long YBCO bridges for various linewidth.

In the case of 300 K, a dramatic decrease of j_c is observed for bridges narrower than 1.5 μm. For w = 0.4 μm only $4x10^4$ A/cm^2 can be reported. According to the pedestals in the resistive transition, 10 μm long lines with w < 0.4 μm do not become completely superconducting above 77 K. As the sputter temperature is gradually reduced, significant improvements in j_c(77 K) can be observed. For sputter temperatures of 300, 150 and 77 K, j_c(77 K) in 0.4 μm wide lines increases monotonously from $4x10^4$ A/cm^2 to $5x10^5$ A/cm^2 and $8x10^5$ A/cm^2 respectively. Additionally for 300 nm wide 10 μm long lines j_c(77 K) can be reported to $4x10^5$ A/cm^2. A further reduction of the temperature during the IBE process does not influence on the critical current density any more.

4. CONCLUSION

We performed dry etching studies of YBCO submicron structures over the wide temperature range from 300 K to 20 K. It has been demonstrated that the superconducting properties of YBCO submicron structures critically depend on the substrate temperature during dry etching. Ion beam induced defects such as oxygen effusion have been reduced by LN_2 substrate cooling during dry etching. Critical current density at 77 K has been increased by more than an order of magnitude by reducing the substrate temperature during IBE from 300 to 77 K.

ACKNOWLEDGEMENT

We thank B. Holzapfel, M. Jansen and J. Auge for the excellent YBCO-film preparation, B. Hadam for careful electron beam lithography and H.G. Roskos for useful discussions and careful correction of the manuscript. The authors would like to appreciate the financial support by Veeco Instruments (A. Dorotik) for the modification of the substrate holder. This work was supported by the BMFT under contract No. TK 0335.

REFERENCES

1. C.A. Copetti, U. Gassig, W. Zander, J. Schubert, Ch. Buchal, Appl. Phys. Lett. **61**, 3041 (1992).

2. D.P. Kern, K.Y. Lee, R.B. Laibowitz, A. Gupta, J.Vac.Sci.Technol. B **9**, 2875 (1991).

3. J.R. Wendt, J.S. Martens, C.I.H. Ashby, T.A. Plut, V.M. Hietala, C.P. Tigges, D.S. Ginely, M.P. Siegal, J.M. Phillips, G.K:G: Hohenwarter, Appl. Phsy. Lett. **61**, 1597 (1992).

4. R. Barth, B. Spangenberg, W. Langheinrich, W. Albrecht, H. Kurz, G. Gieres, B. Holzapfel, J.Vac.Sci.Technol. A **10**, 3411 (1992).

5. H. Assink, A.J.M.v.d. Harg, C.M. Schep, N.Y. Chen, D.v.d. Marel, P. Hadeley, E.W.J.M.v.d. Drift, J.E. Mooij, Appl. Supercond. Conf. (1992), Proceedings.

6. J.R. Cooper, S.D. Obertelli, A. Carrington, J.W. Loram, Phsy.Rev.B, **44**, 21, 12086 (1991).

7. R. Barth, B. Spangenberg, C. Jaekel, H.G. Roskos, H. Kurz, B. Holzapfel, Appl. Phys. Lett **63**, 1148 (1993).

8. B. Holzapfel, B. Roas, L. Schultz, P. Bauer, G. Saemann-Ischenko, Appl. Phy. Lett. **61**, 3178 (1992).

9. J. Auge, M. Jansen, H.G. Roskos, H. Kurz, EUCAS' 93 (1993), Proceedings.

10. W. Langheinrich, B. Spangenberg, R. Barth, H. Kurz, Mircroelectron. Eng. **21** 479 (1993).

Vacuum Electron and Ion Technologies
D. S. Karpuzov (Ed.)

ISBN 1-56072-242-8

MORPHOLOGICAL AND COMPOSITIONAL CHANGES OF TARGET SURFACE DURING RF MAGNETRON SPUTTERING OF Y-Ba-Cu-O SYSTEM

R.A. CHAKALOV AND M.V. ABRASHE*
Institute of Electronics, Bulgarian Academy of Sciences, 72, Tzarigradsko chaussee, Sofia, 1784, Bulgaria
*Faculty of Physics, Sofia University, 1126, Sofia, Bulgaria

A B S T R A C T

Morphological and compositional changes of target surface during RF magnetron sputtering of an off-stoichiometric Y-Ba-Cu-O (Y:Ba:Cu = 1:2.6:5) ceramic target have been investigated by optical microscopy, Raman spectroscopy and Energy dispersive analysis by X-rays (EDAX). It has been established that target material contains stoichiometric $YBa_2Cu_3O_{7-\delta}$ microcrystals and cuprum oxide as intermedium. Nearly the same morphology and composition retain during sputtering in the region of erosion zone.

I. I N T R O D U C T I O N

The fabrication of high quality Y-Ba-Cu-O thin films by any deposition techniques requires growth conditions providing correct metallic stoichiometry Y:Ba:Cu = 1:2:3. Sputtering is a deposition method where the film composition is normally equal to the target composition and films prepared by sputtering of single stoichiometric target $YBa_2Cu_3O_{7-\delta}$ are expected to be stoichiometric. But sputter deposition of high-T_c superconducting thin films is complicated by several effects and strong deviations can be observed [1].

Generally, films deposited from a single stoichiometric (Y:Ba:Cu = 1:2:3) target are Ba- and Cu- deficient. Compensated composition targets are often used in order to obtain stoichiometric films (for a review see *Leskela et al.*[2]). If the cation ratio of the target is not 1:2:3, insulating phases are present and the DC sputtering rate from insulating phases is very low. As a consequence, for DC sputtering the film composition can not be adjusted by changing the target composition. By using RF sputtering the problems mentioned can be avoided. In fact, films have been deposited from mechanical mixtures of oxides [3], but they are not good thermal conductors and the heating may cause segregation and bad reproducibility. Although sintered targets are prefered the peculiarities of off-stoichiometric target sputtering are not investigated in details.

In this paper we report the morphology and phase composition on the surface of an off-stoichiometric $YBa_{2.6}Cu_5O_x$ target and their changes observed during RF magnetron sputtering.

II. EXPERIMENTAL

A ceramic target was prepared by mixing, pressing and sintering of fine powders in appropriate amounts in order to achieve a metalic cation ratio of Y : Ba : Cu = 1 : 2.6 : 5. The target in the form of a disc of 51 mm diameter and about 4 mm in thickness was mounted in a planar magnetron sputtering assembly. RF power (13.56 MHz, 50 W) was applied. The off-stoichiometric target was sputtered in argon/oxygen mixture (10:1 respectively) at total pressure of 1 Pa. The planar magnetron construction formed a spatial configuration of the plasma with enhanced sputtering yield over an annular zone on target surface (erosion zone).

Preparation of high-T_c superconducting thin films "in-situ" requires the cooling down of the substrate after deposition to be carried out at high oxygen pressure ($\leq$ 1 atm). It is not clear whether interactions between oxygen atoms and target material take place. The target surface is still hot at this technological stage because the substrate heater is not switched off. That is why we investigated the morphological and compositional changes after several deposition runs but not after long time target sputtering. The procedure of one deposition run was as follows: the vacuum chamber was evacuated to a base pressure of 3.10^{-3} Pa, the substrate was heated up to 700°C for 10 min, the chamber was back-filled with argon/oxygen gas mixture and RF power was applied, the target was presputtered (with shutter) for 10 minutes and sputtered for 60 minutes, the RF power was switched off and oxygen gas was introduced into the chamber, the substrate temperature was decreased to room temperature by a rate of 10°C/min i.e. during 60-70 min.

The morphology of the target surface was investigated by optical microscopy. Raman spectroscopy and Energy dispersive analysis by X-rays (EDAX) were employed to determine the presence of different phases. The micro-Raman spectra were measured using a multichannel triple spectrometer (Microdil-28) equipped with a microscope (λ_L = 514.5 nm, 1 μm diameter of the spot).

III. RESULTS

The observation of the target after preparation under optical microscope showed rough but relatively homogeneous surface (Fig.1a). After several deposition runs the target was again examined. Fig.1b shows the surface morphology in the erosion zone. Microcrystals can be clearly distinguished. To answer the question whether they were formed under sputtering conditions on the surface we investigated the target material in the depth. Polishing revealed a surface consisting of straggling large (20-40 μm) plate-like and

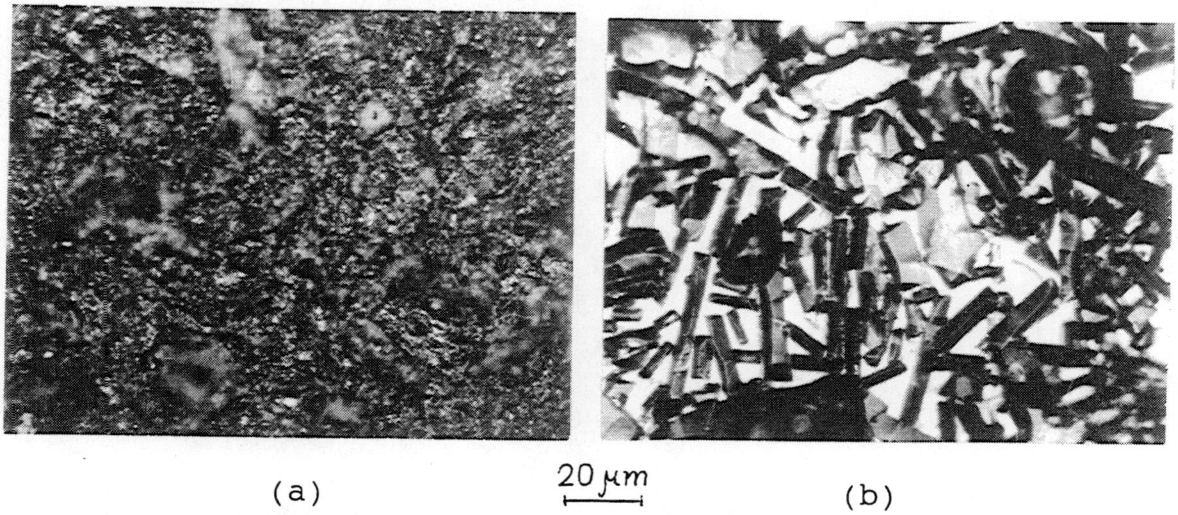

Fig.1. Surface morphology of $YBa_{2.6}Cu_5O_x$ target (a) after preparation and (b) in the erosion zone after several deposition runs.

needle-like dark areas and bright space between them. This picture looked like cross-section view of the morphology shown in Fig.1b.

The EDX analysis indicated that the material of the white fields observed both in the depth and on the sputtered surface, is CuO. The compositions of the dark rectangles and the microcrystals were also identical, respectively 1 : 2.47 : 3.22 and 1 : 2.57 : 3.24 . Some excess of Ba and Cu was detected but no areas with additional phases like $BaCuO_2$ were identified.

The Raman spectra of a dark area on the polished surface and of a microcrystal on the sputtered surface are shown in Fig.2a and Fig.2b, respectively. In both cases all characteristic for $YBa_2Cu_3O_{7-\delta}$ Raman-active A_g modes [4] were observed. From the polished target a perfectly oriented zx-surface of the microcrystal was investigated and $y(xx)\bar{y}$ and $y(xx)\bar{y}$ (using Porto notations) polarized Raman spectra were obtained (Fig.2a). The frequency of the apex oxygen (O4) mode was 502 cm^{-1} and the modes of Cu2 (150 cm^{-1}) and O2-O3 in-face (441 cm^{-1}) vibrations appeared in zz-polarization only. All microcrystals in the erosion zone had tilted faces only, as one can see in Fig.1b. The Raman spectra from their surfaces were with unknown polarization. Analysing the features of the spectra shown in Fig.2a and Fig.2b we conclude that such polarization behaviour is typical for $YBa_2Cu_3O_{7-\delta}$ and oxygen content in both cases is close to 7, i.e. the type of crystal lattice is orthorhombic.

IV. DISCUSSION AND CONCLUSION

By Raman spectroscopy and EDAX it was established that off-stochiometric sintered $YBa_{2.6}Cu_5O_x$ target contains stoichiometric $YBa_2Cu_3O_{7-\delta}$ microcrystals and cuprum oxide as intermedium. Since no additional phase was detected it can be assumed that the excess of Ba is concentrated near the microcrystal walls.

It is well known that the flux of sputtered atoms should precisely reflect the composition of the target only in steady state i.e. when an altered surface layer depleted

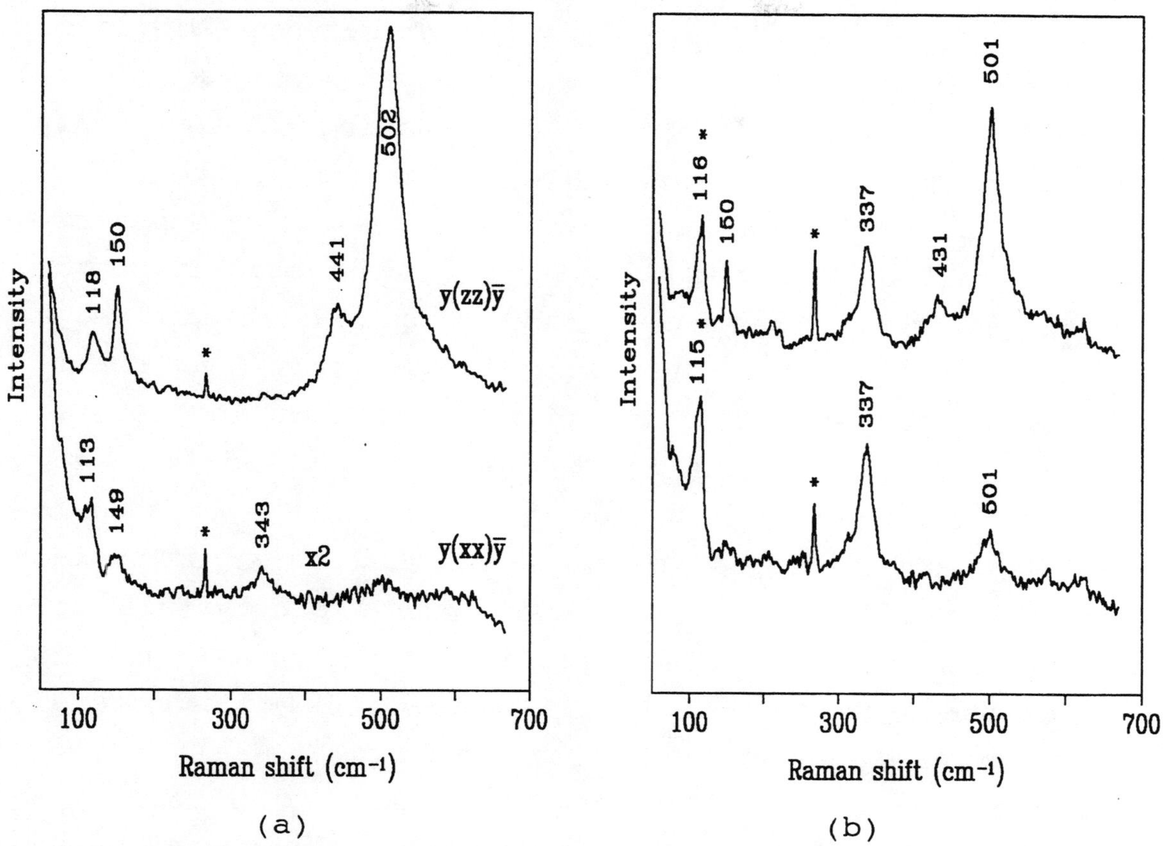

Fig.2. Polarized Raman spectra of $YBa_2Cu_3O_{7-\delta}$ ($\delta \approx 0$) microcrystals in the target: (a) from internal polished surface with $y(zz)\bar{y}$ and $y(xx)\bar{y}$ polarisations; (b) from the erosion zone with two unknown polarizations in two perpendicular directions from tilted face of a microcrystal. The asterisks denote the Ar^+ plasma lines.

in the fast sputtering component/element has developed. *A.M.Kadin et al.* [5] have noted a period of about 10 hours or more to establish steady state when sputtering of stoichiometric $YBa_2Cu_3O_{7-\delta}$ target under nearly the same sputtering conditions. The authors have estimated that such time corresponds to ≈ 10 μm of target material sputtered.

In the case of sintered off-stoichiometric target sputtering it is unlikely that a constant altered surface layer forms. In our case two multi-elemental components are present: $YBa_2Cu_3O_{7-\delta}$ (more than 60 % of the whole area) and CuO. The morphological and compositional changes can be divided into two groups: changes in the multi-elemental components and changes in the multi-component target as a whole. The investigations allow us to estimate the order of importance of the changes. In the erosion zone the microcrystals remained stoichiometric having changed only their form to polyhedrons because of preferable sputtering of the edges. So variations in the multi-elemental components were not detected. The reason could be find if the changes of whole target surface are supposed to be faster and more essential. Really, the $YBa_2Cu_3O_{7-\delta}$ microcrystals are small. Some of them can be sputtered entirely during one or several deposition runs. At the same time unexposed to ion bombardment microcrystals appear from the bulk. It can be assumed from the pictures in Fig.1 that CuO is RF-sputtered more easily than $YBa_2Cu_3O_{7-\delta}$. However, the microcrystals are randomly placed and a part of situated between them CuO is always exposed. So we believe that the flux of sputtered atoms is copper-rich but various in time.

In conclusion, the RF magnetron sputtering of sintered off-stoichiometric $YBa_{2.6}Cu_5O_x$ target have to be considered as sputtering of a two-component surface where each component is multi-elemental: $YBa_2Cu_3O_{7-\delta}$ microcrystals and melt-like CuO.

V. ACKNOWLEGEMENT

We thank E.Dinolova for EDAX and G.Bogachev for technical help. This work was supported by the National Fund for Scientific Investigations under Project F72 and in part by Grant No F1/91 (2001 NIS) of the Bulgarian National Foundation for Science.

REFERENCES

1. T.I.Selinder, G.Larsson, U.Helmersson and S.Rudner, J.Appl. Phys. **69**(1), 390 (1991)

2. M.Leskela, J.K.Truman, C.H.Mueller and P.H.Holloway, J.Vac. Sci.Technol. **A7(6)**, 3147 (1989)

3. S.H.Liou, M.Hong, J.Kwo, B.A.Davidson, H.S.Chen, S.Nakahara, T.Boone and R.J.Felder, Appl.Phys.Lett. **52**, 1735 (1988)

4. R.Feile, Physica C, **159**, 1 (1989)

5. A.M.Kadin, P.H.Ballentine, J.Argana and R.C.Rath, IEEE Trans.Magn. **25(2)**, 2437 (1989)

Vacuum Electron and Ion Technologies
D. S. Karpuzov (Ed.)

ISBN 1-56072-242-8

SUPERCONDUCTING THIN FILMS OF $Y_1Ba_2Cu_3O_{7-X}$ PRODUCED BY ND: YAG LASER ABLATION

Dimitar Valkov[a], Radosting Stoyanov[a], Elena Nazarova[b], Vasil Lovchinov[b] and Rumen Gerasimov[a]

[a]Sofia University, Faculty of Physics, 5 james Bourchier Blvd., 1126 Sofia, Bulgaria
[b] Instutute fo Solid State Physics, 72 Tzarigradsko Chaussee Blvd., 1784 Sofia, Bulgaria

Abstract

Thin films of $Y_1Ba_2Cu_3O_{7-x}$ have been deposited on <100> yttria-stabilized ZrO_2 single crystal substrates by Nd:YAG laser ablation process from $Y_1Ba_2Cu_3O_{7-x}$ superconducting targets. The influence of laser pulse energy density on the superconducting characteristics and microstructure of the thin films has been investigated and discussed.

1. Introduction

Pulsed laser deposition has emerged since late 1987 as the primary means for high throughput deposition of high quality superconducting thin films for research and devices. Generally, excimer lasers were used for the deposition process which yielded thin films with smooth surface, high transition temperatures and critical currents. One alternative of the expensive excimer lasers are the Nd:YAG lasers operating in Q-switched mode at 1064nm. They have some advantages : the initial laser cost and maintenance expenses are lower than for the excimer lasers; ordinary glass optics can be used instead of the quartz optics; the beam from the Nd:YAG laser can be focused sharper than the beam from the excimer laser because of much greater coherence.

The process of pulsed laser deposition of thin films of HTS involves a complicated interplay among the variables of the ablation laser wavelength and pulse energy density delivered to the target, the distance between the target and the substrate, the pressure of the background gas, the temperature of the substrate and all the other parameters characterizing the deposition process.

Intensive theoretical and experimental researches concerning the physics of the excimer laser ablation were carried out. The peculiarities of Nd:YAG laser deposition of thin films of HTS have not been sufficiently studied so far compared to the excimer laser ablation.

In our previous works [1-4] we have presented the results of our investigations concerning the effect of laser pulse duration, the influence of in-situ oxygen heat treatment after deposition, the effect of target composition, the influence of substrate surface polish quality, the role of laser wavelength, the properties of the thin films deposited on different kinds of substrates and the interaction of the plasma species with the substrate surface characterized by the sticking coefficient.

In this work we present a part of our investigations concerning the influence of pulse energy density on the microstructural morphology and superconducting properties of YBCO thin films deposited on $Y:ZrO_2$ substrates.

2. Experimental details

The experimental setup for the deposition of thin films of High Temperature Superconductors (HTS) has been described in details previously [1]. For our experiments we used a Nd:YAG laser operating in Q-switched mode at 1064nm. The duration of the Q-pulse was 30ns. The pulse energy could be varied from 30mJ to 150mJ and produced energy densities of $3 \div 7\ J.cm^{-2}$. The pulse repetition rate was 5-6Hz.

Sintered ceramic targets of stoichiometric composition $Y_1Ba_2Cu_3O_{7-x}$ were employed. All targets were superconducting with a T_c value of 92K and a transition width, ΔT of 2K.

Single $Y:ZrO_2$ crystals 10x10x0.8 mmxmmxmm with <100> orientation were used as substrate materials. The substrate surfaces were ground and polished with successively finer diamond powder down to 0.3μm particle size.

Substrate temperature during laser deposition was maintained at 750°C. The base pressure of the diffusion-pumped vacuum system was 10^{-6}Torr. The partial pressure of the oxygen during the deposition was 100-200 mTorr. Pure oxygen was introduced into the vacuum chamber immediately after the deposition. The samples were kept at 750°C for 2min, after that they were cooled to 470°C in about 6min. The thin films were held at 470°C for 30min in 1 atm of oxygen. The heater was turned off after this treatment and the samples were allowed to cool down to room temperature.

The laser beam was focused onto the target at an angle of incidence of 45° and was scanned across the target at 100μm.s^{-1} during the deposition. The target to substrate distance was 4cm.

The critical temperature of the films was measured using the four-probe technique. The measuring current was 10μA. The morphological, structural and compositional properties of the thin films and the targets were determined using a Philips 515 Scanning Electron Microscope (SEM) and an X-ray diffractometer with a Cu Kα source.

3. Results and discussion

Experiments were carried out with pulse energy densities varying from 3J/cm^2 to 7J/cm^2. The thin films were ablated from $Y_1Ba_2Cu_3O_{7-x}$ superconducting targets.

The SEM micrographs revealed some peculiarities in the morphology of the thin films deposited at high pulse energy densities. In fig.1 is presented a typical micrograph from the surface of a film deposited at 7, 6 or 5J/cm^2. Systems of rod-like features are seen on the top of the film. The samples were observed under optical microscope at transmitted light. Star-shaped systems of "rods" were seen across the whole area of the films.

Some differences appeared between the films deposited at 7, 6 and 5J/cm^2. The length of the "rods" shortened with the decrease of the pulse energy density and the distance between them was filled with material more homogeneously.

In fig.2 is presented a micrograph of a "rod" as it appears on the top of the film. A micrograph from an area between the "rods" is shown in fig.3. It is seen that the film is formed by two-dimensional grains with size of few microns and a random plate-like morphology.

The "rods" disappeared in the films deposited at 4J/cm^2. In fig.4 and fig.5 are presented micrographs from such sample at different magnifications. It is seen that the film is formed by small two-dimensional grains with random orientation in the plane of the film. On the top of this primary layer are formed areas with a diameter of few hundred microns, filled with large lumps-like grains. Those areas were separated from each other by thin dark boundaries.

When the films were deposited at 3J/cm^2 a change in sample's morphology occurred. Fig.6 and fig.7 present micrographs from such film at different magnifications. A continuous YBCO film was formed although growth pits and numerous outgrowths disturbed the smoothness of the sample's surface.

The peculiarities observed in the morphology of the thin films deposited at high pulse energy densities reflected in their superconducting properties. In fig.8 (a, b, c, and d) the typical R(T) curves for samples deposited at 3,4,5 and $7J/cm^2$ are presented respectively. All the films produced at pulse energy densities above $4J/cm^2$ showed poor superconducting properties. The samples had residual resistances at 78K with varying values and no correlation with the pulse energy density could be observed.

We suppose that the large pulse energy densities causes removal of big quantities of the target material and the evaporated species sweep along large droplets of melted substance. Those droplets splash on the substrate and disturb the growth of the thin film. As the dimensions and velocities of the droplets depend not only on the pulse energy density, but also on geometrical factors and target density, the superconducting properties of the films produced at high energy densities may vary.

The samples grown at energy densities 3 and $4J/cm^2$ showed higher $T_c(R=0)$ - 84K and 88K respectively. Using AC susceptibility technique we estimated that the superconducting thin films had critical current densities $J_c = 10^3 \div 10^4 A/cm^2$.

Fig.9 shows the typical intensity versus angle (2θ) plots for such films. The X-ray diffraction spectrum shows only the presence of [oon] Bragg peaks, where n = 2÷7, indicating a strong preferential orientation of the film with the "c" axis bearing a normal relation to the surface of the substrate.

It was recently shown [5,7] that in "c" axis oriented films on $Y:ZrO_2$, epitaxy may occur with the YBCO "a" axis parallel to the substrate <100> or <110> directions, causing 45° grain boundaries. In addition, $Y:ZrO_2$ frequently supports epitaxy with the YBCO "a" axis making an angle of $\approx 9^o$ with the substrate <100>.

The reduction of J_c at a tilt or twist grain boundary in YBCO has been studied and reported extensively. Ivanov et al. [6] have shown that in in-situ processed "c" axis YBCO films on $Y:ZrO_2$ bicrystals, the J_c at 77K varies from $1\div5x10^6A/cm^2$ at 0° tilt to $1\div2x10^3A/cm^2$ at 40^otilt. Those results show that texturing in the film plane can lower J_c by almost four orders of magnitude.

From the SEM micrographs it was seen that our samples had random orientation of the grains in the plane of the film and this is the cause for the low values of the critical current densities.

As was shown by leading authors in this field, in-plane epitaxy of YBCO films on $Y:ZrO_2$ depends on various parameters, such as substrate temperature, substrate surface preparation, the use of homoepitaxial $Y:ZrO_2$ and heteroepitaxial CeO_2 layers on $Y:ZrO_2$ single crystal substrates [5-8].

Conclusions

The influence of pulse energy density on the morphological structure and the superconducting properties of thin YBCO films was investigated. The optimal values for pulse energy densities were found to be $3\div4 J/cm^2$. Although the films were "c" oriented and had high T_c(R=0) they had low critical current densities caused by the random in-plane orientation of the YBCO grains. Special attention should be paid to the parameters influencing in-plane epitaxial alignment of "c" axis YBCO films on $Y:ZrO_2$ substrates for achieving optimal transport properties.

References

1.D.Valkov, E.Nazarova, S.Savtchenko, V.Lovchinov and R.Gerasimov, Mat.Sc.Eng.B14 (1992) 11-16

2.S.Savtchenko, M.Ilieva, D.Valkov, R.Gerasimov and V.Gaidarova, III International School "New Materials and Technologies", 3-8 September 1988, Varna, Bulgaria

3.D.Valkov, E.Nazarova, R.Gerasimov and V.Lovchinov, submitted to Materials Science & Engineering B

4.E.Nazarova, D.Valkov and V.Lovchinov, ISSP Jubil. Collect., Scientific World, Singapoure, 1993

5.D.K.Fork, S.M.Garrison, M.Hawley, and T.Gebale, J.Mater.Res., Vol.7, No7, 1992, 1641-1651

6.Z.Ivanov, P.Nilsson, D.Winkler, J.Alarco, T.Claeson, E.Stepantsov and A.Tzalenchuk, Appl.Phys.Lett. 59, 3030 (1991)

7.S.M.Garrison, N.Newman, B.F.Cole, K.Char and R.W.Barton, Appl.Phys.Lett., 58, 2168 (1991)

8.B.H.Moeckly, S.E.Russek, D.K.Lathrop, R.A.Buhrman, Jian Li and J.W.Mayer, Appl.Phys.Lett., 57 (16), 1687 (1990)

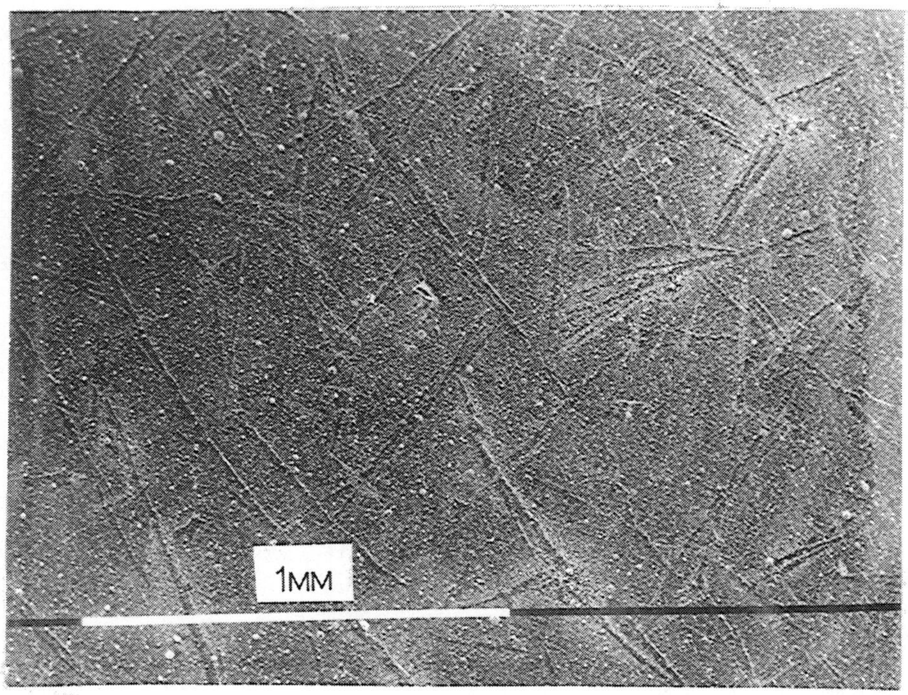

Fig.1 Typical micrograph from the surface of a film deposited at 7, 6 or 5J/cm^2

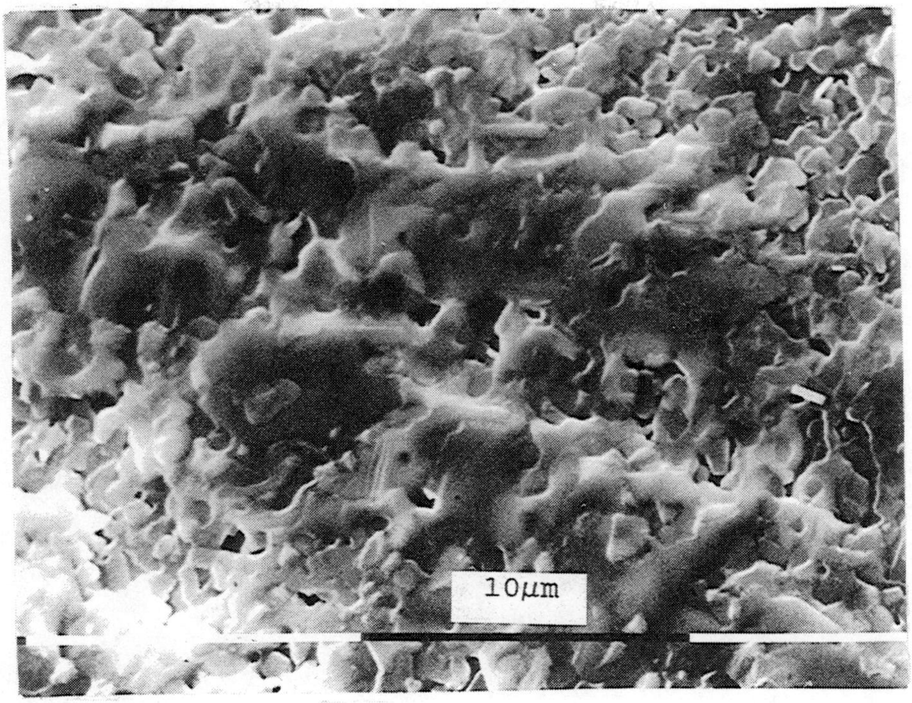

Fig.2 Micrograph of a "rod"-like feature on the top of a film deposited at high pulse energy density

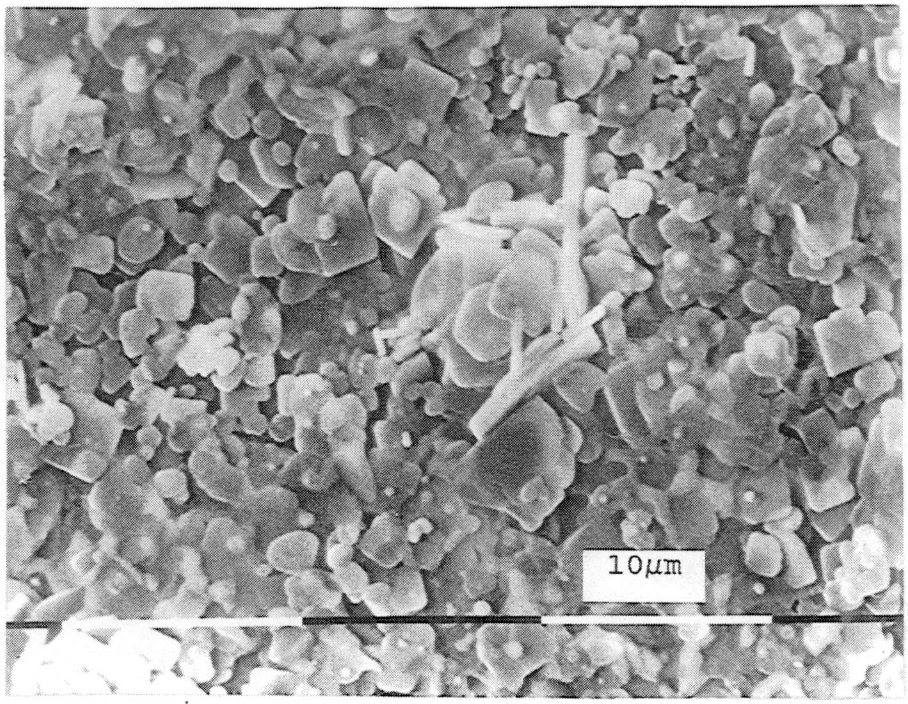

Fig.3 Micrograph from an area between the "rod"-like features on the top of a film deposited at high energy density

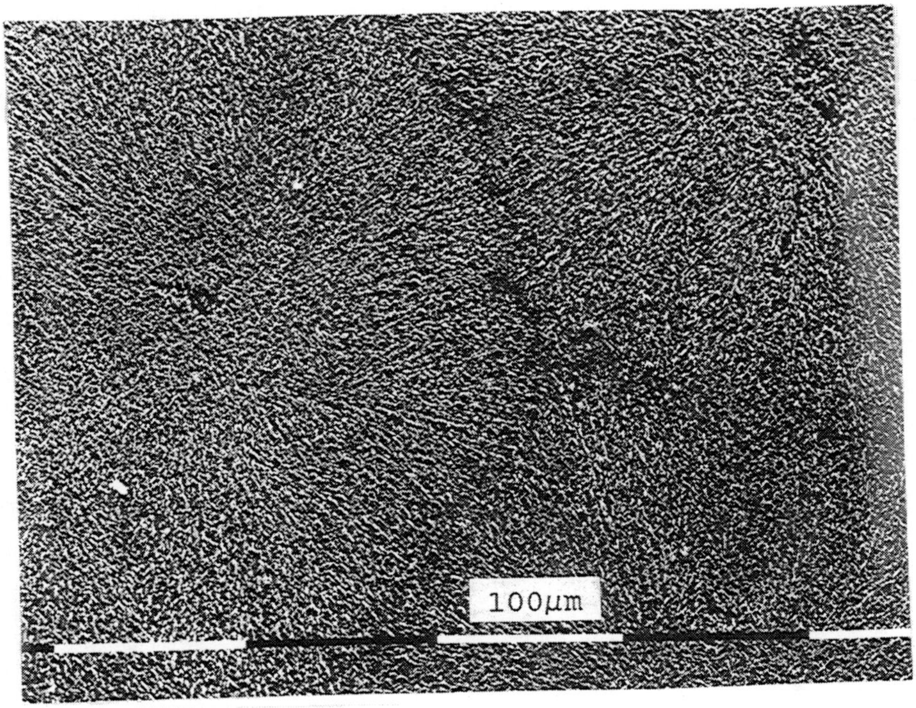

Fig.4 Micrograph from the surface of a sample deposited at $4J/cm^2$

Fig.5 Micrograph from the surface of a sample deposited at $4J/cm^2$

Fig.6 Micrograph from the surface of a film deposited at $3J/cm^2$

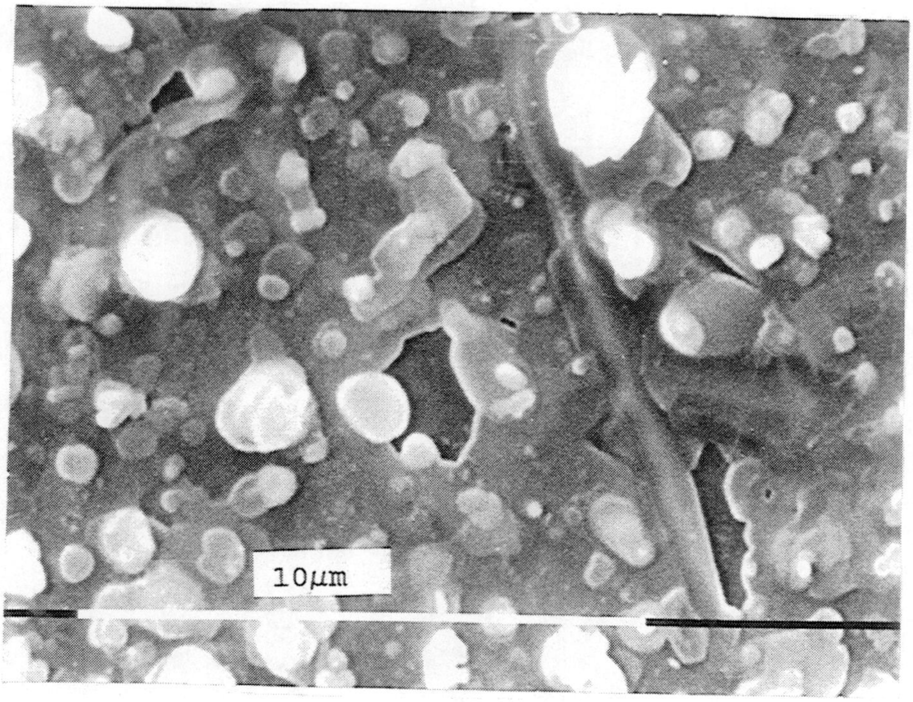

Fig.7 Micrograph from the surface of a film deposited at $3J/cm^2$

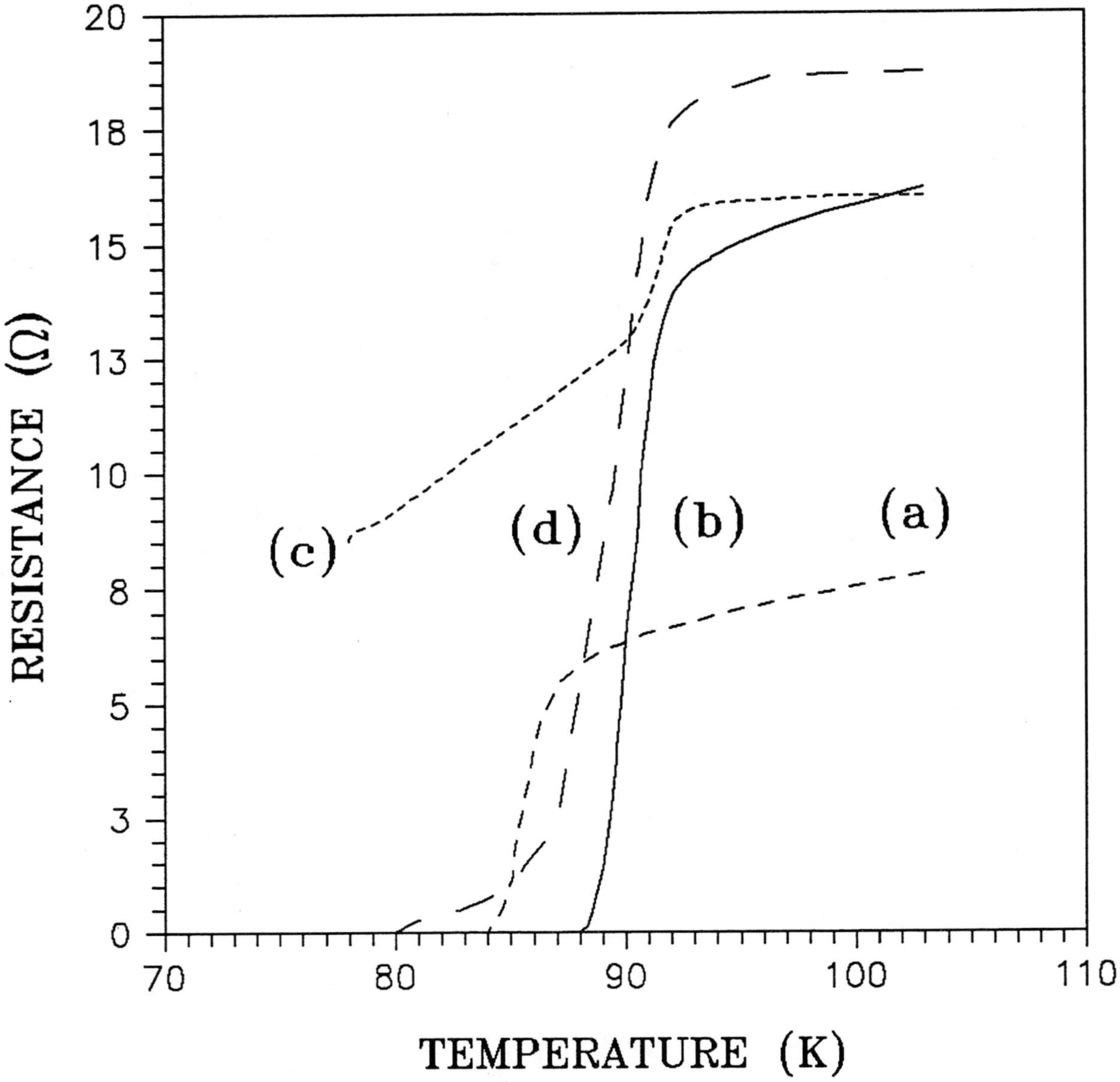

Fig.8 R(T) curves of samples deposited at different pulse energy densities: (a) $3J/cm^2$, (b) $4J/cm^2$, (c) $5J/cm^2$, and (d) $7J/cm^2$

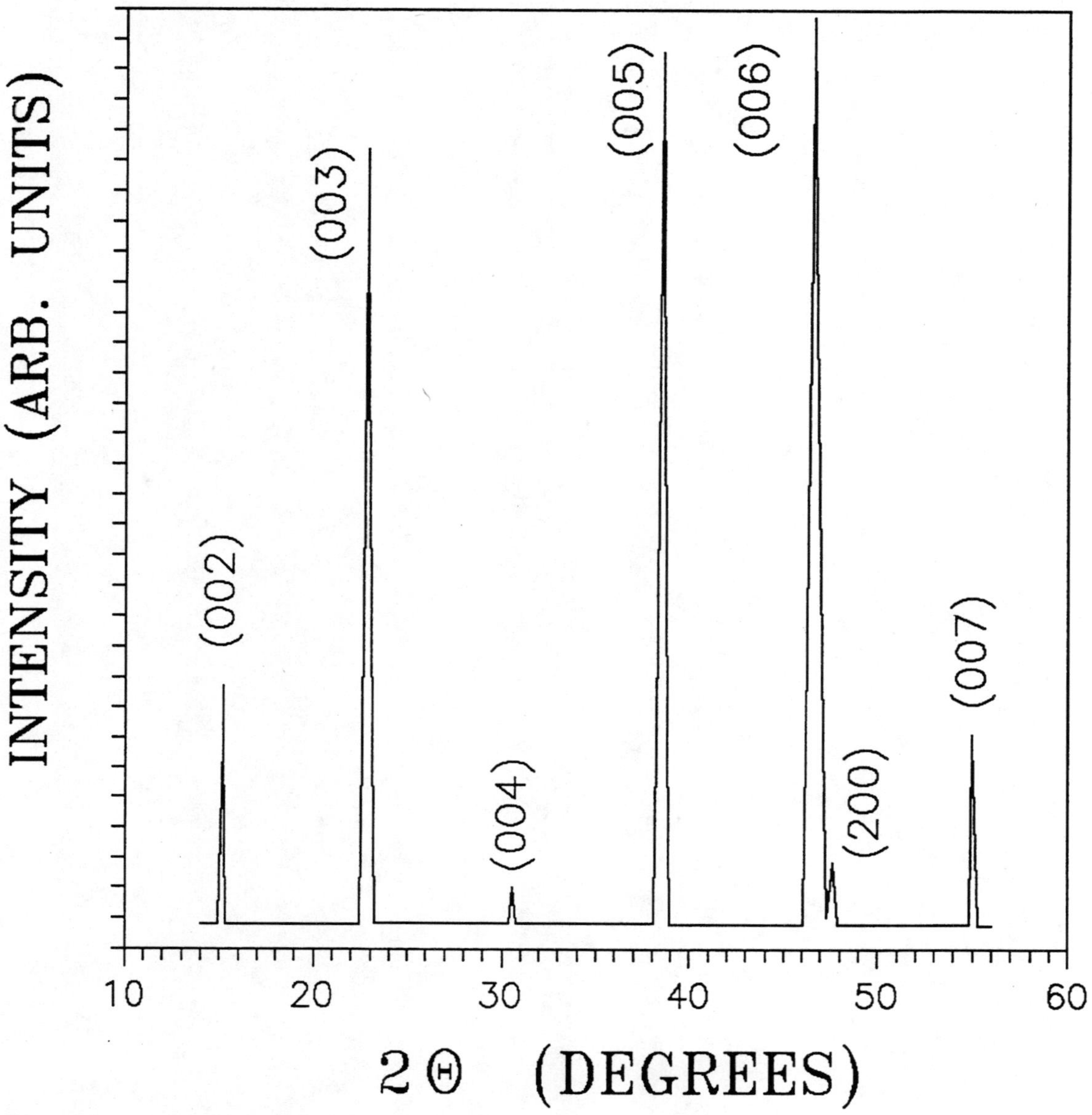

Fig.9 Typical X-ray diffraction spectrum (Cu Kα source) of thin superconducting YBCO film

Vacuum Electron and Ion Technologies
D. S. Karpuzov (Ed.)

ISBN 1-56072-242-8

INVESTIGATION OF YBCO SUPERCONDUCTING FILMS IN LAYERED STRUCTURES BY RAMAN SCATTERING

M.V. Belousov, R.A. Chakalov*, V.Yu. Davydov and S.F. Karmanenko*****
Institute of Physics, St. Petersburg University, 198904, Petrodvorets, Russia
*Institute of Electronics, Bulgarian Academy of Sciences, 1784, Sofia, Bulgaria
**A.F. Ioffe Physico-Technical Institute, Russian Academy of Sciences, 194021, St. Petersburg, Russia
***Electrotechnical University of St. Petersburg, 1973776, St. Petersburg, Russia

INTRODUCTION

The layered structure consisting of superconducting films separated by barrier layer is a prospective object for designing of various devices such as Josephson junctions [1, 2], strip-lines for microwave electronics [3], active control devices [4] using the nonlinear properties of barrier layer.

Along high-T_c materials $YBa_2Cu_3O_{7-\delta}$ (YBCO) compound is most appropriate for practical applications nowadays. To prepare layered structures with YBCO thin films one must employ effective diagnostic methods which would allow to control the properties of the separate layers. Raman spectroscopy is adopted as a convenient method for investigation of oxygen sublattice in YBCO crystal structure [5]. Since scattered light spectrum may be obtained from the film/substrate interface as well as from the film surface [6], we were able to use this technique for characterization of the bottom film in trilayer structure. As a material of the intermediate layer ferroelectric compound $Ba_xSr_{1-x}TiO_3$ (BSTO) was chosen due to the possible applications of microwave devices based on superconductors and non-linear dielectrics [7].

Presenting the results of Raman scattering analysis of layered $YBa_2Cu_3O_{7-\delta}/Ba_xSr_{1-x}TiO_3/YBa_2Cu_3O_{7-\delta}$ structures we demonstrate the efficiency of this method by obtaining information necessary for film quality improvement.

FORMATION OF YBCO / BSTO / YBCO LAYERED STRUCTURES

The layered structures were deposited by DC and RF sputtering in a vacuum chamber with two magnetron sources where sintered ceramic targets of YBCO and BSTO compounds were mounted. "On-axis" configuration was used and the distance from the target plane to the substrate holder was 20 mm. The magnetron source with corresponding target was moved to the substrate when each layer was deposited. The parameters of sputtering processes during the growth of YBCO/BSTO/YBCO layered structure are listed in Table 1. As a substrate MgO<100> with heteroepitaxial $Ba_{0.6}Sr_{0.4}TiO_3$ buffer layer (~200nm) was used thus getting better matching to the YBCO lattice parameter.

Table 1. Magnetron sputtering parameters during the growth of YBCO/BSTO/YBCO layered structure

	YBCO	BSTO
Target, ϕ 51 mm	$YBa_2Cu_3O_{7-\delta}$	$Ba_{0.2}Sr_{0.8}TiO_3$
Sputtering	DC, 140 V	RF, 50 W
Gas composition	$Ar+O_2$ (70%:30%)	$Ar+O_2$ (70%:30%)
Gas pressure, Pa	5	5
Substrate temperature, °C	650	670-720
Cooling down	O_2, 10 °C/min	O_2, 10 °C/min
Presputtering time	5 min	10 min
Growth rate, nm/min	5	1-2
Film thickness, nm	200-250	40-50

Fig.1. shows the sketch of the layered structure. During the deposition of BSTO and top YBCO layers the underlying film surface was partly screened by masks so the uncovered areas ensured free access to each layer. Six silver ohmic contacts for electrical measurements were prepared by thermal evaporation. Two couples of contacts (1-1 and 2-2) were

placed on bottom YBCO layer and third couple (3-3) was located on the upper superconducting film.

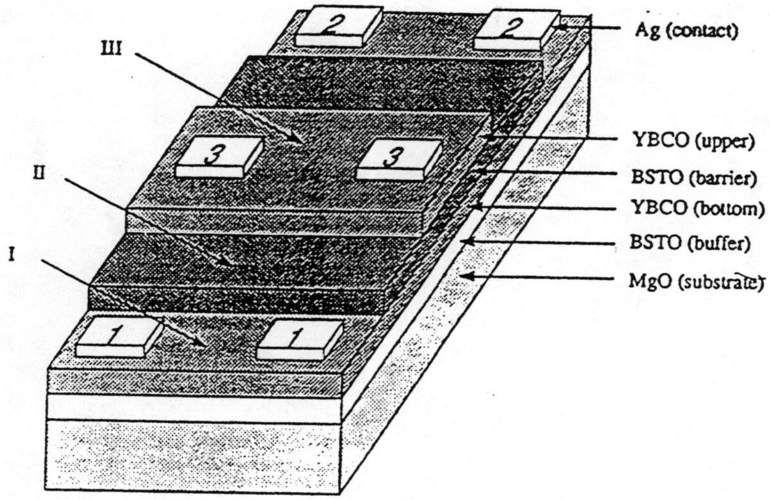

Figure 1. Sketch of layered YBCO/BSTO/YBCO structure used for optical and electrical measurements. The letters I, II and III indicate the spots of Raman scattering analysis.

Since both targets are mounted in the vacuum chamber material transfer from the target sputtered to the other one is probable. To investigate the influence of this feature we realized two sputtering procedures. In the first one named as Process A the presputtering of each target was carried out in a position where the target was simply located outside the substrate holder. In the second process named as Process B movable shields were used to separate the volume where the target was presputtered and sputtered from the rest space in the vacuum chamber. The growth regimes presented in the Table 1 were the same for both processes. We supposed that the additional separation used in Process B would lead to sharp interface boundary between neighbour layers.

RESULTS AND DISCUSSION

The crystal lattice of trilayer structures was studied by Raman scattering. Raman spectra were exited by Ar laser (λ = 514.5 nm, $P \sim$ 30 mW in a spot of about 50 μm in diameter) and registered by double monochromator DFS-24. The intensity of scattered light was measured in the frequency range of 200-700 cm^{-1}. There are three active vibration modes hitting in this measuring range. For YBCO films with $\delta \sim 0.05-0.07$ the frequencies of the modes are as follows: $\nu_5 = 502$, $\nu_4 = 435$ and $\nu_3 = 338$ cm^{-1}. The decreasing of oxygen content in YBCO lattice to $\delta \sim 0.5$ leads to the change of the frequencies ν_5 and ν_4 to 470 cm^{-1} and 455 cm^{-1} respectively. The intensity of ν_5, ν_4 and ν_3 Raman modes is strongly dependent on the directions of light polarization and crystallographic axes of $YBa_2Cu_3O_{7-\delta}$ lattice and this property allows to determine the degree of the film orientation.

Raman spectra presented in Fig. 2(a) and Fig. 2(b) were measured from the samples grown in the Process A and in the Process B respectively. These spectra were obtained in (xx) - polarization, i.e. at parallel polarizations of exited and scattered light. The ratio of ν_3 and ν_5 line intensities was ~ 0.25 and using the data in [6] it was possible to establish that both upper and bottom YBCO layers have practically pure *a*-orientation. In this case the *a* axis of YBCO lattice is

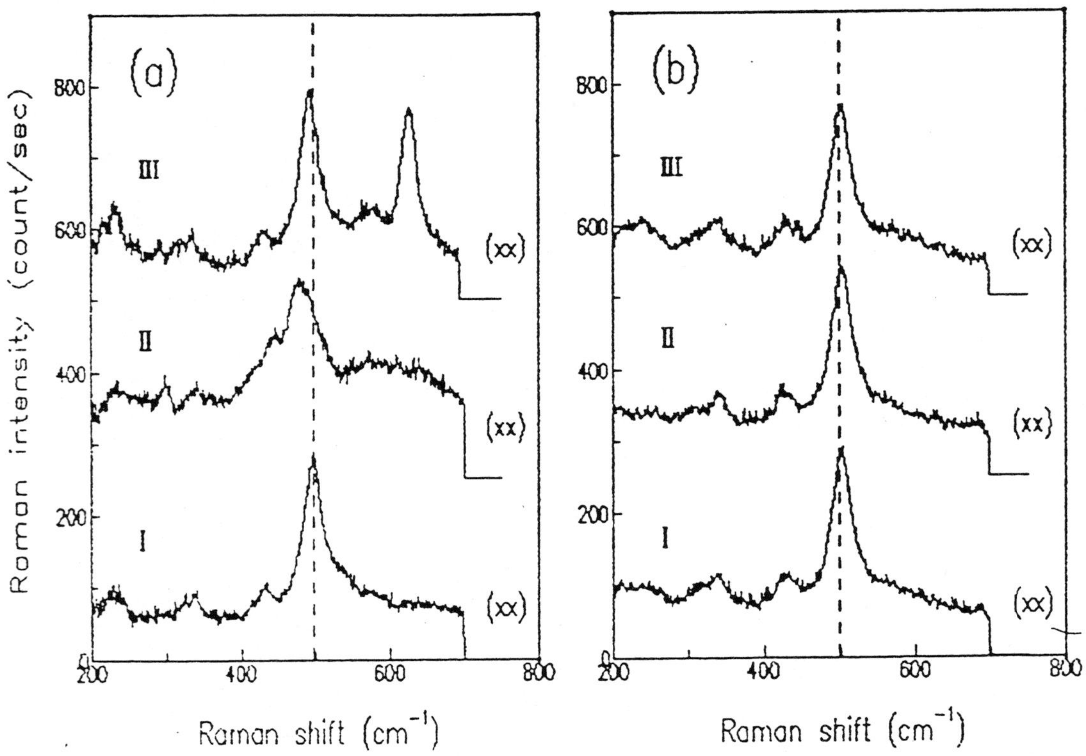

Figure 2. Raman spectra of trilayer YBCO/BSTO/YBCO structures measured in different spots indicated in Fig.1;
(a)-spectra obtained from a structure grown in the Process A;
(b)-spectra obtained from a structure grown in the Process B.

perpendicular to the film surface and *b* and *c* axes lay in the plane. We suppose that the growth regimes used determine high film growth rate ($V_g \simeq 5$ nm/min) due to relatively low gas pressure and substrate temperature. Under these conditions *a*-oriented YBCO films are formed.

For the bottom layer of structures grown in the Process A the frequencies of ν_5 and ν_4 lines shifted to the values of 490 and 445 cm^{-1}, respectively (Fig.2a), which indicates that the deposition of BSTO layer has led to oxygen loss ($\delta \sim 0.4$). The YBCO film grown on the intermediate BSTO layer was also not perfect – the oxygen concentration ($\delta \sim 0.15$) differed from the maximum value and, in addition, an essential fracture of impurity phases, mainly $BaCuO_2$, was included in the film (frequency modes > 600 cm^{-1}).

The spectra of trilayer structures grown after some improvement of technological facilities in the Process B are presented in Fig.2(b). Both bottom and upper YBCO layers were *a*-oriented and well formed structurally; the oxygen content was close to the maximum value ($\delta \sim 0.05$). So the deposition of BSTO film did not influence the properties of YBCO films.

The results of Raman scattering investigations were compared with resistivity measurements of YBCO films in the layered structures presented in Fig.3(a) and (b). The resistance of uncovered bottom YBCO layers was measured between contacts 1-1 (curves 1). The curves 2 correspond to

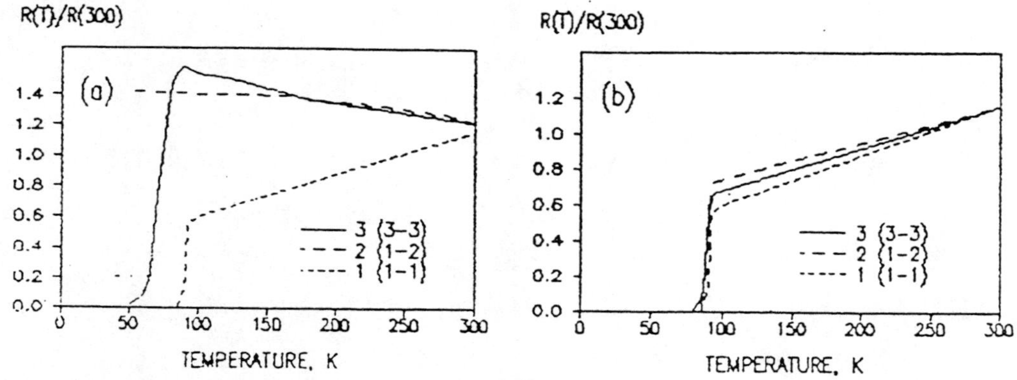

Figure 3. Temperature dependences of YBCO film relative resistance R(T)/R(300) measured between different contacts depickted in Figure 1 on YBCO/BSTO/YBCO structures grown in the Process A (a) and Process B (b).

the temperature dependences of the parts of bottom YBCO films lying under BSTO layer between contacts 1-2. The dependences drawn as curves 3 were obtained from the upper YBCO layers between contacts 3-3. One can see that bottom YBCO layer lying under BSTO prepared in the Process A was non-superconducting (low oxygen content) as distinct from the similar layer in the Process B. The upper YBCO in the Process A (containing impurity phases) showed semiconductor behaviour of the R(T) dependence unlike usual metallic character for the upper layer in the Process B. The critical (R=0) temperatures measured in all parts of YBCO films grown in Process B only were high: $T_{C(1-1)}$ = 85.1 K, $T_{C(2-2)}$ = 85.1 K, $T_{C(3-3)}$ = 83.7 K. Therefore joint structural and electrical study allow to establish that trilayer structures grown in the Process B demonstrate essentially higher quality.

CONCLUSION

The intersected sputtering and mixing of components from different targets during deposition of trilayer YBCO/BSTO/ YBCO structures do not allow sharp interfaces between separate films to form. As result, bottom YBCO layer lose oxygen atoms and superconducting properties are destroyed. The lattice growth is disturbed and top YBCO contains impurity phases. When an improved technological process preventing material transfer is realized high superconducting properties in the strustures are achieved.The high efficiency of Raman scattering diagnostics for analysis of superconducting films in layered structures has been demonstrated.

REFERENCES

1. J.Gao et al., Appl. Surface Science, **46** (1990), 74.

2. H.Akoh et al., Physica C, **180** (1991), 227.

3. Yu.Boikov et al., Appl.Phys.Lett., **59**/20 (1991), 2606.

4. P.H.Carr et al., IEEE Trans.on MTT, **39**/9 (1991), 1445.

5. M.V.Belousov et al., Proc. of the Europ. Conf. HTSC films and Single Structure, World Sci.Singapour, 1990, p.731.

6. M.V.Belousov et al., Sverkhprov., Fiz., Khim., Tekhn., 6/1 (1993), 104.

7. O.G.Vendik et al., Ferroelectrics (1993).

Vacuum Electron and Ion Technologies
D. S. Karpuzov (Ed.)

ISBN 1-56072-242-8

PREPARATION AND PROPERTIES OF $Tl_2Ba_2CaCu_2O_8$ THIN FILMS

L-G. Johansson
Department of Inorganic Chemistry, Chalmers University of Technology and University of Gothenburg, S-41296 Göteborg, Sweden
T. Claeson, Z.G. Ivanov, H. Olin, And E. Olsson
Department of Physics, Chalmers University of Technology and University of Gothenburg, S-41296 Göteborg, Sweden
D. Erts
Department of Condensed Matter Chemical Physics, University of Latvia, LV1098 Riga, Latvia

Abstract

An ex situ process has been developed to produce thin superconducting $Tl_2Ba_2CaCu_2O_8$ films. Properties of films grown on different substrates using different annealing regimes were studied. Critical temperatures of 103-107K were measured on films prepared in a broad range of annealing temperatures on $SrTiO_3$, $LaAlO_3$ and Y-ZrO_2 substrates. A critical current density, J_c, of $2 \cdot 10^6$ A/cm^2 at 77Kwas measured on $LaAlO_3$. Film morphology was studied by SEM, AFM and STM.

Key words: Tl cuprates, thin films, ex-situ thin film processing

1. Introduction

In the area of Tl-cuprate superconducting thin films much interest is directed towards, e.g., $Tl_2Ba_2Ca_2Cu_3O_{10}$ (Tl-2223) and $Tl_2Ba_2CaCu_2O_8$ (Tl-2212). The former compound has a maximum T_c of about 125K. Tl-2212 is characterised by greater ease of preparation while the optimum T_c is slightly lower compared to Tl-2223. Typical values of some important parameters for Tl-2212 and Tl-2223 films include T_c of about 107 and 115K, respectively, critical current density, J_c, at 77K of 5-10$\cdot 10^5$ A/cm^2 and microwave surface resistance, R_s, of 0.5 mΩ at 77K and 10 GHz [1,2]. Thus, the thallium cuprate thin films offer several advantages compared to $YBa_2Cu_3O_x$ (YBCO) thin films including higher critical temperature, low surface resistance and low 1/f noise [3]. Epitaxial superconducting Tl-cuprate thin films and a low noise Tl-cuprate

SQUID were reported already in 1989 [4]. Nevertheless, progress has been slow compared to the YBCO system because of the difficulty in preparing pure and homogenous films with good superconducting properties and by a lack of reproducibility of the process. In this paper we report on a simple and reliable method for the preparation of superconducting $Tl_2Ba_2CaCu_2O_8$ thin films.

2. Experimental

The preparation of superconducting Tl-2212 films was carried out in two consecutive steps:

(i) Tl-free precursor films with nominal composition Ba:Ca:Cu=2:1:2 were deposited on different $5x5mm^2$ substrates by pulsed laser deposition. The beam of an KrF eximer laser (λ=248 nm) was focused on a rotating oxide target with the elemental composition $Ba_2CaCu_2O_x$. The beam energy density was 1-1.5 J/cm^2 in a spot of 1x3 mm^2. The substrates were kept at room temperature and the films were amorphous after deposition.

(ii) formation of the superconducting film by annealing the precursor film from 30 minutes to several hours at 835-895°C in Tl_2O vapour generated by ceramic Tl-2212:

$$Tl_2Ba_2CaCu_2O_8 \Leftrightarrow Tl_{2-x}Ba_2CaCu_2O_{8-3x/2} + x/2\ Tl_2O(g) + x/2\ O_2(g)$$

Ex-situ processes for preparing thallium cuprate thin films usually employ Tl_2O sources that are more thallium-rich than the ideal film stoichiometry (see for example ref.5). The extra thallium oxide is added to compensate for thallium losses from the system or because the thallium source has a lower temperature than the film. We present a method that involves the annealing of the film under near-equilibrium conditions, avoiding the difficulties of controlling vapour phase transport of Tl_2O. Figure 1 shows a schematic view of the experimental set-up. The Tl-free precursor film is annealed inside a crucible made from ceramic Tl-2212 with a lid of the same material. The Tl-2212 crucible is in turn enclosed in a gold-sealed alumina crucible. Using this method the precursor film is exposed to a vapour pressure of thallium oxide corresponding to the equilibrium value over pure Tl-2212. Thallium losses from the film are thus avoided and the thallium content of the film can be kept at the optimum value.

The loss of thallium from the crucible was monitored by weighing the closed alumina crucible before and after each annealing. In a typical run, less than two mg of thallium oxide was lost during annealing. The thallium oxide loss mainly occurred from the outer surface of the Tl-2212 crucible and consequently did not affect the film. Due to the minimal thallium loss and to the small amount of thallium absorbed by the film the same Tl-2212 crucible could be used repeatedly with no change of film characteristics.

The films were deposited on (001) $LaAlO_3$, $SrTiO_3$, and Y-ZrO_2 single crystal substrates. The effect of annealing temperature in the range 835-895°C was studied using T_c measurement (standard four probe resistive measurements), θ-2θ X-ray diffractometry, Scanning Electron Microscopy (SEM), Atomic Force Microscopy (AFM) and Scanning Tunneling Microscopy (STM). The in-plane orientation was studied using Φ-scan X-ray diffractometry.

3. Results and Discussion

The films on Y-stabilised ZrO_2, $SrTiO_3$, and $LaAlO_3$ all showed similar transitions to the superconducting state, see Fig. 2 for films on $SrTiO_3$, and $LaAlO_3$. Films deposited on different substrate materials, but otherwise under identical deposition conditions, exhibited different surface morphologies. A comparison between films deposited at 855°C on (001) $LaAlO_3$ and (001) Y-ZrO_2 substrates is shown in Fig. 3. The films on $LaAlO_3$ were in general smoother on a local scale with relatively large flat regions. This was confirmed by STM and AFM which showed that the films on $LaAlO_3$ were locally atomically smooth, an AFM picture is shown in Fig. 4. Using the Φ-scan technique the films grown on $SrTiO_3$ and $LaAlO_3$ were shown to be epitaxial (see fig.5 for a Φ-scan of Tl-2212 on $LaAlO_3$). Double side coated films were prepared on $LaAlO_3$ substrates, the two sides exhibiting identical properties.

The annealing temperature affected phase purity, T_c, surface resistance as well as the surface morphology. The optimal annealing temperature was found to be 875°C giving a T_c of 107 K for films on $LaAlO_3$. Films treated at 835°C were insulating while treatments at temperatures above 875°C caused T_c to decrease by 1-2 K (see Fig.6).

Films treated with thallium oxide vapour in the appropriate temperature range and during sufficient time were single phase c-axis oriented Tl-2212, no crystalline impurities being detected by X-ray diffraction. However, T_c as well as phase purity was affected by the annealing time, longer annealing times being necessary in the lower part of the temperature range. Annealing at 845°C initially produced films with Tl-1212 as the dominating phase and with Tl-2212 and small amounts of Tl-2201 also present, all three compounds being perfectly c-axis oriented, see Fig 7a. Re-annealing the film for three hours at the same temperature converted the film to Tl-2212 (with traces of Tl-1212), see Fig 7b. The formation of Tl-1212 after short annealing times at relatively low temperature shows that the film has not yet reached its equilibrium composition with respect to Tl. The prolonged exposure of the film to Tl_2O vapour at high temperature apparently converts the single TlO-layer Tl-1212 into its double layer counterpart. It is suggested that the formation of double layer thallium cuprate thin films by ex-situ methods may in general be preceded by the formation of the corresponding single-layer compounds. The formation of Tl-2201 and its subsequent disappearance after prolonged Tl_2O annealing indicates the presence of a compound on the surface that contains calcium and copper and that is able to react with Tl-2201 to form Tl-2212.

The measured T_c values of our films were typical of bulk Tl-2212, as synthesised. It has been reported that treatment in argon atmosphere at 300-400°C may increase the T_c of bulk Tl-2212 to about 112K [6]. Annealing in Ar at 300°C for 15 hours caused the T_c of the film to drop to 90K from its starting value of 104K (Fig. 8). Re-annealing this film at 855°C for 30 min. resulted in an increase of T_c to 105K. Tl losses as well as changes in oxygen concentration may be responsible for the observed behaviour. At this point we cannot resolve this question and further studies are needed. The observed reversibility of the annealing process may have important implications for Tl-2212 films since it allows the modification of film properties in a controlled way.

To measure the critical current densities, microbridges with width 4- 8 µm and length of 50 µm were patterned by two different methods. In the first method the precursor film was deposited through a lift-off mask of photoresist and thallium-annealed afterwards. In the second method the film was patterned <u>after</u> annealing using a photoresist mask and Ar ion beam etching. The first method was simple, avoiding Tl processing but resulted in microbridges with needle-like outgrowths at the edges. The measured T_c:s of 80-90K and J_c:s of 10^3-10^4 A/cm^2 were very low and we related that to residues of photoresist on the substrate. For films annealed in the temperature interval 855-895°C, the microbridges patterned by the second method, showed a decrease of 3-5K in T_c compared to the unpatterned film while the critical current densities were $2 \cdot 10^6$ A/cm^2 at 77K. At the high end of the temperature range, holes developed in the films which in turn increased the proportion of defect bridges.

The microwave surface resistance was measured at 21.5 GHz by the parallel plate resonator method and was found to be 1.5 mΩ at 4.2K. Films on double-side coated substrates exhibited identical surface resistance.

4 Conclusions

An annealing procedure is presented that allows ex-situ double side coated epitaxial Tl-2212 films to be prepared routinely on (001) $SrTiO_3$, Y-ZrO_2 and $LaAlO_3$ substrates. The films on $LaAlO_3$ were epitaxial and locally smooth on an atomic scale. T_c ranged from 103 to 107K depending on annealing time and temperature (835-895°C), J_c reaching $2 \cdot 10^6$ A/cm^2 at 77K for films on $LaAlO_3$. Short annealing times at low temperature resulted in the formation of single-layer Tl-1212 that was converted to Tl-2212 by prolonged annealing at the same temperature. It is suggested that the formation of single layer Tl-cuprates may generally precede the formation of the corresponding double layer compounds. High annealing temperature (895°C) resulted in films containing holes down to the substrate. Argon annealing of Tl-2212 films at 300°C resulted in a deterioration of the superconducting properties. Re-annealing in Tl_2O vapour at high temperature lead to a recovery of T_c.

Acknowledgements: The work was supported by the Swedish Superconductivity Consortium. The help from M. Löfgren with the microwave surface resistance measurements and from Pedro Berastegui with the Φ-scans is gratefully acknowledged.

References

[1] D. S. Ginley, J. S. Martens, E. L. Venturini, C. P. Tigges, C. Ashby and S. Volk, *IEEE Trans. Appl. Superconductivity*, **3,** 1201 (1993).

[2] W. Y. Lee, S. M. Garrison, M. Kawasaki, E. L. Venturini, B. T. Ahn, R. Boyers, J. Salem, R. Savoy, and J. Vazquez, *Appl. Phys. Lett.*; **60**, 772 (1992).

[3] A. Misra, Yi Song, P.P. Crooker, and J. R. Gaines, *Appl. Phys. Lett.*, **59**, 863 (1991).

[4] V. Foglietti, R. H. Koch, W. J. Gallagher, B. Oh, and B. Bumble, *Appl. Phys. Lett.*, **54**, 2259 (1989).

[5] W. L. Holstein, L. A. Parisi, C. Wilker and R. B. Flippen, *IEEE Transactions on applied superconductivity*, **3**, 1, 1197 (1993)

[6] C. Martin, A. Maignan, J. Provost, C. Michel, M. Hervieu, R. Tournier, and B. Raveau, *Physica C*, **168**, 8 (1990)

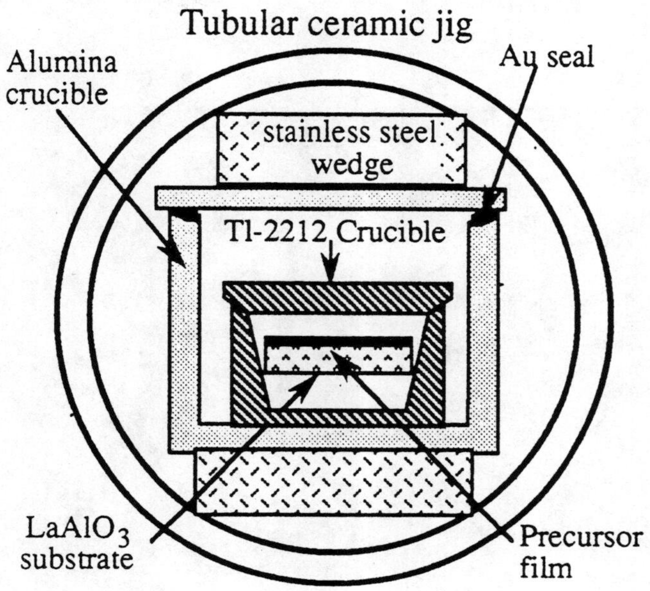

Fig.1. Schematic cross-section of the experimental set-up used in the annealing of precursor films. The gold seal is put under compression during annealing using the stainless steel wedges.

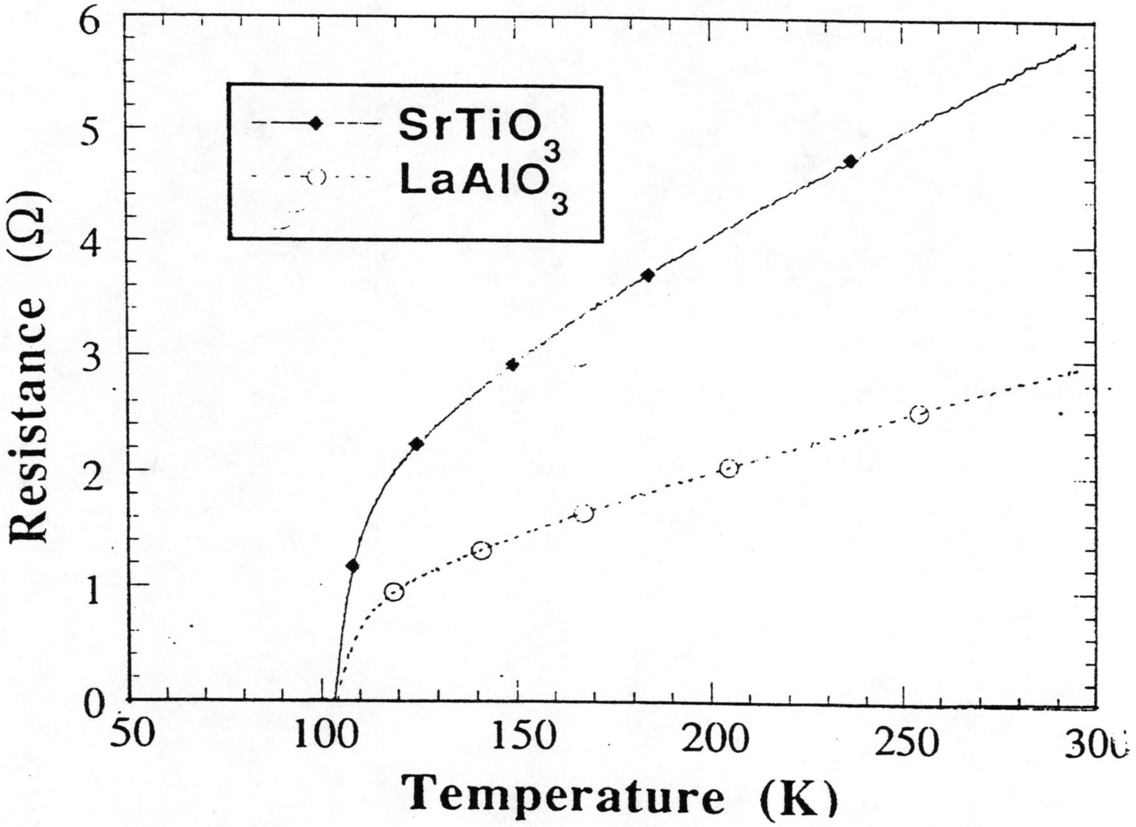

Fig.2. Resistance vs. temperature for films grown on (001) $LaAlO_3$, and $SrTiO_3$. Annealing time = 30 minutes. Annealing temperature = 895°C.

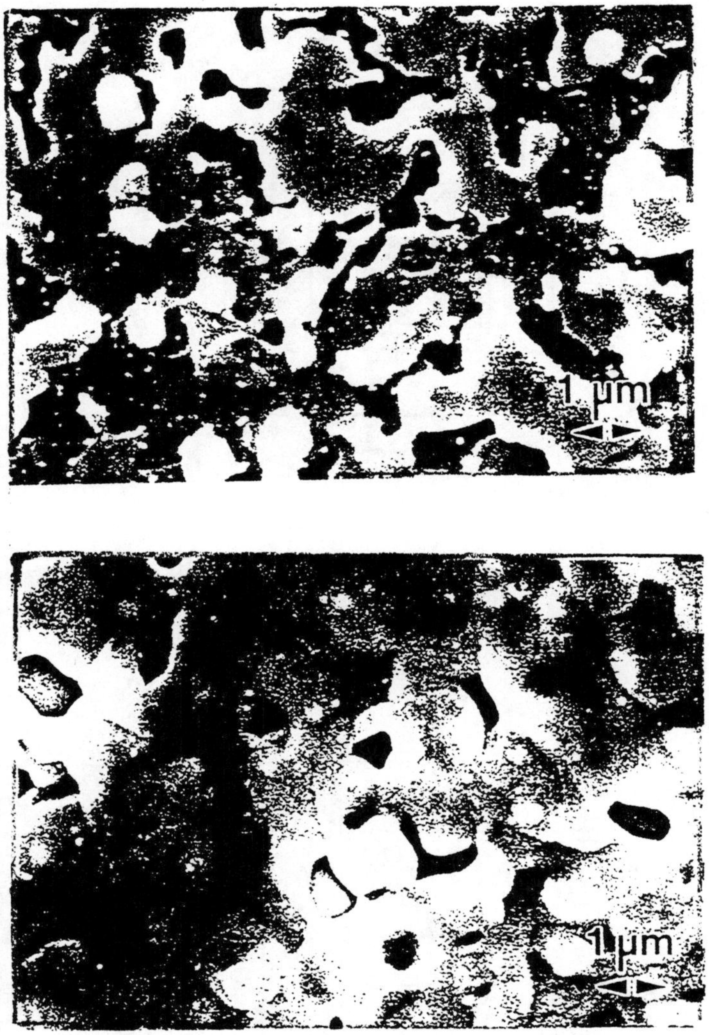

Fig.3. SEM microphotographs showing the surface morphology of $Tl_2Ba_2CaCu_2O_8$ films on Y-ZrO_2 (below) and $LaAlO_3$ (above). The films were annealed at 855°C for 30 minutes.

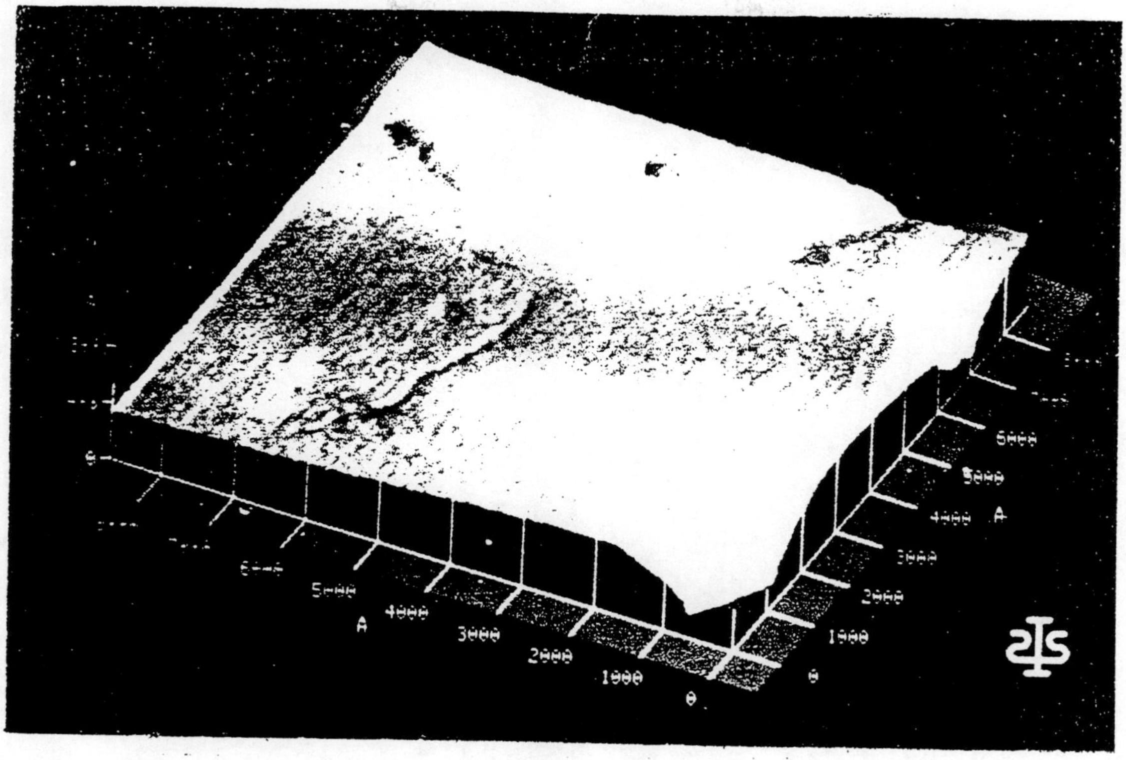

Fig.4. An AFM image of the surface morphology of a Tl-2212 film on a $LaAlO_3$ substrate. Note the terrace-like structure with step height corresponding to one to several unit cells.

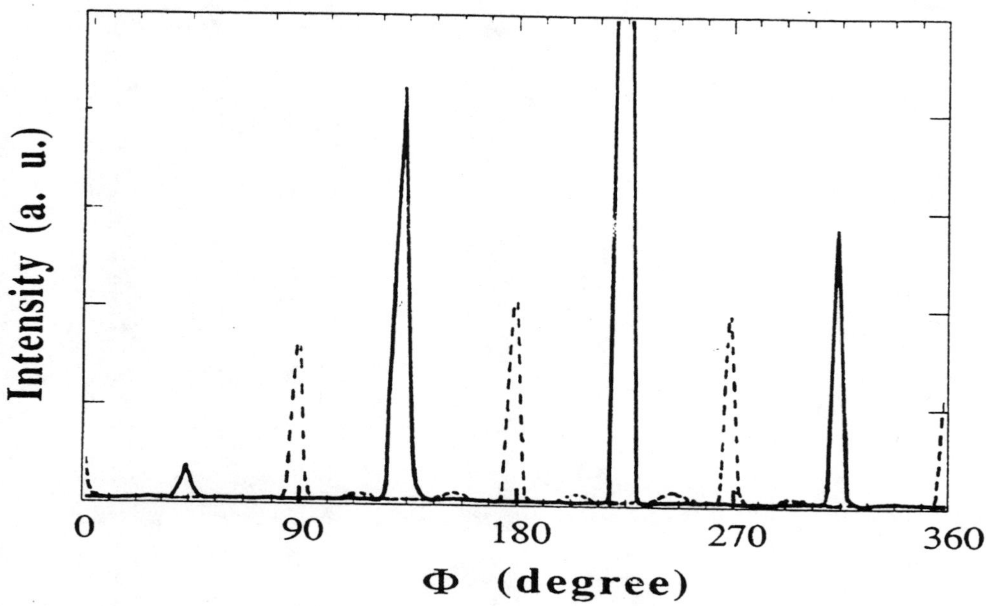

Fig. 5 Φ-scans of 300 nm thick $Tl_2Ba_2CaCu_2O_8$ film on $LaAlO_3$ substrate treated at 875°C for 6 minutes. The Tl-2212 (1, 1, 18) family of peaks (dotted line) (2θ=66.8°), indicating the [110] direction in Tl-2212, is shifted by 45° in relation to the pseudo-cubic (1, 0, 2) peaks of the $LaAlO_3$ substrate (full line) (2θ=54.1°), indicating cube on cube epitaxy. Intensity differences are due to mounting inaccuracy.

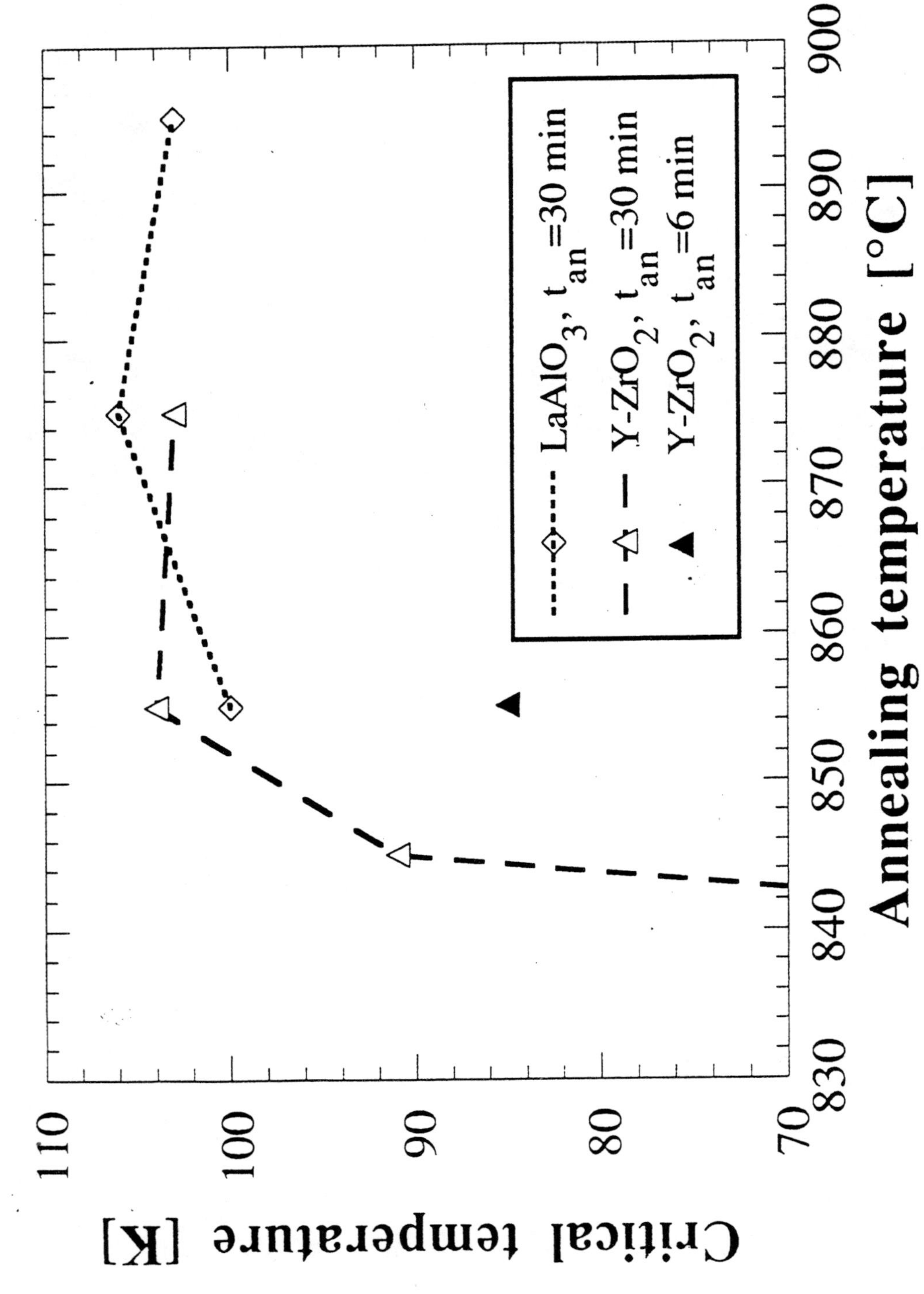

Fig.6. T_c as a function of annealing temperature for films grown on $LaAlO_3$ and $Y\text{-}ZrO_2$.

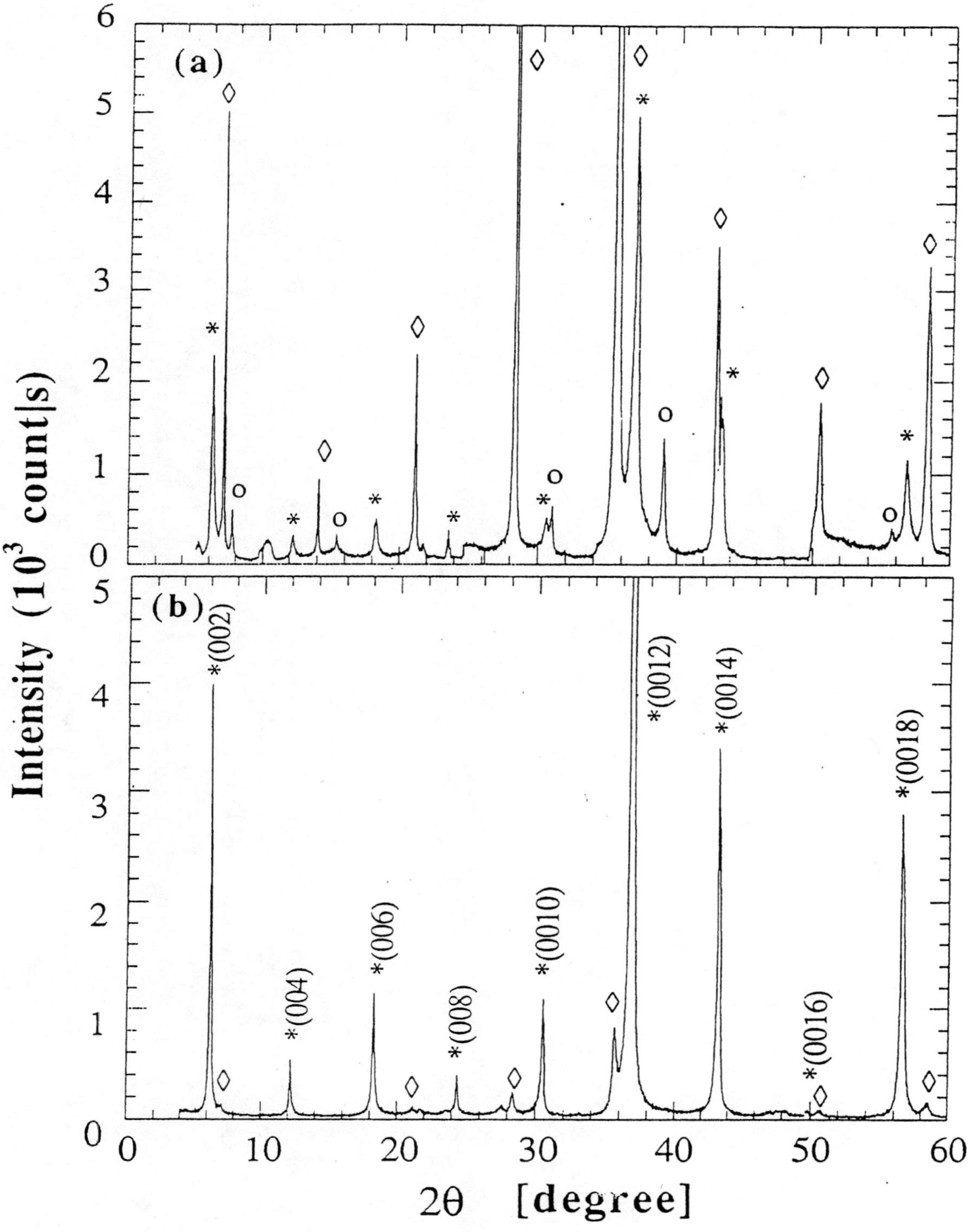

Fig.7 XRD θ-2θ scan of 240 nm thick Tl-2212 film on $LaAlO_3$ substrate annealed at 855°C for (a) 30 minutes and (b) another 3h.(00l) peaks are identified by (*) for 2212 phase, (◊) for 1212 phase (o) for 2201 phase. The substrate peaks have been subtracted.

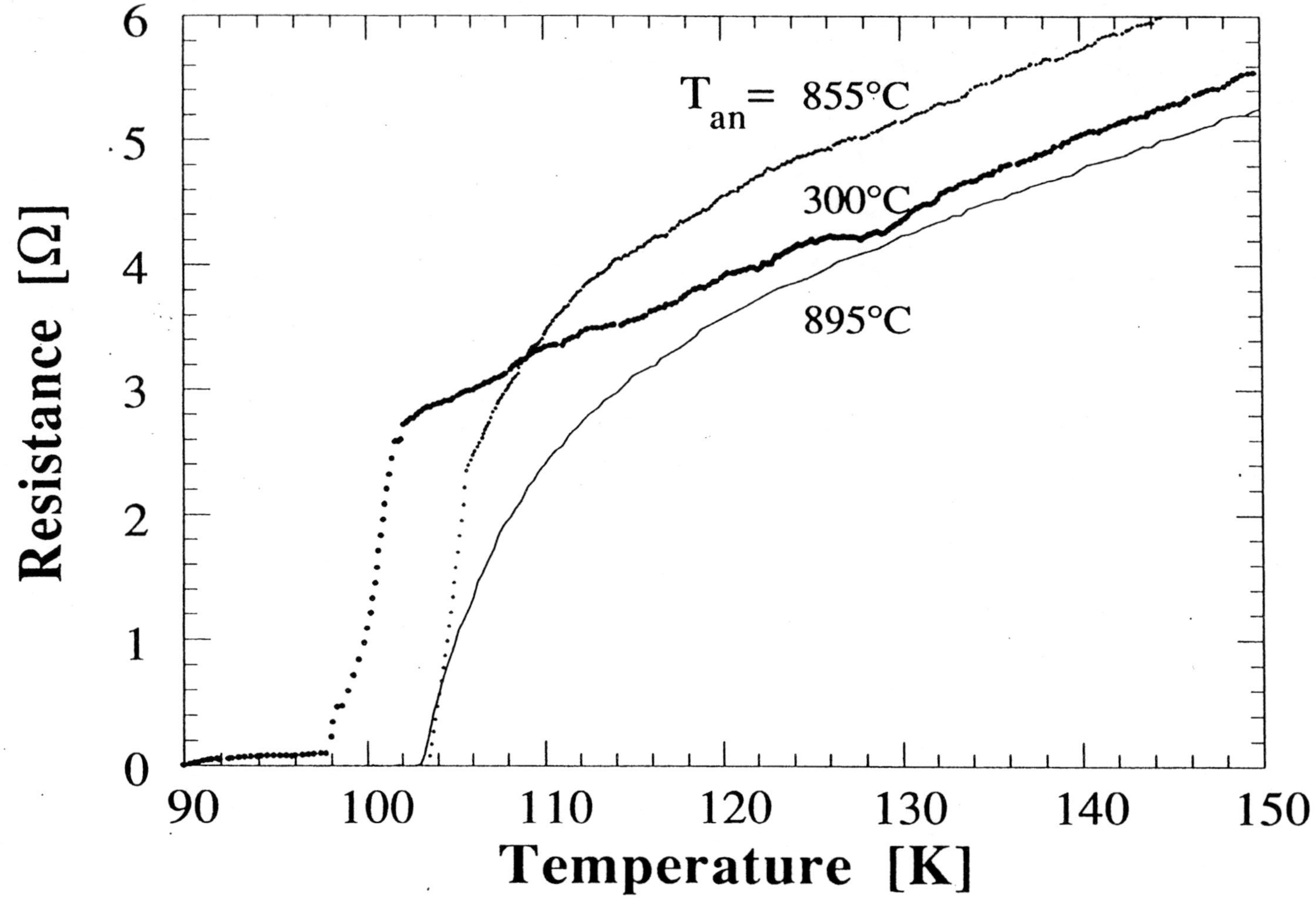

Fig.8. Resistance vs. temperature for Tl-2212 film on $LaAlO_3$ after different annealing procedures. Note the reversible behaviour of T_c of the film after annealing at 855°C for 30 min.

Vacuum Electron and Ion Technologies
D. S. Karpuzov (Ed.)

ISBN 1-56072-242-8

THE FORMATION OF Y-Ba-Cu-O THING FILMS BY PULSED LASER DEPOSITION

R.I. Tomov, P.A. Atanasov
Institute of Electronics, Bulgarian Academy of Sciences,
72, Tzarigradsko chaussee, Sofia, 1784, Bulgaria

Superconducting $Y_1Ba_2Cu_3O_7$ thin films on $SrTiO_3$ and MgO with zero transition temperatures T_{c0}= 82-89 K were prepared by in-situ laser deposition (XeCl- 308nm, 20ns) at different laser energy densities. On the top of the relatively smooth film surface, splashes of spherical droplets are observed, worsening the superconducting properties. It was revealed that cw CO_2 laser heating of substrate surface may be a method of particulate density reduction. The buffer layer such as sol gel deposited YSZ(111) on Si(100) also were successfully experimented for preparation of $Y_1Ba_2Cu_3O_7$ superconducting thin films. The 400-600nm thick films were characterized by EDAX, SEM and XRD.

1. Introduction

The discovery of the high-temperature superconducting oxides by Bednorz J. and Muller K.[1] has excited extraordinary interest and research activity in the area of superconducting thin films technologies. The low critical current, the brittleness and humidity degradation of bulk high-temperature superconductors (HTSC), make their use in thin film form closer to practical application.

The lasers have been widely used for superconducting thin film processing. The advantages of laser beam application have been reported in many works : UV excimer laser ablation ensures strict stoichiometry transfer from the bulk ceramic targets to the substrate [2]and high rate of material transfer[3]; cw CO_2 laser heating of the substrate ensures higher degree of purity compared to other heating techniques - resistive, lamp etc. [4,5,6]; "in situ" excimer laser irradiation on growing film surface improves the superconducting properties and the surface morphology at relatively low substrate temperature[7]; last but not least this technique is relatively simple and inexpensive.

Films with good superconducting properties were obtained "in-situ" at T_s= 650-700°C and by establishing the oxygen partial pressure at 150-200 mTorr. Slow cooling and annealing at 400-500°C ensured oxygen incorporation[8,9] and formation of orthorombic phase. Even at these temperatures certain interdiffusion and worsening of the superconducting properties are observed, especially on more reactive substrates[10] like Si and Al_2O_3. The "in situ" deposition in the presence of activated oxygen allowed reduction of T_s down to 450-550°C and smooth superconducting films with very sharp interfaces were obtained[11,12]. The buffer layers gave promissing results in the deposition of HTSC on substrates like Si and GaAs, widely used in conventional microelectronic.

Generation of particulates is a problematic feature of the laser deposition technique that is difficult to solve. Particulates are 0.1-10 μm in size and consist of solidified melt drops or irregularly shaped granular matter. Higher particulate densities are generally found under laser ablation conditions that produce practical deposition rates ($0.1\text{-}10 A.s^{-1}$) and under power densities where desirable energetic species are generated. Particulates that are incorporated in the film strongly degrade its superconducting properties.

The aim of this work was to overcome some of the above problems using oxygen plasma-assisted XeCl excimer laser deposition combined with CO_2-laser heating and annealing We report the development of an "in situ" method of laser deposition of $Y_1Ba_2Cu_3O_7$ thin films on the $SrTiO_3$, MgO as well as on silicon with YSZ buffer layer at substrate temperatures 450-550°C based on the "ex situ" technique proposed previously by us [4].

2. Experimental

The schematic diagrams of our set-up for pulsed laser deposition of HTSC thin films are shown on FIG.1 (a and b).

The XeCl excimer laser radiation (308nm, 20ns, 5 Hz) focused by a quartz spherical lens is introduced into the vacuum chamber through a plane-parallel quartz window. The lens mounting ensured adjustment of the beam-focus position on the target surface. In our experiments, the

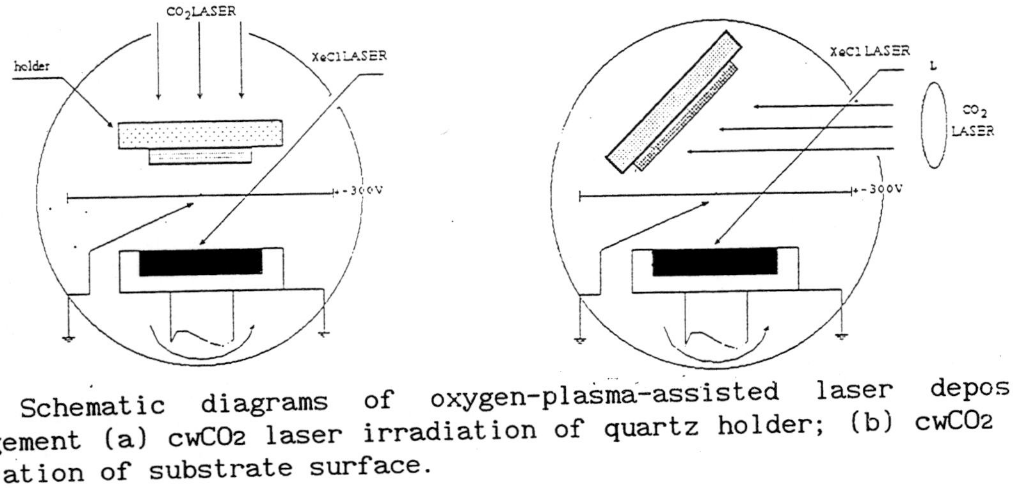

FIG.1 Schematic diagrams of oxygen-plasma-assisted laser deposition arrangement (a) cwCO_2 laser irradiation of quartz holder; (b) cwCO_2 laser irradiation of substrate surface.

laser-energy density on the target surface was 2.0-3.8 $J.cm^{-2}$. The distance between the target and the substrate was 30-50mm. The substrate was heated by a cw CO_2 laser with output power of up to 70 W entering the vacuum chamber through a NaCl window. The substrate temperature was selected by adjusting the CO_2-laser power directed on the back of substrate silica holder (Fig.1a) or on the substrate surface (FIG.1b). The substrate temperature was controlled by a thermocouple in direct contact with the backside of the substrate. The dimensions of the CO_2 laser beam spot on the substrate surface was controlled by moving the GaAs spherical lens. The texturing of the pellet surface during the deposition process was avoided by rotation using a dc motor situated outside the vacuum chamber.

A dc bias voltage (U_{dc}=+300 V) was applied on an intermediate ring electrode placed between the substrate and the grounded target. In addition, an auxiliary needle electrode (electrically grounded) has been located at the ring-electrode center ensuring stable electrical discharge at oxygen pressure of up to 200 mTorr. The substrate was kept at floating potential.

A pellet of superconducting material with a nominal composition of $Y_1Ba_2Cu_3O_x$ prepared by a standard ceramic technology was used as a target. During the deposition, the ambient oxygen pressure was kept in the interval 50-200 mTorr. The substrate temperature (T_s) was optimized in the interval 500 - 600°C. After the film deposition, the samples were slowly cooled down to 400°C and held at that temperature for 30 min in about 1 bar of O_2. Finally, the temperature of the samples was gradually decreased to room temperature.

3. Results and discussion

Typical experimental results are summarized in TABLE 1. The films deposited on $SrTiO_3$ substrates at 50 mTorr oxygen pressure and substrate temperature of 500°C (Sample 1) showed onset transition temperature T_c=89K, relatively broad transition - ΔT_c=5K and low ratio r=R(300K)/R(100K)= 2,1. The SEM analysis showed a granular morphology of the films with high particulate density (FIG.2.a). Although the films were relatively homogeneous and without defects, obviously nonsuperconductive uncrystallized particles existed and optimization of the oxygen content was still needed. The use of cw CO_2 laser irradiation directed on the substrate surface (FIG.1.b) and higher partial oxygen pressure (up to 200 mTorr) have improved the superconducting properties of the films deposited on $SrTiO_3$(100) substrate considerably (see Sample2) - T_c= 90K, ΔT_c= 2K and r up to 2,8. The SEM analysis (FIG.2b) showed substantially lower particulate density. The SEM analysis of the film deposited on MgO (100) substrate at the identical condition (Sample 3) showed the same effect of particulate density reduction (FIG.2c).

TABLE 1 Superconducting transition temperatures and process parameters of $Y_1Ba_2Cu_3O_7$ films deposited at two different arrangements of cw CO_2 laser heating of the substrate.

substrate	E $J.cm^{-2}$	Ts C^o	$P(O_2)$ mTorr	Tc onset K	ΔT_c K	R(300)/R(100)
			Schematic diagram (a)			
1-$SrTiO_3$	2.0	500	50	89	5	2 13
			Schematic diagram (b)			
2-$SrTiO_3$	3.2	550	200	90	2	2.90
3-MgO	3.8	550	150	89	6	1.20
4-YSZ/Si	2.0	550	100	88	11	semiconduct.
5-YSZ/Si	2.0	600	150	87	-	semiconduct.
6-YSZ/Si	2.8	500	100	83	-	semiconduct.

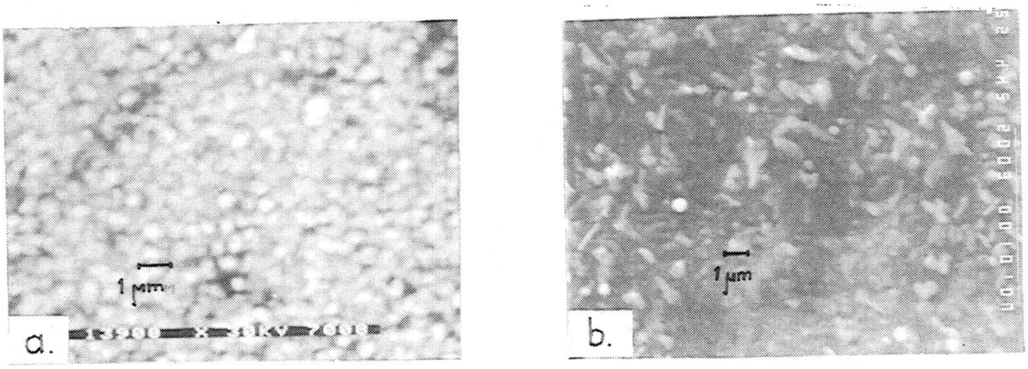

FIG.2 SEM - pictures of $Y_1Ba_2Cu_3O_{7-x}$ films (a) on $SrTiO_3$ (100) - Sample 1; (b) on $SrTiO_3$ (100) - Sample 2.

Various techniques have been used by many groups to eliminate or reduce the density of particulates: deflector vane, electrostatic bias, colliding gas streams, variable laser flounce, spinning disk source, second laser. At power densities below 10 $J.cm^{-2}$ near the evaporation thresholds, the particulate density increased with power density[13]. Decreasing the evaporation laser wavelength results in lower number and size of particulates on the film substrate. Koren et al.[14] gave an explanation in the stronger absorption of shorter wavelength photons by the clusters ejected from the target and in the products of their subsequent fragmentation. The linear absorption coefficients are 1.2, 1.5, and $1.7x10^5 cm^{-1}$ for 1064, 532, and 355 nm, respectively [14]. An explanation of the particulate densities reduction in the case of samples deposited by substrate surface cw CO_2 laser heating (scheme b) with respect to those deposited with quartz holder heating (scheme a) probably is the absorption of IR photons by the photofragments in the excimer laser-produced plasma and on the film surface and their subsequent decomposition, nevertheless of very low absorption coefficient of the fragments at 10.6μm ($\sim 10^3 cm^{-1}$)[15].

In order to avoid the strong substrate-film interaction and migration in the case of Si substrates we used a buffer layer of YSZ(111) deposited on Si(100) by sol-gel method. The XRD investigations showed that these buffer layers were polycristalline, textured with (111) orientation. The thicknesses of YSZ layers were in the range of 150-200nm. The best results were obtain at substrate temperature Ts=550°C (Sample4)- T_{con}= 88K, ΔT_c=11K. The increasing of Ts to 600°C (Sample 5) causes broad transition bellow liquid nitrogen temperature probably because of the stronger film-substrate interaction. By reduction of Ts to 500°C (Sample 6) the worsening of the superconducting properties was observed - bad crystallization of the film and lowest T_c=83K with very broad transition.

TABLE 2. EDAX analyses of YBCO thin films deposited on $SrTiO_3$ and YSZ/Si.

		YBCO/$SrTiO_3$		YBCO/YSZ/Si	
Element	Line	Weight %	Atomic %	Weight %	Atomic %
Si	K	0.00	0.00	5.10	7.89
O	-	15.99	52.22	20.62	55.95
Y	K	15.10	8.87	15.20	7.42
Ba	L	40.21	15.30	31.29	9.89
Cu	K	28.71	23.61	27.15	18.55
Zr	K	0.00	0.00	0.63	0.30

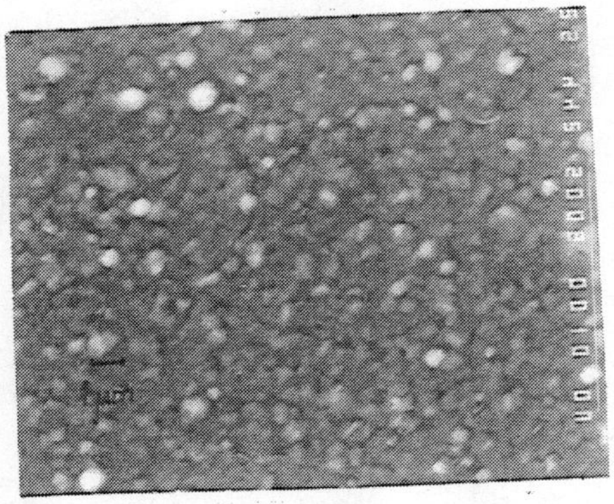

FIG.3 SEM - pictures of $Y_1Ba_2Cu_3O_{7-x}$ film on YSZ/Si - Sample 3.

The EDAX analysis of $Y_1Ba_2Cu_3O_{7-x}$ films deposited on $SrTiO_3$(100) and YSZ(111)/Si(100) at laser energy densities of 2.0-2.8 $J.cm^{-2}$ (see TABLE 2) showed the strict stoichiometric transfer of the target composition to the substrate, but in the case of the silicon substrate the strong film-substrate interdiffusion was observed. The SEM picture of Sample 3 (Fig.3) shows clearly the cracks of the film. Obviously, the packing density of YSZ buffer layers was too low to prevent interdiffusion and stress even at substrate temperature as low as 550°C.

4. Conclusions

The results described in this paper show the possibility of successful application of "in situ" oxygen plasma-assisted laser deposition of $Y_1Ba_2Cu_3O_{7-x}$ thin films at low temperature (500 - 600°C) combined with cw CO_2-laser heating of the substrate and subsequent annealing in oxygen.

The direct cw CO_2 laser irradiation of substrate surface and excimer laser plume have been used successfully to reduce the particulate density in order to obtain smooth superconducting films.

The results of the use of YSZ buffer layer on Si show the need of optimization of the buffer layer packing density and the conditions of laser deposition, heating, and annealing.

5. References

1. J.Bednorz and K.Muller, **Z.Phys. B64**, (1986) 189.
2. E.Fogarassy, C.Fuchs, F.Kerherve, G.Hauchecorne, and J.Perriere, **J.Appl.Phys. 66**, (1989) 457.
3. L.Lynds, B.R.Weinberger, G.G.Paterson, and H.A.Kranski, **Appl.Phys.Lett. 52**, (1988) 320.
4. V.Serbezov, S.Benacka, D.Hadgiev, P.Atanasov, N.Elektronov, V.Smatko, and V.Stribik, **J.Appl.Phys. 67**, (1990) 6953.
5. K.H.Wu, C.L.Lee, J.Y.Jang, T.M.Uen, and Y.S.Gou, **Appl.Phys.Lett. 58**, (1990) 1089.
6. P.Dyer, A.Issa, P.Key, P.Monk, **Supercond.Sci.Tech. 3**, (1990) 472.
7. S.Otsubo, T.Minimakawa, Y.Yonezawa, T.Maeda, A.Morimoto, and T.Shimizu, **Japan.J.Appl.Phys. 28**, (1989) 2211.
8. B.Roas and L.Schultz, **Appl.Phys.Lett. 53**, (1988) 1557.
9. G.Koren, A.Gupta, E.A.Giess, A.Segmuller, and R.B.Labowitz, **Appl.Phys.Lett. 54**, (1989) 1054.
10. T.Venkatesan, E.W.Chase, X.D.Wu, A.Inam, C.C.Chang, and F.K.Shokoohi, **Appl.Phys.Lett. 53**, (1988) 243.
11. S.Witanachchi, H.S.Kwock, X.W.Wang, and D.T.Shaw, **Appl.Phys.Lett. 53** (1988) 234.
12. R.K.Singh, J.Narayan, A.K.Singh and J.Krishnaswamy, **Appl.Phys.Lett. 54**, (1989) 2271.
13. H.Sankur and J.T.Cheung, **Appl.Phys.A 47**, (1988) 271.
14. G.Koren, A.Gupta, R.J.Baseman, M.I.Lutwyche and R.B.Laibowitz, **Appl.Phys.Lett. 55**, (1989) 2450.
15. S.Otssubo, T.Minimakawa, Y.Yonezawa, T.Maeda, A.Morimoto and T.Shimitzu, **Japan.J.Appl.Phys. 28**, (1989) 2211.

Vacuum Electron and Ion Technologies
D. S. Karpuzov (Ed.)

ISBN 1-56072-242-8

EFFECTS OF LATENT TRACKS FORMED AT HIGH ENERGY I ON IMPLANTATION IN CRYSTALS

F.F. Komarov
Institute of Applied Physics Problems, Belorussi an State
University, 7 Kurchatov Str., Minsk, 220064, Belarus

Abstract. The influence of various effects inherent to the high energy ion bombardment of crystals on the latent track formation is treated. The nature of tracks is dependent on a type of solids. The relaxation of strong electron excitation arrisen under irradiation of materials by energetic ions is the main parameter which defines a nature of track regions. The track migration of implanted atoms as well as track channeling of incident ions are discussed for ion implantation in semiconductors.

I. Introduction

For energetic ions with a kinetic energy of E $\gtrsim$ 1 MeV/amu intensity of energy deposition in the electron subsystem is 10^3 - 10^4 times more higher than in nuclear subsystem. It may be a few KeV/Å for heavy ions (for example, M $\simeq$ 100 amu, E $\simeq$ 1 MeV/amu, dE/dx $\simeq$ 2 $\div$ 3 KeV/Å). High energy deposition in the electron subsystem increases the role of electronic excitations in structure defects generation, intensive inelastic sputtering of materials, and promates a range of specific effects such as track formation, local melting, amorphization, creation of unusual solid phases (high pressure phases and fullerenes), generation of shock waves and material destruction. The most important result of high energy ion implantation in solids is the formation of specific stretched macrodefects-latent tracks.

First latent tracks have been observed already in 1959 in mica by electron microscopy [1]. Price and Walker [2] have used chemical etching and optical microscopy to detect nuclear tracks induced by fission fragments in mica. Later, the track morphology was investigated by different methods which revealed spherical and discontinuous or continuos cylindrical regions.

Lately, new classes of materials such as magnetic insulators ($Y_3Fe_5O_{12}$, $BaFe_{12}O_{12}$, $NiFeO_4$, $ZnFe_2O_4$ etc.), high temperature superconductors (HTS) ($YBa_2Cu_3O_7$, $Bi_2Sr_2Ca_1Cu_2O_8$), some metallic alloys illustrated a behaviour to form tracks at irradiation by energetic ions [3-6].

The radiation effects in the electronic stopping power regime have been investigated using the heavy ion beams delivered by the GANIL (Caen, France), Unilac (Darmstadt, Germany), U - 300 (Dubna, Russia) and TASCC (Chalk River Laboratories, Canada) accelerators.

It has been found recently that this cylindrical track structure in HTS causes larger increases in magnetization and crytical current at high temperature and high magnetic field than other types of defects in these materials [3-4]. This can be understood as a consequence of the effective pinning of magnetic vortex lines provided by the cylindrical (columnar) defects. By 1-GeV lead ions irradiation of the $YBa_2Cu_3O_{7-\delta}$ ceramics, an enhancement of the critical current density up to a factor 5 is obtained [3]. Measurements confirm that the pinning is strongest when the magnetic field is aligned with the ion tracks.

II. Track Nature and Relaxation of Strong Electronic Excitations.

The nature of tracks is dependent on a type of solids. In a case of insulators such as dielectrics, magnetic insulators as well as HTS ceramics and some conducting alloys (for example Ni - Zr - alloy) [7] that tracks consist of amorphous material.

In metals these observations involve the mechanism of point defect clustering resulting either from thermally activated migration or from athermal processes occuring in a wake of the incident ions. There is now much experimental evidence that high electronic excitations can stimulate a modification of the damage processes resulting from nuclear elastic collisions in metals. Moreover the observed effects are very complex [6] since in some metals and alloys such as Cu and Cu_3Al no effect has been revealed; on the contrary, in some materials such as Ni, Ni_3Fe electronic excitations promote a partial radiation annealing of defects produced in elastic collisions. In a third category of metals [6, 7] such as Ga, Fe, Co, Ti and Zr a strong enhancement of damage production accurs above an electronic stopping power threshold. The value of electronic energy loss above which the defect generation increases is about 0,5 keV for the FeCo alloy [6] and ~ 4 keV/Å for Fe [8].

The relaxation of strong electron excitations arrisen under irradiation of materials by energetic ions is the main factor which defines a nature of track regions. In many perfect metals the relaxation is determined by electron thermal conductivity. Thus the

$$t_1 = R_o^2 / 4K_{min}$$

approximately equals to 10^{-15} s. Here $R_o \approx 1$ nm is the cylindrical excited region and K is the electron thermal diffusivity, $K_{min} \simeq 1$ cm^2/s [9]. The electron-phonon relaxation time is approximately 10^{-12} - 10^{-13} s [9, 10]. The electron cooling time in such metals is much less than the electron-phonon interaction time.

A qualitative analysis of these processes can be done using the treatment of Davydov and Kalinichenko [11] for penetration of fission fragments in a matter. When an energetic ion penetrates a metallic target, the cylindrical region of overheated electron gas with the effective radius $R_{eo} \simeq I^2/4\pi Ze^2n$ (I is the ionization potential, Z is the atomic number, n is the atomic density of a target) is formed after the processes of ionization and electron-electron relaxation. An electron temperature decreases, in main, due to diffusion of hot electrons. At this time stage a heat transfer from electrons to the crystal lattice is negligible, i.e.

$$\beta / C_e (T_e - T_i) \ll 1 / C_e \, \mathrm{div} (\varkappa \nabla T_e) \qquad (1)$$

where β is the coefficient of heat transfer from electrons to the lattice, T_e and T_i are the temperatures of electron gas and lattice, respectively; C_e and $\varkappa_e$ are the electron heat capacity and heat conductivity. The mentioned mechanism of cooling is changed to another one when

$$1 / C_e \left| \mathrm{div}\,(\chi_e \nabla T_e) \right| \simeq \beta / C_e\,(T_e - T_i), \qquad (2)$$

i.e. the electron gas temperature is decreased essentially and heat transfer to the lattice is the more important process of cooling. The Eq. 2 can be used for estimation of a track radius R_c taking into account the electron diffusion. For an approximate estimation of R_c a few simple expressions can be involved: $\mathrm{div} \sim \nabla \sim 1 / R$, $T_e \simeq 1 / R \sqrt{(dE/dx) / \pi \gamma}$, $\gamma = C_e / T_e$ and $T_i \ll T_e$ and then R_c is defined as

$$R_c = \left(\frac{\gamma}{\beta} K_e \sqrt{\frac{dE/dx}{\pi \gamma}} \right)^{1/3}. \qquad (3)$$

Here $K_e = \chi_e / C_e$ is the temperature conductivity coefficient (thermal diffusivity) for electron gas which is determined by the length of electron-phonon collisions. Time of the heat transfer from electrons to lattice

$$\tau_{ei} = \frac{1}{R} \sqrt{\frac{dE/dx}{\pi \gamma}} \Big/ \frac{\beta}{\gamma} = \frac{R_c^2}{K_e} \qquad (4)$$

is of the order of a duration of the first stage (diffusion of "hot" electrons). According to such approach a change of the overheated region dimension at the second stage is neglected, that is why the radius of energetic track is about R_c and the its temperature is described as

$$T_e = \frac{dE/dx}{\pi R_c^2 C_i} \qquad (5)$$

After equilibration of the electron and lattice temperatures a stage of relatively slow heat spreading from the energetic track takes place. This stage is characterized by the coefficient of heat conductivity $K = \mathcal{X}_e / C_i$. As a life time for the energetic track the characteristic time of heat spreading $\tau_T = R_c^2 / K$ can be chosed. The strict theory of temperature evolution in tracks formed by high energy ions has been represented in Ref. [12].

The results (Table 1), deduced according to this treatment, explain the lack of destructive tracks in metallic targets with high electroconductivity (for example: Cu, Al, Ag, Zn, Cu_3Al and perfect structure. On the contrary crystal structure damages can be observed in tracks of heavy ions or fission fragments penetrating such targets as Fe, Ti, Zr, Pl, U. In this case the temperature in tracks exceeds the melting temperature (Table 1).

However, we have to mention that this discription is relevant only for metals with the perfect crystal structure. If a crystal structure has microdefects then a cross section of the energetic track can be described by characteristic dimensions of substructure elements such as grains in polycrystalline targets or interfaces (if its dimension is less then R_c). Many experimental results on efficient inelastic sputtering of fine-grained metallic targets [13] as well as insulating materials [14] support this idea. If the electron subsystem energy is closed up in the special region of a crystal, strong heating of the lattice and even target destroy can happen. Of course, the electron-phonon relaxation time τ_{ei} is strongly dependent on other structure microdefects: dopands (in semiconductors), dislocations, microtwins etc.

In insulators free electrons are absent outside the excitation region. The double electric layer is formed on the excited region boundary [14]. A width of this layer is approximately equal D, where D is the Debye length. The boundary electric field prevents from escaping the electrons which can ionize lattice atoms. Therefore the lattice atoms receive the energy enough for melting of some region in a crystal.

In dielectrics (gypsum, mica, diamond, water) a radius of energetic tracks can be about $R_{eo} \sim 1 - 10$ nm. Corresponding overheatings with $T_e > 10^4$ K should promote boiling up and chemical decomposition of a matter in tracks as well as nonlinear generation of powerful mechanical pulses [11].

At high dose irradiation $D \simeq 5 \times 10^{13} - 10^{14}$ ion/cm^2 (taking into account a diameter of tracks) a buried layer or a layer extended from the surface to the definite depth, which consists of amorphous material can be formed during high energy ion implantation. This effect has been observed at irradiation of U_3Si by fission fragments [15]. Overlapping of a specimen cross-section by tracks having an amorphous phase in its volume could explain the mentioned phenomenon.

The gradual accomulation of amorphous phase till the total amorphization is revealed also in a case of 235 MeV ^{84}Kr, 50 MeV ^{32}S and 50 MeV ^{63}Cu ions penetrating yttrium ion garnet $Y_3Fe_5O_{12}$ or $Cd_3Ga_5O_{12}$ targets [16]. The high resolution transmission electron microscopy (HRTEM) micrograph and diffraction patterns show total phase transition of these materials for high fluences ($7,5x10^{13}$ Kr/cm^2 and $2x10^{14}$ S/cm^2). The track core diameters were about 6,5 - 7,0 nm here. Note that all these amorphization cross sections of YIG or GGG, i.e. the track-core cross sections, are 6 orders of magnitude higher than those of the nuclear collision regime [17]. Thus, due to very high quenching rates (> 10^{11} K/s), complete amorphization can be achieved in a case of dielectrics. For metals the final picture can be characterized by high amount of "quenched" defects in a recrystallized volume.

Recrystallization of the 50 MeV ^{63}Cu tracks for higher fluences was observed in YIG, leading to grain sizes around 10 nm, while no recrystallization was found for the 235 MeV ^{84}Kr tracks [16].

Following the "thermal-spike" model already used for the recrystallization of a - Si [18] and a - Ge [19], the process must go through two steps. The first step means energy deposition through primary and secondary electrons, and exchange of energy by electron-electron collisions, occuring between 10^{-16} and 10^{-13} s. The second step comes when atomic motion can occur after electron-phonon collisions during a few Debye periods between 10^{-13} and 10^{-12} s, in which a high non-equilibrium density of defects is produced and then quenched in.

The release of this stored energy through the thermally-activated processes of recovery and recrystallization can occur later on. The grain sizes are lower than 10 nm i.e. its value is close to the track-core diameter.

In this scheme, recrystallization can occur when an ion hits a track-core of radius R_c. The probability of such "second-hit" events P can be approximated [16] as

$$P = 4{,}7\pi R^2 \Phi t \exp(-4{,}7\pi R^2 \Phi t),$$

where Φt stands for the fluence. This recrystallization process is thought to be triggered by the dense electronic excitations produced by energetic ions.

Chromatographic analyses of solvent extracts from graphitc bombarded with 13 MeV/amu $^{161}Dy^{22+}$ ions give clear evidence for the in situ synthesis of fullerene - C_{60} in track regions [20]. This is extremely interesting phase exciting many scientists through the world at present. These results of Chadderton with coworkers [9] will be published soon.

For semiconductors no reliable proof for track formation existed yet.

III. Track Morphologies and Threshold Energies for Track Formation

To correlate damage efficiencies and morphologies one needs a direct observation of the defects. High resolution electron microscopy appeared soon as a valuable tool for this purpose. A systematic study of the defects created by heavy ion irradiations in magnetic insulator was underdaken by Houpert et al. [21], Studer et al. [5] and Constantini et al. [16, 17]. Five specified regions can be introduced according to these results (fig.1):

(1) In range I the damage results only from the nuclear collisions. The amorphization cross section (A) in the high energy regime will be in the range of 10^{-18} - 10^{-16} cm^2.

If one considers the damage efficiency $\mathcal{E}$, as defined by the ratio $\mathcal{E} = A / (dE/dx)$, the shape of the curve as a function of dE/dx can be understood as the amorphization low of each material. This saturation curve, shown here for the ittrium garnet (fig.1 upper part) is characterized by an electronic stopping power threshold T_S (T_S = 0.8 keV/Å), above which the damage efficiency increases suddenly, and a saturation behaviour in the domain of high electronic stopping (above 2.2 keV/Å in the YIG) where the damage effeciency stárts being independent (dE/dx). In the saturation region a damage cross section A is proportional to (dE/dx).

(2) At $T_e \leq dE/dx \leq T_S$ the damage induced by inelastic collisions overcomes the nuclear elastic collisions one. The defect morphology is then spherical. The A values are ranging between 10^{-16} to 10^{-14} cm^2.

(3) At $T_S \leq dE/dx \leq T_C$ a percolation of the extended sperical defects leads to discontinuous cylindrical tracks. A is ranging between 10^{-14} to 10^{-13} cm^2.

(4) Between T_C and T_m, the cylindrical defects become continuous but the damage is inhomogeneous inside this cylinder. Etchable tracks appeared with a distribution of diameters observed by optical microscopy, which narrows as (dE/dx) increases (10^{-13} < A < $3x10^{-13}$ cm^2).

(5) At $dE/dx > T_m$ the defects are cylindrical and homogeneus inside the cylinder of defects ($A \gtrsim 5x10^{-13}$ cm^2). In this region the radii of latent tracks deduced from the damage cross sections ($\pi R^2 = A$) are in agreement with the HREM observations. The values of T_i(i = e, s, c, m) are dependent on the specific materials while amorphization cross section A will have the some values for all materials in the different quated ranges.

In the case of amorphous creation in $Y_3Fe_5O_{12}$ the corresponding values are [5, 16] T_s = 0.8 keV/Å; T_c = 1.5 keV/Å; T_m = 2.2 keV/Å while T_e is unknown. The threshold energy losses (T) at which elongated extended defects appear deduced for other magnetic insulators are: 1.0 keV/Å in $BaFe_{12}O_{10}$ and 1.4 keV/Å in $NiFe_2O_4$ [5].

The dependence of track morphology on the rate of electronic energy loss in the HTS is illustrated in Table 2. According to these experimental data the above mentioned threshold energy for $YBaCu_3O_7$ are $T_s \approx$ 1.9 keV/Å and $T_c \approx$ 3 keV/Å.

IV. Track Formation Mechanisms

The experimental results reveal the nonuniform distribution of the radiation defects generated by energetic ions in solids along the ion pathway. It results in a multilayer defect structure of the irradiated target. The target can contain five specific layers at the most [22] (fig.2):

(i) A slightly damaged layer occurs at the surface. The layer contains, in the main, homogeneously distributed point radiation defects. The layer is formed when $\sigma \gg \sigma_i$. Here

υ is the ion velocity and υ_i is the velocity of the deep shell electrons of the target atoms.

(ii) A layer of increased concentration of point defects which are distributed non-homogeneously is formed when $\upsilon \approx \upsilon_i$. If the electronic energy loss is high enough (i.e. $(dE/dx)_e > E_T$), this layer contains the tracks. In some cases it can be the upper layer in a multilayer structure.

(iii) The disordered layer lies at the depth of the maximum nuclear stopping power.

(iv) The ion-implanted layer is doped with the implanted atoms.

(v) The layer of defects produced by secondary irradiation caused by nuclear revetions of ions interacted with target atoms. This damaged layer can be produced by channeled ions as well [24 - 27].

In these cases the radiation defects or implanted atoms spreads much deeper then the ion-implanted layer predicted according to standard calculations (fig. 3 and 4).

The HRTEM studies concluded that the tracks are produced by the very high electronic stopping power of ions, which causes permanent damage if the rate of energy loss is above a certain threshold. Mechanisms of enhanced damage production in different materials are still under investigation. Nevertheless a list of possible mechanisms initiating the tracks by high energy ions contains the following [22, 23].

(1) The "Coulomb explosion" mechanism is very effective and takes place when the velocity of ions becomes equal to the velocity of K - shell electrons of target atoms. This mechanism is very effective for light targets (carbon, boron nitride, silicon etc.).

(2) A high density of ionization leads to a significant reduction in the threshold energy E_d for removing atoms out of the lattice sites. The smaller value of E_d increases the probability of the displacement of those matrix atoms which have obtained small amounts of energy from the penetrating ion.

(3) The thermal mechanism is very important in many cases as it was discussed in Sec. II.

There is a very scarce information on acoustic pulses (or shock ware) generation and its influence on the track formation [28], yet.

In general case, a few mechanisms can contribute to a total effect of track formation simultaneously. The most probable mechanism of the track production in light insulators (mica, gypsum, polymers etc.) is proved to be Coulomb explosion [29, 30] of positively multiply charged zones. The origin of this phenomenon is ionization of the deep shells of target atoms with the subsequent Auger recombination process [31, 32] when the neighbouring atoms to the ion lose almost all the electrons. The probability of the Auger recombination is about unity for light and medium mass atoms. For carbon and silicon these values are of 0.9 and 0.96 respectively [33].

The track diameter strongly depends on the $(dE/dx)_e$ value, since the electronic stopping governs the deep shell ionization process. Every substance has its own definite threshold E_T of the $(dE/dx)_e$ value. When exceeding this value the track formation takes place [34]. If the electronic stopping power only slightly top the E_T value the ion track appeares like a chain of spherical somewhat stretched ionization defect spikes. In this case the track diameter is proportional to $(dE/dx)_e^{1/3}$. When the inelastic stopping is great enough to produce a continuous cylindrical (column type) track, the track diameter linearly follows the $(dE/dx)_e$ value. Such a dependence strongly points that ionization process plays the leading role in the track formation mechanism.

In order to check this assumption, we have estimated the probability of K- and L- shells ionization for target atoms in the onefold collision depending on the ion energy E and impact parameter b for the various ion-target combinations in accordance with the model from Ref. [35]. On the basis of this model the numerical simulation of the atom ionization processes in the lattice by the moving ion has been performed using the Monte Carlo procedure [36]. The energy of the interparticle pushing-apart during ionization of atoms (per atom) was calculated according to the expression:

$$E_i = -\sum_{j \neq i} \frac{g_i g_j l^2}{r_{ij}} ,$$

where the summation is performed over the nearest neighbouring atoms.

Fig.1 (a,b) represents the dependence of the average atom energy in the diamond lattice acquired due to ionization on the ion impact parameter. This dependence E_i (b) is compared with the dependence of the threshold energy E_d for defect formation on the impact parameter P of the ion passing near the treated atom. The latter quantity is calculated from the approximate formula $E_d \simeq E_{do}^{x}$ x(1-P((b)). For diamond crystals the value $E_{do} \simeq 50$ eV was used. The intersection of both curves b_m can be interpreted as the lower limit value of the impact parameter at which an atom can be knocked out the lattice position, i.e. the approximate track radius. In the cases of implantation of the 60 MeV Ni^+ and 130 MeV Xe^+ ions into diamond crystals, the b_m values come to 2 and 3 Å respectively, which are relatively consistent with the track radius values as it will be discussed in Sect. V. All these data speak in favour that the throwing out of the atoms from the track core during its formation takes place and strongly depends on the matrix parameters.

In fact many experiments observed the somewhat lower density within the core region then that of regular matrix. For example, in zircon [37] the atom density inside the 14 MeV/amu Pb ion track has been established at about 10% of the undamaged zircon crystal atom density. In gypsum crystal [38] the lack of atom density within the 1 MeV/amu Xe ion track is up to 100%, that is an empty hole is formed along the high energy ion path. In polymers this value is about of 12% [39] and the track densities in $YBa_2Cu_3O_7$ produced by the 1.0 GeV Au irradiation are within 20% of the nominal values [3]. The reduced atom density in the center of tracks appearing in garnets and spinels was also reported in Ref. [40].

According to these results one can imagine the track as a hole or a multivacancy cord surrounded by an interstitialrich envelope. The crystalline envelope surrounding the ion track can sbject to high compression stresses. This effect was revealed lately [37, 41 -44].

The observed geometry of tracks is very favourable for formation of solid fullerenes in carbon containing materials (diamond, graphite or polymers). The buckyball type of fullerenes (a molecule diameter is about 7 Å) can be formed in discontinuous tracks ("beads" like) or continuous columnar tracks. As a matter of fact this effect was revealed by Chadderton and coworkers [20]. To my mind, the buckytube type of solid fullerenes (a molecule diameter is about 20 Å) could be produced in the mentioned above materials at penetration of heavy energetic ions with E $\gtrsim$ 1 GeV, when a diameter of low atomic density cores of continuous cylindrical tracks is about a few nanometers. High energy conditions in the cores of energetic ion tracks in carbon-rich solids might support formation of the highly stable fullerene structures through radiation-enhanced chemical reactions, enhanced carbon atom mobility and clustering of supra-thermal atomic carbon [20]. The dense ionization plasma generated in tracks may be a much more efficient environment for this phase production in the geometry of tightly bound energetically favoured quasi-spherical or cylindrical structures.

There are not any obvious manifestations of track formation in semiconductors up to now. Nevertheless there are experimental data obtained on semiconductors which are quite analogous to those obtained on typical insulators and which can be accounted well for as a display of high energy ion tracks. For instance, the high pressure local regions were observed in wide-gap semiconductors such as diamond and cubic boron nitride irradiated with high energy ions [42, 45]. The shape of the compression value distribution through depth correlates well with the electronic stopping power distribution through ion irradiated layer.

On the basis of EPR investigations of diamonds irradiated by 63 MeV Co ions and 26.7 MeV Ne ions the authors of Ref. [46] have showed that the formation of a quesi-one-dimensional structure (track like) with nontetrahedral long-range atom order took place.

V. Track migration and track channeling.

There are other specific effects relating to the ion tracks in semiconductors. They are migration of atomic species along the tracks [45] and channeling of ions in tracks formed at a previous stage of irradiation [36, 47]. The "tube" like structure of tracks with low atomic density can result in acceleration of diffusion.

It is well known that there is no impurity diffusion in diamond until the temperature above 2000 °C. So the appearance of any diffusion through long distance at lower temperatures can be explained as a result of increasing the mobility of atoms by some specific structure formed during penetration of swift ions.

This is confirmed in the experiments on irradiation of diamond and cubic boron nitride by energetic ions [45]. The effect of "track migration" is illustrated in fig.6. Two profiles of the distribution of helium atoms implanted into diamond with energy of 300 keV (fig.6a) and of 200 keV (fig.6b) and subsequently annealed at 1000 °C are compared.

If the diamond sample was preliminary irradiated with 1 MeV/amu Xe ions then helium atoms migrate at a temperature of 1000

°C to a depth of 6 um. According to fig.6 the helium migration is effective only in the track-containing layer so far as the penetration of helium atoms does not take place at a depth below the beginning of the xenon-containing layer which is rich in defects. The first peak in the helium distribution profile means the initial helium-implanted distribution, the second one reveals the distribution of helium atoms captured into the tracks and then migrated along them. The mechanism of "track migration" can be used for explaining the appearance of a long tail to the surface in the xenon atom distribution when irradiation was performed at 1350 °C. If xenon irradiation of the sample is carried out at room temperature the subsequent heating does not redistribute xenon atoms out of the ion-implanted layer. The authors of the paper [45] suppose that some xenon atoms implanted into the heated lattice do not have enough time to get into stable position and hence migrate through the tracks back to the surface.

As a posible manifestation of tracks at high energy ion implantation can also be the effect of deep penetration of a portion of ions captured into tracks already created. Indeed , one can hardly imagine the absence of the influence of the channels with the reduced atom density (or quite hollow) on the path of fast ions moving in such channels, if the channels are perfectly alligned with the ion penetration direction. Obviously the movement of high energy ion through ion track is characterized by a lower electronic stopping. As a result a deeper penetration of the ions captured into the tracks as compared with the ions moving through undamaged matrix occures. This deeper ion penetration leads to a considerable broadening of the final implanted atom distribution. If the ion tracks are emty holes (as in Ref. [38]) one can even expect doubling the standard ion penetration depth R_p .

The following features are inherent to this effect. First, the deeper penetration of impuritis appeares at doses exceeding a critical one. Second, the deeper penetration forms the specific deep tale in the distribution of implanted atoms. When ion dose reaches a value about 10^{14} - 10^{15} cm^{-2} all the irradiated surface is covered with the tracks and the probability for the following ions to be captured into tracks becomes considerable.

Computer simulation of high energy ion penetration in solds performed recently [48-51] has shown that among the effects characteristic of high energy ion implantation, the fluctuations of ion charge state and crystallographic channeling influence the implanted atoms and radiation defects profiles considerably. Nevertheless, the account of only these factors can not explain a lot of experimental data [36, 47]. More adequate discription of these data has been reached when the effect of ion channeling through the tracks is also taken into account.

The channeling through the ion tacks is quite similar to the common axial crystallographic channeling. For description of the movement of energetic ions in such tracks the low energy version of the Fokker-Plank equation developed by Beloshitsky and Kumakhov (see: [52]) for proton and α- particles channeling in crystals was adopted for track channeling [36, 47], namely

$$\frac{\partial F(x,E_\alpha)}{\partial x} = \frac{\partial}{\partial E_\perp}\left[S(E_\alpha)D(E_\perp)\frac{\partial}{\partial E_\perp}\left(\frac{F(x,E_\perp)}{S(E_\perp)}\right) - \right.$$

$$- \frac{\partial}{\partial E_\perp}\left[\left(\frac{dE_\perp}{dx}\right)_{loss} F(x,E_\perp)\right],$$

where $F(x,E_\perp)$ is the distribution of channeled ion through the transversal energy $E_\perp$ at the depth x. D(E) and $(dE_\perp/dx)_{loss}$ are the diffusivity and the drift velocity for the transversal energy relatively. $S(E_\perp)$ is the channel cross section accessible for the ions of $E_\perp$ energy. The penetration of dechanneled from track ions was described by the Boltzmann type transport equation [50].

The simulation results are represented on fig.7-9. The track channeling effect as well as the charge state fluctuation effect have to be taken into account for realistic deseription of the experimental results (fig.7). The double peak structure is observed on the final distribution. The second peak is caused by the track channeling. The numerical calculations by the formula mentioned above have shown that the effect of ion channeling through tracks strongly depends on the track radius r_o, the defect (scattering centres) concentration inside track N_d and the length of channeling within track L_{tr}. In order to evaluate the parameters r_o, N_d and L_{tr} only one way is to simulate experementally measured implanted impurity and radiation defects depth distributions being used as the compared data. The following values of parameters were used for obtaining the results of fig.7: r_o = 6Å; N_d = 0.02 N_a; L_{tr} = 3 μm (N_a is the atom density of diamond). One can see that the theoretical profiles of implanted atoms are very close to the experimental points.

The depth distribution of the optically active radiation damage in silicon caused by 335 MeV Ni-ions is presented in fig.8. The optical reflectivity profile is much broader than the nuclear stopping power profile given by TRIM (fig. 8). Much better agreement can be reached when taking into account the influence of the electronic stopping and the ion channeling through tracks. The $(dE/dx)_e$ $(dE/dx)_n$- dependence describes very well the slope of the optically active damage profile towords the surface. The shape of the profile towards a greater depth is in good agreement with the $(dE/dx)_n$ distribution when the channeling through tracks is taken into consideration. The following values of the parameters have been used: r_o = 3.5 Å; N_d = 0.01 N_a; L_{tr}= 20 µm. It should be marked that the theoretical prediction describes the unusual shoulder appearing in the experiment at a depth of 54 µm.

The radiation damage depth profile obtained for silicon irradiated with 377 MeV Kr-ions at a dose of $4x10^{13}$ cm^{-2} has relealed the analogous peculiarities [47]. The most pronounced feature of the damage distribution is the shoulder at a depth of 50 um. This shoulder spreads up to 10 µm from the main distribution peak. Such a shoulder one can discribe theoretically well involving the effect of ion channeling through tracks [47]. The parameters used in these calculations were: r_o = 9 Å; N_d = 0.02 N_a; L_{tr} = 30 µm.

In the same way the depth distribution of nickel atoms in diamond implanted with energy of about 1 MeV/amu [47] have been simulated (fig.9). In this case the following parameters were used: r_o = 3.3 Å; N_d = 0.02 N_a; L_{tr} = 3 µm. One can see a good agreement between the theoretical profiles and experimental points.

VI. Conclusions

The relaxation of strong electron excitations in solid induced by high energy ion bombardment is a parameter which define a final structure of the treck region. In metals the relaxation is determined by electron thermal conductivity. In insulators the elecronic heat is localized, and lattice atoms receive the energy enough for melting.

A proposed structure of the track as a multivacancy cord surrounded by an interstitial-rich envelope is confirmed in many experiments on the formation of stable local areas under high pressure and extremely efficient migration of impurities and defects through the tracks.

Hot carbon atoms in tracks can shape large, compete units-such as the fullerene-through radiation enhanced chemical reactivity. The threshold energy loss for track formation is dependent on a type of solids (metals, semiconductors or insulators) and structure of a target.

The principal result of the ion channeling through latent tracks is the deeper penetration of a part of ions resulting in a considerable tail (in some cases double-peaked picture) in the implanted impurity and radiation defect depth distributions toward a greater depth. The effect of ion charge fluctuations results in the considerable broading of implantation profiles near the ion projected range.

The discussed simulation results on energetic ion implantation in diamond and silicon allows tentative conclusions that the average path of ions channeling through the tracks in semiconductors is about of a half of the R_p-value of randomly moving ions and that the defect density within the tracks is about 0.5 - 2% of the initial matrix atom density. The dechanneling probability for ligt ions out of the tracks appears to be 5-10 times larger than for heavy ions. Such a considerable difference results probably from different structures of the latent tracks created by light and heavy ions. In silicon for light ions a good agreement between theory and experiment has been achieved by increasing the defect concentration in tracks up to 0.1 the normal density.

Using the tracks, one can find new possibilities for controlling doping of nanometric areas in crystals. Unique photoelectronic and superconductive properties of fullerenes could well lead to a new generation of electronic and optoelectronic circuit elements in graphite, diamond and polymers.

References

1. E.C.M. Silk and R.S. Barnes, Philos. Mag. 4 (1959) 970
2. P.B. Price and R.M. Walker, Phys. Rev. Lett. 8 (1962) 217
3. Th. Schuster, H. Kuhn, M.R. Koblischka, T. Theuss and H. Kronmuller, Phys. Rev. B47 (1993) 373.
4. A.D. Marwick, L. Civale, L. Krusin-Elbaum, R. Wheeler, J.R. Thompson, T.K. Worthington, M.A. Kirk, Y.R. Sun, H.R. Kerchner and F. Holtzberg, Processing High-T- Superconductors with GeV Heavy Ions, Nucl. Instr. Meth. (B), to be published.
5. F. Studer and M. Toulemonde, Nucl. Instr. and Meth. B65 (1992) 560.
6. J.P. Riviere, J.F. Dinhut, E. Paumier and J. Dural, Nucl. Instr. and Meth. B80/81 (1993) 65.
7. A. Barbu, A. Dunlop, D. Lesueur and R.S. Averback, Europhys. Lett. 15 (1991) 37.
8. A. Dunlop, D. Leseuer, J. Morillo, J. Dural, R. Spohr and J. Vetter, Nucl. Instr. and Meth. B48 (1990) 419.
9. V.V. Katin, Yu.V. Martynenko, Yu.N. Yavlinskii, in: Interaction of Corpuscular Flows with Solid Surfaces, ed M.S. Sondov (FA N. Tashkent, 1992) p.147.
10. E.L. Yoffa, Phys. Rev. B21 (1980) 2415.
11. A.A. Dauydov, A.I. Kalinichenko. Problems of Nuclear Physics and Cosmic Rays 26 (1986).
12. A.I. Glushzov, F.F. Komarov, A.P. Novikov, A.I. Urbanovich and A.G. Hoang, Reports of Belorussian Acad. Sci. 31 (1987) 609.
13. YU.V. Martynenko, Yu.N. Yavlinskii, Surface (USSR) 6 (1988) 5.
14. I.A. Baranov, S..Tsepelevich, Surface (USSR) 1 (1989) 73.
15. I.A. Naskidashvily, H.E. Maile in: Radiation Effects in Solids (Naukova Dmka, Kiev, 1977), p.24.

16. J.M. Costantini, F.Rovel, F.Brisard, M.Caput and C.Cluzeau, Structural Modifications of Yttrium Iron Garnet after High-Energy Heavy Ion Irradiations, to be published in Nucl. Instr. and Meth.

17. J.M. Costantini, F.Brisard, J.L.Flament, A.Mftah, M.Toulemonde and M.Hage-Ali, Nucl. Instr. and Meth. B65 (1992) 568.

18. K.Izui, J.Phys. Soc. Jpn. 20 (1965) 915.

19. C.Dufour, B.Lesellier de Chezelles, V.Delignon, M.Toulemonde and E.Paumier, in: Modifications Induced by Irradiation in Glasses, Gd.P.Mazzoldi (Elsevier, Amsterdam, 1992).

20. L.M.Chadderton, D.Fink, H.Mockel, K.K.Dwivedi and A.Hammoudi, Fullerene Formation in the Tracks of Energetic Ions, to be published in Nucl. Instr. and Meth.

21. C.Houpert, F.Studer, D.Groult and M.Toulemonde, Nucl. Instr. and Meth. B39 (1989) 720.

22. A.M.Zaitsev, Mater. Sci. and Eng. B11 (1992) 179-190.

23. L.T.Chadderton and I.M.Torrens, Fission Damage in Crystals, (Methuen, London, 1969).

24. A.F.Burenkov, V.S.Varychenko, A.M.Zaitsev, F.F.Komarov, G.G.Konoplyanik, A.A.Melnikov, V.S.Stelmakh and V.V.Tkachev, Phys. Status Solidi A115 (1989) 427.

25. F.F.Komarov, .F.Burenkov, G.G.Konoplyanik,V.S.Varichenko, A.M.Zaitsev, N.M.Kazyuchits and V.V.Karpovich, Nucl. Instr. and Meth. B48 (1990) 439.

26. A.F.Burenkov, F.F.Komarov, S.A.Fedotov, V.S.Varichenko, A.M.Zaitsev and N..Kazyutchitz, Rad. Eff. and Def. in Solids 25 (1993) 169.

27. A.F.Burenkov, F.F.Komarov and S.A.Fedotov, Nucl. Instr. and Meth. B67 (1992) 30.

28. L.M.Ljamshev and B.I.Chelnokov, in: Radiation Acoustics (Science, Moscow, 1987) p.8.

29. S.Klaumunzer, M.-D.Hou, G.Schumacher, Phys. Rev. Lett. 57 (1986) 850.

30. R.L.Fleischer, P.B.Price, R.M.Walker, Nuclear Tracks in Solids (University of California Press, Berkeley, 1978) p.14.

31. T.A.Tombrello, Nucl. Instr. and Meth. B1 (1984) 23.

32. T.A.Tombrello, C.R.Wie, N.Itoh, T.Nakayama, Phys. Letters A100 (1987) 42.

33. J.D.Garcia, R.J.Fortner, T.M.Kavanach, Rev. Mod. Phys. 45/2 (1973) III.

34. F.Studer, D.Groult, N.Nguen, M.Toulemonde, Nucl. Instr. and Meth. B19/20 (1987) 806.

35. F.F.Komarov, Rad. Eff. 46 (1979) 39.

36. F.F.Komarov, S.A.Fedotov, V.S.Varichenko, A.M.Zaitsev and N.M.Kazyuchitz, Influence of the Fluctuations of Ion Charges and Latent Tracks on the Process of the High-Energy Ion Implantation, Nucl. Instr. and Meth., to be published.

37. L.A.Bursill, G.Braunshausen, Phil. Mag. Ab2 (1990) 395.

38. I.V.Vorobjeva, V.E.Monastyrenko and V.P.Perelygin, Solid State Physics 31 (1989) 1, in russian.

39. R.Spohr, P.Armbruster and K.Schaupert, Rad. Eff. and Defects in Solids 110 (1989) 27.

40. C.Houpert, D.Groult, F.Studer and M.Toulemonde, Rad. Eff. and Defects in Solids 110 (1989) 189.

41. C.Houpert, M.Hervien, D.Groult, F.Studer, M.Toulemonde, Nucl. Instr. and Meth. B32 (1988) 393.

42. V.S.Varichenko, A.M.Zaitsev, A.A.Melnikov, V.F.Stelmakh, Superhard Materials 1 (1989) 3, in russian.

43. R.B. Guimaraes, L. Amaral, M. Behar, D. Fink and F.C. Zawislak, J. Mater. Research 3 (1988) 1422.

44. R.B. Guimaraes, L. Amaral, M. Behar, D.F.P. Fichtner, F.C. Zawislak and D. Fink, J. Appl. Phys. 63 (1988) 2086.

45. A.M. Zaitsev, Mater. Sci. and Engineering B11 (1992) 179.

46. D.P. Erchak, V.G. Efimov, A.M. Zaitsev, V.F. Stelmakh, N.M. Penina, V.S. Varichenko and V.P. Tolstykh, Nucl. Instr. and Meth. B69 (1992) 443.

47. A.M. Zaitsev, S.A. Fedotov, A.A. Melnikov, F.F. Komarov, W.R. Fahrner, V.S. Varichenko, and E.H. Te Kaat, Influence of Tracks on Penetration of High Energy Ions in Semiconductors. Simulation with Transport Equations, Nucl. Instr. and Meth., to be published.

48. A.F. Burenkov, F.F. Komarov and S.A. Fedotov, Phys. Stat. Sol. B169 (1992) 33.

49. A.F. Burenkov, F.F. Komarov, S.A. Fedotov, Nucl. Instr. and Meth. B67 (1992) 30.

50. A.F. Burenkov, F.F. Komarov, S.A. Fedotov, Nucl. Instr. and Meth. B67 (1992) 35.

51. A.F. Burenkov, F.F. Komarov, S.A. Fedotov, V.S. Varichenko, A.M. Zaitsev and N.M. Kazyutchitz, Rad. Eff. and Def. in Solids 25 (1993) 169.

52. M.A. Kumakhov, G. Shirmer, Atomic Collisions in Solids (Gordon and Breach, NY) 1989, 272 pp.

Table 1

Dependence of track radious R_c, overheating temperatures T_c, formation time τ_{ei} and life time τ_T for tracks formed by fission fragments (light group) an a type of metals [11]

Metal	R_c, 10^{-6} cm	T_c, °C	τ_{ei}, 10^{-12} c	τ_T, 10^{-12} c
Cu	8.0	25	0.3	65
Al	2.5	130	0.06	7.5
Zn	2.3	300	0.1	13
Pb	1.6	1300	0.2	11
Fe	0.5	6000	0.06	1

Table 2

Dependence of track morphology on the rate of electronic energy loss in $YBaCu_3O_7$ and $Bi_2SrCa_2O_8$ [4]

Ion	Energy (GeV)	dE/dx (keV/Å)	Morphology
		$YBaCu_3O_7$	
Br	1.4	1.1	none
Xe	3.5	1.9	discontinuous
Sn	0.58	2.8	discontinuous
Pb	5.3	3.5	continuous
Au	1.0	3.8	continuous
		$Bi_2Sr_2Ca_1Cu_2O_8$	
I	0.5	2.7	continuous

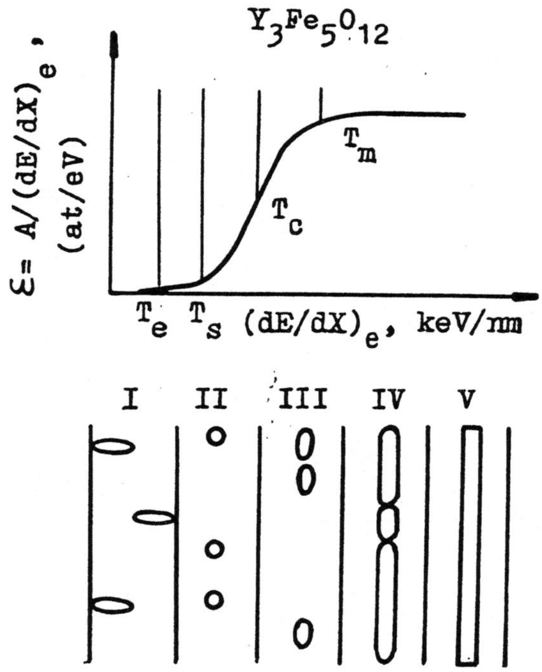

Fig. 1. General behaviour of the track formation and damage efficiency as observed in magnetic insulators [5]. In the upper part of the figure the evolution of the damage efficiency versus (dE/dx) is represented.

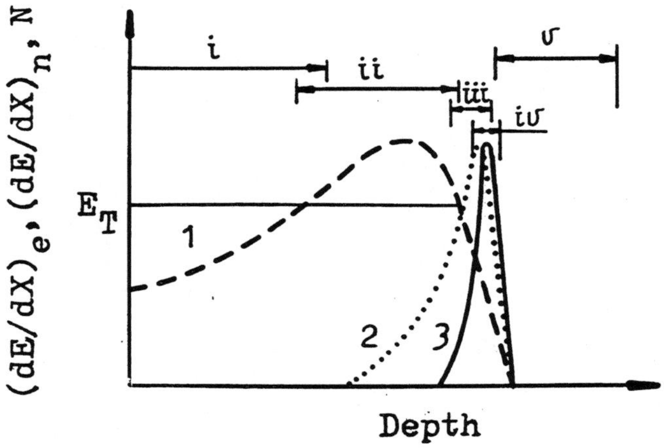

Fig. 2. The outline of the electronic (curve 1), nuclear (curve 2) stopping power curves as well as distribution of implanted impurity concentration N (curve 3) for ions with energy of several MeV/amu. E_T denotes the threshold (dE/dx) value for the track formation. i, ii, iii, iv and v mark the layers according to the text.

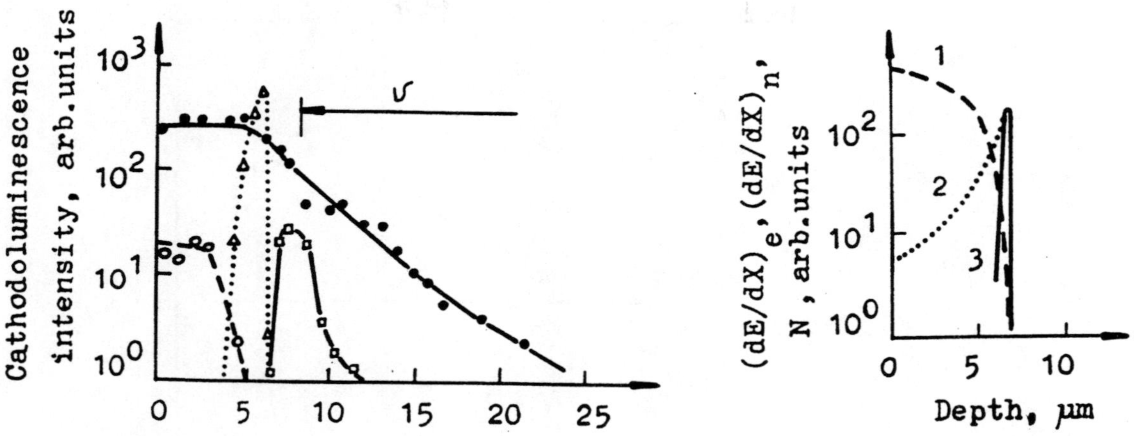

Fig. 3 (a). The depth distribution of cathodoluminescence intensity (reduced to the A band intensity) of the centres at 575 nm (●), 638 nm (○), 415 nm (□) and the B band (Δ) in type Ia diamond implanted with 58 MeV copper ions and annealed at 1400 °C [22].

(b). As for fig. 2.

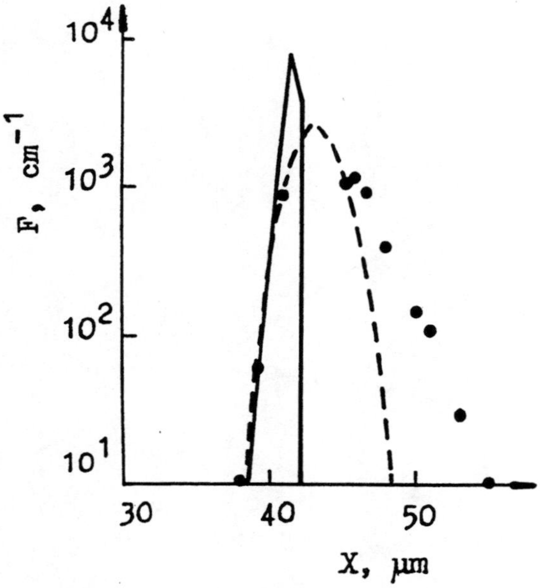

Fig. 4. The experimental depth distribution profile for the impurity in the diamond crystal implanted by nitrogen ions with an energy of 60 MeV [24] (circles) is compared to calculated stopped ion distribution profiles (broken line) obtained taking account of the ion charge state fluctuations [27] and without it (solid line). All distributions are normalized to unity.

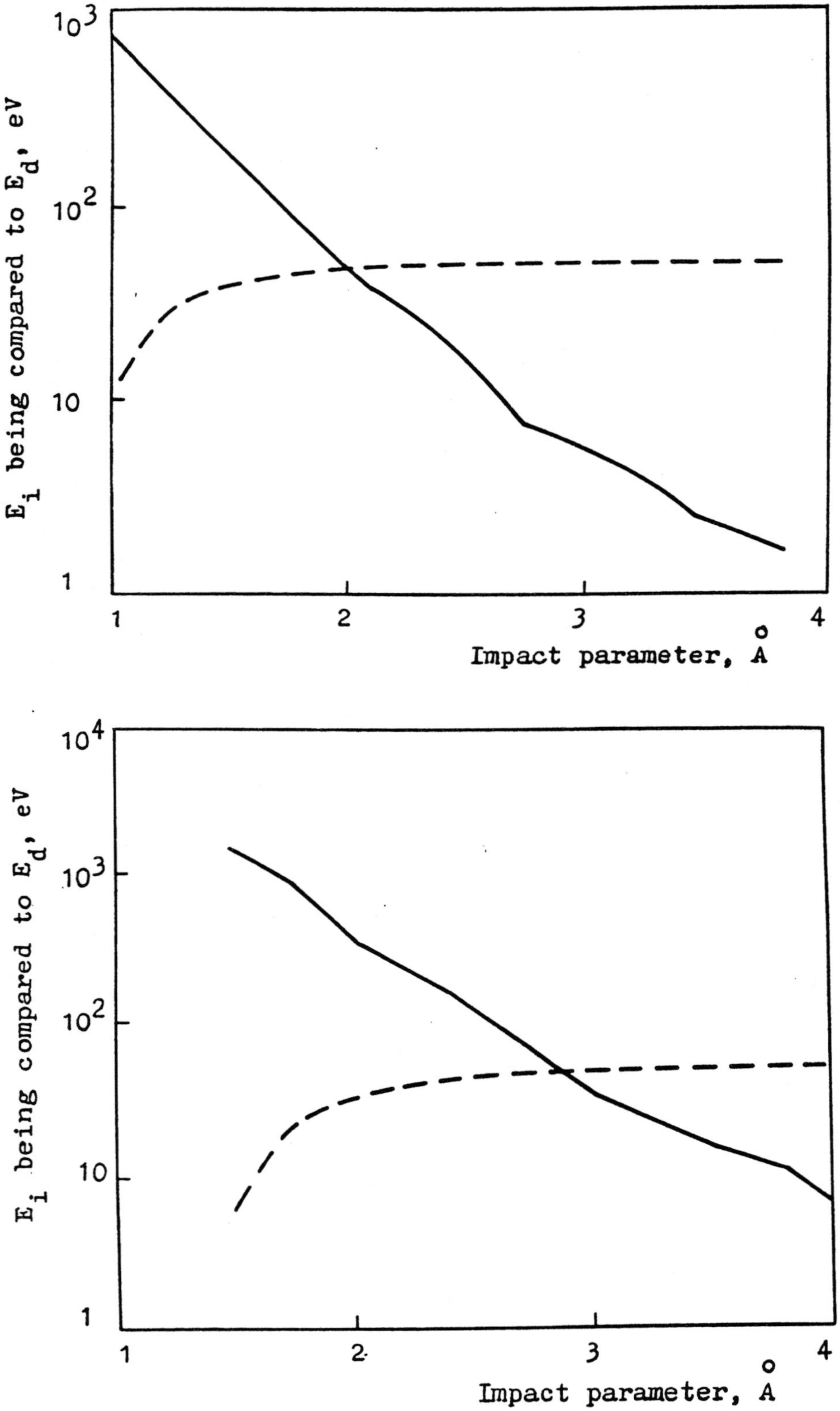

Fig. 5. Impact parameter dependence of the energies $E_d(b)$ (dashed line) and $E_i(b)$ (solid line) for diamond crystals implanted with 60 MeV Ni^+-ions (a) and 130 MeV Xe^+-ions (b).

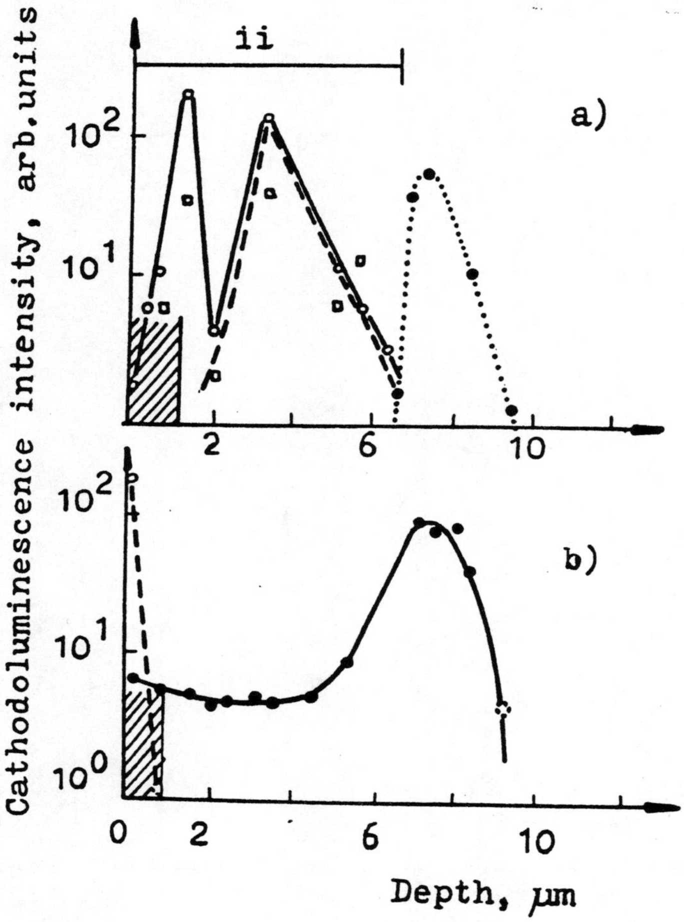

Fig. 6. The migration through the tracks in diamond irradiated with 124 MeV Xe^+-ions [45], showing the depth distribution of helium-containing centres at 536.5 nm (o) and 560.5 nm (□) and a xenon-containing centre at 816 nm (•) registered by the luminescence method: (a) irradiated at room temperature with xenon ions and then with 300 keV helium ions and subsequently annealed at 1000 °C and (b) irradiated at 1350 °C with xenon ions and than at room temperature with 200 keV helium ions and subsequently annealed at 1000 °C: ii is the layer containing the tracks; --- calculated helium-ion -implanted layers.

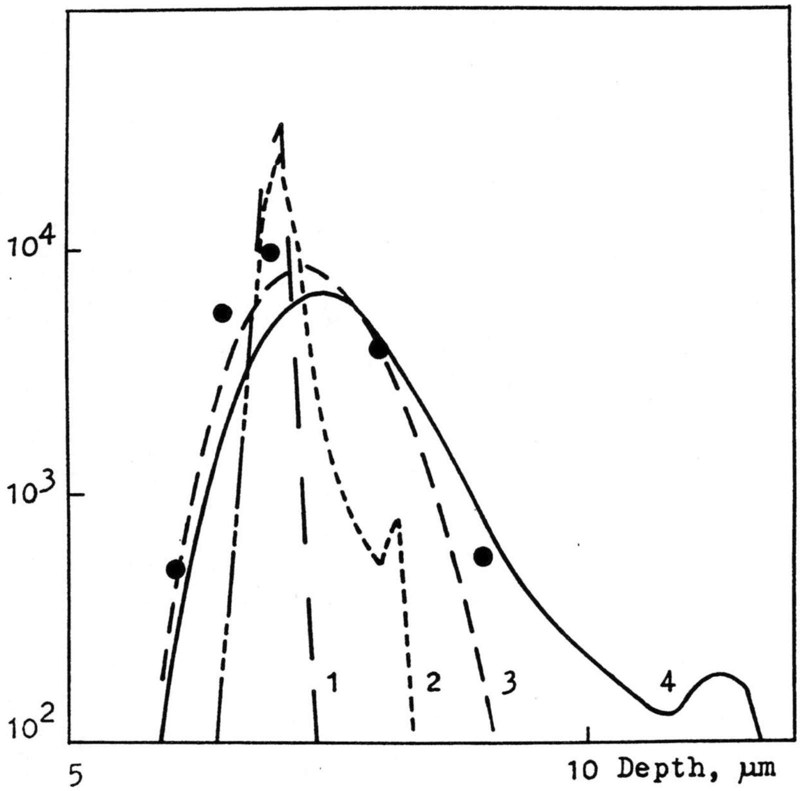

Fig. 7. Depth distributions of implanted atoms in diamond irradiated with 130 MeV Xe^+ -ions:
1 - standard calculation; 2 - using the Boltzmann type transport equation and taking into account particle motion in tracks; 3 - the charge state fluctuation effect is treated alone; 4 - both effects are taken into account; o - experiment [24]. All distributions are normalized to 1.

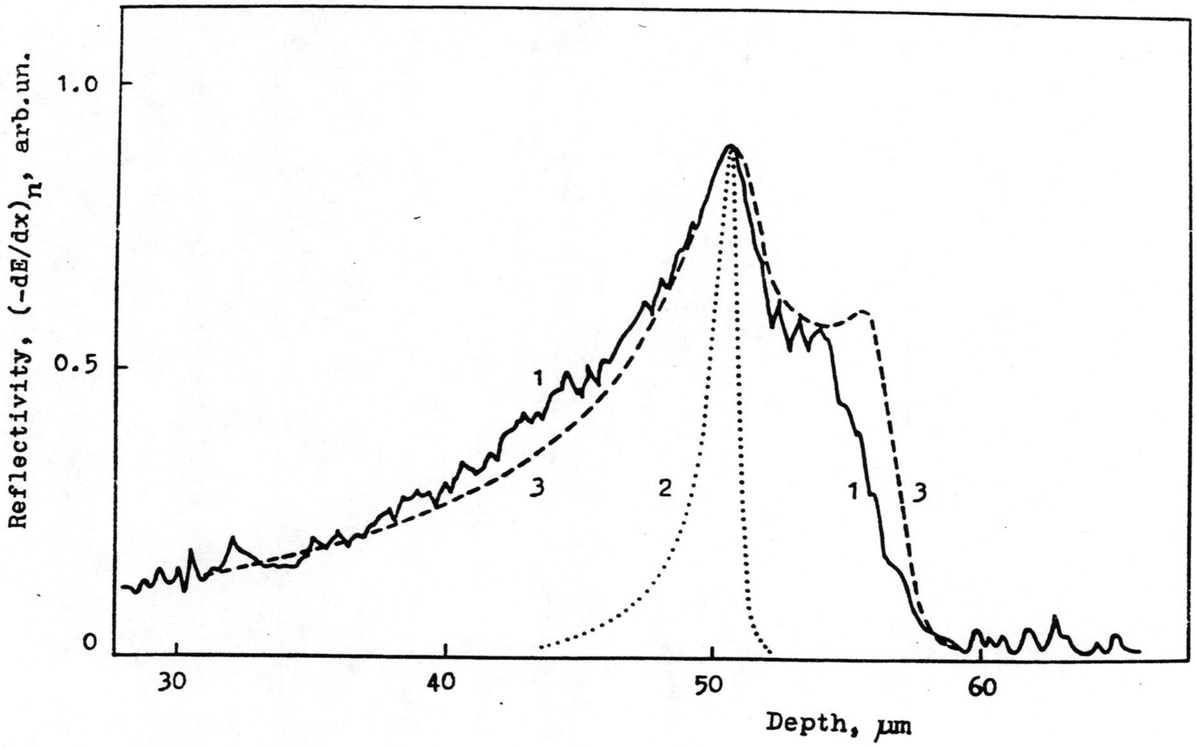

Fig. 8. Depth distribution of optical reflectivity at a wave length of 650 nm (1) on silicon implanted with 335 MeV Ni-ions at a dose of $5x10^{14}$ cm^{-2} [47]. The bevel angle is 10°. Nuclear stopping power depth distribution (2) and radiation defect depth distribution calculated taking into account both the influence of ionization on the amorphization damage production and the ion channelingl through tracks (3).

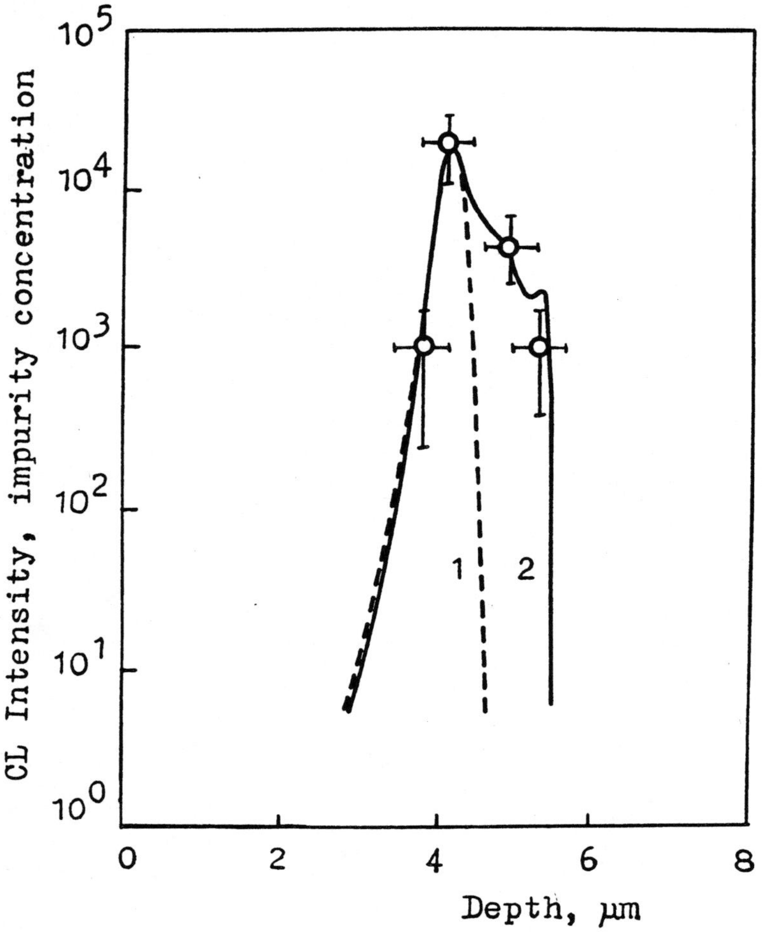

Fig. 9. Depth distribution profiles of nuclear stopping power for 60 MeV Ni^+ -ions in diamond (1), taking into account the ion channeling through tracks (2). The circles are the experimental values of cathodoluminescence intensity of the implanted nickel centres [47].

Vacuum Electron and Ion Technologies
D. S. Karpuzov (Ed.)

ISBN 1-56072-242-8

ENHANCED INTERDIFFUSION IN GaAs/AlGaAs SINGLE QUANTUM WELLS DUE TO ARSENIC IMPLANTION

I.V. Bradley, W.P. Gillin, K.P. Homewood and R. Grey*.
Department of Electronic and Electrical Engineering, University of Surrey, Guildford, Surrey, GU2 5XH, U.K.
*Department of Electronic and Electrical Engineering, University of Sheffield, Mapping Street, Sheffield, S1 3JD, U.K.

Abstract

Implants of arsenic and gallium ions have been made into GaAs/AlGaAs single quantum wells to a dose of 10^{13} atoms/cm^2. In each case the implantation energies are chosen so as to get the peak of the implantation profile at the centre of the well. Using repetitive anneals and photoluminescence from the n=1 electron to heavy hole transition, it has been possible to follow the evolution of the diffusion coefficients for the interdiffusion with anneal time. There was little difference between the behaviour of implanted and unimplanted materials except for an initial broadening of the wells for the implanted samples which is thought to be due to collisional mixing or recovery from gross damage caused by the ion beam. However some enhancement of the diffusion coefficient was found for the sample implanted with 10^{13} cm^{-2} arsenic. This enhancement was not as dramatic as that previously discovered in the InGaAs/GaAs system.

Introduction

Since Laidig (1) discovered that GaAs/AlGaAs quantum well heterostructures were unstable against zinc diffused from a surface source, many authors have studied the effects of different impurities on III/V semiconductor heterojunctions. Ion implantation is the preferred technique used to introduce impurities into semiconductors. If these two technologies could be married it should be possible to create passive and active optoelectronic devices on planar quantum well structures, using implants and anneals only. The effect on interdiffusion coefficients of implantation of ions such as Si(2) and Zn(3) have been studied. However both Zn and Si are dopants and can introduce significant carrier densities into the structures, which is undesirable for many applications. Many other authors have studied the effects ion implantation of the lattice constituents (such as gallium(4), arsenic(5) and aluminium(6)) as well as a range of other ions. Gallium has been particularly widely studied since it can be used in liquid metal ion sources, producing a very stable, finely focused beam. This technique can be used in direct beam writing, enabling implantation of features less that 100nm to be produced without use of a mask. Authors have claimed that both gallium(7) and arsenic(5) produce enhanced diffusion in the ion implanted regions, however the effect of the implanted ion and implant damage have never been separated.

In this work we make a comparative study of the effects of implants of lattice constituents gallium and arsenic into AlGaAs/GaAs single quantum wells, (SQWs). By the use of repetitive anneals and photoluminescence measurements on individuals samples it is possible to follow the evolution of the diffusion coefficient with anneal time. Using a standard method for modelling the shift in the photoluminescence peak, in order to convert the shift into a diffusion length(6), and then by plotting the calculated diffusion lengths squared as a function of the total anneal time it has been possible to observe changes in the diffusion coefficients with time and ascribe different regions to possible diffusion mechanisms(8).

Experimental method

The samples used in this work were 100Å GaAs SQWs in $Al_{0.2}Ga_{0.8}As$ barriers, with the overlayer thickness of 1000Å, a further 500Å cap of GaAs was grown to protect the upper AlGaAs barrier which would oxidise on contact with the air. All samples were grown by molecular beam epitaxy (MBE) on (100) orientated GaAs substrates. The growth temperature was calibrated against the sharp C(4x4) to C(4x2) transition in the surface reconstruction, monitored using reflection high energy electron diffraction (RHEED). This transition (T_t) occurs at 530°C at the 5:1 arsenic to gallium ratio used throughout this work. For the samples used in this study the GaAs was grown at T_t + 49°C and the AlGaAs at T_t + 105°C, the wells being grown at the optimum temperature for that material, with the temperature being ramped up or down during the growth of the barriers. A growth rate of approximately one monolayer per second was used for the well region, enabling accurate control of well widths.

After growth, the photoluminescence peak energy of each wafer was measured in order to determine the initial points from which to measure the shift in photoluminescence peak energy. Quarter wafers were implanted with gallium or arsenic doses of 10^{13}

atoms/cm^2 respectively, while a further quarter wafer was used as a control sample. In addition, dual implants were also made. One quarter wafer was implanted with 7.5 x 10^{12} atoms/cm^2 arsenic and 2.5 x 10^{12} atoms/cm^3 gallium and a further quarter wafer was implanted with equal doses (5 x 10^{12} atoms/cm^2) of the two ions. The total dose in all case was thus 10^{13} atoms/cm^2. In order to keep the peak of the implant range profile at the well centre the implant energies were varied in accordance with calculations using SUSPRE (9). The implant energies used were 255 KeV and 260 KeV for the gallium and arsenic implants respectively.

All wafers were encapsulated on both front and back surfaces with 500Å and 1000Å respectively of silicon nitride in a plasma enhanced chemical vapour deposition system at 300°C. All anneals were performed in a double graphite strip annealer (10) with a nitrogen ambient, and with the samples placed between the two strips which were resistively heated. Temperature control was produced by use of a two colour optical pyrometer as part of a feedback loop linked to the transformer supplying power to the graphite strips. The system was calibrated against the melting point of gold and silver and was always found to be within ±5°C The furnace has a rise time of ~3 seconds between 700°C and 900°C and a fall time of ~5 seconds over the same range. As the ramp phases are short, their effects may be neglected even for the shortest anneals used. All anneal times were measured as the time from when the anneal temperature is reached until the current to the strip is switched off. In order to reduce the scatter, on the results of the repetitive anneals and photoluminescence measurements, a full set of data was obtained at a given temperature before the furnace temperature was reset. By averaging over many data points from a single sample, the accuracy with which the value of the diffusion coefficients can be determined is improved. Moreover, errors caused, for example by implant damage transiently enhancing intermixing, can be identified, this is not possible if only one data point was taken.

To convert the photoluminescence peak shift to diffusion length upon intermixing, a model is required to estimate the well shape after interdiffusion. In this work it is assumed that Fick's second law is obeyed. For the case of a GaAs/AlGaAs single quantum well with infinitely thick barriers, the aluminium concentration profile in the growth direction in the growth direction C(z) is given by,

$$C_z = C_o \left\{1 - \frac{1}{2}\left[erf\left(\frac{h-z}{L_D}\right) + erf\left(\frac{h+z}{L_D}\right)\right]\right\}$$

were C_o is the initial aluminium concentration of barriers, 2h is the well width, z is the distance in the growth direction with z = 0 at the well centre and $L_D = 2(Dt)^{1/2}$ the diffusion length, where t is the time and D is the diffusion coefficient. This solution for Fick's law is adequate if the well is separated from the surface, or a second source of aluminium atoms, by at least three diffusion lengths, conditions which were satisfied in all these experiments.

Once the interdiffused well shape was determined, the conduction and valence band profiles can be calculated from the empirical relationship between the band gap E_g and the concentration x. For AlGaAs the relationship obtained by Bosio (11) from photoluminescence of single quantum wells was used,

$$E_g = 1.516 + 1.36x + 0.22x^2$$

The band offset ratio in all cases was taken to be 60:40 (12) and the electron effective mass was calculated using a linear extrapolation between the value for GaAs (0.067) and AlAs (0.150). The heavy hole effective mass was taken to be 0.62 in all cases(13).

Consequently, when the band structure profiles for the diffused well have been determined, the Schrödinger's equation can be solved, using a shooting method. Thus the confined energy levels for both conduction and valance bands can be determined and the photoluminescence transition energy calculated. From this theoretical data, the measured shift in the photoluminescence with annealing can be can be converted to a characteristic diffusion length, L_D.

Plotting the square of the measured diffusion lengths (L_D^2) against the total anneal time for a single sample, the assumption that Fick's law diffusion process is correct if the data points lie on a straight line. The gradient of this line is 4D where D is the diffusion coefficient for intermixing. If the diffusion coefficient changes with anneal time, as is possible particularly in the case of implanted samples, this is evident from this plot. These results should provide some insight as to the microscopic processes controlling the interdiffusion.

Results and discussion

The effects on interdiffusion of the implantation of gallium, arsenic and dual implants of gallium and arsenic into GaAs/$Al_{0.2}Ga_{0.8}As$ SQWs, to a total dose in each case of 10^{13} atoms/cm^2 for samples annealed at 900°C are shown in figure 1. For the unimplanted control sample it can be seen that the fits to the data of the L^2_D against anneal

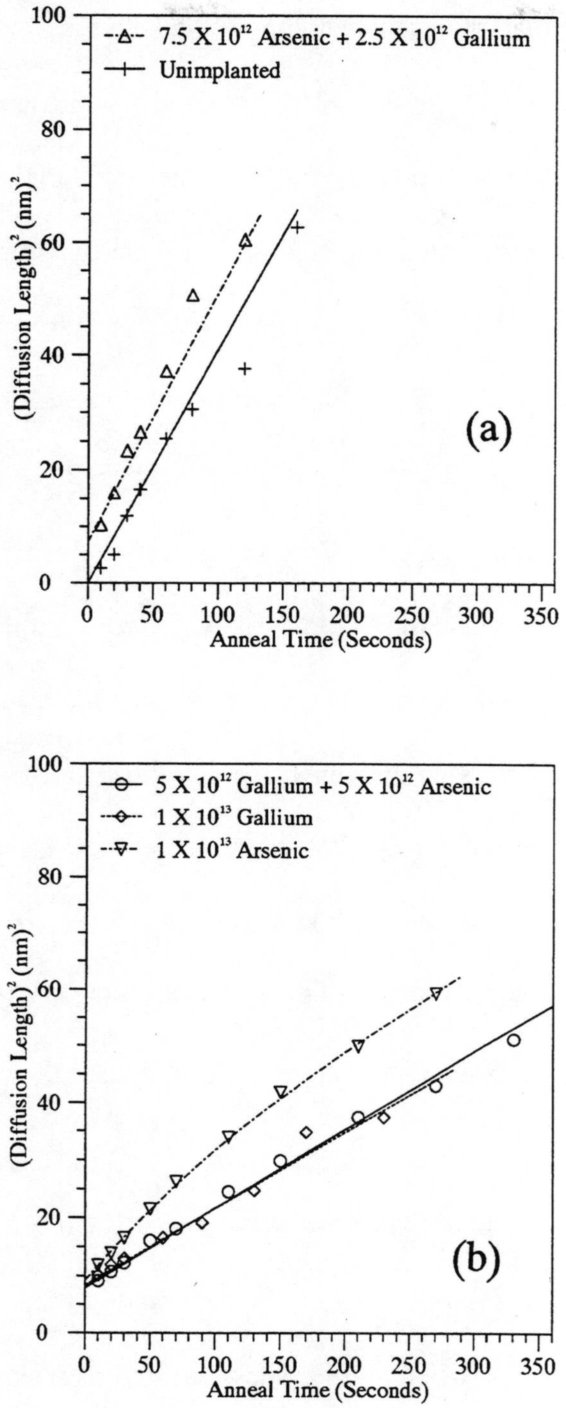

FIGURE 1. The variation of L_D^2 with anneal time for a 100Å GaAs/$Ga_{0.2}Ga_{0.8}$As single well after dual implants with gallium and arsenic, annealed at 900°C. The results shown in figure (a) were measured from samples taken from one wafer and (b) from samples taken from a second nominally identical wafer.

time plots are straight lines that pass through the origin. This shows that our initial assumption of a Fick's law diffusion process is correct and the diffusion proceeds at a constant D which can be determined for the gradient. The plots for the ion implanted samples all show similar behaviour and this is different for that of the unimplanted sample. The graphs of the implanted samples can be divided up into two regions, firstly the intercept with the y axis (at t = 0), this is not zero as in the case of the unimplanted samples but were equivalent to diffusion lengths in the range 28 Å to 32 Å. These intercepts suggest that a significant degree of intermixing occurs during the implantation or the first anneal. However from simulations(8), it has been predicted that probably only 2Å of broadening was caused directly by collisional intermixing due to the ion beam. It is thus suggested that the initial intermixing may either be due to rapid diffusion of the point defects, created by the implant, as the crystal damage regrows or by a radiation enhanced process during implantation. The intercept on the y axis for all the implants are almost identical and are also similar to those found previously in the InGaAs/GaAs system (13), as would be expected since the damage profiles are similar in all cases.

In the second region, for all implanted samples the diffusion obeys Fick's law. Figure 1 is divided up into graphs (a) and (b), with the samples shown in figure (a) being taken from one wafer and (b) from a different one. The measured diffusion coefficients on each wafer were approximately equal, with the gradient of the L_D^2 versus anneal time for the unimplanted and 7.5×10^{12} atom/cm^2 arsenic + 2.5×10^{12} atoms/cm^2 gallium implant being the same, (figure a). Also, the gallium single implant and dual implant with equal doses again give the same diffusion coefficients, (figure b). However, there was a factor of four difference in the measured diffusion coefficients between the two wafers. From previous experiments on samples taken from nominally identical wafers, annealed under identical conditions, variations of up to a factor of four in the measured diffusion coefficients were observed. Thus this variation in measured diffusion coefficients is believed not to be due to implantation effects.

It has been shown that, for the InGaAs/GaAs system, for arsenic implanted samples(8), the diffusion behaviour is greatly modified. In that case the graph can be divided into three regions. Again there was an intercept with the with the y axis. However, following this initial intermixing there was a region where the diffusion coefficient was greatly enhanced by approximately 20 times that measured for the unimplanted samples; then, at a diffusion length of approximately 85 Å, the diffusion coefficient reduced back to approximately that of the unimplanted control. In this study the arsenic 1×10^{13} cm^{-2}, also produced similar behaviour with a three stage process evident, however the enhancement in the diffusion is not as pronounced. Moreover, there seemed to be no enhancement in the cases of the dual implants with an excess of arsenic, as would be expected from the InGaAs/GaAs study. This effect however may be masked, in part, by the higher intrinsic diffusion coefficients in these samples.

Conclusion.

We have shown that using repetitive annealing and photoluminescence measurements it was possible to monitor the interdiffusion coefficients of III/V compounds accurately. It has been discovered that Fick's law was obeyed in all cases, and that it was possible to deconvolute the effect of ion implantation damage from the intrinsic diffusion and impurity effects.

It was found that samples taken from one wafer produced similar measured diffusion coefficients but there was a factor of four difference in the measured diffusion coefficients between two nominally identical wafers. It has been shown that gallium implantation has no effect on the interdiffusion coefficient for intermixing of GaAs/AlGaAs single quantum wells. However, an initial addition shift was found for all implanted samples. This initial interdiffusion is believed to be linked to recovery from implant damage. After this initial diffusion, the diffusion was found to proceed at a constant rate with no enhancement found over the unimplanted control for the dual or gallium implanted samples. For the 10^{13} cm^{-2} arsenic implanted sample some evidence of an enhanced diffusion region was found, however this enhancement was much less that observed for the arsenic implants into the InGaAs/GaAs system.

References

1 W.D. Laidig, N. Holonyak, Jr., M.D. Camras, K. Hess, J.J. Coleman, P.D. Dapkus and J. Bardeen.
Appl. Phys. Letts., **38**, *p. 776*, 1981.

2 M.D. Camras, J.J. Coleman, N. Holonyak, Jr., K. Hess, P.D. Dapkus and C.G. Kirkpatrick,
Gallium Arsenide and Related Compounds, Vol **9**, edited G. E. Stillman (IOP, London), *p. 233*, 1983.

3 J.J. Coleman, P.D. Dapkus, C.G. Kirkpatrick, M.D. Camras and N. Holonyak, Jr.,
Appl. Phys. Letts., **40**, *p. 904*, 1982.

4 Y. Hirayama, Y. Suzuki and H. Okamoto,
Jpn. J. Appl. Phys., **24**, *p. 1498*, 1985.

5 B. Elman, E.S. Koteles, P. Melman, and C.A. Armiento,

Mat. Res. Soc. Symp. Proc., **144**, *p. 445*, 1989.

6 K. Kash, B Tell, P. Grabbe, E.A. Dobisz, H.G. Craighead and M.C. Tamargo, J. Appl. Phys., **63**, *p. 190*, 1988.

7 F. Laruelle, P. Hu, R. Simes, R. Kubena, W. Robinson, J. Metz and P.M. Petroff, J. Vac. Sci. Technol., **B7**, *p. 2034*, 1989.

8 I.V. Bradley, W.P. Gillin, K.P. Homewood and R.P. Webb, J. Appl. Phys., **73**, *p. 1686*, 1993.

9 See VG IONEX, Charles Ave., Burgess Hill, West Susses. RH15 9TO.

10 R. Gwilliam, R. Bensalem, B.J. Sealy and K.G. Stephens, Physica B, **129**, *p. 440*, 1985.

11 C. Bosio, J.L. Staehli, M. Guzzi, G. Burri, and R.A. Login, Phys. Phys. Rev., **B38**, *p. 3263*, 1988.

12 V.A. Wilkinson, A.D. Prins, D.J. Dunstan, L.K. Howard and M.T. Emeny, J. Electron. Mat., **20**, *p. 509*, 1991.

13 S. Adachi, J. Appl. Phys. **53**, *p. 8775*, 1982.

Acknowledgements

The authors would like to thank the staff of the D.K. Chick laboratory for their help with the ion implantation.

Vacuum Electron and Ion Technologies
D. S. Karpuzov (Ed.)

ISBN 1-56072-242-8

CRYSTAL - BINARY COLLISION SIMULATION OF ATOMIC COLLISIONS AND DYNAMIC DAMAGE BUILDUP IN CRYSTALLINE SILICON

IR Chakarov and RP Webb
Dept. of EE Eng., Univ. of Surrey
Guildford, Surrey GU2 5XH, UK
(September 15, 1993)

A beam of fast ions (energy range approx. 1 eV/amu to 25 keV/amu) entering a crystalline silicon target is slowed down and scattered due to nuclear collisions and electronic interaction. Along its path, an individual projectile may create fast recoil atoms which in turn may initiate collision cascades of moving target atoms. These may either leave the surface (be sputtered) or deposited on a site different from their original one. Together with the projectiles being deposited in the substrate, this results in local compositional changes, damage creation and finally amorphisation of the target. Depending on the crystal orientation and/or the direction of the beam, the implanted projectiles and the damage created by them has different spatial distribution. With even more higher fluences, these fenomena will cause collisional mixing in a layered substances, changes of the surface composition due to preferential sputtering and the establishment of a stationary range profile of the implanted ions. The paths of the individual moving particles and their collisions are modeled by means of the binary collision approximation for a crystalline, polycrystalline and amorphous substance, using a screened Coulomb potential for the nuclear collisions and local or nonlocal free-electron-gas approximation for the electronic energy loss. For each nuclear collision, the impact parameter and the azimuthal deflection angle are determined according to the crystal structure using its translational symmetry. For amorphous materials the impact parameter and the azimuthal deflection angle are determined from random numbers. A proper scaling is chosen so that each incident projectile ("pseudoprojectile") represents an interval of implantation fluence. Subsequent to the termination of each pseudoprojectile and its associated collision cascades, the local concentrations of the implanted species, created vacancies and interstitials are calculated according to the density of the matrix. The running time depends strongly on the problem chosen and is mainly influenced by the number of pseudoprojectiles, their energy and their atomic species. The typical values are between 1 and 40 hours on Sparc2 workstation.

INTRODUCTION

The first computer simulations using binary collision approximation was used for studying collision cascades [1] and penetration of the projectiles in structureless materials, [2,3]. *It is worth mentioning that by means of such methods for calculating the penetration of particles into crystals, [4-6], the effect of channeling was discovered in 1963.* Nevertheless, the first calculations were performed by Beeler et al. [7]. Earlier and recent reviews in this field can be found in refs. [8-11].

A number of different codes have been developed applicable for a crystal structures. It is the MARLOWE code [12], which has international popularity and is used by many groups for studying ion implantation, sputtering and surface scattering in crystal materials. Among these programs, the CRYSTAL code was written explicitly to meet the requirements for reliable modeling of channeling implantation and damage accumulation in crystal silicon. The program is highly parallelized and vectorized, which typically reduces the computing time by a factor of 2 to 6 compared to initial version. The above reduction is believed to be quite important in view of computing times of an order of one hour to several days.

COMPUTATIONAL PROCEDURE

The Simulation Model in Binary Collision Approximation

To calculate the penetration and depth distribution of bombarding particles, the model uses so called Binary Collision Approximation (BCA), in which the deflection of the trajectories of moving particles is calculated in strict binary way - between the moving atom and closest atom from the lattice. This gives approximation to the real dynamic of the penetration of energetic particles in the matter, in which the real trajectory of the moving particle is replaced by straight lines between deflection points. The real benefit of this approach is the moderate speed of calculation combined with the possibility of including single crystal structures in the calculation.

The inelastic atomic collisions are considered to be composed of a quasielastic part and of an essentially separate electron excitation part. The trajectories of two particles interacting according to a conservative central repulsive force are shown in Fig. 1.

The barycentric scattering angle is

$$\theta = \pi - 2s \int_R^\infty dr[r^2 g(r)]^{-1}, \tag{1}$$

where

$$g(r) = [1 - \frac{s^2}{r^2} - \frac{V(r)}{E_r}]^{1/2}, \tag{2}$$

s is the impact parameter, E_r is the relative kinetic energy, r is the (variable) interatomic separation, $V(r)$ is the potential of interatomic force, and R is the apsis of the collision defined by $g(r) = 0$. The relative kinetic energy is

$$E_r = AE_0/(1+A), \tag{3}$$

where E_0 is the incident kinetic energy of the projectile and A is the ratio of the mass of the target atom to that of the projectile atom. The scattering angle θ in eq. 1 is evaluated following the formalism described in [13].

The laboratory scattering angles of the projectile and the target are calculated using the expressions

$$\tan\vartheta = Af\sin\theta/(1+Af\cos\theta), \tag{4}$$

$$\tan\phi = f\sin\theta/(1-f\cos\theta), \tag{5}$$

where

$$f = (1 - Q/E_r)^{1/2} \tag{6}$$

and Q is the energy lost by the electron excitation. The energy transferred to the target atom in the collision is

$$T = [f\sin^2\frac{1}{2}\theta + \frac{1}{4}(1-f)^2]T_m, \tag{7}$$

where the maximum transferred energy is

$$T_m = 4E_r/(1+A). \tag{8}$$

The energy of the projectile after scattering is

$$E_{recoil} = E_0 - T - Q. \tag{9}$$

Location of the asymptotic trajectories in laboratory co-ordinate system requires the evaluation of the quantities τ and τ_{rec} in Fig. 1. Because, the formalism used in [13] approximates the time integral by its hard-core value $s\tan\frac{1}{2}\theta$, $\tau_{rec} = 0$ which, results in targets moving directly off their lattice sites

$$\tau = s\tan\frac{1}{2}\theta, \tag{10}$$

$$\tau_{rec} = s\tan\frac{1}{2}\theta - \tau = 0. \tag{11}$$

Interatomic Potentials

The present program uses screened Coulomb potential

$$V(r) = (Z_1 Z_2 e^2/r)\Phi(r/a_{12}), \tag{12}$$

where Z_1e and Z_2e are the nuclear charges of the projectile and target, respectively, a_s is the screening length, which depends on the properties of the collision partners, $\Phi(x)$ is either the universal screening function described in [13]

$$\Phi(x) = 0.1818e^{-3.2x} + 0.5099e^{-0.9423x} + 0.2802e^{-0.4029x} + 0.02817e^{-0.2016x} \tag{13}$$

or the Molière approximation to the Thomas-Fermi screening function [14],

$$\Phi(x) = 0.35e^{-0.3x} + 0.55e^{-1.2x} + 0.10e^{-6.0x} \tag{14}$$

The screening lengths in the calculation are either Firsov's [15], which was delivered on the basis of approximate numerical solutions of Thomas-Fermi diatomic molecule problem

$$a_s = 0.88534 \times a_B(Z_1^{1/2} + Z_2^{1/2})^{-2/3}, \tag{15}$$

or the universal screening length proposed in [13]

$$a_s = 0.88534 \times a_B(Z_1^{0.23} + Z_2^{0.23})^{-1}, \tag{16}$$

where a_B is the Bohr radius. The screening length can be estimated in other ways - for example by matching the Molière potential at the nearest neighbor distance to a Born-Mayer potential whose parameters were obtained mainly from equilibrium crystal data [12].

Inelastic Energy Losses

In order to simulate the channeling, CRYSTAL uses nonlocal and local inelastic models. The local i.e. impact dependent inelastic energy loss model, $Q = Q(s, E_0)$, is either Firsov's [15] 5th power

$$Q(s, E_0) = \alpha E_0^{1/2}/[1 + \beta R(s, E_0)]^5, \tag{17}$$

where

$$\alpha = 0.05973 \times (Z_1 + Z_2)^{5/3}/\sqrt{M_1}, \tag{18}$$

$$\beta = 0.31 \times (Z_1 + Z_2)^{1/3}, \tag{19}$$

M_1 being the mass of the projectile or Oen-Robinson's [16] exponential model

$$Q(s, E_0) = (0.045K E_0^{1/2}/\pi a_s^2)\exp[-0.3R(s, E_0)/a_s], \tag{20}$$

where s is the the impact parameter in the collision, E_0 is the kinetic energy of the projectile, a_s is the screening length in the potential, R is the apsis of the trajectory, and K is a parameter.

The theory of Lindhard et al. [17] is used for the value of K. The electronic cross section associated with eqs. 17, 20 is

$$S_e(E_0) = 2\pi \int_0^{s_{max}} sQ(s, E_0)ds, \tag{21}$$

where s_{max} is the maximum value of the impact parameter determined by the density and crystal structure of the target. For high energies, where $R(s, E_0) \approx s$, eq. 21 reduces to

$$S_e(E_0) = KE^{1/2}(1 - \gamma), \tag{22}$$

where the value of $\gamma = (1 + 0.3s_{max}/a_s)\exp(-0.3s_{max}/a_s)$ is typically less than 0.1.

Crystal Descriprion and Cascade Development

The scheme used in CRYSTAL to describe the target crystal, uses the translational symmetry of the crystal. A list is made of the position of silicon atoms in the crystal, using one of the lattice sites as the origin of co-ordinates. The procedure for searching the crystal is illustrated in Fig. 2.

The projectile P, near which is the associated lattice site R, moves in the direction $\vec{\lambda_0}$. Using the list of atoms neighboring R, the position of the site T is generated and the following quantities are checked:

$$\zeta = \vec{\lambda_0} \cdot \vec{\Delta x}, \tag{23}$$

$$s^2 = |\vec{\Delta x}|^2 - \zeta^2. \tag{24}$$

If $\zeta > \zeta_{min} \geq 0$, the site T is in the correct direction to be partner of P in a collision. If the impact parameter $s > s_{max}$, then T is regarded as too far from the path of P to produce a significant deflection. Compared to MARLOWE [12], there is difference in the way CRYSTAL finds the lattice site R. The searching procedure, using the free-flight-distance ($\propto n^{-1/3}$, n being the target density), makes guess in direction $\vec{\lambda_0}$ for the next position of the projectile, near which with translation places the reference center R. Detailed comparisons between CRYSTAL and MARLOWE (under equally impact points) show nearly identical collision histories i.e. the co-ordinates of the nearest partner for interaction, scattering angles, transferred kinetic energy and inelastic local losses. The value of s_{max} is chosen large enough so that in no part of the crystal the projectile be free of the influence of surrounding atoms, but small enough to avoid too many collisions involving very small energy transfers. It is roughly equal to $n^{-1/3}$, where n is the target density. Additional option is available so that s_{max} depends on the kinetic energy of moving particle and minimum transfer energy $T_{min} = 5$ eV [13].

A simple model of thermal vibrations is included in the program as an option. The vibrations are regarded as producing randomly displaced but static lattice atoms. The displacement parallel to each of the three Cartesian axes is distributed according to a Gaussian. The mean square displacement is based on the Debye model, the only parameters required being the temperature of the target and the Debye temperature for the silicon.

Displacement Model and Damage Buildup

Since the binary collision procedure used in CRYSTAL cannot deal directly with questions of defect stability and interactions, a simple model was included describing the conditions under which atoms may become permanently displaced from lattice sites. Consider a collision from which the original projectile emerges with kinetic energy E_{recoil}, after transferring kinetic energy T to the target atom [cf. eqs. 7 and 9]. The target is displaced if its energy exceeds a sharp threshold energy E_d. It may be required at the same time to overcome a binding energy $E_b \leq E_d$. Thus, if $T > E_d$, the target is added to the cascade with kinetic energy

$$E_{target} = T - E_b. \tag{25}$$

The atoms in the cascade are followed as long as their energies exceed a predefined cut-off energy E_{co}. No limitation is placed on the relation between E_{co} and E_d. But in order to obtain reflection and sputtering yields correctly, E_{co} has to be chosen equal to or smaller than the surface binding energy, E_{sb}. The displacement model corresponds to the familiar one of Kinchin and Pease [18] if $E_b = 0$, $E_{co} = E_d$, and E_d is chosen to represent the threshold energy for producing permanent displacements, about $16eV$ for silicon.

To be able to make direct comparisons to the experimental results, CRYSTAL takes into account damage accumulation with the dose. The amount of damage is estimated by the equation

$$\wp_{new} = \wp_{old} \times [1 - \frac{\Delta N_V}{\Delta N}\Phi], \tag{26}$$

where $\wp$ is a probability representing the amount of crystal not yet damaged, ΔN & ΔN_V the target and vacancy concentrations respectively, and Φ is the amount dose per pseudoprojectile normalized to the unit length used throughout the calculation. Estimated in this way damage, corrected for the "openness" of the channel, is used by CRYSTAL for choosing whether the projectile moves in crystal or random media.

3D Simulation Including Surface Topography

Additional provision is made in the program for simulation of ion implantation through three-dimensional surface structure.

The program can perform ion implantation in targets with any initial surface topography and takes automatically into account the re-implantation and deposition because of scattering (**B**, **C**) or sputtering (**D**) from the different surface planes. Also, the program is capable to consider any number of inclusions (α - ϵ) with different structure and density. Note that case **C** in Fig. 3 is significant in producing shallow implants because of redirection of the impulse in the cascade and the high flux of emitted particles through the walls.

TEST RUNS

Comparison with 1D experimental range profiles

Careful examination of all available experimental data shows big discrepancy in depth profiles among different research groups. The main differences can be divided into three groups:

- implantation geometry (beam divergence, beam alignment along crystallographic channels) and proper control to maintain it;
- the quality of the target surface;
- the depth profile measurements.

Any kind of consolidation between experiment and simulation should take into account these factors.

For comparison we have chosen several sets of experimental data, which in our opinion account the remarks made above. Good example of how the misalignment with the channel change the profiles is given in [19]. Therefore we selected published experiments with estimated misalignment, beam divergence, quality of the surface.

All comparisons are presented in Fig. 5-15. The most unsatisfactory result is in Fig. 6 where the surface concentration in the simulation is higher than in the experiment. In our opinion this is due to the fact the CRYSTAL uses one collisional partner for scattering, which increases the probability for dechannelling. Also, the inelastic losses has to be replaced with better model. In Fig. 7 the inelastic losses need either to be reduced a bit or the impact parameter dependence has to be changed.

Spatial Distributions with a 2D Surface Topography

Only as a test runs, three simulations were performed - 100 keV boron into Si_3N_4 mask on $(100)Si$ at angle -7° and fluence $10^{15}cm^{-2}$. The simulation set-up is shown on Fig.4.

CONCLUSIONS

The present model developed was especially designed for heavy ion implantation into crystal Silicon for energies not higher than the energy corresponding to Bohr's velocity. The basic limitations consists in the inelastic energy loss model, which is valid for velocities smaller than Bohr velocity. Being into allowed energy range when doing simulation and satisfying the experimental setup, the program gives reasonable and reliable profiles when compared to the experiment. The second set of limitations occur when simulating the damage buildup during the implantation. The modeling is restricted with in an could implants, where no significant diffusion exists and damage cannot be altered.

In summary, the overall match between present simulation and the experiment proofs that the used in CRYSTAL scattering model and impact parameter dependent inelastic energy loss (Molière screening function and Oen-Robinson or Firsov local energy losses) are adequate for describing channeling ion implantation in silicon. Such parameters as beam divergence and alignment to the channel, the surface oxide layer are absolutely necessary to be specified for good agreement in the simulation and the experiment as well. The disorder buildup plays key role in exact reproduction of the channeled profiles well below the amorphisation threshold.

[1] M. Yoshida. *Journal Soc. Japan*, 16:44, 1961.

[2] O. S. Oen, D. K. Holmes, and M. T. Robinson. *Journ. Appl. Phys.*, 34:302, 1963.

[3] O. S. Oen and M. T. Robinson. *Journ. Appl. Phys.*, 34:2515, 1964.

[4] M. T. Robinson and O. S. Oen. *Appl. Phys. Lett.*, 2:30, 1963.

[5] M. T. Robinson and O. S. Oen. Computer studies of the slowing down of energetic atoms in crystals. *Physical Review*, 132:2385–2397, 1963.

[6] J. A. Davies. *Nuclear Physics Applications on Materials Science*, volume E144, pages 1–23. Kluwer Academic Publishers, Netherlands, 1988.

[7] J. R. Beeler and Besco D. G. *radiation Damage in Solids*, volume 1, page 43. Intern. Atomic Energy Agency, Vienna, 1962.

[8] D. P. Jackson and D. V. Morgan. *Contemp. Phys.*, 14:24, 1974.

[9] M. T. Robinson. *Sputtering by Particle Bombardment I*, volume 47, pages 73–144. Springer Verlag, Berlin, 1981.

[10] H. H. Andersen. *Nucl. Instr. and Meth.*, B18:321, 1987.

[11] W. Moeller. *Simulation of Stopping and Ranges*, volume 155 of *NATO ASI Series E: Applied Sciences*. Kluwer Academic Publishers, Dordrecht, 1989.

[12] M. T. Robinson and I. M. Torrens. Computer simulation of atomic-displacement cascades in solids in the binary-collision approximation. *Physical Review B*, 9:5008–5024, 1974.

[13] J.F. Ziegler, J.P. Biersack, and U. Littmark. *The Stopping and Range of Ions in Solids*, volume 1 of *The Stopping and Ranges of Ions in Matter*. Pergamon Press, Oxford, 1985.

[14] G. Molière. *Z. Naturforsh. A*, 2:133, 1947.

[15] O. B. Firsov. A qualitative interpretation of the mean electron exitation energy in atomic collisions. *Sov. Phys. - JETP*, 36:1076, 1959.

[16] O. S. Oen and M.T. Robinson. Computer studies of the reflection of light ions from solids. *Nuclear Instruments and Methods*, 132:647–653, 1976.

[17] J. Lindhard, M. Scharff, and H. E. Schiott. Unknown. *Mat.-Fys. Medd. Kgl. Dan. Vid. Selsk.*, 33(14), 1963.

[18] G. H. Kinchin and R. S. Pease. *Rep. Prog. Phys.*, 18:1, 1955.

[19] E. H. A. Dekempeneer, P. C. Zalm, G. van Hoften, and J. Politiek. Light- and heavy-ion channeling profiles in silicon. *Nuclear Instruments and Methods*, B48:224–230, 1990.

[20] A. E. Michel, R. H. Kastl, S. R. Mader, B. J. Masters, and J. A. Gardner. Channeling in low energy ion implantation. *Appl. Phys. Lett.*, 44:404–406, 1984.

[21] R.G. Wilson. Channeling of 20-800 kev arsenic ions in the $< 110 >$ and $< 100 >$ directions of silicon, and the roles of electronic and nuclear stopping. *J. Appl. Phys.*, 52:3985–3988, 1981.

[22] R. J. Schreutelkamp, V. Raineri, F. W. Saris, R. E. Kaim, J. F. M. Westendorp, P. F. H. M. van der Meulen, and K. T. F. Janssen. Channeling implantation of b and p in silicon. *Nuclear Instriments and Methods*, B55:615–619, 1991.

[23] V. Raineri, A. Cacciato, F. Benyaich, F. Priolo, E. Rimini, G Galvagno, and S. Capizzi. High energy p implants in silicon. *Nucl. Instrum. Methods*, B62:331–337, 1992.

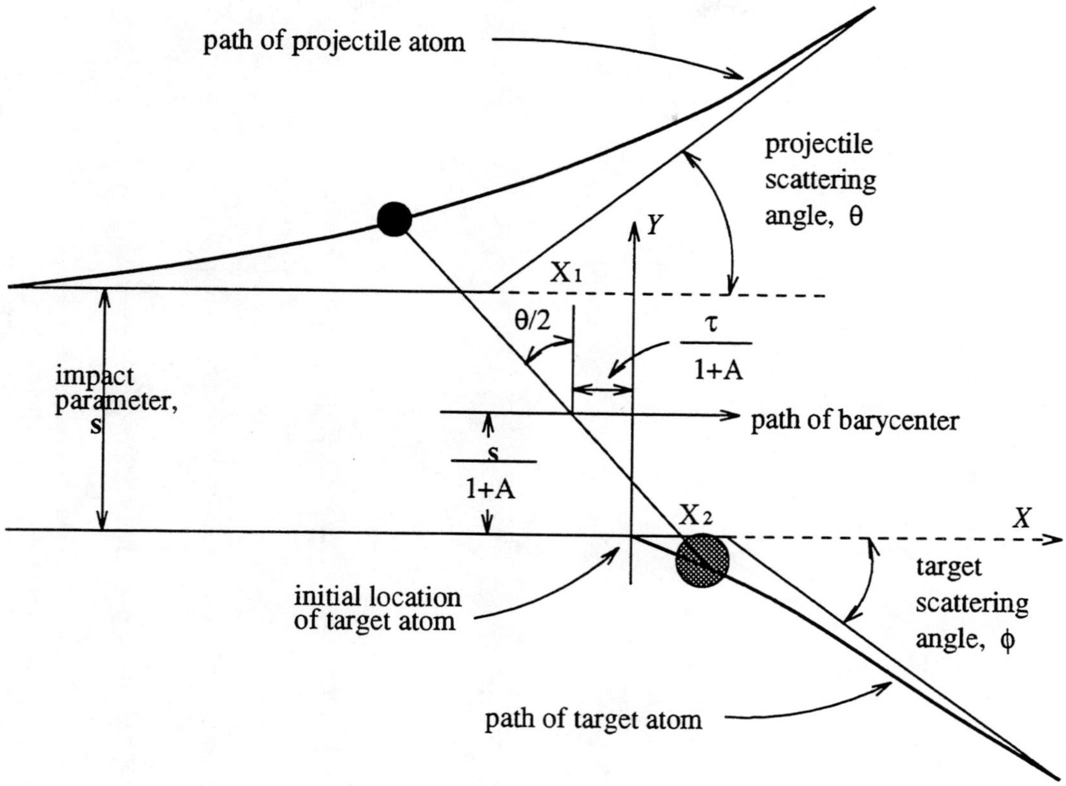

FIG. 1. Trajectories in laboratory co-ordinate system of two particles interacting according to a conservative central repulsive force. The positions of the particles and of the barycenter are shown at the apsis of the collision. All notations are according to [12]. Note, $X_2 \equiv 0$ because of using MAGIC-formula approximation [13] to the scattering integral.

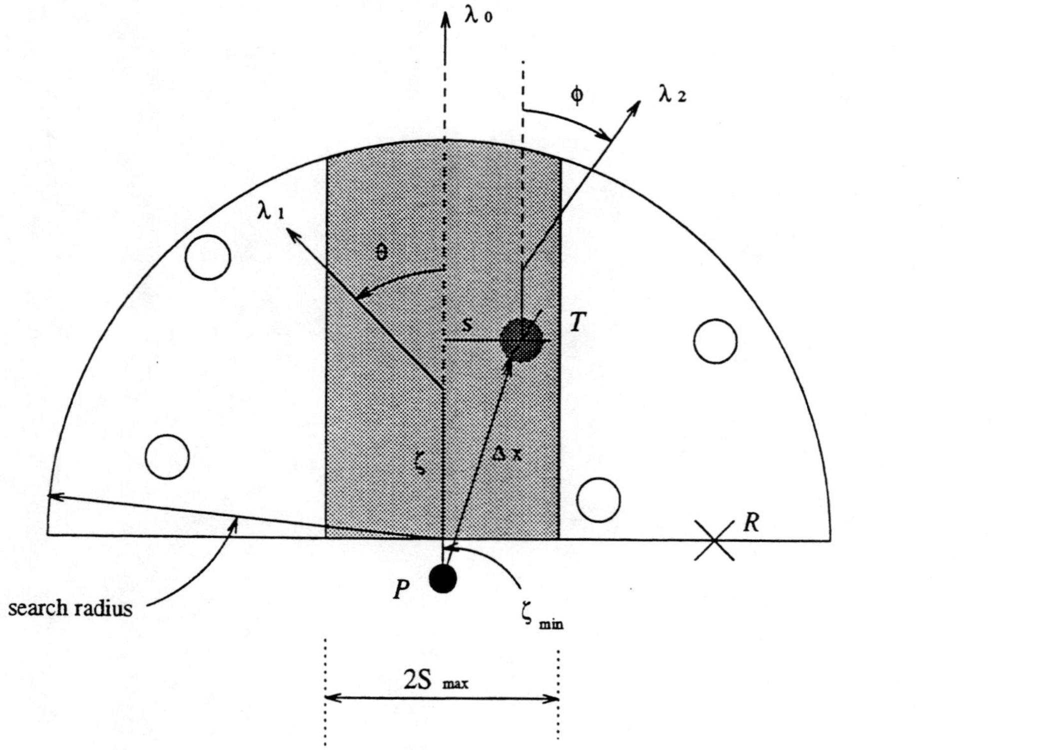

FIG. 2. Schematic representation of the CRYSTAL searching procedure. s_{max} depends on density and energy of the projectile.

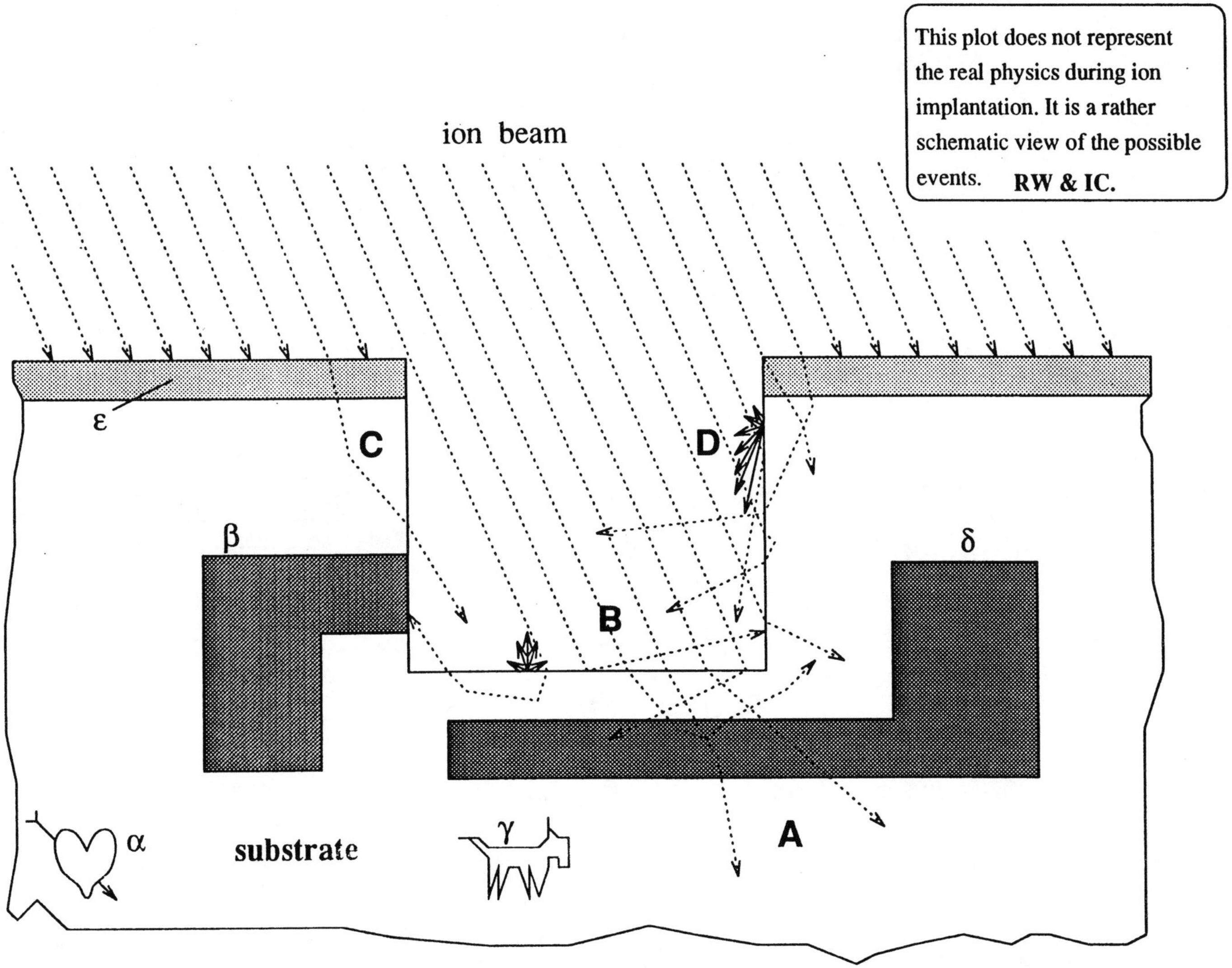

FIG. 3. Schematic diagram of what may happen during ion implantation through two and three dimension surface topography. α, β, γ, δ and ϵ are different phases of the bulk material or entirely different inclusions. **A** are ions which will be directly implanted, **B** are those which after scattering probably will be re-implanted, **C** is another way for implantation with transmitted particles through the walls and **D** are the sputtered from the surfaces particles which partly will be deposited on the opposite surfaces.

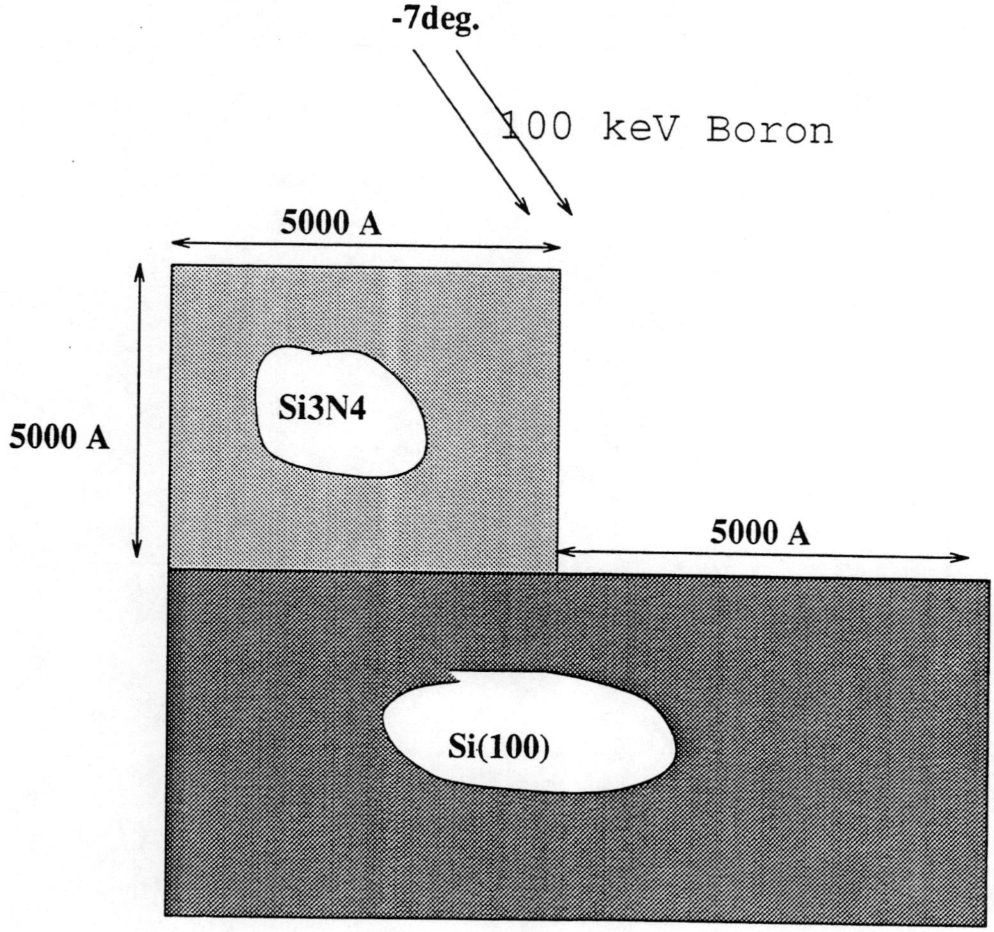

FIG. 4. Simulation setup for the results presented in Fig. 16

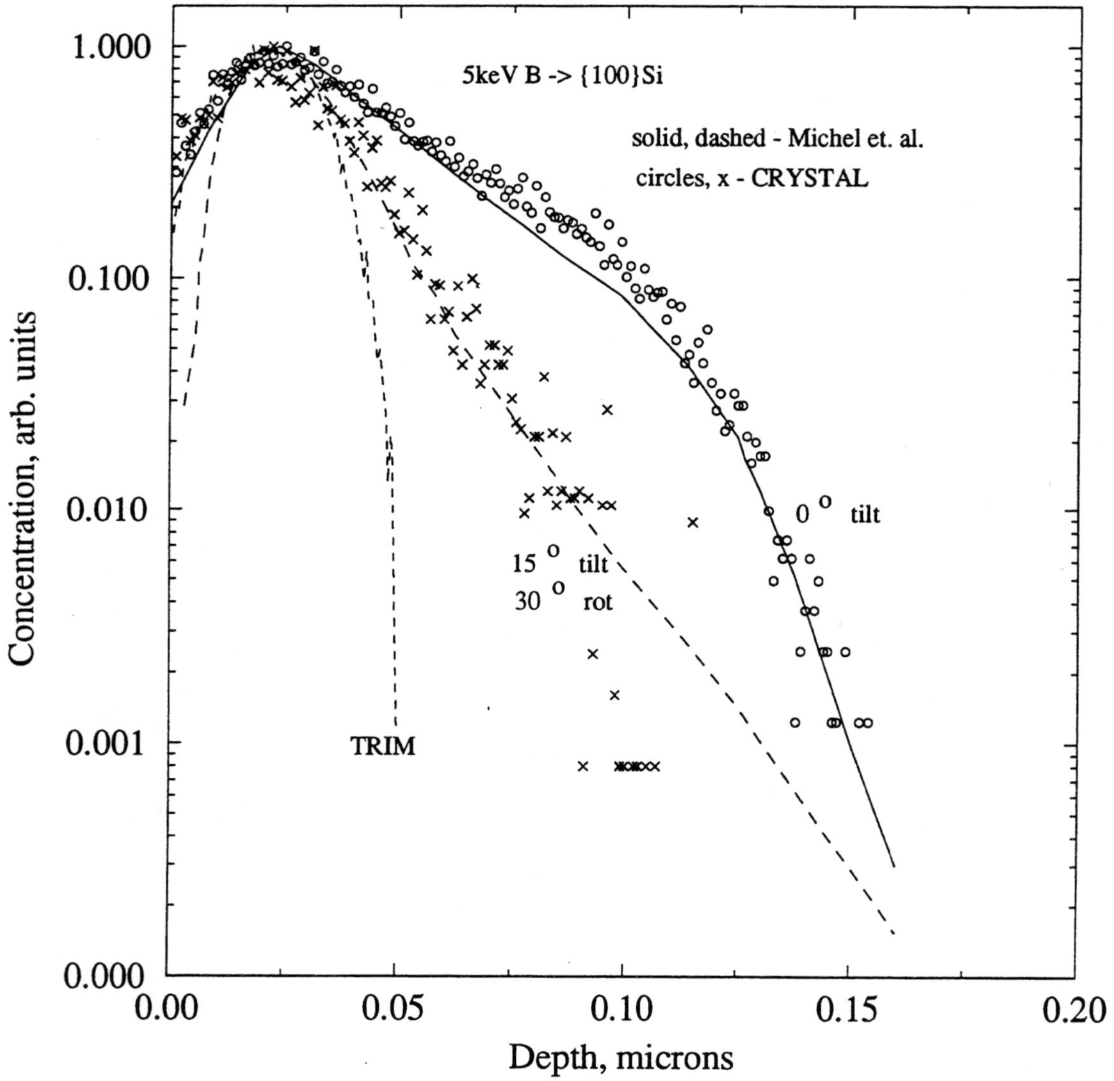

FIG. 5. Profiles of 5 keV implanted boron along < 001 > in silicon. The experimental profiles are taken from [20] (crystallographic orientation checked by x-ray analysis, ion beam adjusted to within 0.5° using laser reflection, minimized beam divergence - unscanned beam). Computer simulation was done with 10 Å surface oxide layer, 0.5° divergence and slightly tilted, 0.3°. Molière screening function, Firsov's local energy loss only.

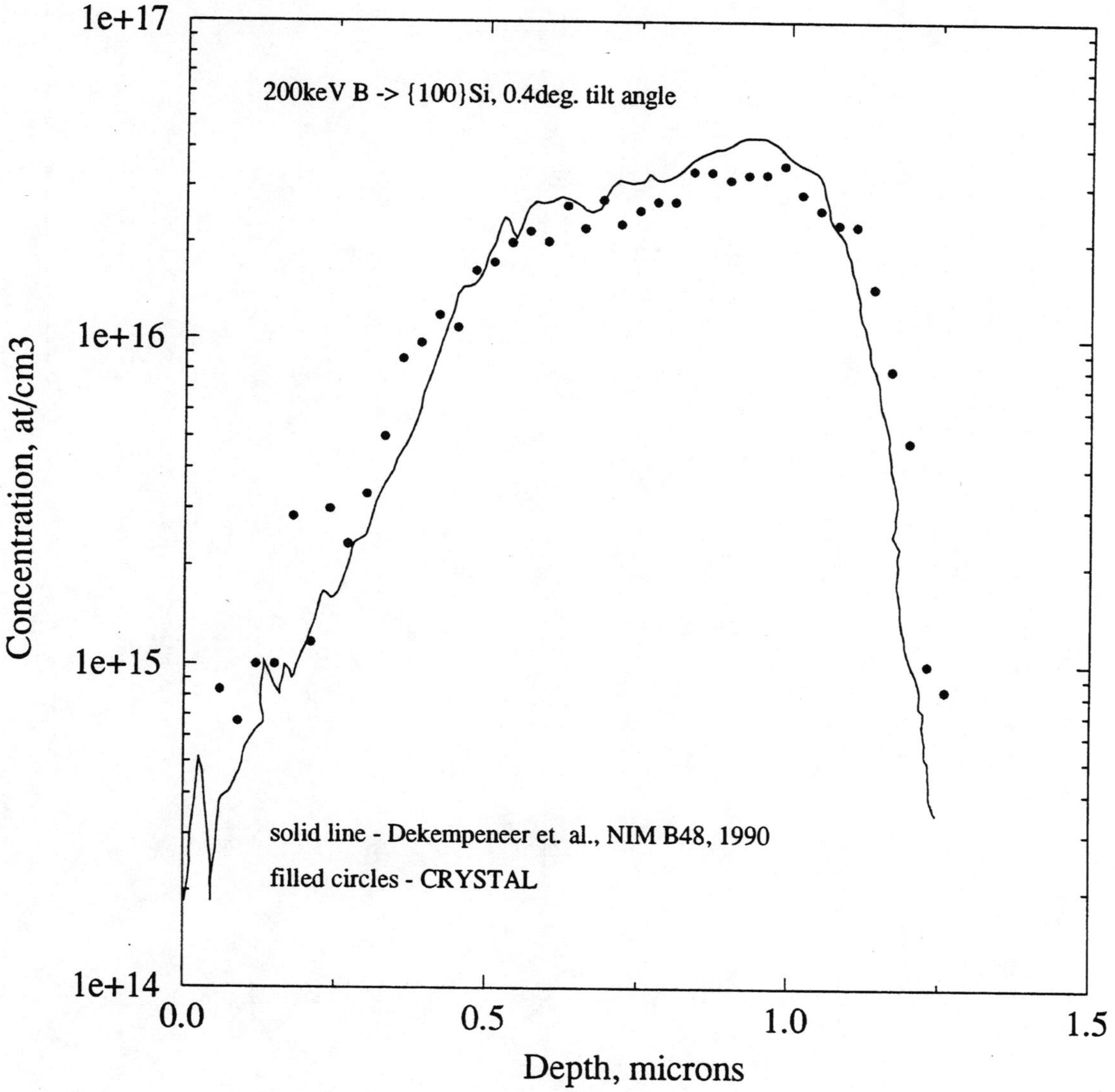

FIG. 6. Profiles of 200 keV implanted boron along < 001 > (0.4° tilt angle) in silicon. The experimental profile is taken from [19] (alignment to the < 001 > - by monitoring the backscattering yield of protons near the center of the target, the estimated uncertainty is ±0.2°, estimated error in depth scale is 1.5%). The simulation is performed with 10 Å surface oxide layer, 1.25° divergence and tilt angle 0.4°. Molière screening function, Firsov's local energy loss only. Ion dose in both simulation and experiment - $2 \times 10^{12} cm^{-2}$, dynamic accumulation of the damage.

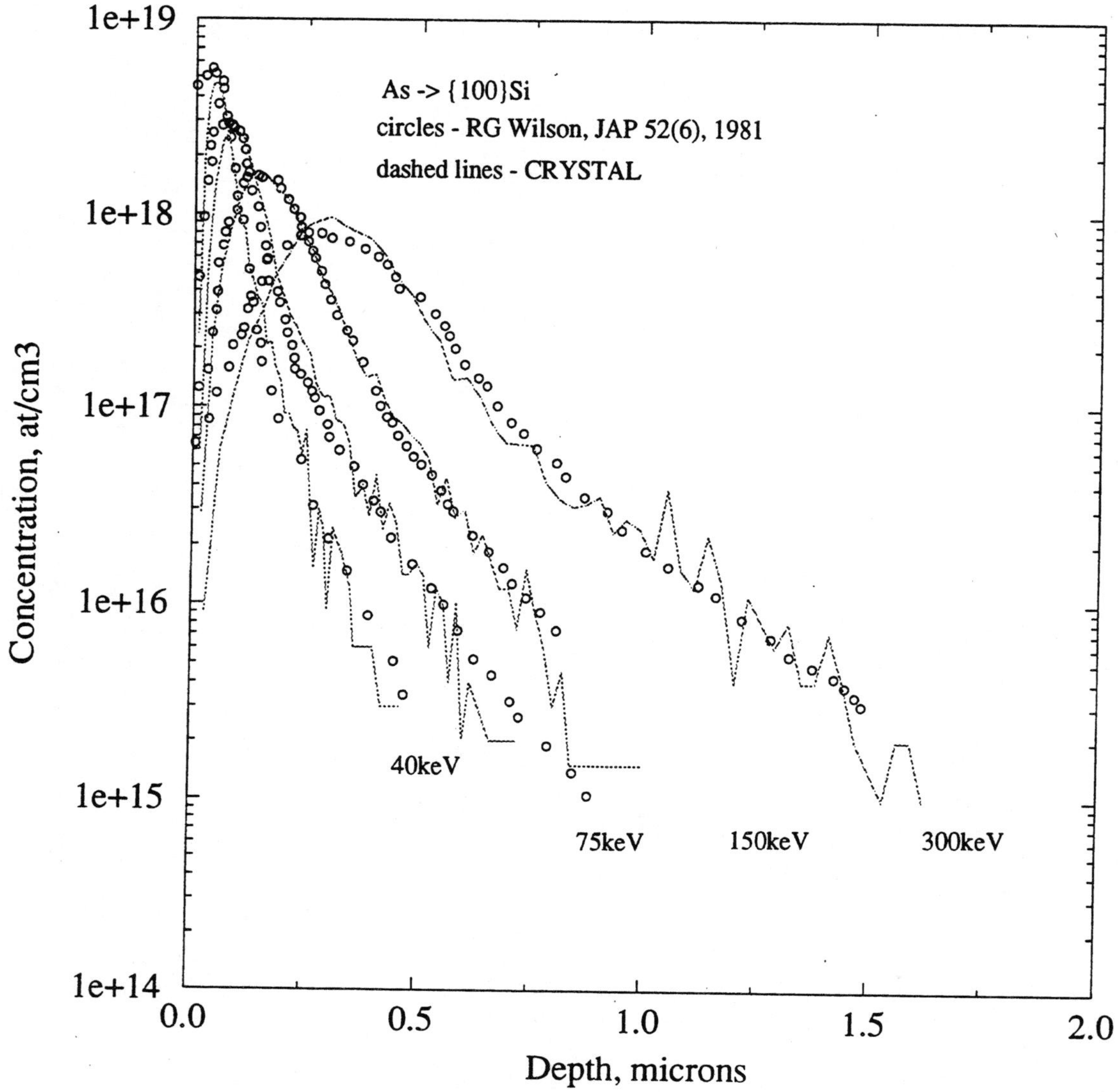

FIG. 7. Depth profiles for 40, 74, 150 and 300 keV arsenic implantation along < 001 > in silicon. The experimental profiles are taken from [21] (hyper-pure, low oxygen, float-zoned Si, 0.05° misalignment to the channels, 7% error in depth scale). The simulation - 0.05° tilt angle, 1° divergence, 10Å oxide layer. Molière screening function, Firsov's local energy loss only. Ion dose in both - $3 \times 10^{13} cm^{-2}$, dynamic accumulation of the damage.

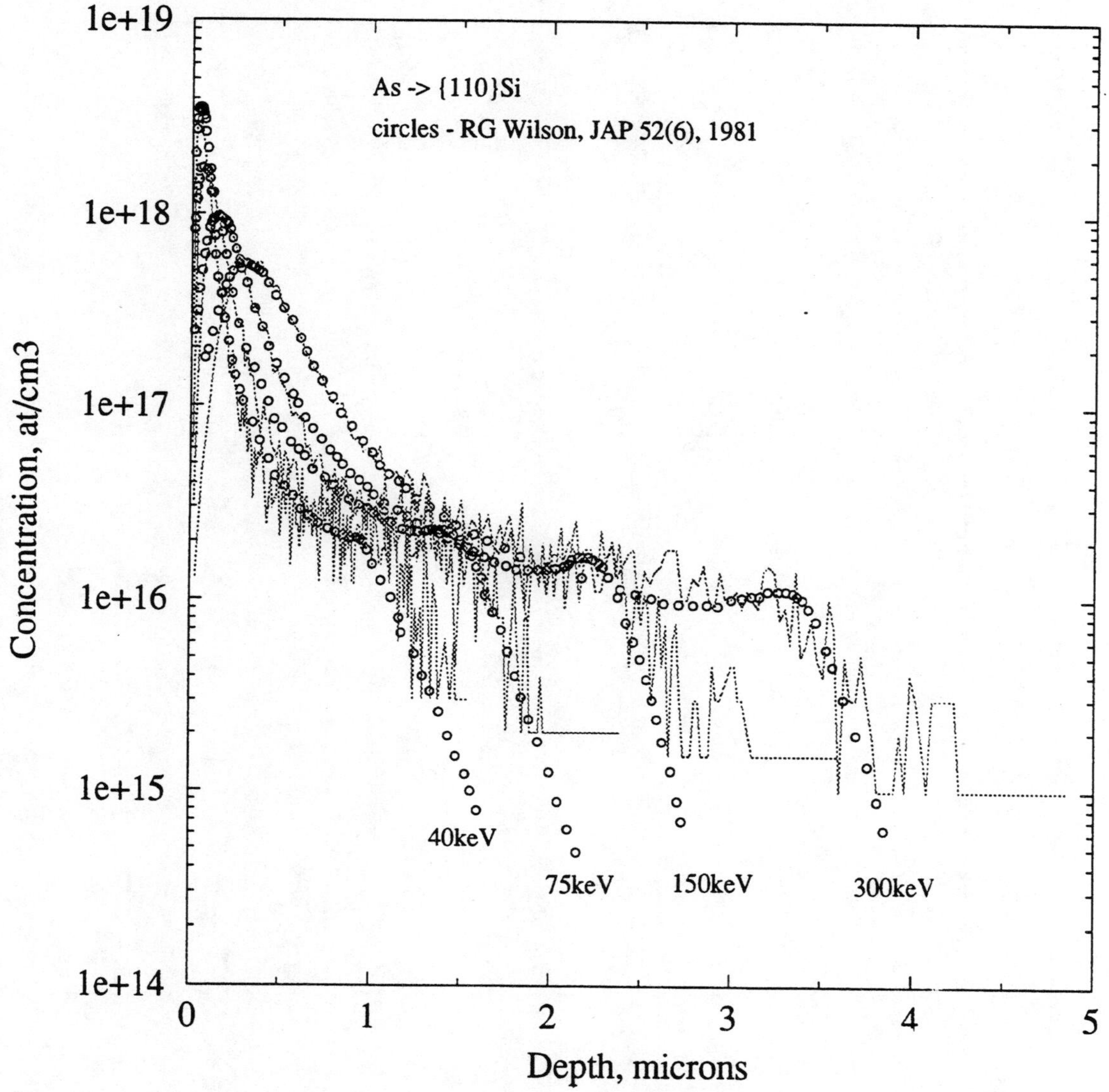

FIG. 8. The same as in Fig. 7 but for < 110 > channel. The tilt angle in the computer simulation is in the range 0.3 – 0.5°.

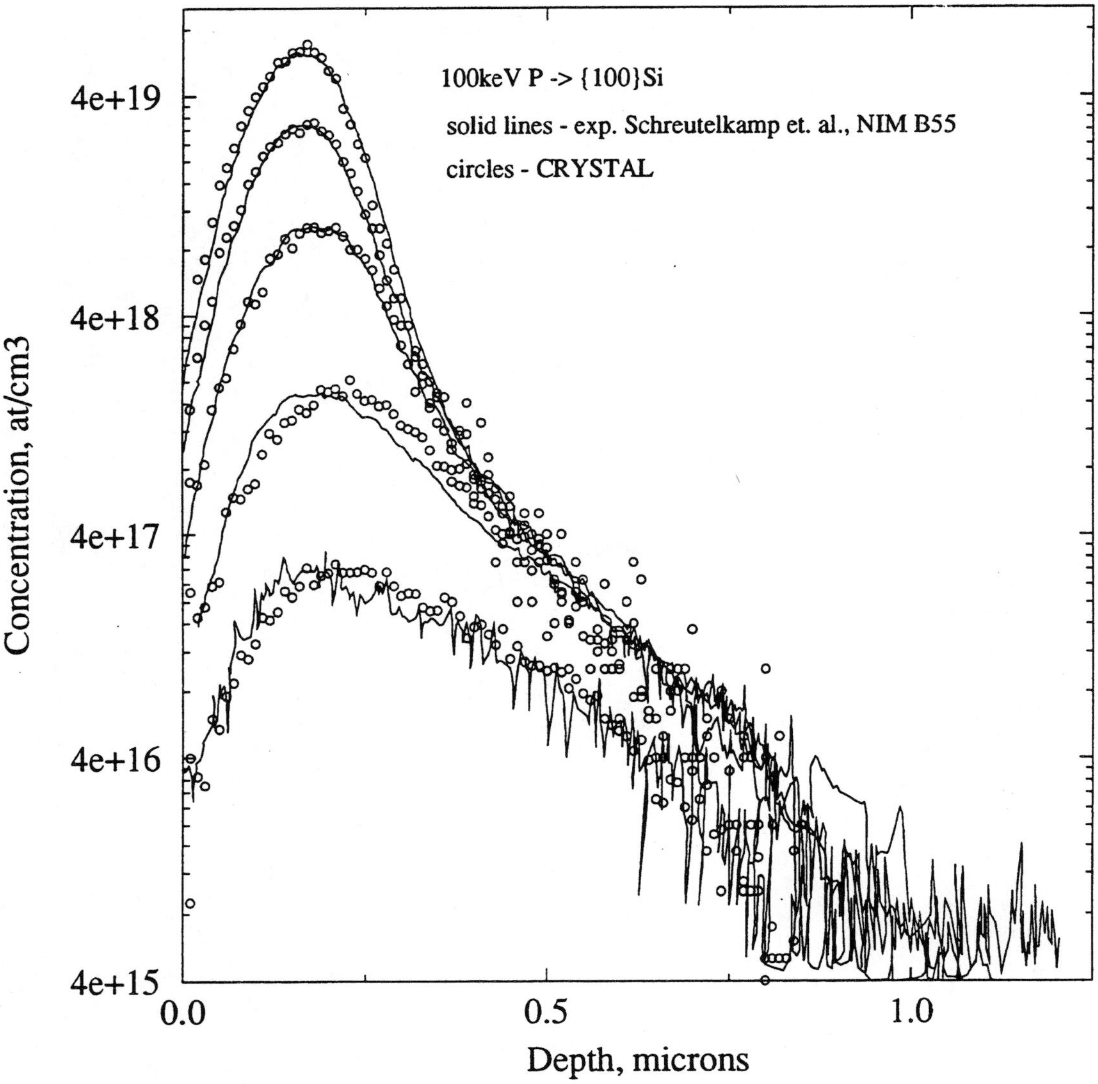

FIG. 9. Phosphorus depth profiles for implantations at 100 keV of 1×10^{13}, 5×10^{13}, 2×10^{14}, 5×10^{14} and 1×10^{15} along < 001 > channel. The experimental profiles are collected from [22] (experiments were performed with VARIAN/EXTRION 220 implanter, "well channeled" conditions were maintained across the wafer by comparing SIMS depth profiles from different spots). The simulation is with 0.05° tilt angle, 1.5° divergence and 12Å surface oxide layer. Molière screening function, Firsov's local energy loss only, dynamic accumulation of the damage.

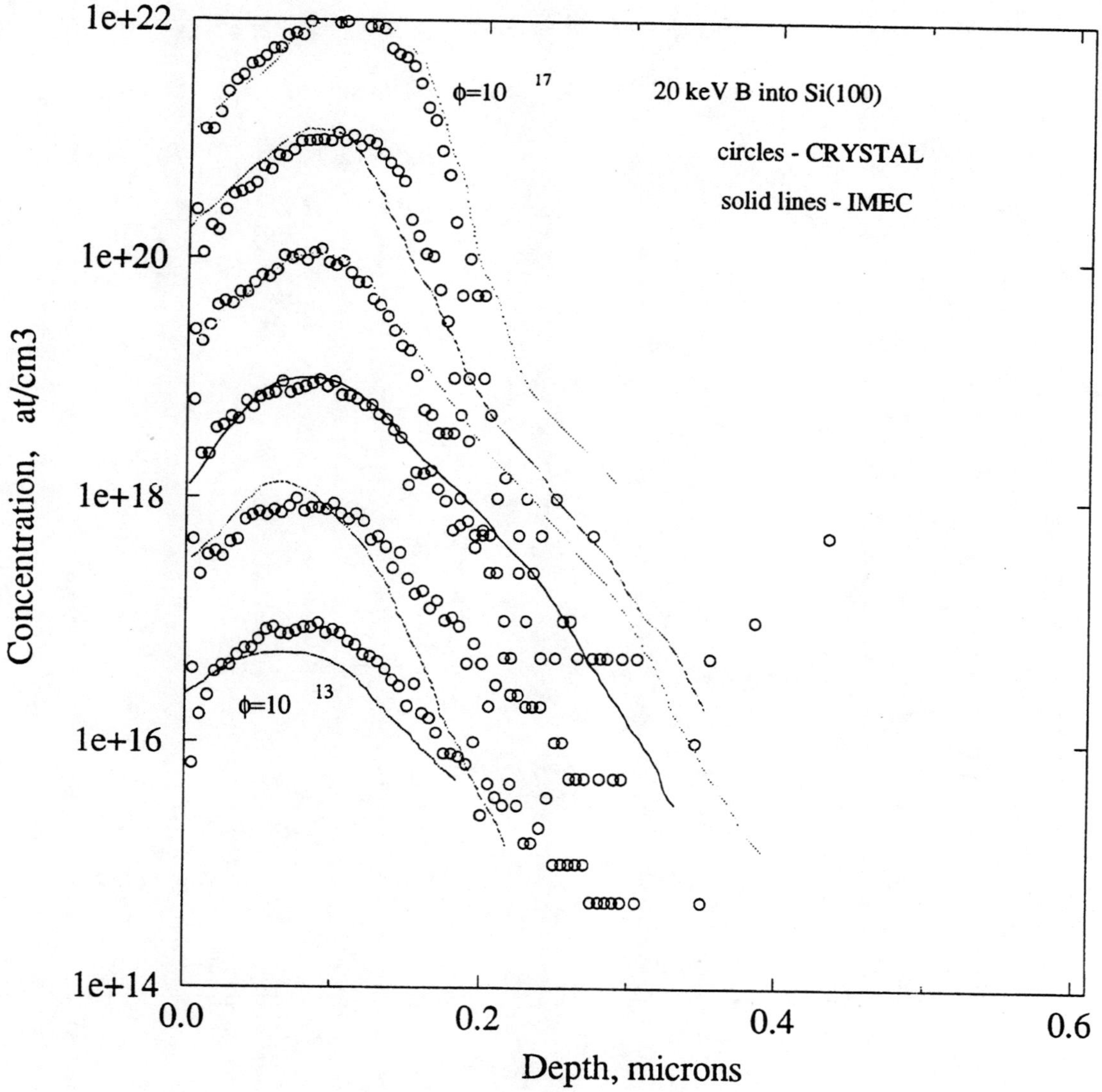

FIG. 10. 20 keV boron implantation into crystal silicon. IMEC data were performed at 7° tilt and 22° rotate angles, then analyzed by SIMS. Ion doses range from $10^{13} cm^{-2}$ to $10^{18} cm^{-2}$.

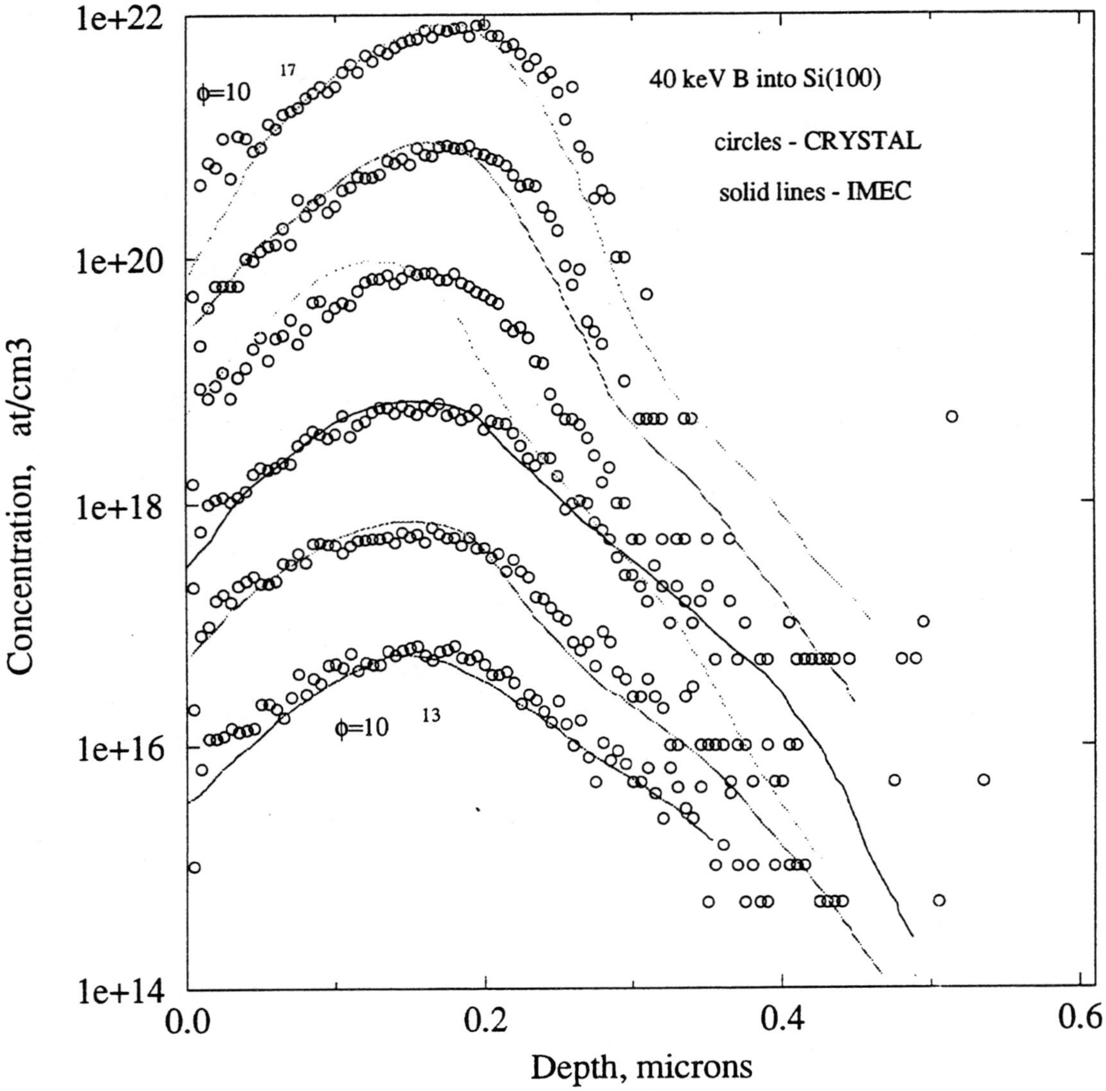

FIG. 11. The same as on 10 but for 40 keV.

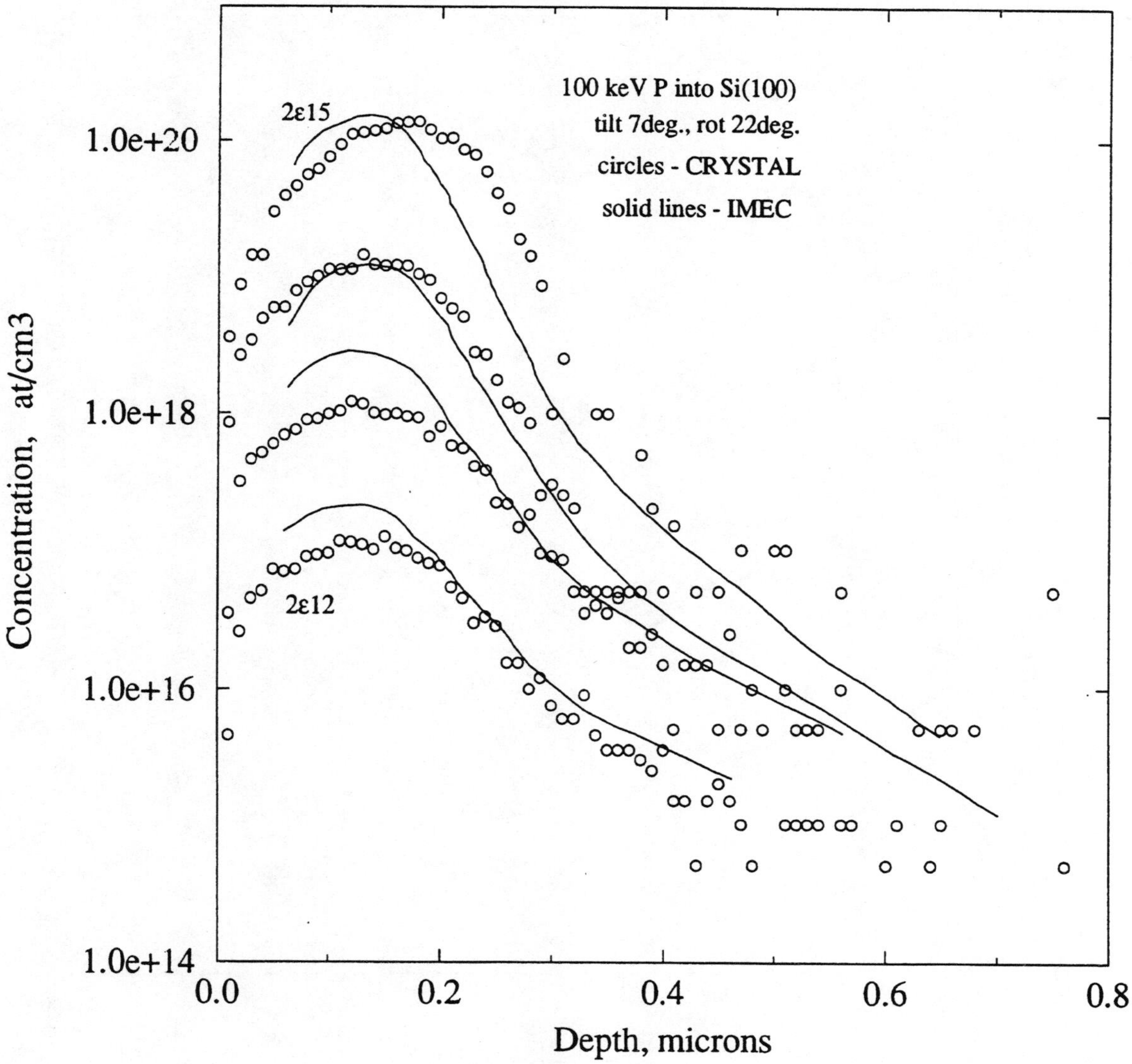

FIG. 12. 100 keV phosphorus implantation into crystal silicon. IMEC data were performed at 7° tilt and 22° rotate angles, then analyzed by SIMS. Ion doses range from $10^{12} cm^{-2}$ to $10^{15} cm^{-2}$.

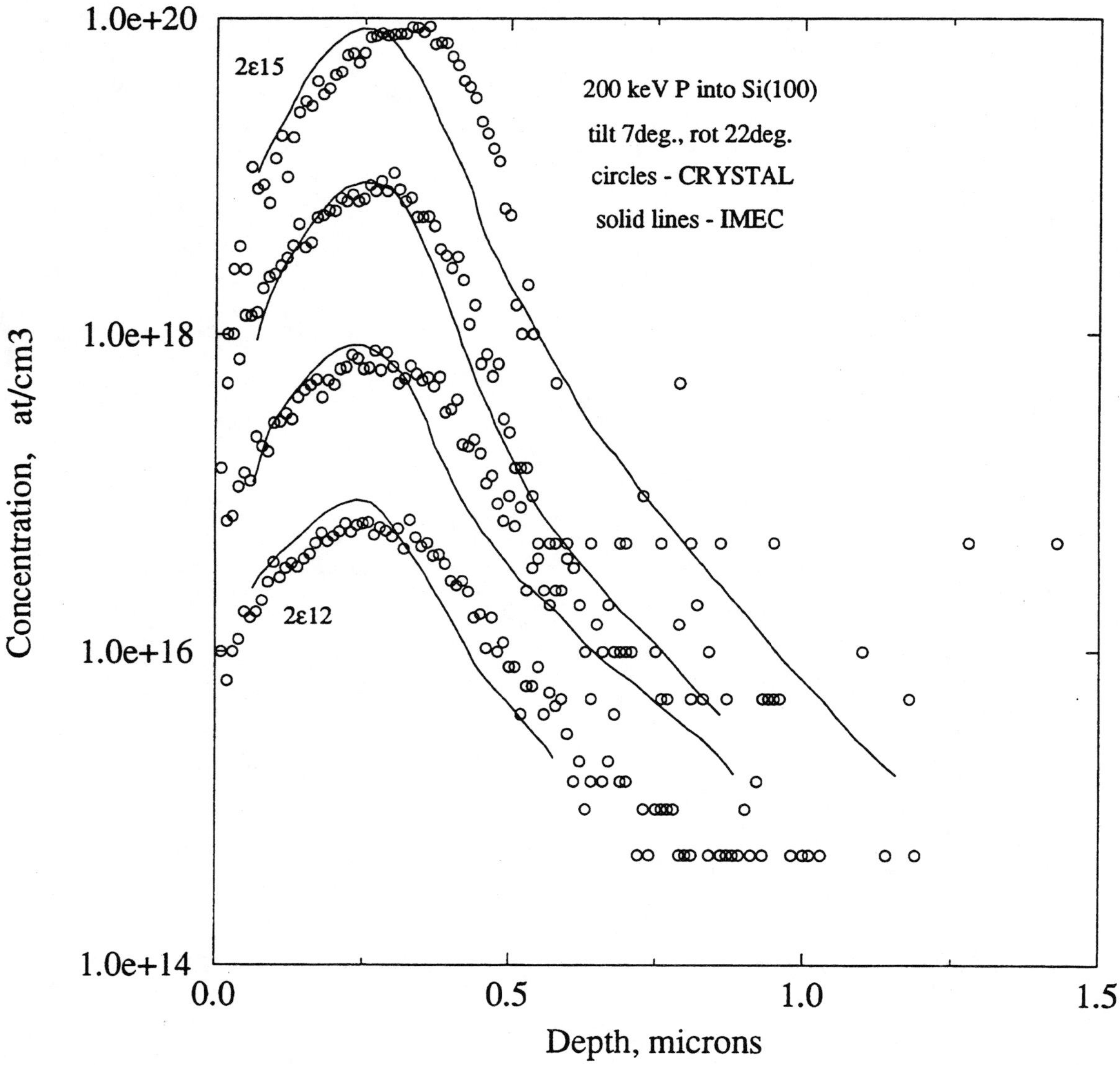

FIG. 13. The same as on 12 but for 200 keV.

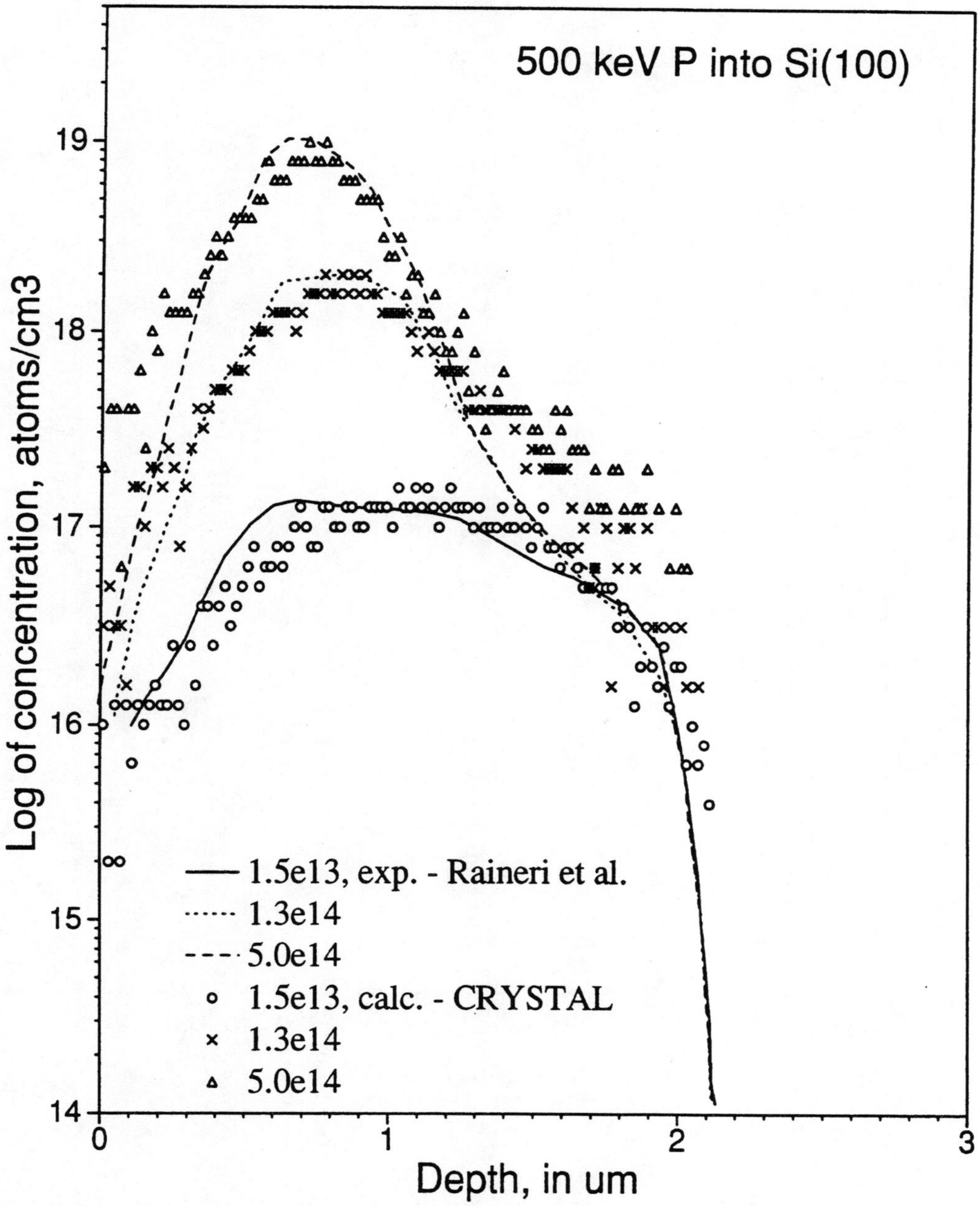

FIG. 14. 500 keV phosphorus implantation into a crystal silicon. The experimental data are taken from [23]

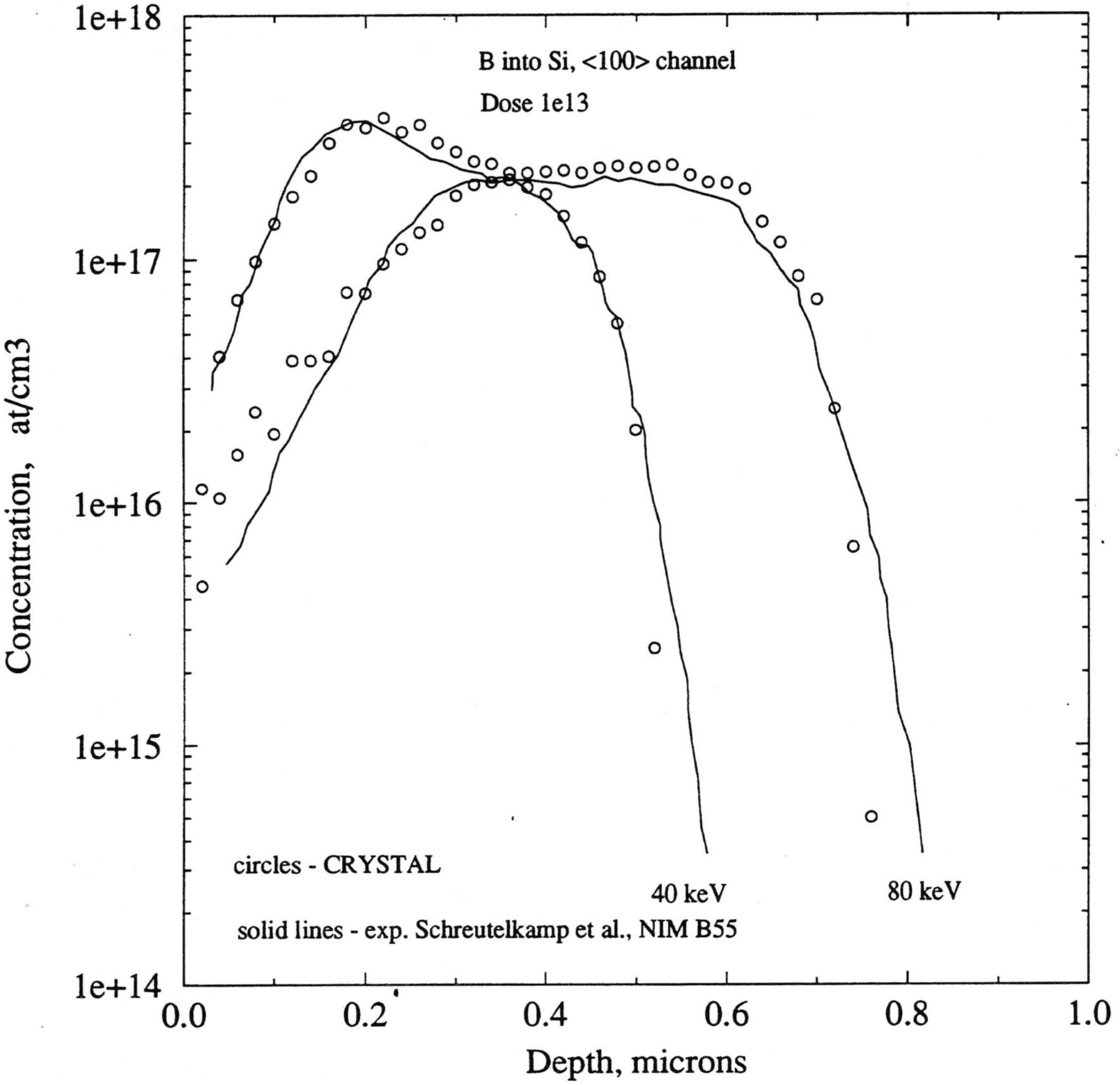

FIG. 15. Boron depth profiles for implantations at 40 and 80 keV of $1 \times 10^{13} cm^{-2}$, along < 001 > channel. The experimental profiles are collected from [22] (experiments were performed with VARIAN/EXTRION 220 implanter, "well channeled" conditions were maintained across the wafer by comparing SIMS depth profiles from different spots). The simulation is with 0.05° tilt angle, 1.5° divergence and 12Å surface oxide layer. Molière screening function, Firsov's local energy loss only, dynamic accumulation of the damage.

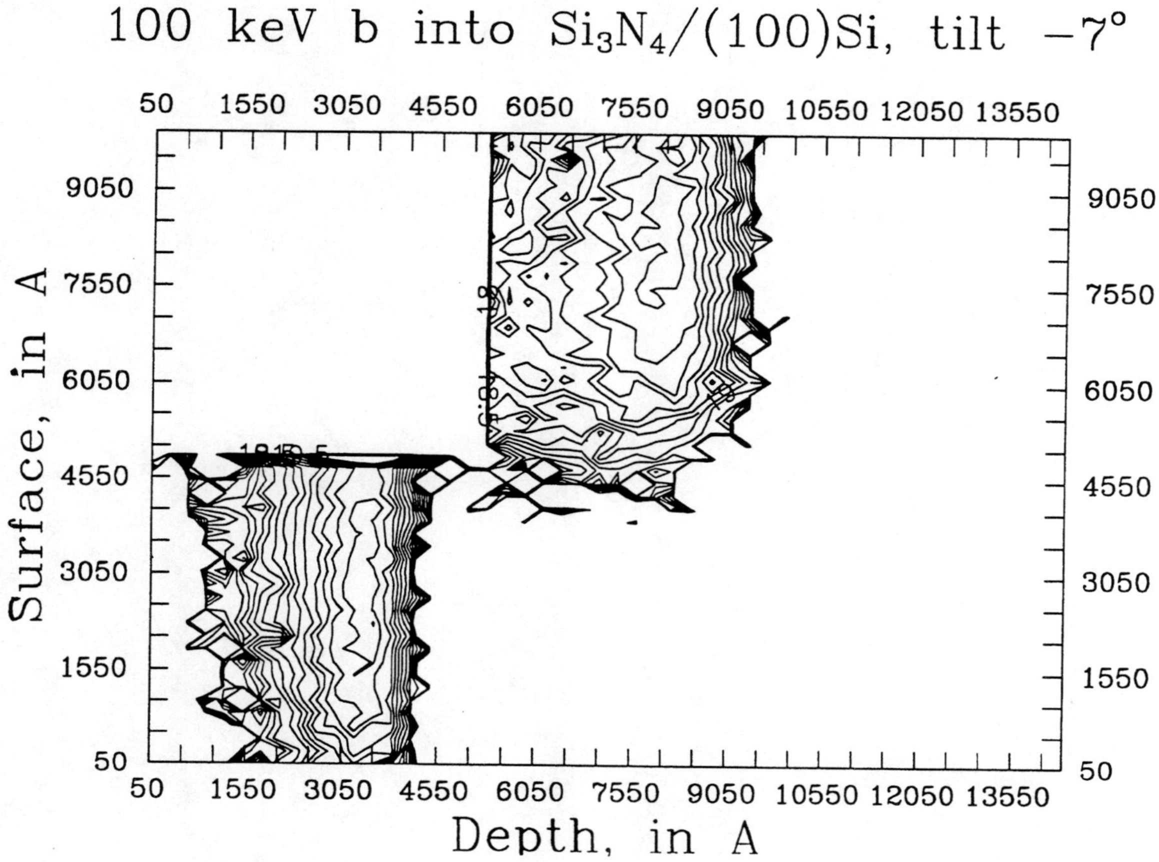

FIG. 16. 2D distribution of 100 keV Boron implant with dose $10^{15} cm^{-2}$ at tilt angle $-7°$ (ions are coming from the left and make shadow at the $(5000\AA, 5000\AA)$ corner of the structure) into $Si_3N_4/(100)Si$.

Vacuum Electron and Ion Technologies
D. S. Karpuzov (Ed.)

ISBN 1-56072-242-8

ION BEAM MIXING IN DENSE COLLISION CASCADES IN METALS

M. Hautala and I. Kopenen,
Department of Physics, P.O. Box
SF-00014 University of Helsinki, Finland

1 Preliminary remarks

We have recently studied ion bombardment induced transient phenomena in solids [1,2]. The basic problem we have been considering is the following one: An energetic ion implanted in metallic solid dissipates its energy rapidly to ions and electrons of the medium. At the end of the collisional stage the ion and electron systems are far out of equilibrium. In dense and small cascades a local equipartition of energy in the ionic system is attained rapidly and all energy in the electronic system is dissipated to large amount of the surrounding atoms. Electrons act only as a heat sink and the cascade thermalization can be described by the thermal spike model.

The cooling of a thermal spike is described by coupled heat conduction equations for conduction electrons and lattice. Atomic transport takes place above the glass transition temperature of a metallic matrix and the transport is treated as diffusion in liquids. The heat exchange between conduction electrons and lattice allows rapid quenching of thermal spikes in metals with high d-electron density. Mass and heat transport coefficients are calculated on the basis of elementary solid state models.

We have treated the problem numerically for spherical spikes [1] and analytically for cylindrical spikes [2]. The model explains the results of high energy ion beam mixing when the mixed atoms are chemically similar. We have also made SIMS-measurements to test the model at low energies. We have been able to calculate the resolution of SIMS in the case of Pt/Au-interfaces [3]. Recently we have extended calculations for a big sparse cascade with comparable amounts of deposited electronic and nuclear energy [4]. In some cases part of the electronic energy seems to warm up the lattice significantly.

The calculations contain a number of approximations and badly known material constants. The ultimate reasons for these difficulties are cascade structure and the strongly nonequilibrium and exotic conditions within the cascade. In the following we try to enlighten these difficulties by as tranparent as possible calculation of the amount of mixing. An analytical thermal spike model is used.

2 Terminology of ion beam mixing

When ion beam mixing is studied experimentally there are two possible configurations. Either a thin marker layer of extra material is used in otherwise homogenous material or one uses bilayer sample. After ion beam bombardment the marker profile is normally found to be of gaussian shape and the interface attains a shape of complementary error function. These profile shapes are characteristic for diffusion. It is thus natural to discuss ion beam mixing in terminology of diffusion. The solution for the diffusion equation for a narrow profile is gaussian with variance $\sigma^2 = 2Dt$ where D is the diffusion coefficient and t the time duration. In mixing experiment the variance is usually proportional to the fluence Φ. Therefore the mixing is characterised by mixing rate $< Dt > /\Phi = \sigma^2/2\Phi$. We write $< Dt >$ to indicate that we do not have normal diffusion and that t is not a physical quantity like bombarding time or the duration of spike cooling.

Let us then look the process microscopically. For clarity we assume that mixing is due to a number of jumps of only one length r_c. There are two ways of thinking. One can either think that each single ion bombardment creates η_c jumps per depth interval or that there is a certain probability (characterized by the cross-section $d\sigma$) for creating η jumps at certain depth. Then $\eta_c = \eta d\sigma n_o$ where n_o is the number density of atoms. The number of jumps per atom after bombarding the matrix with dose Φ (ions/cm^2) is $\eta_c\Phi/n_o = \eta d\sigma\Phi$. Then the mean-square displacement $< R^2 >$ of the atoms during the irradiation is $\eta_c\Phi r_c^2/n_o = \eta d\sigma\Phi r_c^2$ since $< R^2 >= (nr_c)^2/n$ where n is the total number of jumps. In normal diffusion with diffusion coefficient D and time duration t there is a connection $Dt = 1/6 < R^2 >$. Accordingly, $< Dt > /\Phi = \sigma^2/2\Phi = R^2/6\Phi = \eta_c r_c^2/6n_o = \eta d\sigma r_c^2/6$. One thus needs to know the jumping distance and the number of jumps per unit length or the number of jumps in one cascade and its cross-section.

3 Collisional mixing

As long as the energies of recoils in the cascade are above few eV their motion is more or less ballistic with occasional hard collisions.

In collisional mixing r_c is the projected range of recoils. Since there are recoils of several energies and therefore several jumping distances we have a random walk problem with several lengths with various weights. The several processes are independent from each other and can be summed together. In practise it is easiest to calculate first the number of recoils (i.e. the cross-sections $d\sigma$) with specific ranges using Monte Carlo techniques and then separately the random walk process. The major uncertainty comes from low energy recoils which being numerous may cause appreciable mixing. The binary collision calculation is doubtful below of the order of 100 eV. We therefore made some test calculations with Molecular Dynamics simulation in order to know how one should make the binary collision calculation to achieve the best accuracy. We then looked how much mixing is affected by the very low energy recoils. The conclusion of these studies is that the probable uncertainty in the calculation of collisional mixing is about 20 % [5] . This is not serious since the mixing rates due to collisional mixing are occasionally found to be less than 10 % of those experimentally measured although there are cases where mixing can be explained in terms of collisional mixing. For these other cases one must seek for some extra mechanism.

4 Spherical spike

At least when we have a cascade in the matrix of heavy atoms a peculiar thing should occur. If one calculates the volume of the cascade and its energy content one notices all the atoms are in motion having energies of the order of 1 eV. We might speculate this volume to be locally thermalized. The natural conclusion is that there is a hot spot in the matrix, a thermal spike. In case of 1 eV/atom the temperature would be 3400 K if the heat capacity per atom is $3k_B$. Under these circumstances matter is certainly in a strongly nonquilibrium state. Is it liquid, gas or overheated amorphous solid? Presumably most of the spike volume behaves as superheated solid. In any case this high temperature should lead to fast diffusion and efficient mixing of atoms. Let us estimate analytically the amount of mixing we should have.

We first assume that the energy E of the spike is initially introduced at a point in the medium. The equation of heat conduction in an isotropic, uniform medium with thermal conductivity κ and heat capacity per unit volume C_L is

$$\nabla \cdot (\kappa_L \nabla T) = C_L \frac{\partial T}{\partial t} \tag{1}$$

According to Debye-model the lattice heat capacity is $C_L = 3n_o k_B$. It is to be noted that for liquid it would be $2n_o k_B$ and for gas $\frac{3}{2} n_o k_B$. We here assume that heat conduction occurs only via lattice atoms, there is no energy transfer to electrons. It is good to remember that electronic, i.e. bulk metallic, heat conductivity would be two orders of magnitude faster than lattice heat conductivity.

The equation of heat conduction can be analytically solved if one neglects the temperature dependence of κ_L giving (with zero initial temperature) [6]

$$T(r,t) = \frac{E\sqrt{C_L}}{(4\pi\kappa_L t)^{3/2}} \exp(-C_L r^2/4\kappa_L t). \tag{2}$$

In order to calculate mixing rate we need to assume that there is a rate process such as the jumping of atoms of the material into new lattice sites, or the jumping of uniformly distributed impurity atoms to new sites. Assume that the rate of jumping (the number of jumps per unit volume per unit time) in the material when its temperature is T is given by $A\exp(-Q/k_BT)$ where A and Q are independent of temperature. The total number of jumps η in this spike is then

$$\eta = \int_0^\infty 4\pi r^2 dr \int_0^\infty A\exp(-Q/k_B T(r,t))dt = \frac{0.00226AE^{5/3}}{\kappa_L C_L^{2/3}(Q/k_B)^{5/3}} \tag{3}$$

with T(r,t) given by (2).

The final result is (we define $< Dt >_{eff} = < Dt > /d\sigma$)

$$< Dt >_{eff} = \frac{\eta r_c^2}{6} = \frac{0.00226AE^{5/3}r_c^2}{6\kappa_L C_L^{2/3}(Q/k_B)^{5/3}} = \frac{0.00226D_o E^{5/3}}{\alpha_L(3n_o Q)^{5/3}} \tag{4}$$

where $\alpha_L = \kappa/C_L$ is heat diffusivity. We have further assumed that $Ar_c^2 = 6n_o D_o$ (since $Dt = 1/6 < R^2 >$) where D_o is the prefactor of apparent diffusion coefficient.

D_o/α_L should be of the order 0.1 since $\alpha_L = 60$ Å^2/ps and $D_o = 6 \cdot 10^{-4}$ cm^2/s just above melting point. D_o/α_L depends mainly on the repulsive part of the interatomic potential and should therefore be unsensitive on temperature and the material considered. We estimate the appropriate activation energy to be $Q = 0.15\Delta H_{coh}$. It can be justified both by mixing experiments and by theoretical arguments. For the discussion of this choice see Cheng [7]. Also the choice of $D_o/\alpha_L = 0.1$ is in agreement with experimentally determined mixing rates.

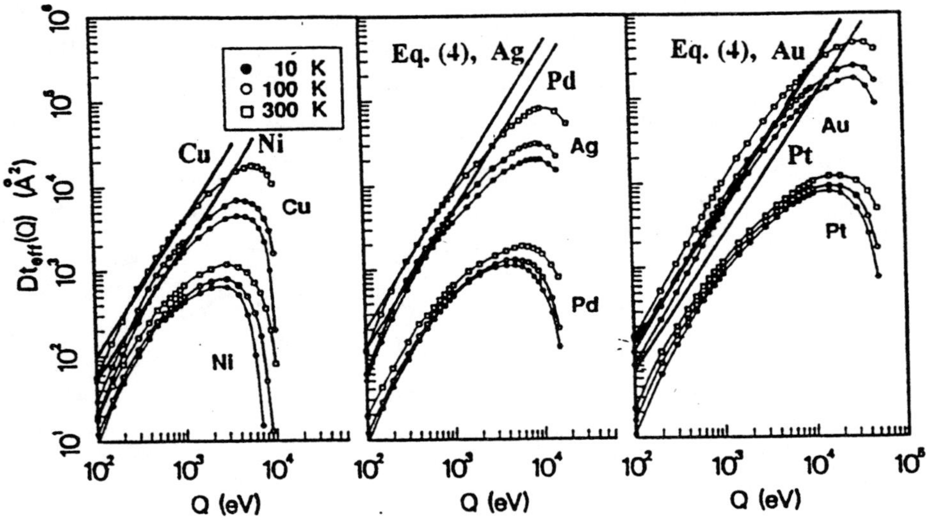

Fig.1. Eq. (4) compared with numerical calculation [1] for matrix(marker) systems Cu(Au), Ag(Au), Au(Cu) and Ni(Pt), Pd(Pt) and Au(Cu) at three ambient temperatures as a function of spike energy E.

In Fig. 1. estimates based on Eq.(4) are compared with more accurate calculations [1]. These numerical calculations for the heat-conduction equation (1) contain somewhat different material constants and they also include the possibility of energy transfer from ions to electrons. This corresponds to the case that Eq.(1) has an extra term which is proportional to the difference between lattice and electronic temperatures. Further the results depend on lattice temperature; in our simple derivation lattice was at 0 K. The overall variations in the results should be rather understood as the uncertainty in the models themselves, not so much to be due to the simplicity of our present estimate. Instead of looking at absolute numbers it is more proper to look at trends from one material to another. From Fig. 1 it is clearly seen that mixing in high resistivity transition metals Ni, Pd and Pt should be much less than in the almost free electron metals Cu, Ag and Au and consequently the spike model presented here is inadequate to describe these high resistivity materials.

5 Cascade structure

When do we actually have thermal spikes? Fig.1 gives a partial answer. In Ni, Pd and Pt there are almost no spikes. The electron-phonon coupling should be small enough that heat is carried away mainly via lattice. However, there must initially be a dense enough collision cascade with enough energy. The occurrence of this depends on the masses of bombarding ion and matrix atoms and the bombarding energy. These determine the cascade structure and its energy content. In the case of a few hundred keV heavy ion bombardment of heavy matrix the slowing down trajectory should be more or less straight line leading to single cylindrical spike. Lower energy bombardment (SIMS) leads more like a single spherical cascade. If the matrix atoms are lighter the cascade is no longer space filling and we might have several individual spherical/cylindrical cascades. An extra complication arises from the possibility that also the energy lost to target electrons may via electron-phonon coupling contribute to the heating of target.

We have taken the structure into account by calculating the cross-sections for creating recoils, each of which creates a separate space filling spherical cascade [1]. Geometric consideration leads to a threshold energy of order 10 keV in order this to be fullfilled.

We made some attempts to take into account the structure of the cascades in more detail. Those cascades were grouped together which were spatially close enough from each other. Each group of cascades was then treated as a single spherical spike. We found some improvement of agreement with experimental data. Unfortunately the results were sensitive on how the clusters were formed. Thus it is questionable whether it is a useful procedure. One reason might be that the cascades were in fact more cylindrical than spherical. The most accurate way would be to calculate each cascade cooling separately but it is necessary to use a numerical procedure and not much useful information may be gained. We are at the level where we have qualitative understanding and of the order of factor two quantitative agreement. It is very difficult to do much better.

6 Discussion

Our present result Eq.(4) is calculated for one spherical spike with energy E. In order to calculate mixing rate we need the statistics of the cascade energies. It turns out that the mixing rate diverges as a function of spike energy. We need a cut-off energy which evidently is the threshold energy for sub-cascade formation. The bending of the curves in Fig. 1 indicates that there is another way. The bending is due to the initial energy sharing. Large cascades turn out to be less dense, so they cool faster and there is less mixing. However, neither the threshold energy or the initial conditions are in our opinion satisfactory final solutions.

An alternative way is to take cylindrical geometry with deposited nuclear energy per depth interval F_D. It has the nice solution even when electron- phonon coupling is included [2]

$$\frac{Dt}{\Phi} = \frac{D_o}{\alpha_L} \frac{(F_D/Q)^2}{8\pi(C_L/k_B)^2} i(\chi) \tag{5}$$

with the parameter $\chi = \tau 12\pi n_o \alpha_L Q/F_D$ and $i(\chi)$ is given in Fig. 2. τ is the temperature relaxation time for the lattice. This model has the pleasant features that we need no assumption for the cascade structure and that the most dramatic differences between materials are in τ. The relative changes in it can be calculated reliably. The influence of τ (0.4, 24, 0.5, 56, 0.7 and 64 ps for Ni, Cu, Pd, Ag, Pt and Au, respectively) on mixing rate is seen from Fig.2. As Fig.2 shows a factor two inaccuracy in the other less well known parameters does not alter the relative differences between the materials.

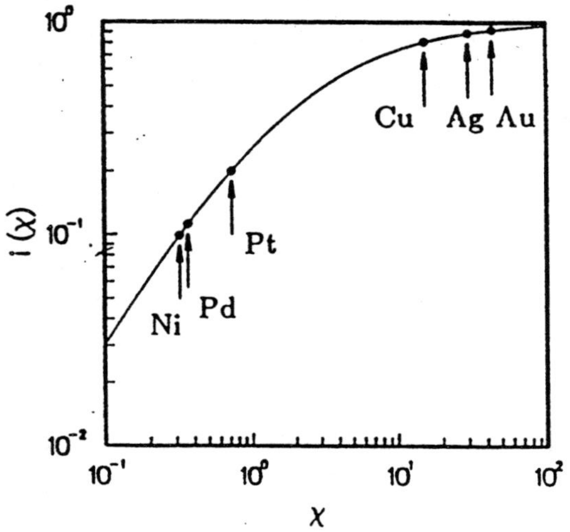

Fig.2. Function $i(\chi) = \tau 12\pi n_o \alpha_L Q/F_D$. The values corresponding spikes with $F_D =$ 200 eV/Å in Ni, Pd and Pt and in Cu, Ag and Au are indicated.

This discussion only treated the case of two similar metallic substances. There exists a lot of work in more complicated cases where chemical effects should be included. Regretably up to now energy transfer to electrons has been ignored in all cases when the experimental data has been analyzed. Further the same material parameters are used in each case. One should therefore be somewhat cautious about the conclusions drawn.

In all, we now have a crude overview of what is going on in cascades. Good example of the accuracy which may achieved can be seen from Eq.(4). There it was assumed that $C_L = 3n_o k_B$ as in lattice. If we instead assume the spike to be melted, i.e. $C_L = 2n_o k_B$ we have a factor two increase in mixing rate. This factor two accuracy is probably what we have now. A more quantitative understanding is very difficult to obtain and of questionable interest.

The most interesting open questions are probably connected to electron-phonon coupling. We do not know how to calculate the coupling coeffient in the nonequlibrium circumstances. We do not know whether it is possible that electronic heat may significantly warm up the lattice. These topics are discussed in [8].

References

[1] I. Koponen and M. Hautala, Nucl. Instrum. Meth. B66 (1992) 182-192.

[2] I. Koponen, J. Appl. Phys. 72 (1992) 1194-1196.

[3] J. Likonen, M. Hautala and I. Koponen, J. Appl. Phys. 72(1992) 5898 - 5904.

[4] I. Koponen and M. Hautala, Nucl. Instrum. Meth. B (in print).

[5] M. Hautala, I. Koponen and H.-P. Kaukonen, Nucl. Instr. Meth. B66 (1992) 508.

[6] G. H. Vineyard, Rad. Effects 29 (1976) 245.

[7] Y.-T. Cheng, Mater. Science Reports 5 (1990) no2.

[8] I. Koponen, Phys. Rev. B47 (1993) 14011.

Vacuum Electron and Ion Technologies
D. S. Karpuzov (Ed.)

ISBN 1-56072-242-8

TEMPORAL ASPECTS OF COLLISION CASCADE DEVELOPMENT IN AMORPHOUS TARGETS

R.G. Vichev, and D.S Karpuzov
Institute of Elctronics, Bulgarian Academy of Sciences,
Blvd. Tzarigradsko Chaussee 72, 1784 Sofia, Bulgaria

Time was introduced in the well-known TRIM code without changing the "event-driven" logic of the program. This modified TRIM code was used to simulate the collision cascades caused by Ar^+ primary ions in amorphous Cu targets. Several time-dependent characteristics of the cascades were obtained such as the mean number and energy of moving target atoms, the number of collisions and vacancies created, the sputtering yield and the mean energy and angle of ejection of sputtered atoms.

INTRODUCTION

Knowledge of the development of collision cascades in time is very important for better understanding of the process of ion-surface interaction. However, the collision cascades are too short in time (some hundreds of femtoseconds) and their experimental study is impossible at present. Computer simulation is therefore the only way to investigate the cascade time-dependent characteristics. Molecular dynamics (MD) codes are usually used for this purpose since they inherently include the time. However, they are restricted to low primary ion energies due to consuming huge amounts of computer time. There are few attempts to include time variable in BCA models [1-3]. We have recently modified the well-known TRIM code [4-6] to investigate the time-dependent characteristics of the slowing-down process of ions in amorphous media [8]. In the present work this modified TRIM version is further extended to yield time-dependent characteristics of development of linear collision cascades in which the moving particles interact only with target atoms at rest.

MODEL

The model used is based on a modified 3-dimensional TRIM code [4-6]. The latter is widely distributed so we shall give mostly details or parameters specific to the present study. The target surface lies at depth P_{max} below the starting point of the primary ion (at X = 0) where $P_{max} = \pi^{-1/2} N^{-1/3}$ is the maximum impact parameter. Here the X-axis is normal to the surface. The free flight path L is equal to $N^{-1/3}$ where N is the target atomic density. The first free flight path of the ion above the surface is randomized. The inelastic energy loss approximations used were the standard nonlocal losses [5] or the local energy losses according to the model of Oen and Robinson [7]. Primary ions start at time t=0. The projectile (ion or recoil atom) moves along a free flight path L. The impact parameter P is determined and the scattering angle in the center-of-mass system ϑ is computed by the MAGIC formula. The point of deflection of the moving particle is shifted back by the quantity $\tau = P \tan(\vartheta/2)$, a value related to the so-called "time-integral", and obtained in the hard-sphere approximation. The nonlocal inelastic losses along the path between the two successive collisions $(L-\tau)$ are calculated and the time for propagating that distance is determined assuming uniform deceleration of the particle. If the local inelastic energy loss model is implemented we assume uniform motion of the particle between the collisions. The elastic losses (the energy transferred to the atom at rest) and the inelastic losses are then subtracted from the energy of the moving particle. The latter is followed in the same way until its energy drops below a given value E_c. The same procedure is also applied to the target atoms involved in the cascade without changing the "event-driven" logic of the program. Target atoms are treated as sputtered if they cross the plane X=0. The surface binding energy is then subtracted from their kinetic energy. The effect of the surface potential barrier on the time of ejection is taken into account.

RESULTS AND DISCUSSION

The code described above was applied to the sputtering of amorphous Cu target by 2 keV Ar^+ ions impinging normal to the surface The universal potential was used. The Ar^+ ions started from a point at the target surface (at X=0). Nonlocal inelastic losses were implemented. The values of the displacement energy E_d was taken to be 19 eV, both the cut-off energy E_c and the surface-binding energy E_{sb} were chosen equal to the heat of sublimation of Cu (3.5 eV). 52500 cascades were followed and distributions of several output time-dependent parameters were obtained.

The mean number of atoms in motion $\langle N_{mov} \rangle$ and the mean number of collisions between target atoms in a cascade per femtosecond $\langle N_{col} \rangle$ as functions of time are shown in Fig. 1a. Collisions of primary ions with target atoms are not included in $\langle N_{col} \rangle$. It is seen that both curves have the same shape. The maximum of the $\langle N_{mov} \rangle$ curve is situated at 67 fs. The peak position of the $\langle N_{col} \rangle$ curve is shifted by about 5 fs towards larger values. It is clear that it takes some 70 fs of the cascade to develop. The full width at half maximum (FWHM) of both curves is 70 fs.

The mean number of vacancies created in a cascade per femtosecond $\langle N_{vac} \rangle$ and the mean number of vacancies created only by a primary ion per femtosecond $\langle N^i_{vac} \rangle$ as functions of time are shown in Fig. 1b. $\langle N_{vac} \rangle$ and $\langle N^i_{vac} \rangle$ have maxima at 50 fs and 5 fs and their FWHM are 60 fs and 40 fs. respectively. The peak positions of $\langle N_{vac} \rangle$ and $\langle N_{col} \rangle$ are shifted by 20 fs. It is seen that at the beginning of the cascade almost all the vacancies are created by the primary ions. As the cascade develops, $\langle N^i_{vac} \rangle$ decreases due to the decrease of ion kinetic energy. $\langle N_{vac} \rangle$ increases with the increase of $\langle N_{mov} \rangle$ and then decreases with the decrease of the mean energy of the recoils which is insufficient to create a vacancy (See Fig. 1a and Fig. 1c).

The dependences of the mean energy of recoil atoms in a cascade $\langle E_{rec} \rangle$ and the mean energy of sputtered target atoms $\langle E_{sput} \rangle$ on time are shown in Fig. 1c. At the very beginning

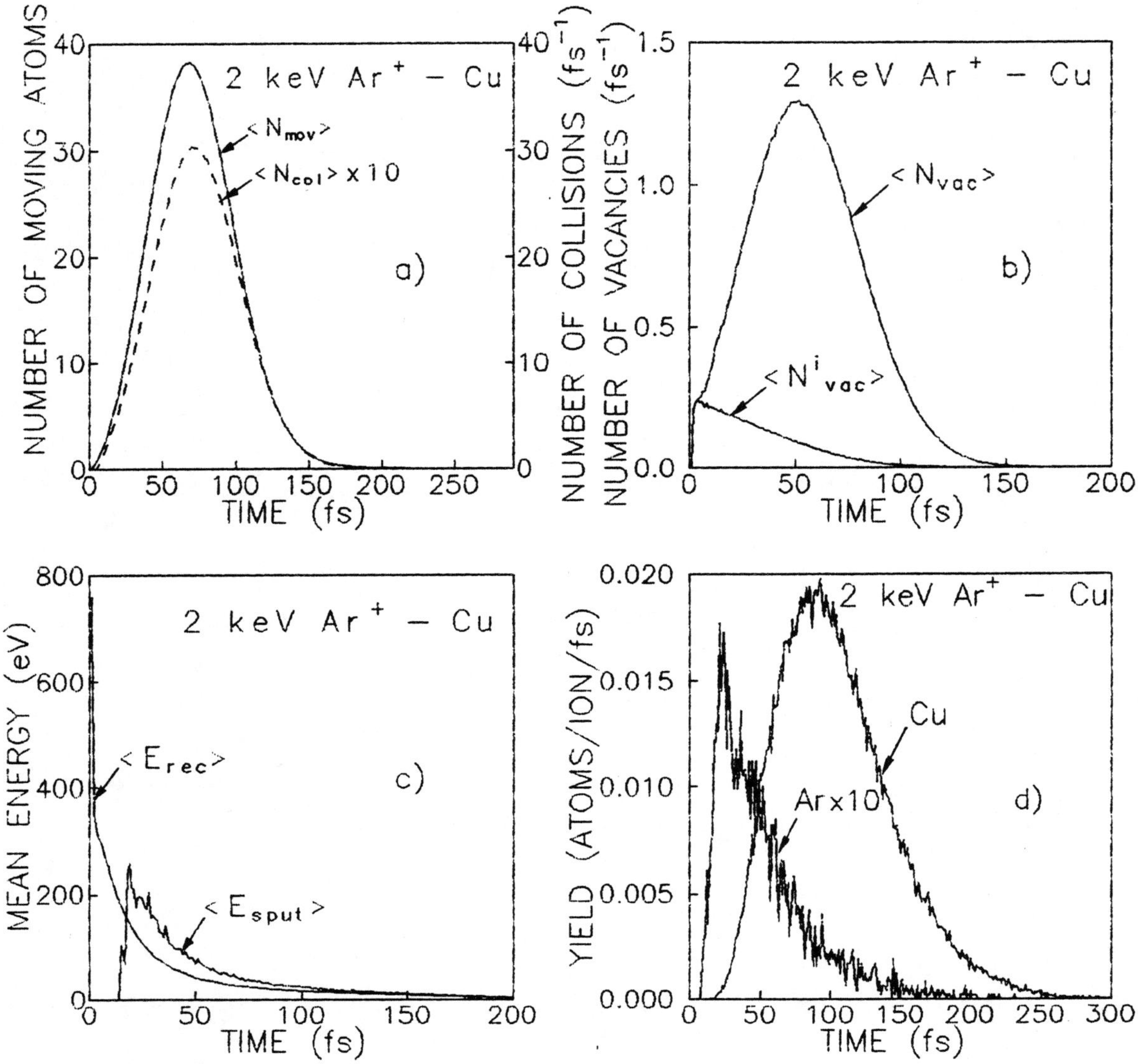

Fig. 1. (a) The mean number of moving recoil atoms $\langle N_{mov} \rangle$ and the mean number of collisions between target atoms per femtosecond in a cascade $\langle N_{col} \rangle$ as functions of time. (b) The mean number of vacancies created in a cascade per femtosecond $\langle N_{vac} \rangle$ and only by the primary ions per femtosecond $\langle N^{i}_{vac} \rangle$ as functions of time. (c) The mean energy of the recoil atoms involved in a cascade $\langle E_{rec} \rangle$ and the mean energy of sputtered Cu atoms $\langle E_{sput} \rangle$ as functions of time. (d) Yields of sputtered Cu atoms and reflected Ar ions plotted as functions of time. The time step is 1 fs.

of the cascade there are few recoils with very high energy. $\langle E_{rec} \rangle$ decreases quickly and has a value of 28 eV at t=67 fs when $\langle N_{mov} \rangle$ has a maximum. The first sputtered atoms are ejected after 15 fs from the starting moment of the ion and have very high energy. $\langle E_{sput} \rangle$ decreases with time as the mean energy of the recoils decreases. Yamamura has found using the DYACAT code that the peak position of the energy spectra of sputtered atoms shift towards lower values in agreement with our results for $\langle E_{sput} \rangle$ [2].

The mean number of sputtered target atoms in a cascade per femtosecond $\langle N_{sput} \rangle$ and the probability of ion scattering $\langle N_{scat} \rangle$ as functions of time are shown in Fig. 1d. The sputtering yield is maximum at t=90 fs, which is 20 fs later than the maximum of the number of moving atoms in a cascade. At this moment $\langle E_{sput} \rangle$ and $\langle E_{rec} \rangle$ are equal to 27 eV and 19 eV, respectively. The number of reflected ions is greatest in the initial stage of the cascade development (at t=25 fs). Similar dependences of $\langle N_{sput} \rangle$ and $\langle N_{scat} \rangle$ on time are reported by Yamamura [2].

CONCLUSIONS

A modified version of the TRIM computer code was developed by introducing the time reference of the collisions. It is shown that the peak positions of the time dependences of the number of moving atoms, the rate of defect generation and sputtering yield do not coincide for the model parameters chosen. The mean energy of recoil atoms and sputtered atoms decreases rapidly with time. Most of the scattered ions are reflected in the early stage of the cascade. The results obtained are consistent with those calculated by the DYACAT code.

REFERENCES:

1. Y. Yamamura, Nucl. Instr. and Meth. B33 (1988) 493.

2. Y. Yamamura, Nucl. Instr. and Meth. B45 (1990) 582.

3. M.T. Robinson, Phys. Rev. B40 (1989) 10717;
M.T. Robinson, Nucl. Instr. and Meth. B48 (1990) 408;
Nucl. Instr. and Meth. B67 (1992) 396.

4. J.P. Biersack and W. Eckstein, Appl. Phys. A34 (1984) 73.

5. J.F. Ziegler, J.P. Biersack and U. Littmark, The Stopping and Range of Ions in Solids (Pergamon, Oxford, 1985).

7. O.S. Oen and M.T. Robinson, Nucl. Instr. and Meth. 132 (1976) 647.

8. R.G. Vichev, D.S. Karpuzov, Nucl. Instr. and Meth. B (1993) (in proofs)

Vacuum Electron and Ion Technologies
D. S. Karpuzov (Ed.)

ISBN 1-56072-242-8

HIGH ENERGY IONS SENSITIVE CHALCOGENIDE GLASSES

J. Assa, R. Stoycheva, A. Buroff
Central Laboratory of Photoprocesses,
Bulgarian Academy of Sciences, 1040 Sofia

It is found that similar to exposure to light and X-ray, the interaction with Kr^{86} ions changes the solubility of evaporated layers of chalcogenide glasses. This phenomenon could be used for ions registration and for the development of ion beam resist as well. The behaviour of vacuum deposited layers of As_2S_3 exposed to Kr^{86} ions with energies 13.01 MeV is studied. The development procedure of layers 0.1 - 1.0 μm thick is optimized. The image obtained could be transferred by wet or dry etching into a Cr layer for pattern production. The correlation between the exposure energy and the thickness of the registration layer is established. It is shown, that the highest sensitivity and resolution are achieved with layers 0.1 μm thick. The available mask is reproduced with sensitivity of 2 mJ/cm^2 and high precision.

1. Introduction

The continuous increase of the integration density of the elements produced in electronic imposes the registration and lithographic transfer of structures of still smaller < (100 A) size. As a possible approach to the solution of this problem, ion beam lithography offers the following advantages [1]:

- resists are more sensitive to ions due to the greater mass of ions. Being heavier, ions cannot penetrate as deeply as lighter electrons in the resist films;
- the lateral scattering of the ions is insignificant, thus enhancing the exact reproduction of the width of the registered structures;
- a large number of ions with various mass and energies could be used.

Generally the resists used so far for ion beam lithographic purposes could be classified as follows: ion-implanted resists, inorganic resists on the basis of Ge_xSe_{1-x}/Ag_2Se and self developing resists. Inorganic resists based on Ge_xSe_{1-x}/Ag_2Se, Ge_xSe_{1-x}/Ag offer substantial advantages due to their high resolution and relatively simple processing. Structures 0.5 μm wide are registered by exposure to ions [2] and 0.3 μm ones are achieved by exposure to electrons [3]. Their potentialities at dry development are also studied [4]. The sensitivity of these resists to H^+ions is 14 - 1.4 μC/cm^2 [2] while in the case of He^+ ions this value amounts to 16 μC/cm^2 [5].

The inorganic resist for photolithographic purposes developed in our Laboratory on the basis of thin As_2S_3 layers is characterized by the following advantages: extremely high resolution and broad exposure and processing latitude [6,7]. Structures smaller than 0.5 μm are successfully reproduced under light exposure [7] and details bellow 0.1 μm - by exposure to electrons [8,9]. As_2S_3 could find application in X-ray lithography as well [10,11]. It is interesting to note that the optimal exposure energy for this resist is 0.5- 0.7 J/cm^2 regardless of the type of radiation - light, electrons or X-ray [10,11].

Here a study of the sensitivity of As_2S_3 layers to some high energy ions is reported. The applicability of these systems to ion beam lithography as well as their use as a medium for the registration of ion beams and traces of single high energy heavy ions are also investigated.

2. Experimental

Experimental samples were prepared by vacuum deposition of a Cr layer with optical density of 2.5 - 3.5 onto glass substrates followed by deposition of As_2S_3 layers with thickness ranging from 0.1 - 1.0 μm. Irradiation with Au^{197} and Kr^{86} ions with energies of 11.47 MeV/N and 13.2 MeV/N, respectively was carried out on the UNILAG accelerator, GSI, Germany. The irradiation dose was varied from 10^7 to 10^{12} N/cm^2. Fig.1 shows the principle scheme of sample irradiation and a micrograph of the mask used is presented in Fig.2.

The samples were treated with a developer optimized for As_2S_3 layers [12] which dissolves the irradiated areas only, without affecting the non-irradiated ones and then the structures were transferred into a Cr layer. The sensitivity is defined as the minimal irradiation dose required for the irradiation-chemical change of the entire resist thickness in the irradiated areas leading to the obtaining of usable lithographic image.

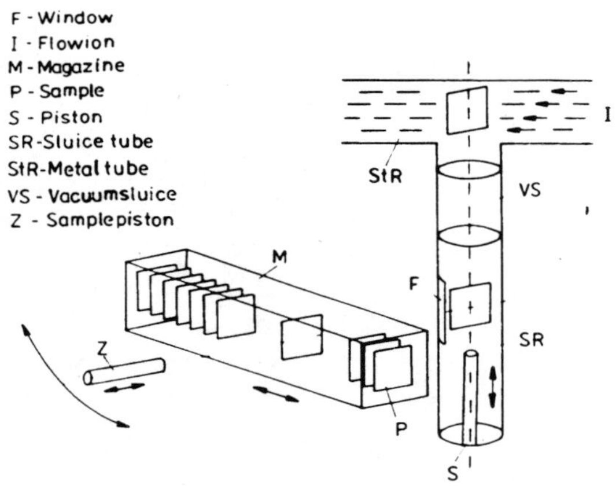

Fig. 1. Principle scheme of sample irradiation.

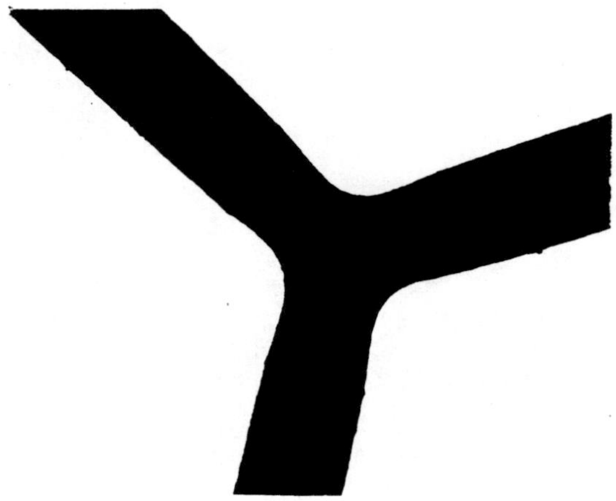

Fig. 2. Micrographs in transmitted light of mask used for samples exposure. Amplification 160x.

3. Results and Discussion

Fig.3 shows the minimal doses irradiation with Au^{197} and Kr^{86} ions depending on the As_2S_3 layer thickness, required for the achievement of

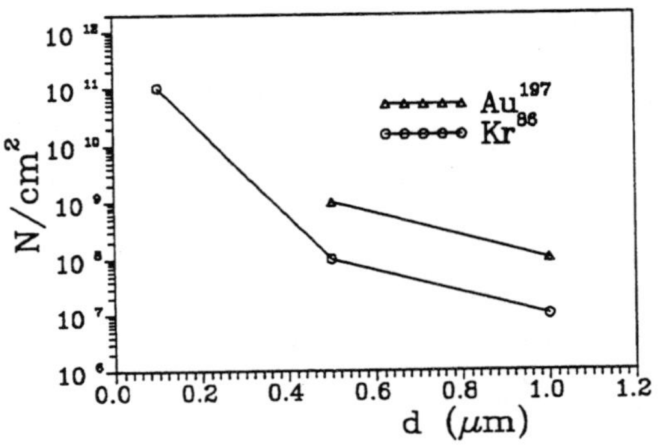

Fig. 3. Minimal doses of irradiation with Au^{197} and Kr^{86} depending on the As_2S_3 layers thickness.

complete dissolution of the irradiated areas. Experimental results show that the ion - target interaction depends mostly on the velocity rather than on the ion mass. Kr^{86} ions have a smaller mass but their energy is higher by 18% and for this reason they interact more actively with As_2S_3 and a higher sensitivity is achieved. This observation leads to the assumption that the results of irradiation with lighter ions used in ion - beam lithography such as He^+, Ar^+, H^+, etc., would be similar.

The results presented in Fig.3 also suggest that the sensitivity increases with the As_2S_3 film thickness. Probably the ion absorption varies in layers with different thickness. Due to their high energies, the bombarding ions attain considerable depths in the targets. Consequently the ions are capable to cause irreversible changes in a greater mass of the target substance of the thicker layers and for this reason the latter are more sensitive. However, high sensitivity is just one of the requirements imposed to resists. As it will be shown later, the quality of the image recorded in thicker layers is lower than that of the structures formed in thinner layers. Furthermore, in order to obtain a relatively good image, thicker layers should be irradiated with doses that are higher by 2 - 3 orders of magnitude than the minimal ones. In the case of thin (0.1 μm) layers the minimal dose coincides with that required for the obtaining of a high quality image.

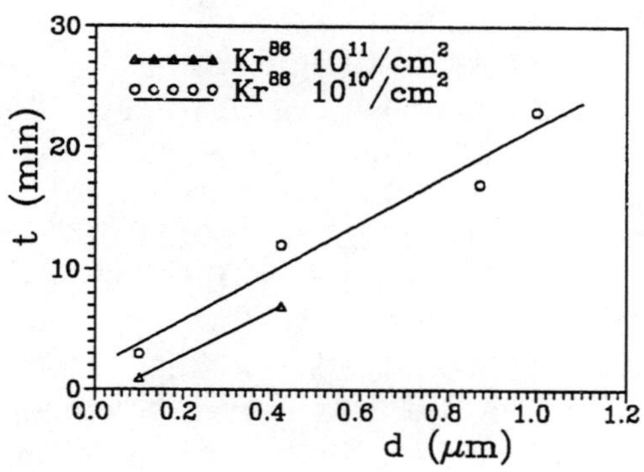

Fig. 4. Development time optimized with respect to the resolution depending on the thickness of As_2S_3 layers, exposed to Kr^{86} ions.

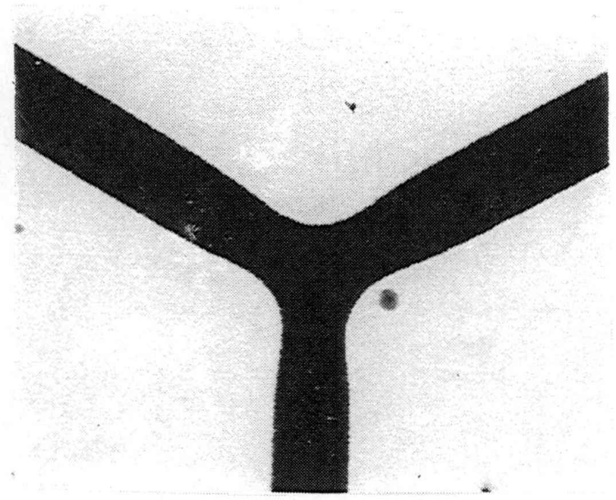

a

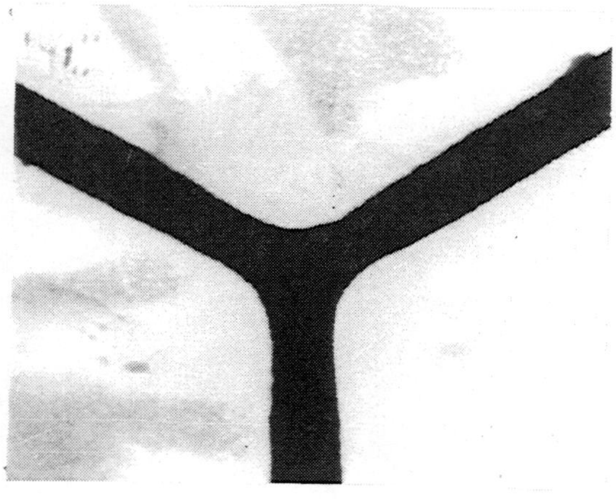

b

Fig. 5. Micrographs of patterns produced by irradiation of As_2S_3 layers 0.1 /a/ and 1.0 μm /b/ with Kr^{86} 10^{10} ions/cm^2 and transferred to Cr layers.

In order to achieve an image with sharper edges, the layer processing is optimized by varying pH of the developer and the development time. Fig.4 shows the time optimization with respect to the resolution of As_2S_3 layers exposed to Kr^{86} ions. Micrographs of As_2S_3 layers exposed to Kr^{86} 10^{10} ions/cm^2 are shown in Fig. 5.

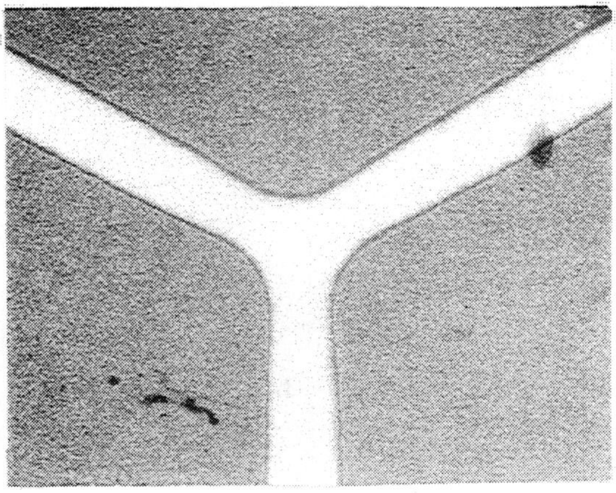

a

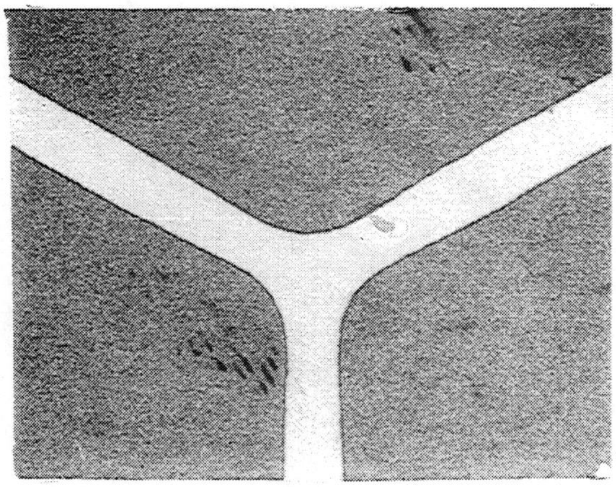

b

Fig. 6. Micrographs in reflected light of patterns produced by irradiation of 1.0 μm As_2S_3 layer with Kr^{86} 10^{10} ions/cm^2 and additional exposure to light - 0.2 /a/ and 0.4 J/cm^2 /b/.

A deterioration of the edge sharpness and of the microimage quality is observed with the thicker layers; it is expressed by incomplete dissolution and reduced protection of the chromium image. Additional exposure to light of layers 1.0 μm thick does not contribute substantially to the improvement of the microimage quality - fig. 6 [13] and that of the thinner layers cannot be attained (fig. 5a). It is interesting to note that better but still insufficiently good results can be obtained with 1.0 μm thick layers by additional exposure to lower energy light - 0.2 J/cm^2 (fig.6.a.). Additional exposure to higher energy light (0.4 J/cm^2 - fig.6.b.) causes substantial changes in the non-irradiated areas, thus lowering the protective properties of the resist.

Since exposure to 10^{10} Kr^{86} ions is insufficient for the formation of a high quality image, Fig.7 shows photographs of layers exposed to 10^{11} Kr^{86} ions.

The best microimage is obtained by using resist layers 0.1 μm thick. The required dose of irradiation with Kr^{86} is about 10^{11}ions/cm^2. It corresponds to a sensitivity of about 2 mJ/cm^2. The latter is comparable to that of PMMA/ethyl acetate ion resist by exposure to H^1 and is by 2 - 3 orders of magnitude lower after its irradiation with Ar^{40}.

The study of ion beams requires the determination of their energy and direction. An appropriate medium for the rapid and reliable determination of these parameters is not established so far. It should be stressed that the resist developed by us offers a quite convenient possibility in this respect. Fig.8 shows traces of single ions registered in As_2S_3 layers and transferred into a Cr layer. It is seen that the holes have an almost circular shape. The elliptical shape of some traces could be due to the fact that the angle of hitting the target by ions differs from 90°. By irradiation at a given angle the traces of the ions can be followed up

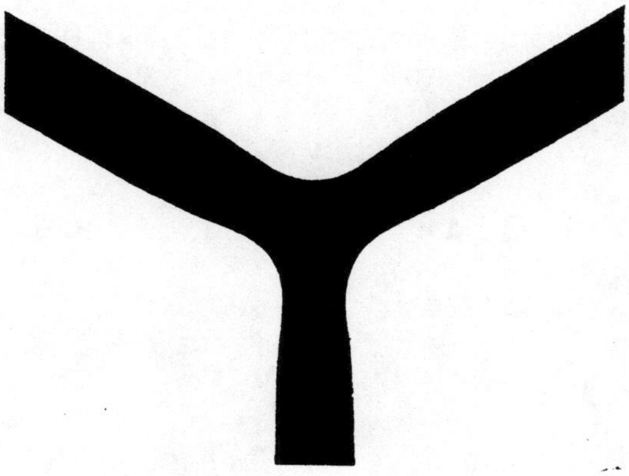

a

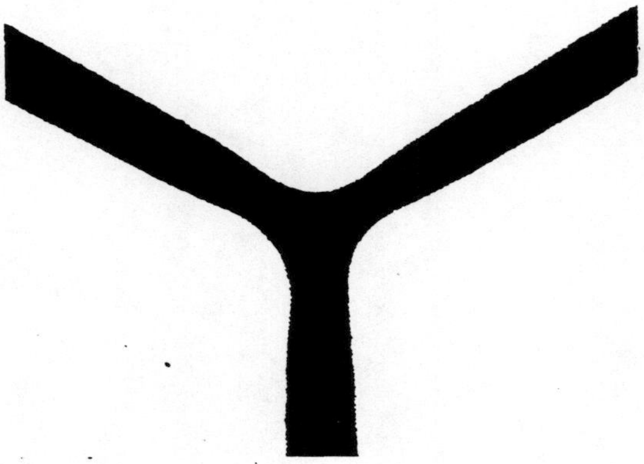

b

Fig. 7. Fig. 5. Micrographs of patterns produced by irradiation of 0.1 and 0.42 μm thick As_2S_3 layers with Kr^{86} 10^{11} ions/cm^2 and transferred to Cr layers.

to their complete absorption. The length of the trace is an indication of the ion energy. The result shown in Fig. 8 suggests that registration of traces of single ions is possible in As_2S_3 layers.

Fig. 8. Traces of single ions, visualized after irradiation of 0.5 μm As^2S_3 with Au^{197} 10^9 ions/cm^2, 160x.

By an electron microscopic study, this technique offers possibilities of determining the penetration depth, the ion path, the trace diameter, the irradiation dose, etc.

4. Conclusion

It is shown that evaporated As_2S_3 layers can be used successfully as a positive ion resist. The layers 0.1 μm thick show the highest resolution. The quality of the structures obtained by irradiation with Kr^{86} 10^{11} ions/cm^2 is quite the same as that of the mask used. On the basis of the previous results [7-9] as the results obtained in this investigation, it could be expected that in the presence of an appropriate mask, the use of the inorganic resist based on As_2S_3 would lead to the successful reproduction of submicron and nanomicron structures. The sensitivity of the resist to Kr^{86} ions is commensurable with that of PMMA and it could be assumed that the inorganic resist based on As_2S_3 will show a good sensitivity to He^+, Ar^+ and H^+ ions applied in ion beam lithography.

The evaporated As_2S_3 layers can be successfully used for registration of single high energy ions thus providing a quite interesting possibility of characterization of ion beams.

5. References

1. U.S.Tandon, Vacuum, 43 (3), 241 (1992).
2. K. Balasubramanyan, J.Adesida, A.Knoff, E.Wolf, J. of Microelectronics 14, 35 (1983).
3. A.Yoshikava, O.Ochi, H. Nagai, Y. Mizushima, Appl. Phys. Lett. 31,161 (1977).
4. A.Yoshikava, O.Ochi, Y. Mizushima, Appl. Phys. Lett., 36, 107 (1980).
5. K.Chopra, L.Malhotra, Harshvardhan, Proc. of the Symposium Inorganic Resist Systems, p.129 (1982).
6. A.Buroff, J. de Physique, 42, (4), 967 (1981).
7. B.Mednikarov, Solid State Technology, 27, 177 (1984).
8. G.Danev, E.Spasova , Vacuum, in press
9. B.Mednikarov - in press
10. A.Buroff, A.Rush, J.Non-Cryst. Solids, 90 (1987) 585
11. E.Spassova, G.Danev, P.Guttman, J.Vac.Sci.Technol. B9 (4), 1939 (1991).
12. Konstantinov et al., Bulgarian Authorship Certificate 46640, 1980.
13. G.Danev, B. Mednikarov, E.Spasova, K. Grancharov, Proc. ICPS, Koln 1986, p.527.

Vacuum Electron and Ion Technologies
D. S. Karpuzov (Ed.)

ISBN 1-56072-242-8

AMORPHIZATION OF ION IMPLANTED GaAsP

J. Krynichi, S. Warchol, H. Rzewuski
Institute of Nuclear Chemistry and Technology
Dorodna 16, 03-195 Warsaw, Poland

Amorphization of P implanted epilayers $Ga_{0.6}P_{0.4}$ at liquid nitrogen temperature has been investigated.

Rutherford backscattering and channeling techniques were used to determine damage in implanted layers.The damage distribution calculated from the RBS spectra have been compared with the results of measurements in implanted GaAs and GaP.

The results show that in ternary compounds the introduction rate of thermally stable lattice defects was much higher than in binary crystals.

INTRODUCTION

Ion implantation doping plays the essential role in semiconductor technology and therefore,is widely applied in the fabrication of many semiconductor devices.

The main drawback of ion implantation technique as a method of semiconductor doping is a heavy crystal damage created during the implantation processes.

Defect production during ion bombardment decrease device performance, therefore, the understanding of the creation of post-implantation damage is of basic importance for semiconductor device technology.

The creation of the crystal damage by ion implantation was studied extensively in basic semiconductor materials like Si [1,2] and GaAs [3-6]. A great number of experimental and theoretical results have shown that total amount and depth distribution of the disorder depend upon many factors, such as mass and energy of ions, type of a crystal, temperature of the implantation process.

The aim of the reported investigations was to obtain the information about the amorphization processes taking place in semiconducting ternary compound GaAsP during ion implantation at liquid nitrogen temperature.

EXPERIMENTAL

In the experiments the epilayers $GaAs_{0.6}P_{0.4}$ and for coparison the GaAs and GaP samples were used.GaAsP epilayers were grown by liquid phase epitaxy method on (100) GaAs:Te substrates.

The samples were implanted with P ions to the dose 1.2×10^{14} ion/cm^2 at liquid nitrogen temperature.The energy of implanted ions was 150 KeV.The temperature of the samples was monitored by copper-constantan thermocouple and kept below 120K.

To avoid axial channeling all the samples were implanted in a misoriented direction.The ion beam intensity was lower than 0.1 μA cm^{-2} to minimize the heating of the target.

The degree of amorphization and the depth distribution of postimplanted damage were measured by means of Rutherford Backscattering (RBS) channeling technique of 2 MeV He^+ions. The analysing He beam intensity was kept below 30 nA. The measurements were performed on the ion Van-de-Graaff accelerator in the Institute of Nuclear Studies (Warsaw).

RESULTS AND DISCUSSION

The RBS spectra from implanted layers of GaAsP,GaAs and GaP have been measured.

Figure 1 shows typical aligned and random RBS spectra of GaAsP implanted with a dose 2×10^{14} of 150 keV P ions at 120K.In the spectrum the post implantation damage peak is present. Below this region ,the yield of RBS-aligned spectra increases with increasing depth within the crystal, as the He ions are dechanneled by the disordered part of the implanted crystal [7].

Figure 2 presentsthe damage density profiles found from the related RBS spectra of P-implanted GaAsP, GaAs and GaP respectively.In Fig. 3 the maxima of amorphization degree for P-implanted GaAsP, GaAs and GaP are presented. The damage depth distribution in the implanted layers were calculated from the related RBS spectra of P-implanted crystals with the following well known formula [8]:

$$k(x)=\frac{X_A(x)-X_D(x)}{X_R(x)-X_D(x)}\times 100\%$$

where: $X_A(x)$ - counts in the aligned RBS spectra at the depth x,
$X_R(x)$ - counts in the random RBS spectra at the depth x,
$X_D(x)$ - counts in the dechanneling fitline.

As it is shown in figure 2 and 3 ,in GaAsP implanted layers damage density reaches about 70 % at 1300 Å from the surface. In GaPmaximum damage density decreases to about 60% and in GaAs is below 30% .

In all cases a long tail of disorder extending up to 1800 Å-GaAs, 2500 Å-GaAs, 3000 Å-GaP can be seen.

This tail can be attributed either to the out-diffusion of point defects from the damaged area or to the channeling of the randomly knock- out matrix atoms by the implanted P ions.

CONCLUSIONS

It was found that there was observable difference between the damage created by implantation in GaAsP and GaAs or GaP.

The total amount of thermally stable lattice defects was much higher for ternary compounds compared to GaAs andGaP.

The experimental results show that in ternary crystals having comparable amount of P and As atoms the recombination and anihilation of defects created during implantation is less efficient than in binary crystals.

REFERENCES

1. J. H. Dennis and E. B. Hale, Rad.Eff.30,219 (1976)
2. S. Prussin, D.I.Margolose and R.N.Tauber,J.Apll.Phys. 57,180(1985)
3. M. Slater, M. J. Nobes and G. Carter, Rad.Eff.83, 219 (1984)
4. D. K. Sadana, Nucl. Instr. Meth. Phys. Research B7/8,375(1985)
5. W.Wesch, E.Wendler, G.Gotz, N.P.Kedelidse J. Apll. Phys. 65,519 (1989)
6. J.Krynicki, H.Rzewuski, R.Groetzschel and Claverie, Phys. Res. 13,197,(1990)
7. J.Krynicki, A.Kozanecki, A.Olszewski and R.Groetzschel Phys.Res. ,8,196(1988)
8. W.K.Chu, J.W.Mayer, M.A.Nicolet, Backscatering Spectroscopy, Acad.Press,New York 1978

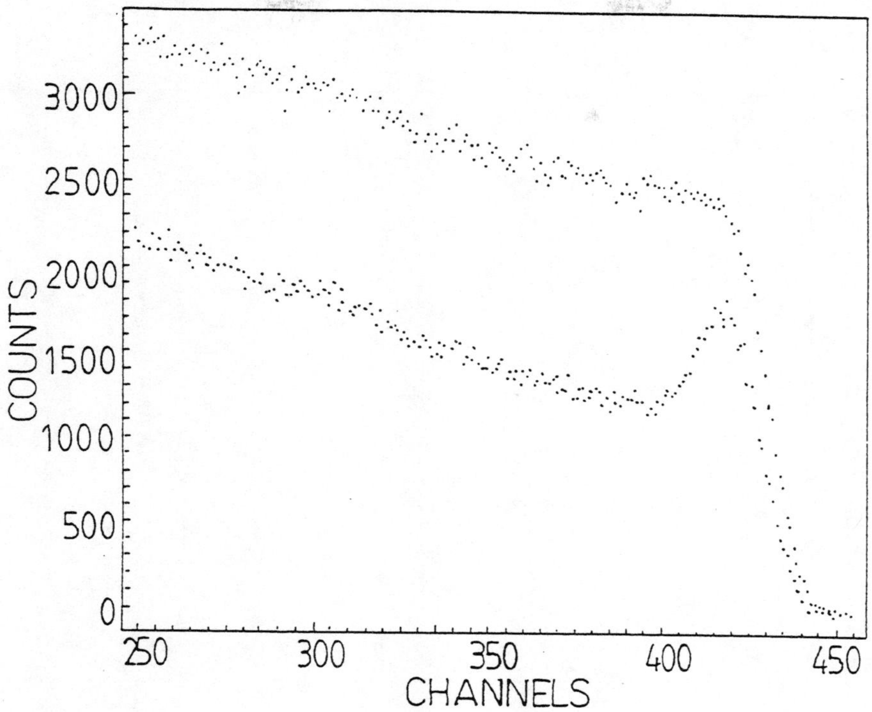

Fig.1 RBS spectra of P ion implanted $GaAs_{0.6}P_{0.4}$ at 150 keV and fluence $1.2 \times 10^{14} cm^{-2}$.

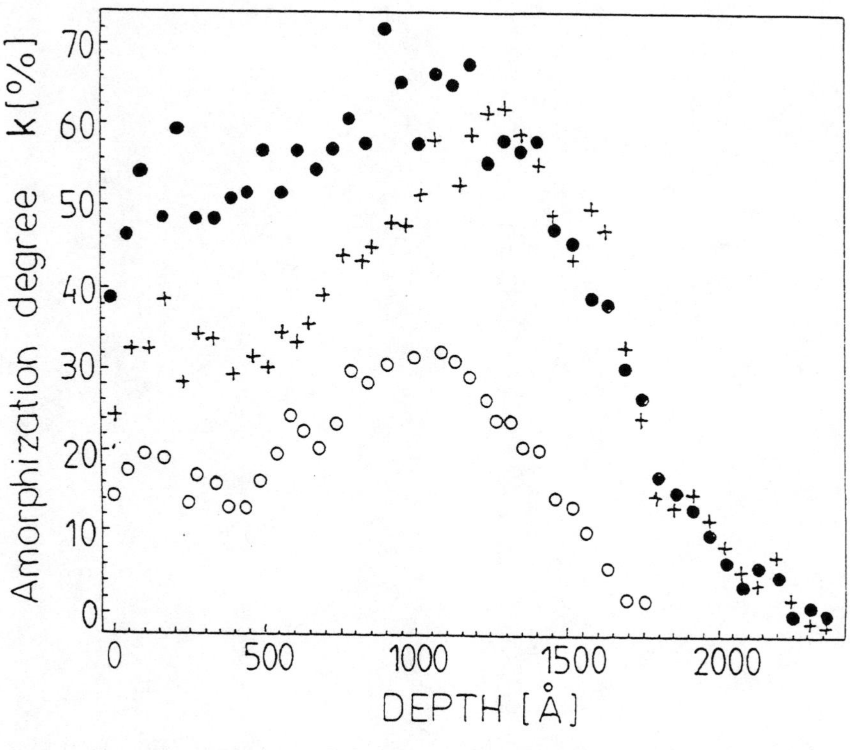

Fig.2 Damage depth distribution for P ion implantation of: ●-$GaAs_{0.6}P_{0.4}$, o- GaAs, +-GaP at 150 keV and fluence $1.2 \times 10^{14} cm^{-2}$.

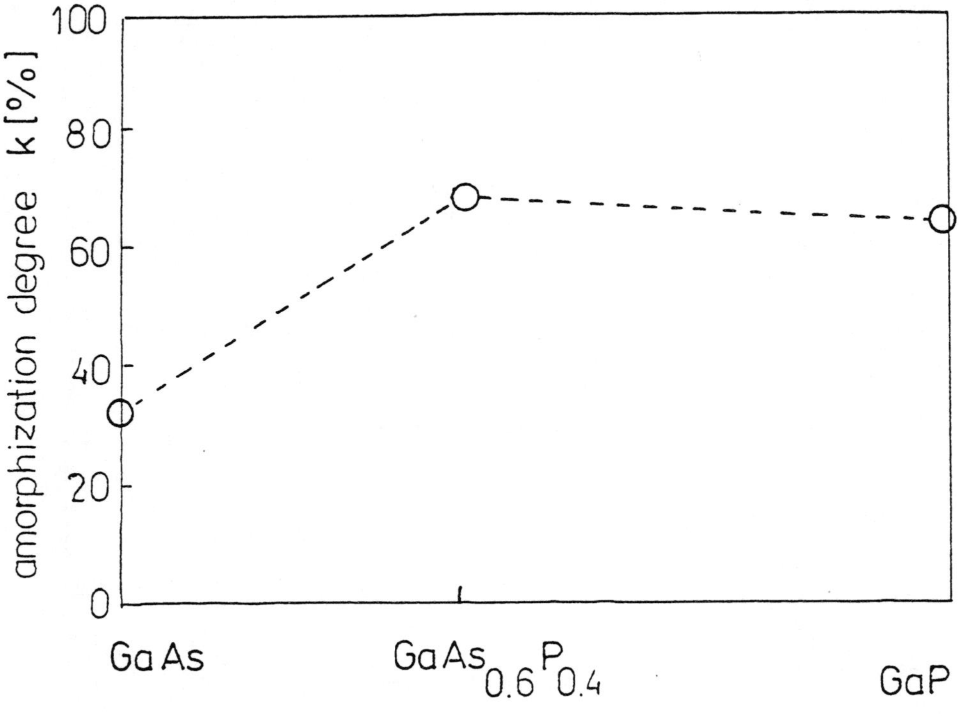

Fig.3 Amorphization degree for P ion implanted $GaAs_{0.6}P_{0.4}$,GaAs and GaP.

Vacuum Electron and Ion Technologies
D. S. Karpuzov (Ed.)

ISBN 1-56072-242-8

HIGH DOSE ION IMPLANTATION OF Sn^+ in a-SiC:H

T. Tsvetkova
Isntitute of Solid State Physics, Bulgarian Academy of Sciences, 72 Tzarigradsko Chaussee Blvd., 1784 Sofia, Bulgaria
N. Tzenov, M. Tzolov and D. Dimova-Malinovska
Central Laboratory for Solar Energy and New Energy Sources, Bulgarian Academy of Sciences, 72 Tzarigradsko Chaussee Blvd., 1784 Sofia, Bulgaria
T. Barancira, H. Pattyn and Ö. Öktü
K.U. Leuven, Department of Physics, B-3001 Heverlee-Leuven, Belgium

With the use of Sn^+ as the implant species there has been achieved a considerable modification effect on the optical properties of thin a-SiC:H alloy films, as demonstrated by optical transmission spectra. High doses of the order of $10^{17}cm^{-2}$ have been used leading to a registered shift of the absorption edge to the lower photon energies. This shift may be attributed both to the introduction of additional disorder and to creation of new bonds between the implanted ions and the atoms of the alloy, confirmed by Raman and IR measurements. Results of ^{119}Sn ion implantation, and investigations with Mössbauer spectroscopy and Rutherford backscattering (RBS) analysis, give additional evidence for the new compound formation, accompanied by the presence of β-Sn precipitates for doses as high as $2x10^{17}cm^{-2}$.

1. Introduction

Ion implantation is commonly used to modify semiconductor materials characteristics via structural and chemical modification, the latter being more clearly manifested when reaching high implantation doses of the order of $10^{17}cm^{-2}$. Using Sn^+ as the implant species for high doses ion implantation of a-SiC:H films has resulted in a considerable modification effect on their optical properties (1). The underlying mechanism of structural and chemical modification has been studied by different optical methods (2).

Supplementary information on the microscopic surrounding and bonding arrangements of the Sn impurity atoms in the matrix has been obtained using Mössbauer spectroscopy on the isotope ^{119}Sn chosen as the implant species. The ^{119}Sn resonance has found widespread application in many fields of science, to be surpassed only by that in ^{57}Fe. Since it exhibits a reasonably high sensitivity to alloying and local bonding rearrangements through its isomer shift S and its quadrupole interaction Δ, it has found application in the study of amorphous semiconductors as well.

In the present work we encounter the same four components that were previously identified in $Si_{1-x}Sn_x$:H (3). We essentially confirm the predominance of a substituted Si site for Sn, while we also detected the presence of a small fraction of Sn^{2+}, presumably due to SnO, starting already from the lowest concentration. This fraction increases at the highest concentration and the increase is accompanied by the formation of Sn^{4+} and of a strong β-Sn component. We have also utilised RBS measurements as a supporting tool for the presented here Mössbauer and optical studies, which have been performed for the elucidation of the observed modification effect.

2.Experimental

The samples were prepared in the form of thin a-SiC:H films deposited on glass and double polished silicon substrates by rf (13.56 MHz) reactive magnetron sputtering (1,2) and rf glow discharge (4) techniques. The carbon content of the as deposited films, obtained by RBS measurements, was 0.18–0.20, and the thickness, measured by the diamond stylus method, was 80–100 nm.

Ion accelerator of the type ILU–4 was used to carry out Sn^+ implantation at an energy E=75 keV, the dose rate being $5\mu A.cm^{-2}$. Typical doses were $D_1=10^{15}cm^{-2}$, $D_2=10^{16}cm^{-2}$ and $D_3=10^{17}cm^{-2}$. Thickness of the films and the ion energy were chosen so as to ensure that almost the whole thickness of the layer was implanted and modified by the ion beam, although inhomogeneously, but possibly in the most optimal manner to allow optical data analysis. The optical transmittance (VIS–NIR and IR region) and Raman measurements were performed as described elsewhere (2).

^{119}Sn was implanted at room temperature, at a dose rate of $2.8\mu A.cm^{-2}$ for the doses of $1x10^{15}cm^{-2}$, $1x10^{16}cm^{-2}$ (sample a), $1x10^{17}cm^{-2}$ (sample b) and at $4.9\mu A.cm^{-2}$ for the dose of $2x10^{17}cm^{-2}$ (sample c). The energy was 50 keV. In view of optimising the statistical information that could be extracted from the Mössbauer spectra collected in a given time period, we opted for the detection of conversion electrons: each sample was mounted inside the electron detector, operating in a flowing He-10% CH_4 gas atmosphere and at a voltage of 850 V. Nonetheless, for the first sample with a dose of $1x10^{15}cm^{-2}$, that the RBS measurement showed to contain about 0.5 at.% Sn in the top 30 nm layer, the effect was negligibly small. All spectra were acquired at room temperature in a constant acceleration mode, using a $Ca^{119m}SnO_3$ source. The spectra were fitted to a sum of Lorentzians using the least squares fitting programme "Fitmos".

3. Results and discussion

A considerable modificattion effect on the optical properties of a-SiC:H films has been achieved by high dose ion implantation of Sn, as demonstrated in fig. 1, the effect being more markedly pronounced with the ion dose. As evident from the figure the transmission edge moves to longer wavelehgths accompanied by a decrease in the transmission coefficient, as a result of ion implantation. This shift of the transmission edge to the lower photon energies is a demonstration of a similar absorption edge shift which signifies a change in the energy

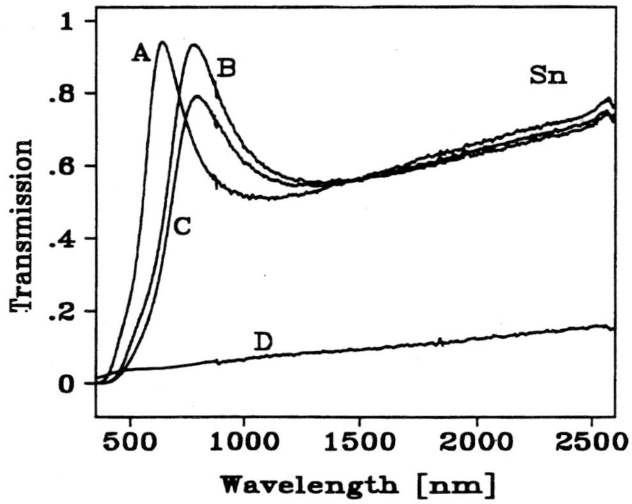

Figure 1. Transmission spectra of a-SiC:H films, implanted with Sn^+ at different doses – as deposited (A), $D_1=10^{15}cm^{-2}$ (B), $D_2=10^{16}cm^{-2}$ (C) and $D_3=10^{17}cm^{-2}$ (D).

spectrum of the electronic states. The observed change could be related to breaking of bonds due to ion bombardment and possible introduction of additional bonds between implanted ions and the atoms of the alloy.

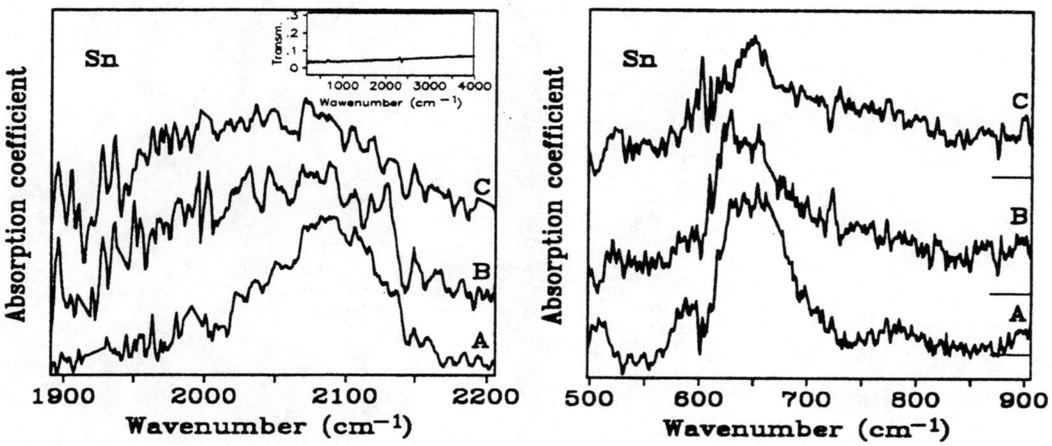

Figure 2. Absorption coefficient spectra in the regions 500–900 cm^{-1} (a) and 1900–2200 cm^{-1} (b) of a-SiC:H films, implanted with Sn^+ at different doses: as deposited (A), $D_1=10^{15}cm^{-2}$ (B), $D_2=10^{16}cm^{-2}$ (C). The zero level for each spectrum is indicated. In the inset of (a) the transmission IR spectrum of an a-SiC:H film, implanted with Sn^+ at dose $D_3=10^{17}cm^{-2}$ is shown.

This assumption is confirmed by the registered changes in the absorption coefficient spectra in the IR region for the implanted a-SiC:H films (fig.2a and b, showing different energy regions). The well defined peak at about 2080 cm^{-1} (fig.2a), related to the stretching vibration mode of Si-H bonds in sputtered a-SiC:H films for the unimplanted material, shows a firm tendency to be moved towards the side of the lower energies when increasing the dose. An underlying mechanism of interaction between the ions of the added elements with the Si atoms in the host material resulting in a change of the Si-H stretching mode could be assumed as the cause for this shift. This could easily be understood bearing in mind the lower electronegativity of the implanted atoms as compared to the host ones. The accompanying decrease of the peak with the dose is a further evidence for the modifying effect on the bonding configuration of the host atoms. Another possible explanation for the registered broadening of the band at ~2080 cm^{-1} could be the formation Sn-H bonds, leading to the appearance of a stretching mode near 1980 cm^{-1} (5).

For the highest ion implantation dose ($D_3=10^{17}cm^{-1}$) the transmission coefficient decreases so drastically that there are no observable features as demonstrated in the inset of fig.2a. This peculiarity together with the similar behaviour observed in the visible range (fig.1, spectrum D) could be attributed to a resulting metallization of the film surface accompanied by a noticeable metallic glistening.

Less explicit but still observable are the changes in the region 500-900 cm^{-1} (fig.2b). An increase of the absorption coefficient at ~740 cm^{-1} with the dose could be related to the emerging of a new backbond configuration of the Si atoms and the resulting shift of the Si-C bond stretching mode at ~780 cm^{-1} (6).

The results of the Raman measurements (fig.3) could be considered also as a justification of the above reasoning. Again there is an effect of increasing disordering of the silicon network with the dose. However, appearence of a Sn-Sn bond vibration mode, which is expected to be at ~150 cm^{-1} (7), is impossible to identify even for the highest dose because of the increased scattering of the laser light due to metallization.

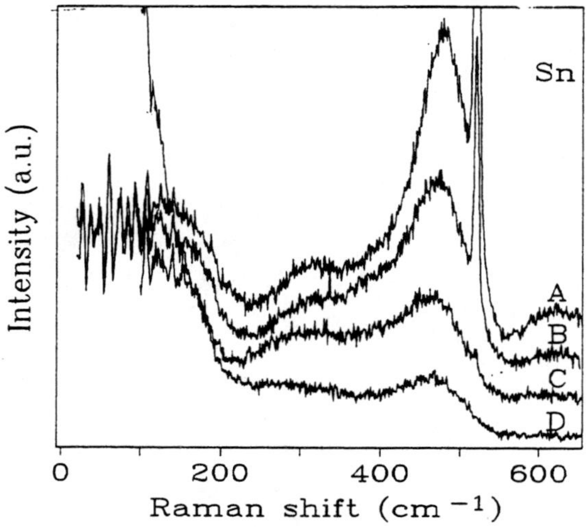

Figure 3. Raman spectra of a-SiC:H films, implanted with Sn^+ at different doses: as deposited (A), $D_1=10^{15}cm^{-2}$ (B), $D_2=10^{16}cm^{-2}$ (C) and $D_3=10^{17}cm^{-2}$ (D). The feature at 520 cm^{-1} in some of the spectra is due to the Raman scattering from the monocrystalline silicon substrate.

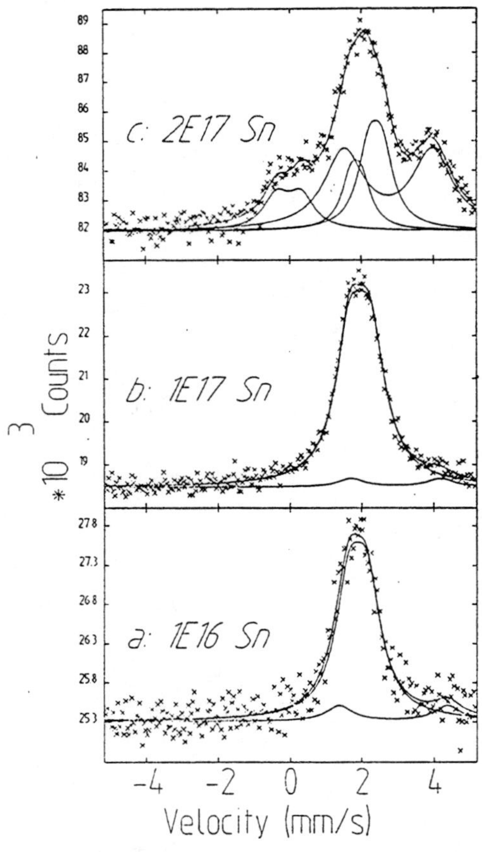

Figure 4. Conversion electron Mössbauer spectra of the a-SiC:H samples **a**, **b** and **c**, respectively, implanted with ^{119}Sn at different doses: (a) $1x10^{16}cm^{-2}$, (b) $1x10^{17}cm^{-2}$ and (c) $2x10^{17}cm^{-2}$. We show the fitted curve together with the decomposition in spectral components.

Supplementary information on the microscopic surrounding and bonding configuration of the Sn impurity atoms in the a-SiC:H matrix was provided by the results of the Mössbauer spectroscopy. Figure 4 a, b and c compare the Mössbauer spectra obtained at a ^{119}Sn dose of $1x10^{16}$, $1x10^{17}$ and $2x10^{17}cm^{-2}$, respectively. The components making up the best fit are also plotted. At the two lower doses, around 90% of the spectral intensity is taken up by a familiar component, site 1, of S value in the neighbourhood of 1.9 mm/s, the remainder of the cpectrum consisting of a very weak contribution of a large S, large Δ component, site 2. The latter component is much stronger in the $2x10^{17}cm^{-2}$ spectrum (fig.4c) as evidenced by the tail at large positive velocity ~4 mm/s. The presence of the former component is now largely obscured by a third component, which is required to fit the right hand side of the main peak in the spectrum, site 3, and a fourth one around 0 mm/s, site 4.

We base the identification of the four sites that we have discerned essentially on their S values, while their Δ values are giving support. We observed site 1 in all three a-SiC:H samples(**a**, **b** and **c**). Its isomer shift lies between the well-documented values of ^{119}Sn in c-Si, close to 1.77 mm/s (8), in a-Si, close to 1.88 mm/s (8) and in semiconducting α-Sn, 1.97 mm/s (9). We therefore clearly attribute that site to sp^3-coordinated Sn impurities, substituting Si atoms in their normal lattice site. The moderate quadrupole interaction experienced by this component very likely finds its origin in the distorted nearest neighbour bond angles or next nearest neighbour dihedral angles. The above interpretation of the Mössbauer data, providing evidence on the quadrupole interaction which affects the sp^3-coordinated site of the Sn impurities, is supported by the results of earlier measurements in which similar quadrupolar splittings in a-Si have been observed. These assumptions are also in good agreement with the Raman spectroscopy results (fig.3) giving evidence for clearly manifested increasing Si-Si bond angle deviation with the dose, and with the data, provided by the IR measurements, explicitly revealing the effect on the local bonding arrangements in the implanted matrix.

The S parameter of site 2, observed in all samples, scatter around those of c-SnO and a-SnO, 2.678 and 2.87 mm/s, respectively (10). The Δ values are larger than the literature values of 1.36 and 1.71 mm/s, respectively (10), which we cannot explain. In the study on $Si_{1-x}Sn_x$:H (3) a spectral intensity around velocity 4 mm/s was also observed in the sample prepared in the absence of H_2 gas. We still find it plausible to attribute this site to Sn^{2+}, presumably in a SnO bond. The S value as Δ value of the latter site are close to those of c-SnO_2, −0.02 and 0.50 mm/s (3), and to the values measured in a-SnO_2, 0.129 and 0.70 mm/s (10). Since the parameters of site 3 are quite close to those of metallic β-Sn, S=2.56 mm/s and Δ=0.3 mm/s (9), we ascribe the presence of this site in the high concentration samples to precipitated β-Sn particles.

Our rather precise determination of S_1, the isomer shift of the substitutional site, permits to corroborate one of the findings of Williamson and Deb (3) where they observed an increase from 1.83 to 1.97 mm/s upon increasing the Sn content of their alloy from 5 to 30 at.%. We deduced 1.87(3) mm/s in sample **a**, which contained about 6 at.% Sn in the top 30nm layer as shown by RBS measurement, and 1.95(1) mm/s in the sample **b**, which contained about 30 at.% Sn. The complexity of the spectrum of sample **c** prevents us in determining its value of S_1 with a good precision. Although the implanted dose is double that of sample **b**, RBS measurements showed it to contain only about 24 at.% Sn. While sputtering will certanly limit the maximum concentration of Sn, the upper limit probably being attained already at $1x10^{17}cm^{-2}$, the decrease observed in sample **c**, upon doubling the dose, can only be explained by a partial peeling off of the a-SiC:H layer. This is confirmed by an RBS measurement that showed the quasi complete disappearance of the a-SiC:H layer on the top of the SiO_2 substrate. The strong concurrent increase of the oxide components 2 and 4, SnO and SnO_2, points in the same direction, although the presence of small fraction of site 2, between 5 and 10 %, already at doses where a-SiC:H layer is still intact, appart of the thinning due to sputtering, seems to weaken this reasoning. Within the limits of the sensitivity of the RBS measurements performed, we detected no oxigen on the a-SiC:H layer. This seems to indicate that Sn forms oxides with trace amount of oxigen, maybe already present right after deposition procedure. From the decrease in fractional population of site 2 in going from $1x10^{16}$ to $1x10^{17}cm^{-2}$, we conclude that this trace amount was not introduced through the implantation process.

While we argued that part of the a-SiC:H layer of sample **c** flaked off, the fact that still about 13% of the spectral intensity belongs to the substitutional site indicates that part of the toplayer should still be present. It remains however unclear why we observe such a large fraction of site 3, which is believed to stem from β-Sn particles. A fraction of 24% of the spectral intensity corresponds in absolute terms to a population of close to 100%, since the recoilless fraction, of around 0.04 at room temperature (11), is about ten times smaller than that of the other sites. The population of that component will be that high, unless the

Debye temperature, and thus the recoilless fraction, are strongly increased by the possibly nanometer sizes of the precipitates, for which no experimental evidence is available now. One would also expect to see some presence of site 3 in sample **b** which we implanted to half the dose of sample **c**: if the recoilles fraction is as low as mentioned and if the absolute population is not high, its presence could however go unobserved in the fitting process. We found some indication for this, from the discrepancy between the observed Sn content in sample **a** (about 6 at.%, from RBS), and that in sample **b** (about 30 at.% – the increase by only a factor of 5, while the dose increases 10 times, is due to sputtering) and the total effect in the Mössbauer spectra which increased by less than factor of 3. In going from sample **b** to **c** RBS observed a slight decrease in Sn content, from about 30 to 24 at.%, while the Mössbauer effect decreased by a factor of 2, again pointing to the low spectral contribution of the large β-Sn component.

4. Conclusions

High dose ion implantation of Sn^+ in a-SiC:H films leads to observable chemical modification of the implanted material, accompanied by considerable changes in the optical properties of the films. These changes could be attributed to breaking of bonds due to ion bombardment and possible introduction of additional bonds between implanted ions and the atoms of the alloy, resulting in decreasing of optical band gap as evidenced by the absorption edge shift to the lower photon energies. The Mössbauer studies on ^{119}Sn implanted a-SiC:H layers deposited on SiO_2 confirm that Sn, as an impurity which is isoelectronic with Si, populates essentially a Si substituted site up to high local concentrations. of 30 atomic %. We detected a small fraction on sites that are Sn^{2+} and Sn^{4+} bonded, probably to oxygen (SnO and SnO_2, respectively). At the highest consentrations we deduced that a rather large amount of the implanted Sn precipitated in the form of β-Sn inclusions.

References

1. T. Tsvetkova, N. Tzenov, M. Tzolov, D. Dimova-Malinovska, G. Adriaenssens, H. Pattyn and W. Lauwerens, *J. Non-Cryst. Solids* (1993), in press.
2. N. Tzenov, M. Tzolov, D. Dimova-Malinovska, T. Tvetkova, C. Angelov, G. Adriaenssens and H. Pattyn, *Nucl. Instr. Meth.* **B** (1993), in press.
3. D. Williamson and S. Deb, *J. Appl. Phys* **54** (1983) 2588.
4. A. Eray, O. Kodolbas, H. Tolunay, Ö. Öktü, W. Lauwerens and G. Adriaenssens, in *"Electronic and Optoelectronic Materials for the 21th Century"*, eds. J. Marshall, N. Kirov and A. Vavrek, (World Scientific, Singapore, 1993), p.423.
5. D. Girginoudi, N. Georgoulas and A. Thanailakis, *J. Appl. Phys.* **66** (1989) 354.
6. W. Mohr, C. Tsai and R. Street, *Mat. Res. Soc. Symp. Proc.*, **v.70** (1986) 319.
7. F. Demichelis, G Kaniadakis, A. Tagliafero, E. Tresso and P. Rava, *Phys. Rev.* **B37** (1988) 1231.
8. G. Weyer, J. Andersen, B. Deutch, J. Golovchenko and A. Nylandsted-Larsen, *Rad. Eff.* **24** (1975) 117.
9. J. Stevens and V. Stevens, eds., *Mössbauer Effect Data Index*, (IFI/Plenum, New York, 1978), p.131.
10. G. Collins, T. Kachnowski, N. Benczer-Koller and M. Pasternak, *Phys. Rev.* **B19** (1979) 1369.
11. C. Hohenemser, *Phys. Rev* **A139** (1965) 185.

Vacuum Electron and Ion Technologies
D. S. Karpuzov (Ed.)

ISBN 1-56072-242-8

ELECTRICAL AND OPTICAL PROPERTIES OF VITREOUS CHALCOGENIDE FILMS AFTER CARBON ION IMPLANTATION

S. Balabanov, E. Skordeva and T. Tsvetkova
Institute of Solid State Physics, Bulg. Acad. Sci
72 Tzarigradsko Chaussee, 1784-Sofia, Bulgaria
B. Amov, C. Angelov and A. Djakov
Institute of Nuclear Research, Bulg. Acad. Sci.
72 Tzarigradsko Chaussee, 1784-Sofia, Bulgaria

In the present work thin amorphous chalcogenide films of the As-Se type have been irradiated with C^+ ions (E=30 KeV), with doses up to 2.10^{17} cm^{-2}. The effect of modification has been evaluated by studying d.c. and a.c. electrical conductivity (100 Hz-100 KHz). Temperature dependence of the d.c. conductivity has been measured in the range 25 to 160°C to allow estimation of the modifying effect on the activation energy. High dose C^+ ion implantation also results in considerable changes of the optical properties of the thin amorphous As-Se films, as registered by optical transmission measurements (0.4-2.5 μm). The obtained results are related to the electrical conductivity data and discussed in the light of existing concepts on amorphous materials modification.

1. Introduction

For a long time conductivity control in chalcogenide vitreous semiconductors has been a challenge for investigators in the field and a widespread belief is that unlike the crystalline semiconductors they are insensitive to doping with impurities. This conception suffered considerable changes with the introduction of the ideas of the so-called "cold methods" of doping by Ovshinsky and co-workers [1]. They were successful in achieving an increase of electrical conductivity over many orders of magnitude for a number of complex chalcogenide materials. Transition metals like Ni proved to be particularly effective when introduced by rf co-sputtering at comparatively low temperatures thus preventing structural and configurational equilibrium to be attained. As a result electrically active impurities were introduced but because of the greater amount needed to control the conductivity in a similar manner as in the case of doping of the crystalline materials, this process was called "chemical modification".

Another highly effective method of electrical conductivity control over many orders of magnitude, i.e. chemical modification of chalcogenide glasses, proved to be the method of ion implantation with Ni and a number of other suitable elements [2,3]. According to this publication of ours electrical conductivity of thin a-As_2Se_3 films could be increased in wide range (reaching 10 orders of magnitude for the highest dose of Ni - $D\approx 10^{17}$ cm^{-2}), while the activation energy change is in the range of 0.9 - 0.2 eV.

In the present work the object of investigation were thin films of the As-Se system modified by carbon ion implantation at comparatively high doses (the highest being $D=2.10^{17}$ cm^{-2}), the carbon impurities being considered as chemically inert towards the host atoms in the chalcogenide material.

2.Experimental

The samples were prepared in the form of thin (0.8-1.8 μm) films of vitreous As_3Se_2 by thermal evaporation in vacuum ($P=10^{-3}$ Pa) onto mica and quartz substrates, for electrical and optical measurements respectively. The film thickness was measured with a Talystep profilometer and proved to be unaffected by the carbon ion implantation up to doses as high as 2.10^{17} cm^{-2}. An ion accelerator (type ILU-4) was used for the implantation of C^+ with energy 30 KeV, dose rate ≈ 5 $\mu A.cm^{-2}$, and different doses ($D=10^{16}$ to 2.10^{17} cm^{-2}). Conductivity σ was measured at planar sample geometry with Au electrodes, using Teraline III electrometer. The temperature dependence of σ was studied in the interval 25°C-165°C. Optical transmission measurements have been carried out with the help of double-beam spectrophotometers of the type Specord 61 NIR and UV-190 Shimadzu. Using the method of R.Swanepoel [5] and suitable computer program,estimations for the optical parameters $n(\lambda)$ and $\alpha(\lambda)$ have been obtained and the modification effect on the Tauc optical gap has been evaluated by using the plot $(\alpha h\nu)^{1/2}=f(h\nu)$.

3.Results and Discussion

a) D.c.-measurements

Fig.1, curve 1, shows the results of the electrical conductivity (σ) changes versus the dose (D) of the implanted C^+ ions. For comparison curve 2 in the same figure illustates the results for fused quartz implanted with carbon ions (see [2]). The attainment of practically the same value of σ at the highest dose of implanted carbon ions ($D=2.10^{17}$ cm^{-2}) could be regarded as an indirect proof that the incorporated carbon in the modified uppermost layer with an approximate thickness 0.1 μm, is not chemically bonded to the host atoms of the As_3Se_2 matrix.

The influence of the dose of the implanted C^+ ions on the temperature dependence of the electrical conductivity for the As_3Se_2 composition is demonstrated in Fig.2. By comparison with the values of E_σ for the unimplanted material (Fig.2, curve 1) it is seen that the activation energy in the higher temperature range undergoes practically no changes during C^+ implantation with a dose $D=10^{16}\,cm^{-2}$ while after $D=1.10^{17}\,cm^{-2}$ and $D=2.10^{17}\,cm^{-2}$, it decreases down to 0.5 eV and 0.45 eV respectively. In the lower temperature region the ion doses $D=10^{16}\,cm^{-2}$ and $D=1.10^{17}\,cm^{-2}$ do not affect noticeably the activation energy, while the highest dose $D=2.10^{17}\,cm^{-2}$ results in a drastic change of E_σ down to ≈ 0.07 eV.

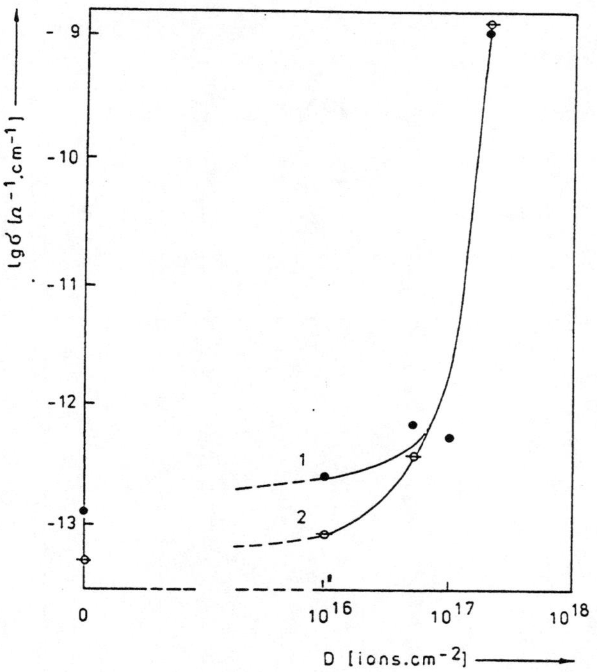

Fig.1.Room temperature d.c. conductivity *vs* dose D for C^+ ion implantation in: (1)-As_3Se_2 and (2)-fused quartz.

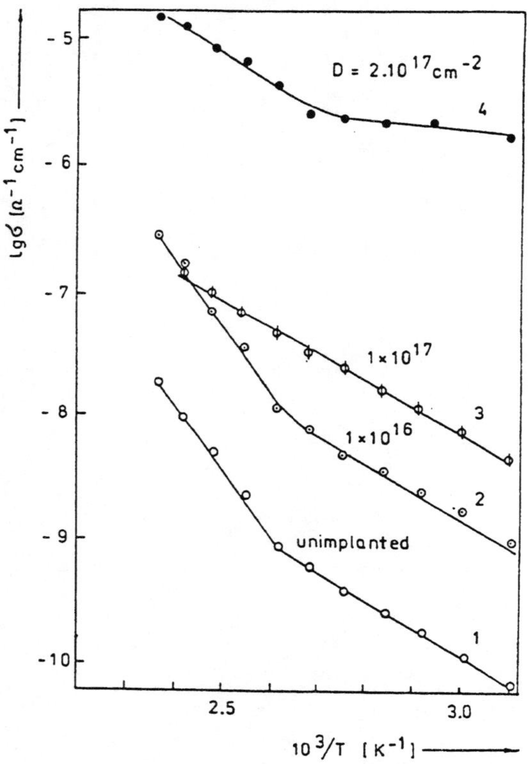

Fig.2. Temperature dependence of d.c. conductivity σ for As_3Se_2 films implanted with C^+ at different doses.

b) A.c.-measurements

Due to the small and not well defined electrical capacity of the investigated thin films samples, as well as the high ohmic resistivity, the parameter of the material,most sensitive to the a.c. frequency proved to be the dielectric losses (tanδ).

In Fig.3 the obtained results for the $\tan\delta=f(\nu)$ dependence is presentted in As_3Se_2 films for $\nu=70-10^5$ Hz (Fig.3, curve 1). In the same figure (Fig.3, curves 2,3 and 4) the changes in this dependence as a result of different doses of C^+ ion implantation are displayed : $D_1=10^{16}$ cm^{-2}, $D_2=5.10^{16}$ cm^{-2} and $D_3=10^{17}$ cm^{-2}. The obtained results for D_1 and D_2 show that the carbon implantation results in a comparatively small increase (excluding the lower frequencies - 100 - 150 Hz) of $\tan\delta$ and emergence of local maxima in the range $3.10^2-1.10^3$ Hz. For the highest C^+ dose, $D=1.10^{17}$ cm^{-2}, the dielectric losses ($\tan\delta$) increase practically by more than an order of magnitude (curve 4), but in the low frequency range only. At 10^3 Hz the dielectric losses drastically decrease (from $2.5.10^{-1}$ to 5.10^{-3}), while at higher frequency $\tan\delta$ surpasses those for the lower doses by a factor of 2. At frequencies above 10^4 Hz this difference is even smaller.

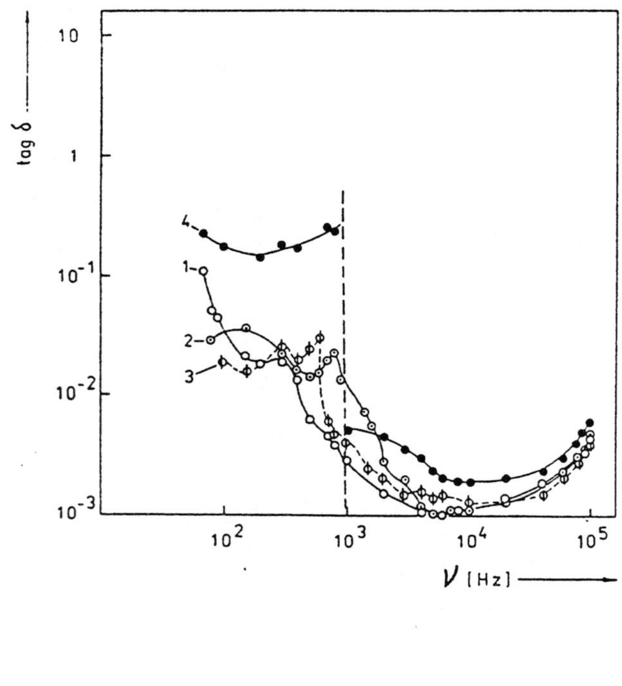

Fig.3. Dielectric losses $\tan\delta$ *vs* frequency ν for As_3Se_2 films implanted with C^+ at different doses D: (1)-0, (2)-10^{16} cm^{-2}, (3)-5.10^{16} cm^{-2}, (4)-10^{17} cm^{-2}.

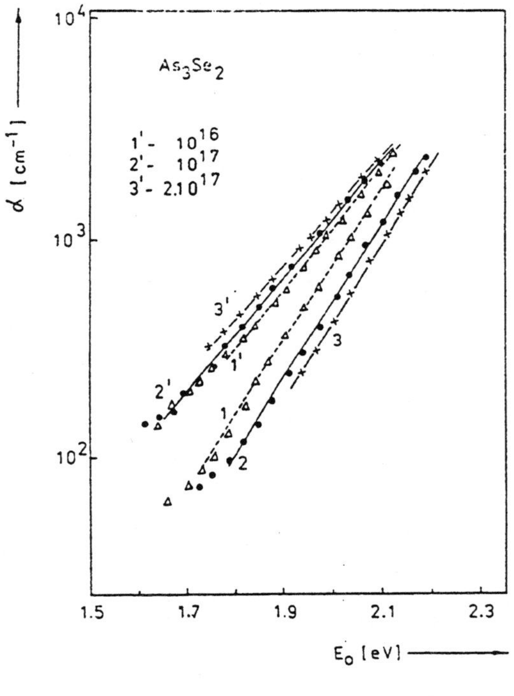

Fig.4. Optical absorption coefficient α *vs* photon energy E_0 for As_3Se_2 films implanted with C^+ at different doses.

c) Optical measurements

Fig.4 presents the results obtained for the absorption coefficient α, where curves 1, 2 and 3 refer to the unimplanted parts of the samples and curves 1′, 2′ and 3′ correspond to their carbon-implanted parts with doses $D_1=10^{16}$ cm^{-2}, $D_2=10^{17}$ cm^{-2} and $D_3=2.10^{17}$ cm^{-2} respectively. The estimated values of the optical gap E_o for the three As_3Se_2 samples, as deduced from the experimental data, are given in Table 1.

Table 1.

Sample №	D	E_o^1 [eV]	E_o^2 [eV]	ΔE_o
T2-32	1.10^{16}	1.846	1.731	0.115
T2-42	1.10^{17}	1.850	1.714	0.136
T2-310	2.10^{17}	1.846	1.689	0.157

In the table by E_o^1 the estimated values of the initial optical gaps are denoted, and by E_o^2 - the corresponding values after the carbon implantation; ΔE_o denotes the differences between E_o^1 and E_o^2.

The analogous measurements after carbon implantation with the same doses for the two other compositions (AsSe and As_2Se_3) show considerably smaller change in the optical gap. However all these data for the optical parameters of the modified chalcogenide films as a result of carbon implantation should be regarded as rather qualitative, because of the optical inhomogeneity of the films due to the comparatively small range of the implanted ions as compared to the whole thickness of the films.

As evidenced by the experimental data, both the absorption edge and the obtained curves of ln(α)=f(E) exhibit well pronounced change in the slope after the implantation. The increase in the slope of the transmission curve in the range of the absorption edge could be related to a resulting higher degree of disorder or a possible formation of clusters in the implanted region of the films.

4. Conclusions

As a result of carbon ion implantation of thin As-Se type vitreous films considerable changes in the electrical conductivity have been registered, reaching over four orders of magnitude for the highest dose of 2.10^{17} cm^{-2}.

Comparison of the results of carbon implantation in As_3Se_2 with those obtained in fused quartz points to insignificant differences, hence it may be concluded that carbon ions practically affect the electronic properties of the chalcogenide material in a similar manner.

The resulting decrease in the slope of the absorption edge with increasing the dose gives further evidence for the modifying effect of carbon ion implantation on the density-of-states distribution in the gap, as could be deduced from the results on the electrical properties.

This work was supported by the National Foundation of Scientific Research, Gr. F-209.

5. References

1. R.B. Ovshinsky, Proc.7th Int.Conf. Amorphous and Liquid Semicond., ed. W.E.Spear, G.G.Stevenson, Dundee, 1977, p. 519.
2. S. Prawer, A. Hoffman and R. Kalish, Nucl.Instrum.and Meth. in Phys.Res. B **80/81** (1993) in press.
3. T.Tsvetkova, B.Amov, E.Vateva and V.Averyanov, Phys.Stat. Sol.(a) **119** (1990) 107.
4. T.Tsvetkova, B.Amov. A.Djakov, S.Balabanov, I.H.Wilson, J.Non-Cryst.Sol., **97&98** (1987) 443.
5. R. Swanepoel, J.Phys.**E 16** (1983) 1214.
6. J.Dresner, K.W.Hang, J.I.Gittelement, P.Kuznetsoff, J.Non-Cryst.Sol., 27 (1978) 39.

Vacuum Electron and Ion Technologies
D. S. Karpuzov (Ed.)

ISBN 1-56072-242-8

MONTE CARLO SIMULATION OF 25-75 keV ELECTRON ENERGY LOSS DISTRIBUTIONS IN PMMA ON TOP OF HIGH-T_c SUPERCONDUCTING THIN FILMS

J. Georgiev, G. Mladenov, and D. Ivanov
Institute of Electronics, Bulgarian Academy of Sciences
Tzarigradsko shoussee 72, 1784 Sofia, Bulgaria

The spatial distributions of absorbed electron energy density in PMMA on top of $YBa_2Cu_3O_7$/substrate were determined using a Monte Carlo Simulation (MCS). The ability of our Monte Carlo program to accurately simulate the radial exposure distributions was demonstrated. The effects of substrate material ($SrTiO_3$ and MgO), beam energy (25, 50 and 75 keV) and high temperature superconducting (HTS) film thickness (100 and 300 nm) on these distributions were investigated.

1. Introduction.

In 1986 Bednorz and Muller[1] reported on a HTS system Ba-La-Cu-O and some months later another HTS material was discovered, namely Y-Ba-Cu-O.[2] Since that time the field of deposition and application of high-T_c thin films has been the object of intensive investigation.

Improvements to the capabilities of high-T_c superconductor electronics depend on the development of versatile lithographic and patterning processes which allow delineation of submicron and even nano- geometries with proper dimensional control and at the same time do not degrade the characteristics of the HTS film.

Electron beam lithography (EBL) is commonly applied for creating patterns of such small dimensions. It is, however, associated with some drawbacks which affect its accuracy and applicability. The main of them is the so called "proximity effect"[3] which is quantitatively described by the spatial distribution of absorbed electron energy density in the resist. Exact data for this distribution are needed for the proper correction for proximity effect and successful application of EBL. The Monte Carlo simulation is most widely used for obtaining such data.[7,8,10-14]

The aim of this work is to study the spatial distributions of the absorbed electron energy density in PMMA on top of $YBa_2Cu_3O_7$/substrate ($SrTiO_3$ and MgO) using Monte Carlo simulation.

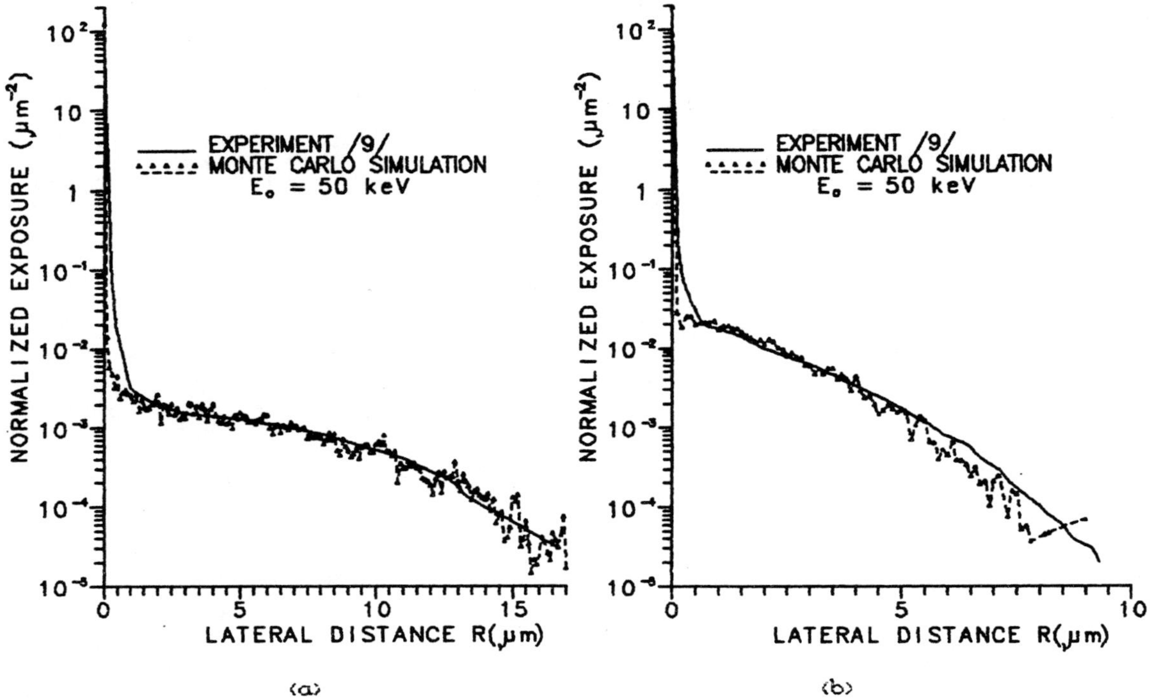

Fig. 1. Comparison of the simulated radial distributions of absorbed energy density on Si and GaAs substrates with the experimental point exposure distributions[9]; E_o=50 keV.

2. Monte Carlo Simulation.

In order to analyze our multilayer systems, an MCS program was developed. It is based on both a single scattering theory and a continuous slowing down approximation. The screened Rutherford formula with the screening parameter of Nigam *et al.*[15] is used for the total elastic scattering cross section and the Bethe equation[16] is assumed for the energy loss between single elastic scattering events. To prevent the failure of this equation at low energies we introduce the parabolic extrapolation of the average energy loss rate $(dE/dx)^{-1}$ derived by Rao-Sahib and Wittry[17] below the certain value of electron energy. Hence, using this extrapolation the trajectory of each electron can be followed until its energy slows down to 50 eV instead of the value of 500 eV which is commonly used as cut-off energy.

A procedure for recalculation of the free path length[14] and of the energy loss is included in order to take into account the differences in both the scattering and the stopping properties at the interfaces between different layers of the target.

In order to evaluate the accuracy of our MCS program, the simulated radial distributions of the absorbed energy density were compared directly with experimental point exposure distributions.[9] As is well seen in Fig. 1 the agreement is rather good. Thus, the distributions obtained from MCS may be used as a reliable bases for the evaluation of actual distributions.

3. Results and Discusion.

In this work calculations were performed on an IBM 4381 machine using a total of 30 000 electron trajectories for each set of parameters. All distributions were obtained for a zero-width δ function.

3.1. Effect of substrate material.

Figure 2 shows a comparison between radial distributions of the absorbed energy density in a 125 nm PMMA resist film on $SrTiO_3$ and MgO substrates as well as on 100 and 300 nm HTS layers of $YBa_2Cu_3O_7$ deposited on the above mentioned substrates at three beam energies - 25 keV (a), 50 keV (b), and 75 keV (c). In this figure three regions are well seen. The first one is in the peaks of the distributions and is usually attributed to the forward scattered electrons. Here there is almost no difference between the compared radial distributions. The second region is from the peaks to the cross-points of the distributions. Here the energy deposited in the resist by the electrons, backscattered from the substrate of higher values of the effective atomic number Z_{ef} and mass density ρ, i.e. from the $SrTiO_3$ substrate, is greater than that deposited by the electrons backscattered from the MgO substrate. The third region is from the cross-

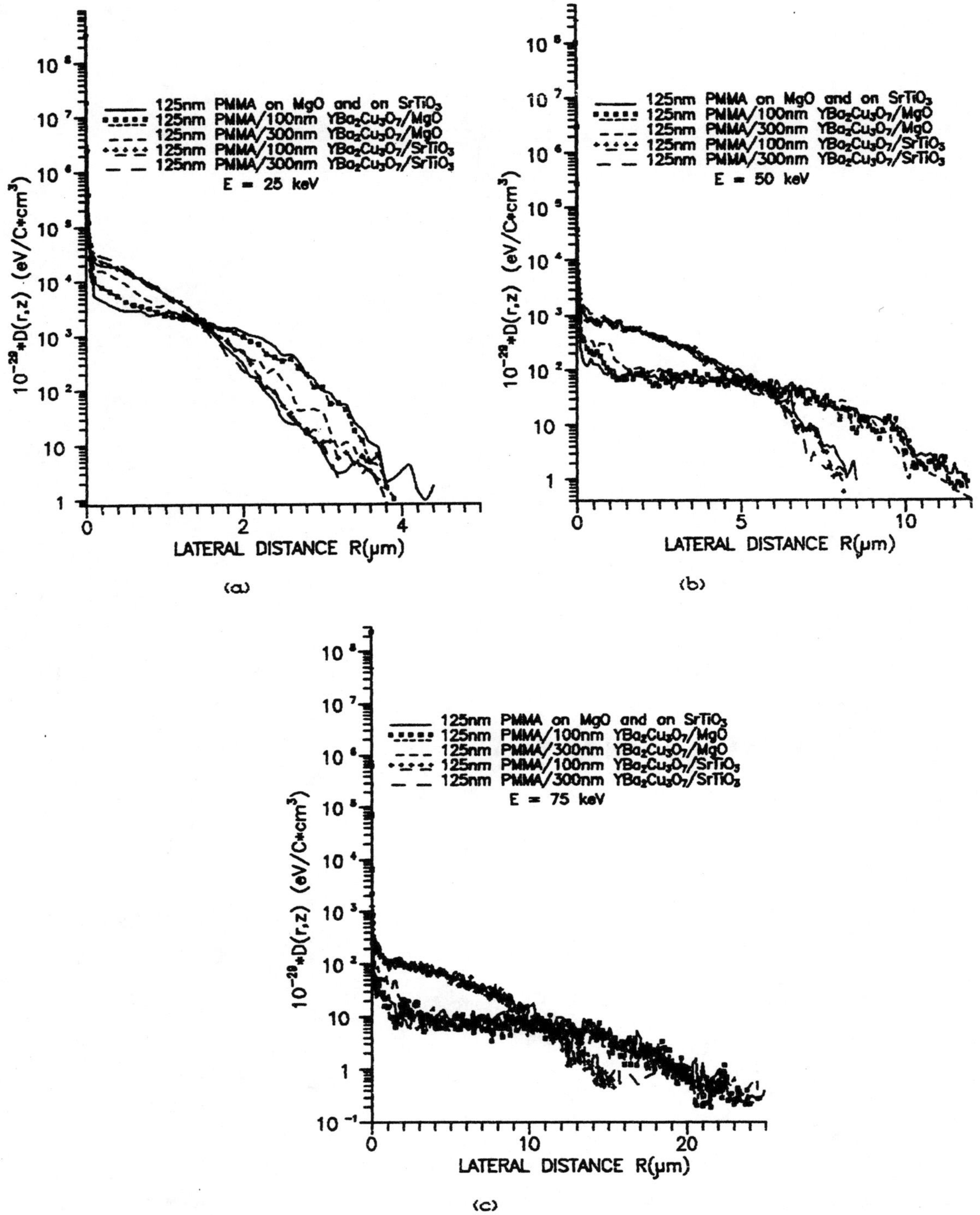

Fig. 2. Comparison between radial distributions of the absorbed energy density in a 125 nm PMMA resist film on $SrTiO_3$ and MgO substrates as well as on 100 and 300 nm HTS layers of $YBa_2Cu_3O_7$ deposited on the substrates: (a) E_o = 25 keV; (b) E_o = 50 keV; (c) E_o = 75 keV.

points to the flanks of the distributions. Here the situation changes - the energy deposited in the resist on the MgO substrate is greater than in that on the $SrTiO_3$ one due to the longer trajectories of the electrons in the MgO substrate.

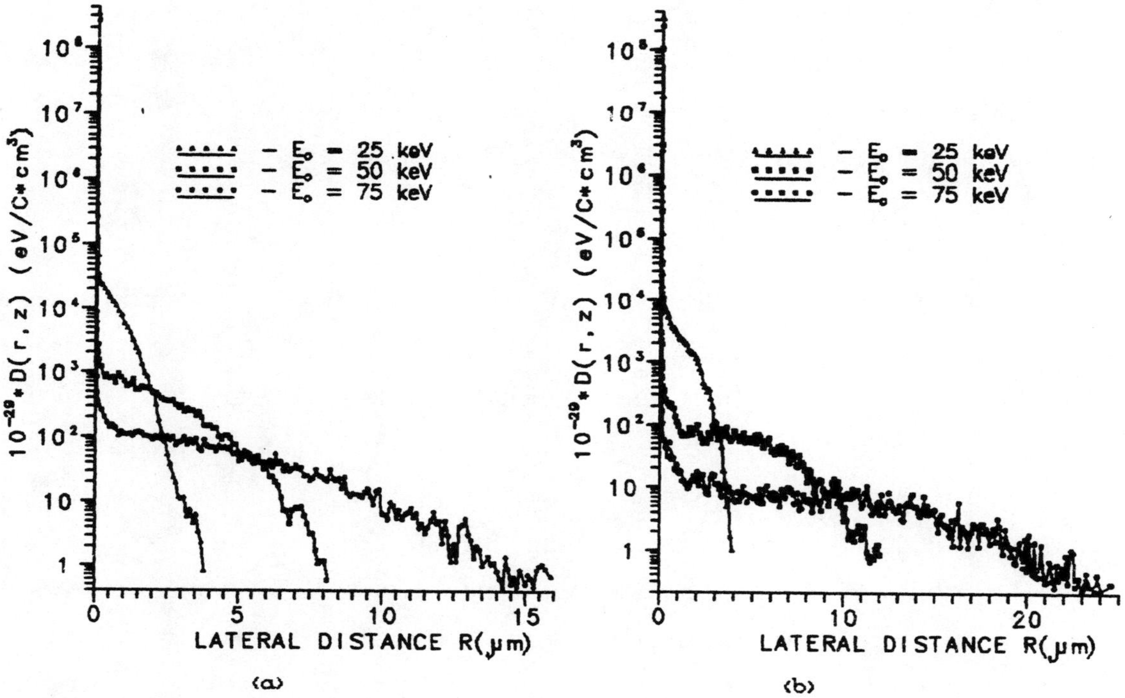

Fig. 3. Comparison of radial distributions of the absorbed energy density for the two structures - (a) 125 nm **PMMA**/ 100 nm $YBa_2Cu_3O_7$/ $SrTiO_3$, and (b) 125 nm **PMMA**/ 100 nm $YBa_2Cu_3O_7$/ MgO at three beam energies - 25, 50 and 75 keV, respectively.

3.2. Effect of beam energy.

Figure 3 compares the radial distributions of absorbed energy density for the two structures - 125 nm PMMA/ 100 nm $YBa_2Cu_3O_7$/ $SrTiO_3$ (a), and 125 nm PMMA/ 100 nm $YBa_2Cu_3O_7$/ MgO (b) at three beam energies - 25, 50 and 75 keV, respectively. Again three regions can be identified when each pair of distributions is considered. This comparison shows that the increase in the beam energy reduces the undesirable exposure by the backscattered electrons only to a certain distance from the incident point. Therefore, if the dimensions of the whole pattern to be created are greater than this distance, and the number of the exposed pixels is large enough (depending on the dimensions of the pattern) it may appear that the proximity effect at a higher accelerating voltage is greater than at a lower one.

3.3. Effect of HTS thin film thickness.

From the comparison made in Fig. 2 it is well seen that the dependence of the spatial distributions of the absorbed energy density on the thickness of HTS film is stronger for the substrate of lower effective atomic number Z_{ef} and mass density ρ (MgO) as well as for lower beam energy.

4. Conclusions.

The spatial distributions of energy density deposited by an electron beam in the resist on $SrTiO_3$ and MgO substrates with $YBa_2Cu_3O_3$ thin films on them were obtained by means of Monte Carlo simulation. The results show that the distributions for the substrate of lower Z_{ef} and ρ (MgO) are wider and lower than those for $SrTiO_3$.

The increase in the beam energy from 25 to 75 keV reduces undesirable exposure by backscattered electrons only to a certain distance from the incident point (the shorter the heavier the substrate is), but causes a fogging exposure beyond this distance.

The influence of HTS film thickness upon the spatial distributions of absorbed energy density is stronger for the substrate of lower Z_{ef} and ρ as well as for lower accelerating voltage. It is strongest for MgO at 25 keV (see Fig. 2 (a)) and is almost negligible for $SrTiO_3$ at 75 keV (see Fig. 2 (c)).

The distributions obtained in this work can be used to correct for the proximity effect or to simulate the profile development in the resist.

Acknowledgments

The authors would like to express their thanks to R. Chakalov of the Institute of Electronics at the BAS for the helpful discussion concerning the technology of deposition of HTS thin films, their properties and application.

References

1 J.G.Bednorz and K.A.Muller, *Z. Phys. B, 64* (1986) 189.

2 T.H.P.Chang, *J. Vac. Sci. Technol., 12* (1975) 1271.

3 M.Parikh and D.F.Kyser, *J. Appl. Phys., 50* (1979) 1104.

4 S.J.Wind, M.G.Rosenfield, G.Pepper, W.W.Molzen, and P.D.Gerber, *J. Vac. Sci. Technol. B*, 7 (1989) 1507.

5 M.G.Rosenfield, S.A.Rishton, D.P.Kern, D.E.Seeger, and C.A.Whiting, *J. Vac. Sci. Technol. B, 8* (1990) 1763.

6 D.F.Kyser and K.Murata, *IBM J. Res. Dev., 18* 352 (1974).

7 K.Murata, M.Kotera, and K.Nagami, *Jpn. J. Appl. Phys., 17* (1978) 1671.

8 K.Murata, D.F.Kyser, and C.H.Ting, *J. Appl. Phys., 52* (1981) 4396.

9 S.A.Rishton and D.P.Kern, *J. Vac. Sci. Technol. B, 5* (1987) 135.

10 K.Murata, H.Kawata, K.Nagami, Y.Hirai, and Y.Mano, *J. Vac Sci. Technol. B, 5* (1987) 124.

11 G.Mladenov, R.Dimitrova, and K.Vutova, *Bulg. J. Phys., 14* (1987) 445.

12 S.Johnson and N.C.MacDonald, *J. Vac. Sci. Technol. B, 7* (1989) 1513.

13 M.Gentili, L.Grella, A.Lucchesini, L.Luciani, L.Mastrogiacomo and P.Musumeci, *J. Vac. Sci. Technol. B, 8* (1990) 1867.

14 B.P.Nigam, M.K.Saunderson, and Ta-You Wu, *Phys. Rev., 115* (1959) 491.

15 H.A.Bethe, *Ann. Phys. (Leipz.), 5* (1930) 325; *Handb. Phys., 24* (1933) 273.

16 T.S.Rao-Sahib and D.B.Wittry, *J. Appl. Phys., 45* (1974) 5060.

Vacuum Electron and Ion Technologies
D. S. Karpuzov (Ed.)

ISBN 1-56072-242-8

ON THE CALCULATION OF ENERGY LOSS IN MONTE CARLO SIMULATION OF SCATTERING OF ACCELERATED ELECTRONS IN SOLIDS

Jordan Georgiev
Institute of Electronics, Bulgarian Academy of Sciences
Tzarigradsko Chaussée 72. 1784 Sofia. Bulgaria

Two schemes of the calculation of energy loss, i.e. the two-points difference scheme and the three-points difference scheme, were applied to the Monte Carlo Simulation (MCS) of penetration and scattering of accelerated electrons in solids. The radial distributions of absorbed energy density in the 125 nm PMMA resist film deposited on two different substrates - Si substrate and GaAs substrate - were obtained numerically by using both schemes. In order to evaluate the accuracy of these schemes the simulated distributions were compared with the normalized point exposure distributions obtained experimentally by Rishton and Kern[1].

1. Introduction.

The Monte Carlo technique is extensively used to simulate the electron scattering in matter[2-15]. In this method the energy loss is typically calculated in the continuous slowing down approximation (CSDA) using the Bethe equation[16] which describes the average energy loss rate dE/dx. Since this equation is derived on the basis of the Born approximation, its accuracy becomes worse in the low energy region or in high atomic number materials. To prevent this failure several other equations were used in the CSDA in conjunction with the Bethe equation as well as instead of it.[11,12] This approach, however, does not take into account the statistical process of inelastic collisions.

A number of techniques have been used in order to account for statistical nature of energy loss.[5-10,12-15] However, most of them do not lead to significant improvement in the energy loss calculation (ELC) but require much more computer time.

Valiev *et al.*[17] have proposed another approach to increase the accuracy of ELC. They have used the Bethe equation in the CSDA but instead of the conventional two-point difference scheme of ELC

$$E_{k+1} = E_k - (dE/dx)_{E=Ek} . \lambda_k, \tag{1}$$

where E_k and E_{k+1} are the electron energies in k-th and (k+1)-th events of elastic scattering, respectively, λ_k is the electron range between these events, and $(dE/dx)_{E=Ek}$ is the average energy loss rate calculated by the Bethe equation for $E = E_k$, they have involved the three-point difference scheme

$$\tilde{E}_{k+1} = E_k - (dE/dx)_{E=Ek} . \lambda_k,$$

$$d\tilde{E}/dx = 1/2[(dE/dx)_{E=Ek} + (dE/dx)_{E=\tilde{E}k+1}] \tag{2}$$

$$E_{k+1} = E_k - (d\tilde{E}/dx) . \lambda_k,$$

which is supposed to increase the accuracy of ELC. However, in that paper there are no data about the influence of this new approach on the output results of calculations, such as beam exposure distributions, backscattering coefficients, etc.

The aim of this work is to study the influence of the application of the three-point difference scheme of ELC to the MCS of electron scattering in solids on the most important output quantity , i.e. the spatial distribution of absorbed energy density

2. Monte Carlo Simulation.

Since the MCS of electron scattering in solids is discussed in details elsewhere[2-4], here only the main features of the present model will be described.

In our program we use the conventional Monte Carlo technique based on a single scattering model and continuous slowing down approximation assuming the screened Rutherford elastic scattering cross section and the Bethe energy loss equation. In order to prevent the failure of the latter one at low electron energies the parabolic extrapolation of $(dE/dx)^{-1}$ derived by Rao-Sahib and Wittry[18] is introduced below a certain value of electron energy. In this way the incident electrons can be pursued until their energy decreases to 50 eV.

To account for the differences in both the scattering and the stopping properties of materials at the interfaces between different layers of the target, a procedure for recalculation of the free path length[12] and of the energy loss is included.

3. Results and Discussion.

The present calculations were performed on an IBM 4381 machine with 30 000 trajectories for each simulation.

In figures 1 and 2 a comparison is made between the numerically obtained by using both schemes of ELC radial distributions of absorbed electron energy density in 125 nm PMMA resist film deposited on bulk Si and GaAs substrates and the experimental point exposure distributions.[1] As seen in these figures the distributions obtained by using (2) go closer to the experimental distributions than that obtained by using (1) particularly in the tails of distributions. However, in Fig. 1 (b) both simulated distributions lie very close to the experimental one and it is difficult to determine which of them is the better. This means that the application of the three-point difference scheme to the ELC increases the accuracy of MCS of electron scattering in solids and that the improvement is greater at lower beam energy for the same substrate (compare figures 1 (a) and 1 (b)) as well as for the heavier substrate at the same beam

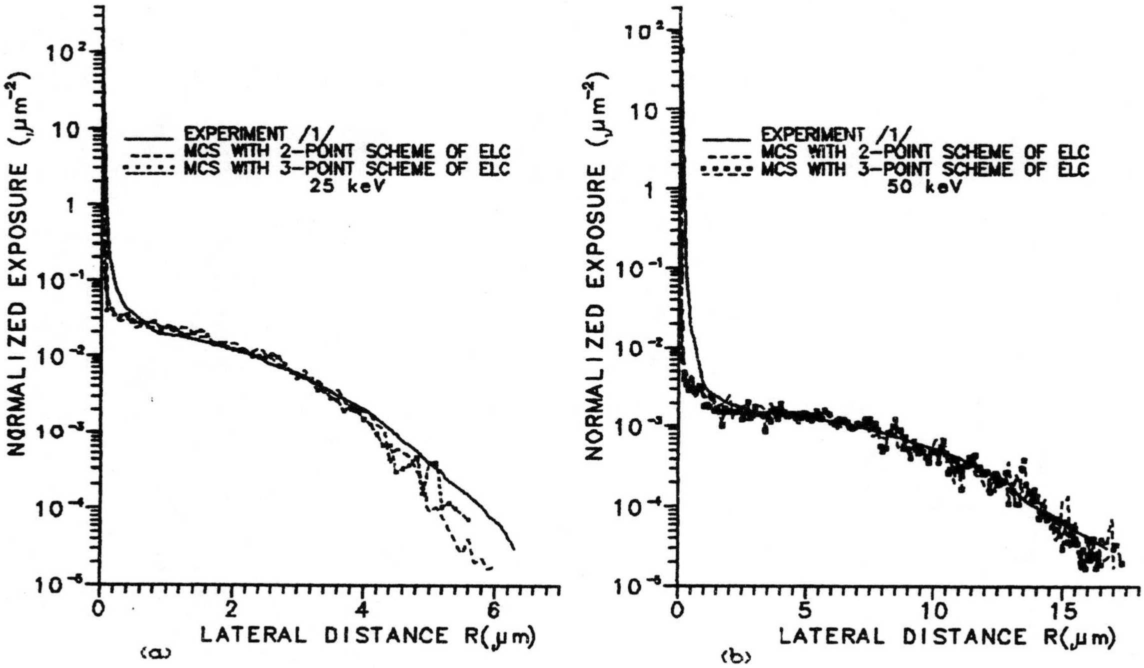

Figure 1. Comparison of simulated by using the 2-points scheme and the 3-points scheme of ELC radial distributions of absorbed energy density with the experimental point exposure distributions[1] in 125 nm PMMA layer on Si substrate: (a) E_o = 25 keV; (b) E_o = 50 keV.

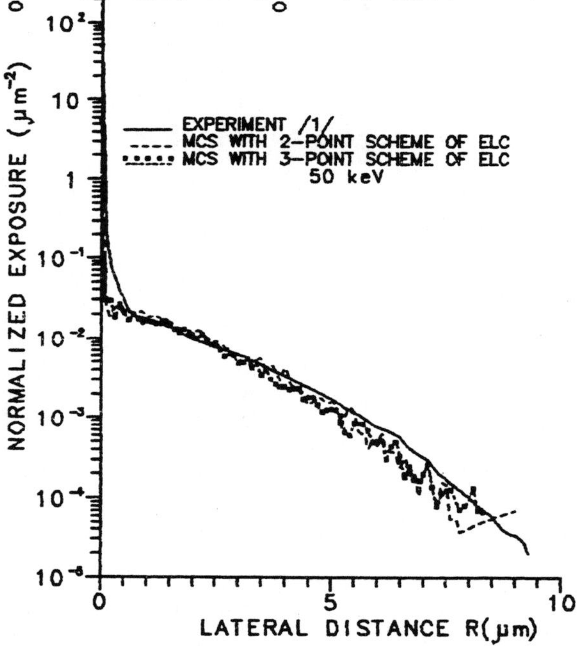

Figure 2. Comparison of simulated by using the 2-points scheme and the 3-points scheme of ELC radial distributions of absorbed energy density with the experimental point exposure distribution[1] in 125 nm PMMA layer on GaAs substrate.

energy (compare figures 1 (b) and 2). Here it must be pointed out that the using of the three-point difference scheme requires extra computery time of about 7%.

4. Conclusions.

The three-points difference scheme of energy loss calculation was applied to the Monte Carlo Simulation of electron scattering in solids instead of the conventional two-points difference scheme. The radial distributions of absorbed energy density in the 125 nm PMMA resist film deposited on Si substrate as well as on GaAs substrate were obtained numerically by using both schemes and were compared with the experimental point exposure distributions in the same structures.[1] The comparison shows that it is worth using the scheme (2) of ELC if the energy of incident electrons is relatively low (E_o < 30 keV) (see Fig. 1 (a)) and/or the atomic number of the target is relatively high (Z > 20) (see Fig. 2).

References

1. S.A.Rishton and D.P.Kern, J. Vac. Sci. Technol. B **5**, 135 (1987).
2. K.Murata, T.Matsukawa, and R.Shimizu, Jap. J. Appl. Phys. **10**, 678 (1971).
3. R.Shimizu and T.E.Everhart, Optik, **36**, 59 (1972).
4. D.F.Kyser and K.Murata, IBM J. Res. Dev. **18**, 352 (1974).
5. R.Shimizu, Y.Kataoka, T.Ikuta, T.Koshikawa, and H.Hashimoto, J. Phys. D: Appl. Phys. **9**, 101 (1976).
6. K.Murata, M.Kotera, and K.Nagami, Jpn. J. Appl. Phys. **17**, 1671 (1978).
7. I.Adesida, R.Shimizu, and T.E.Everhart, Appl. Phys. Lett. **33**, 849 (1978).
8. M.S.C.Chung and K.L.Tai, *Proceedings of the 8th Int. Conf. on Electron and Ion Beam Science and Technology,*

edited by R.Bakish (Electrochemical Society, Pennington, N.J., 1978), p. 242.

9. K.Murata, E.Nomura, K.Nagami, T.Kato, and H.Nakata, J. Vac. Sci. Technol. **16**, 1734 (1979).

10. K.Murata, D.F.Kyser, and C.H.Ting, J. Appl. Phys. **52**, 4396 (1981).

11. M.Kotera, K.Murata, and K.Nagami, J. Appl. Phys. **52**, 7403 (1981).

12. K.Murata, H.Kawata, K.Nagami, Y.Hirai, and Y.Mano, J. Vac. Sci. Technol. B **5**, 124 (1987).

13. S.J.Wind, M.G.Rosenfield, G.Pepper, W.W.Molzen, and P.D.Gerber, J. Vac. Sci. Technol. B **7**, 1507 (1989).

14. M.G.Rosenfield, S.A.Rishton, D.P.Kern, D.E.Seeger, and C.A.Whiting, J. Vac. Sci. Technol. B, **8**, 1763 (1990).

15. M.Gentili, L.Grella, A.Lucchesini, L.Luciani, L.Mastrogiacomo, and P.Musumeci, J. Vac. Sci. Technol. **B 8**, 1867 (1990).

16. H.A.Bethe, Ann. Phys. (Leipz.) **5**, 325 (1930); Handb. Phys. **24**, 273 (1933).

17. K.A.Vliev, A.N.Kirillov, B.N.Kovtun, T.M.Machviladze, and M.M.Mkrtchian, Proceedings of the Institute of General Physics at the Academy of Sciences of the USSR 8, 5 (1987) (in Russian).

18. T.S.Rao-Sahib and D.B.Wittry, J. Appl. Phys. **45**, 5060 (1974).

Vacuum Electron and Ion Technologies
D. S. Karpuzov (Ed.)

ISBN 1-56072-242-8

A UHV SYSTEM FOR ION SURFACE SCATTERING STUDIES WITH A COMPACT HIGH - PRECISION SAMPLE MANIPULATOR

S. Tzanev and I. Neshev*
Institute of Electronics, Bulgarian Academy of Sciences
blvd. Tzarigradsko Chaussée 72. 1784 Sofia. Bulgaria

Abstract

In this paper we describe a UHV system being designed and created for low energy ion beam surface scattering analysis of solids. The system's performance characteristics were estimated by carrying out surface scattering studies of 3 KeV Ar^+ ions on polycrystalline Cu and monocrystalline $Ni(111)$ targets. The experimental set-up comprises a hemispherical electrostatic energy analyzer with corrected fringe field effects, an original high precision sample holder and manipulator mounted in a UHV chamber, and an ion source of duoplasmatron type with cold cathode. The movable energy analyzer allows variation of the scattering angle θ from 0° to 128°. The target can be rotated independently around three axes: **a)** an axis normal to the sample surface for varying the azimuthal angle φ; **b)** a normal to the scattering plane axis lying on the sample surface for varying the incidence angle α, and **c)** an axis determined by the crossection of the sample surface and the scattering plane for changing the tilt angle ψ. The sample is electrically isolated for measuring and integrating the ion beam flux at any position. The target surface can be electron-impact desorption cleaned prior to measurements or heated during the measurements by an original electron gun mounted between the energy analyzer and the target. The whole system is bakeable up to 250° C.

***Present address**: Dept. of Physics, Illinois Institute of Technology, Chicago, IL, USA 60616

1 INTRODUCTION

Ion solid surface scattering spectrometry has experienced rapid development as one of the major surface structural analysis techniques [1], [2], [3]. This is due to the fact that information concerning the structural parameters, i.e. the distance between the scattering centers, is easily obtained [4], [5] together with information about the thermal motion of the scattering centers [6], [7]. The possibility of directly probing the surface periodicity of a single crystal [8], [9] is also an advantage of the above spectroscopy method.

In this paper, we describe a compact UHV system designed and created for low energy ion solid surface scattering studies. The system's performance characteristics have been investigated by carrying out experiments of 3 KeV Ar^+ ions scattering on polycrystalline Cu and monocrystalline $Ni(111)$ targets. A very good reliability, troublefree operation and reproductivity of the experimental results are observed. Good resolution and performance characteristics are achieved by using a home-made compact high-precision sample manipulator and rotatable hemispherical energy analyzer.

2 ANALYZER SYSTEM

A schematic diagram of the scattering system and the target chamber is shown in fig. 1. The ion beam source is of a duoplasmatron type with a cold cathode and is capable of providing a beam of various kinds of ions, e.g. Ar^+, He^+, H^+, N^+. The differentially pumped ion beam forming line consists of two $\mathcal{X}$ - $\mathcal{Y}$ deflectors/scanners, two einzel lenses and apertures. A rotation of the energy analyzer in the scattering plane allows changing of the ion scattering angle. This movement is independent on the other mechanical movemetns, which are provided in the UHV chamber, to the target holder. The original design, creation and performance of a compact target manipulator allows a rotation of the sample around three axes: **a)** an axis normal to the target surface for varying the azimuthal angle φ; **b)** an axis normal to the scattering plane for varying the incidence angle α, and **c**) an axis lying in the scattering plane for tilting the sample face. An electron gun firmly fixed in front of the energy analyzer provides the cleaning of the target by heating it up to 1200° C. In the following paragraphs the ion scattering system and the sample manipulator will be described in detail.

2.1 Primary Ion Beam Source

A duoplasmatron type ion source with a cold cathode is used for production of the primary ions. The ion source and the beam forming line are differentially pumped by a liquid nitrogen trapped diffusion pump. The beam forming line comprises two einzel electrostatic lenses and two pairs of rastering/deflecting plates. The ion source is coupled to the target chamber via a CF50 flange (see fig. 1). The typical working argon gas pressure, as measured in the differentially pumped stage, is about 1×10^{-6} $torr$. The beam spot diameter, i.e. the size of the irradiated target surface area at fixed incidence angle, is variable and could be changed by means of two electrostatic forming lenses placed in the beam line. A typical working size of the beam is a diameter Φ=1 mm at a distance L=120 mm from the source. The beam can be rastered automatically or manually over the crystal surface in two perpendicular directions. This allows ion beam cleaning of the sample surface by sputtering of the few outermost target layers prior to the ion scattering experiments. The typical primary ion beam current, when working with Ar^+ and as measured on the target surface is I_p=10 nA $\div$ 2 μA with corresponding ion energies in the range E_p=0.8 $\div$ 10 keV.

2.2 Vacuum System

The ion scattering system is based on a 250 mm in dia commercially available UHV target chamber. It is oil-free evacuated from atmospheric pressure to the working pressure of a 250 l/s triode ion pump by two sorption pumps. A titanium sublimation pump together with a 1260 cm^2 liquid nitrogen cooled trap are used to reach the base pressure of $p_o = 2 \times 10^{-10}$ $torr$. The target holder and the detector manipulator for the ion scattering studies are placed in the target chamber.

The base pressure in the scattering chamber is reached after several hours of bakeout at 200° C. A typical working pressure after a continuous work on ion surface scattering experiments is $p = 1 \times 10^{-9}$ $torr$.

2.3 Energy Analyzer with an Electron Gun for Sample Heating

A cross sectional view of the energy analyzer, the electron gun mounted in front of it and the target manipulator is shown in fig. 2. The hemispherical electrostatic energy analyzer is based on the proposed by Simpson design, described in ref. [10] and performed by Armour [11]. It consists of convex and concave hemispheres with radii $R_{in} = 22.5$ mm and $R_{out} = 27.5$ mm respectively. The hemispheres are attached to a supporting plate by saphire spheres. The supporting plate incorporates field termination diaphragms to provide Herzog correction for the fringing fields [12] and is connected to a rotatable analyzer holder with a balancing body. The latter provides the centre of the mass of the whole system to lie on the vertical axis, $\mathcal{Z}$ thus loading the stainless steel bearings only axially. The analyzer holder is rotated by a high precision external mechanical feedthru. The resolution of this motion which determines the scattering angle $\theta = 0° \div 128°$ is $\Delta\ \theta = 0.2°$.

The whole assembly is shielded by a soft iron plate which reduces to a negligible level or totally eliminates the surrounding magnetic field. The acceptance angle, specified as the full angle at the top of the cone and determined by a double entrance aperture is 1.5°.

The energy analyzer was calibrated by using a compact solid-state Na^+ ion source as it is known that the energy spread of such types of ion sources is very narrow. The relative energy resolution is found to be ΔE/E=2.5 % see fig. 3.

An electron gun connected firmly to the entrance of the energy analyzer (see fig. 2) provides heating of the sample. It has an original design and consists of a stainless steel cylinder with a diameter Φ=12 mm within which an ○ - shape tungsten wire cathode is located. The electrons emitted from the heated cathode are accelerated to the sample, which could be shifted at positive voltage, thus heating the target up to $T = 1200^\circ\ C$. The $\bigcirc$ - shape form allows avoiding the interaction between the scattered ions and the cathode.

2.4 UHV Compact Sample Manipulator

A schematic cross section of the manipulator design is shown in fig. 2. It is constructed on a commercial positioning stage, allowing $\mathcal{X}$, $\mathcal{Y}$ and $\mathcal{Z}$ linear motions and a rotation around $\mathcal{Z}$ - axis. $\mathcal{X}$ and $\mathcal{Y}$ linear motions are converted into a rotational motion for tilting the target. The $\mathcal{Z}$ linear motion is used for driving the mechanism for azimuthal rotation of the sample. The rotation of the positioning stage around $\mathcal{Z}$ - axis gives the rotation of the sample. All of the above mentioned motions are independent on each other.

2.4.1 Construction Details

A spherical sample holder is used for converting the $\mathcal{X}$ - $\mathcal{Y}$ motions into a tilt motion. This construction decision provides $\psi = \pm 3^\circ$ tilt and $\alpha = 0^\circ - 120^\circ$ rotation of the target by rotating the spherical holder around its center with a precision of $\Delta\psi = 0.1^\circ$ and $\Delta\alpha = 0.2^\circ$ respectively. The stainless steel sphere is precisely located between specially designed BN rings which in turn electrically insulate the sample holder from the other metal parts and also allow its tilt and rotation with very low mechanical friction under UHV conditions (it has to be noted that the spherical surface is shiny polished). The center of the spherical holder lies on the target surface and the axis of the azimuthal rotation which is normal to the sample surface. The target holder is capable of handling cyllindrically shaped samples with diameter $\Phi \leq 10$ mm and thickness $d \leq 2$ mm.

The azimuthal motion is provided by using a rod connected to the $\mathcal{Z}$ drive of the positioning stage (the linear drive for the azimuthal rotation - see fig. 2). Moving the rod along the $\mathcal{Z}$ direction allows azimuthal rotation of the sample around an axis normal to its surface. The possible azimuthal rotation covers a range of $\varphi = 0° - 210°$ with a precision of $\Delta\varphi = 0.2°$. The sample holder with the BN rings is placed in a main supporting body which is unmovable and is firmly connected to a CF150 flange.

The manipulator has functioned for more than 2 years without failure. The base pressures of $p_o = 4 \times 10^{-10}$ *torr* have been reached with the whole system mounted.

2.5 Data Collection Section

The scattered and energy sellected ions impinge the entrance cone of a channeltron which is mounted on the supporting plate of the movable energy analyzer. The electrical impulses created by each of the impinging ions are amplified by a wide-bandpass amplifier and formed by a low- and high-peak rejection system. The energy of the ions together with the number of ions encountered within a preset time interval are stored by a computer. The energy scan time could be varied from few seconds to 60 minutes. The energy scan limits can be varied and set between 0 KeV and up to 5 keV. The scans could desirebly be stored on a computer hard disk or on flopy diskettes. The data could then be processed, i.e. normalized, plotted, etc. in different ways.

3 PERFORMANCE

In the following paragraph the instrumental capabilities of the whole complex will be demonstrated with angle-resolved studies of clean polycrystalline Cu and monocrystalline $Ni(111)$ surfaces.

It was decided the first experiments to be carried out on a polycrystalline target with well established parameters, which is known to have previously been intensively used for such studies. We have chosen polycrystalline Cu as an acceptable one. Series of Ar^+ ion scattering experiments on polycrystalline Cu surface were performed in order to estimate the capabilities of the complex to obtain reproductive data as well as to find the degree of agreement between the experimentally observed peak positions and the theoretically predicted ones. It was found that the peak positions of the energy scans measured at different scattering geometries (several combinations of incident and scattering angles) agree very well with the corresponding calculated ones as it is seen in figures 4 and 5. The experimental conditions were as follows: primary Ar^+ ion energy $E_p = 3$ KeV for all data, scattering angle $\theta = 20°$, incidence angle (measured from the target surface) $\alpha = 10°$ for the results shown in fig. 4 and $\theta = 90°$ and $\alpha = 50°$ for the results shown in fig. 5 respectively. It can be seen from fig. 5 that the energy distribution have only one peak which corresponds to the single scattered Ar^+ ions from Cu atoms and the peak position at the energy scale fits very well the calculated one. The energy distribution on fig. 5 exibits three maxima. The most energetic one corresponds to the single scattered primary ions from Cu atoms. The lowest peak on the energetic scale is from oxygen contaminants on the Cu surface. It has to be noted that no special care and efforts were taken for the polycrystalline Cu face to be atomicaly cleaned prior the experiments in order to obtain higher scattering signal from the surface contaminants such as oxygen. According to the calculations the positions of both peaks fit very well the experimental ones. The small peak on the left shoulder of the single scattering Cu peak corresponds to the Cu recoils.

The second step was to reveal the capabilities of the complex to obtain more sophisticated information by obtaining data of the azimuthal anizotropy of the scattering ions from a monocrystalline target. A typical picture of the scattered ions distributions at different azimuthal angles, as obtained on this apparatus, is shown in figures 6 and 7, where the azimuthal distributions of 3 keV Ar^+ ions scattered from $Ni(111)$ surface are measured at two different scattering energies. A direct determination of the crystal surface periodicity by revealing the main crystallographic axis of the surface can be obtained from it. This could be achieved by monitoring the azimuthal anisotropy of scattered ion intensity in appropriate geometry [13]. The measurements were performed under the following conditions: primary Ar^+ ion beam energy $E_p = 3$ keV, average current intensity $j_p = 3 \times 10^{-5}$ A/cm^2, incidence angle $\alpha = 5°$, scattering angle $\theta = 40°$, tilt angle $\psi = 0°$ and working gas pressure in the target chamber

$p = 8 \times 10^{-10}$ *torr*. The azimuthal distribution shown in fig. 6 corresponds to quasi-single scattered Ar^+ ions, while that shown in fig. 7 – to Ar^+ ions scattered with higher energy (multiple scattering). The comparison of the experimental results shown on fig. 6 and fig. 7 could serve as a basis to say that the anizotropical azimuthal distribution of the multiple scattered Ar^+ ions could provide with more detailed information about the surface directions of the $Ni(111)$ surface as compared to the anizotropical azimuthal distribution of the quasi-single scattered Ar^+ ions.

4 SUMMARY

The design and performance of a compact UHV system for ion scattering with a simple high precision sample manipulator have been described. A very good reliability, troublefree operation of the set-up and also a very good reproductivity of the experimental results are observed. The performed tests on the system and the studies of clean polycrystalline Cu and monocrystaline $Ni(111)$ surfaces show that it is possible to study, on the above described set-up, the fundamental processes associated with low energy ion scattering from solid surfaces. The apparatus described above is currently operating as a research machine for carrying out studies of ion-scattering spectrometry at the Institute of Electronics, Bulgarian Academy of Sciences, Sofia.

5 ACKNOWLEDGMENTS

The authors thank A. Kebedjiev for helping to automate the analytical part of the set-up and D.S. Karpuzov for the valuable discussions. This research was financially supported by contracts F-6 and MT-2 of the Ministry of Science and the National Fund "Scientific Researches".

References

[1] W. Heiland, Vacuum, **39**, (1989), p. 367.

[2] Th. Fauster, Vacuum, bf. 38, (1988), p. 129.

[3] J. W. Rabalais, Science, **250**, (1990), p. 521.

[4] M. Aono, Nucl. Instr. and Meth. **B2**, (1984), p. 374.

[5] H. Niehus, Nucl. Instr. and Meth. **B33**, (1988), p. 876.

[6] R. Souda, M. Aono, C. Oshima, S. Otani and Y. Ishizava, Surf. Sci., **128**, (1983), L236.

[7] E. Taglauer. M. Beckschulte, R. Margraf and D. Mehl, Nucl. Instr. and Meth. **B35**, (1988), p. 404.

[8] H. Bu, M. Shi and J. W. Rabalais, Surf. Sci., **236**, (1990), p. 135.

[9] H. Hetterich and W. Heiland, Surf. Sci., **210**, (1989), p. 129.

[10] J. A. Simpson and C. E. Kuyatt, Rev. Sci. Instrum., **38**, (1967), p. 1.

[11] D. J. Armour, J. J. Jimenez-Rodriguez, C. H. Barber, K. Snowdon and P. Headbavny, Vacuum, **34**, (1984), p. 217.

[12] , Rev. Sci. Instrum., **52**, (1981), p. 1148.

[13] F. Masson, H. Bu, M. Shi and J. W. Rabalais, Surf. Sci., **249**, (1991), p. 313.

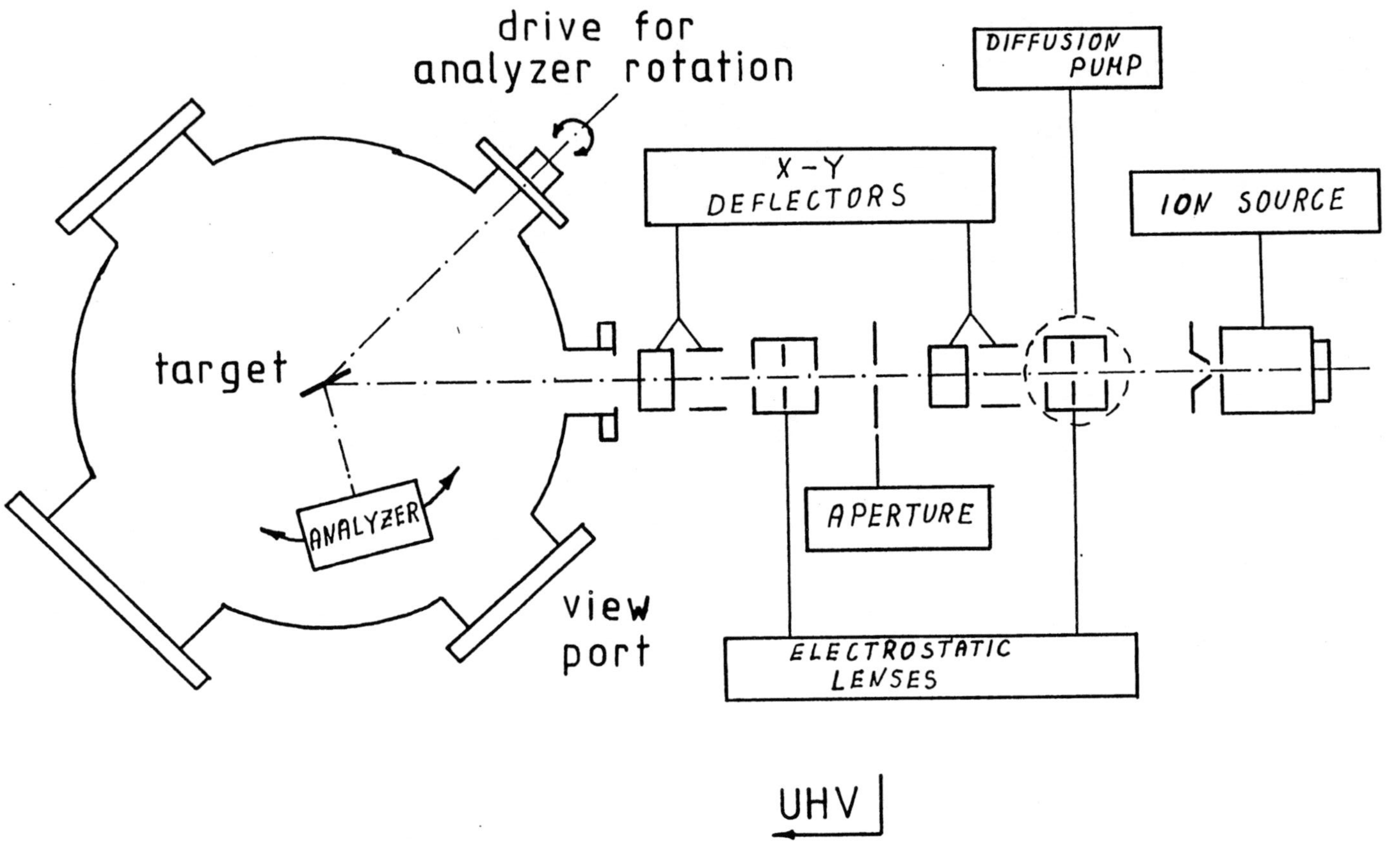

Figure 1: Schematic diagram of the major units of the apparatus.

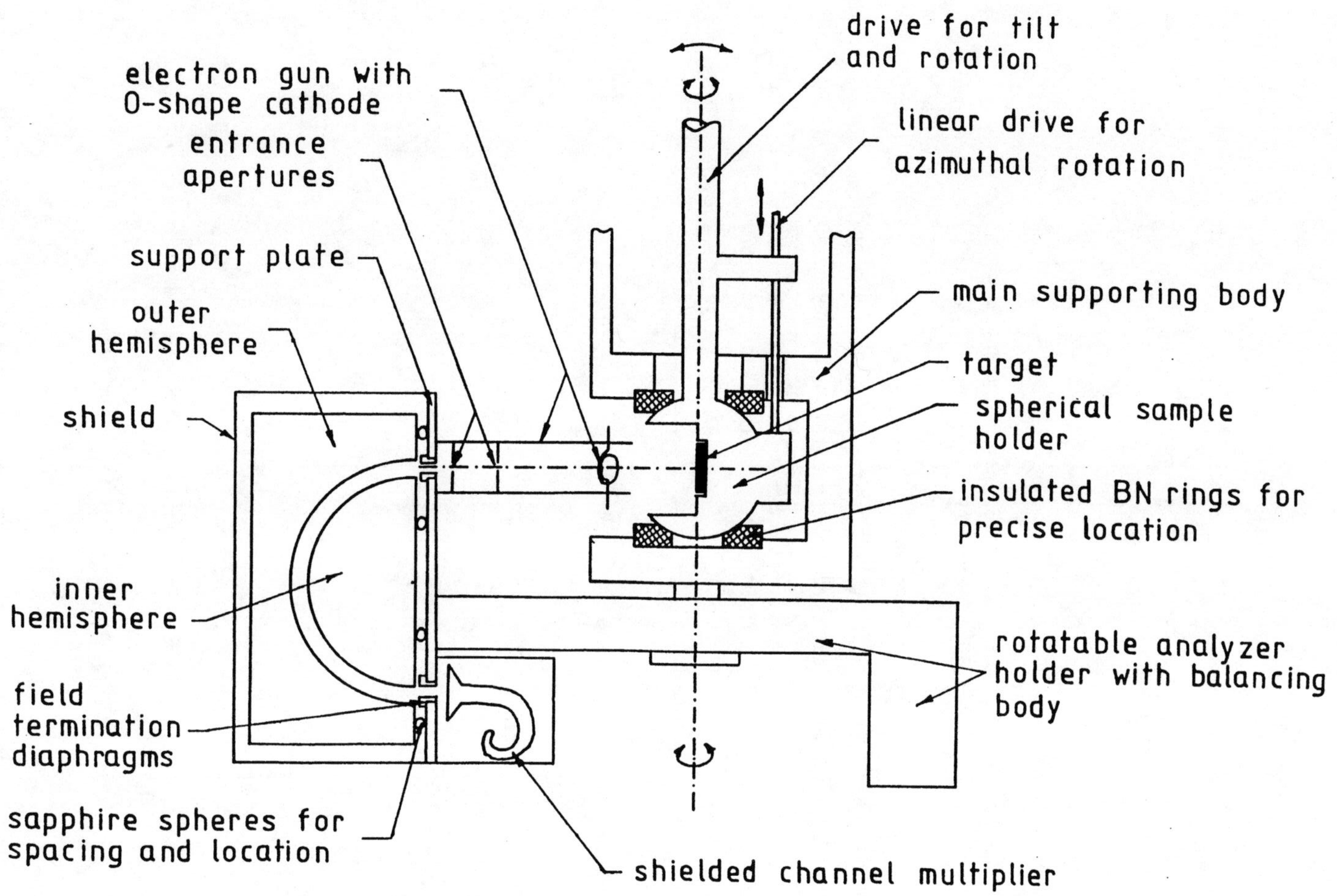

Figure 2: Cross - sectional view of the sample holder, target manipulator, energy analyzer and the electron gun (for more detailes see the text).

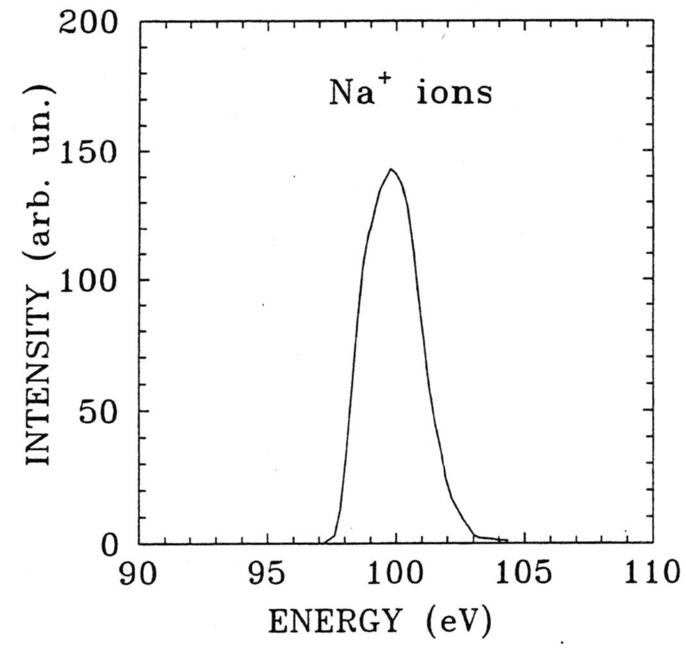

Figure 3: Energy distribution of 100 eV Na^+ ions emitted by a solid-state Na^+ ion source as measured by the hemispherical analizer and used for the calibration of the energy analyzer. The relative energy resolution, $\Delta E/E$ is found to be 2.5 %, where ΔE = FWHM and E is the value of the maximum of the above energy distribution.

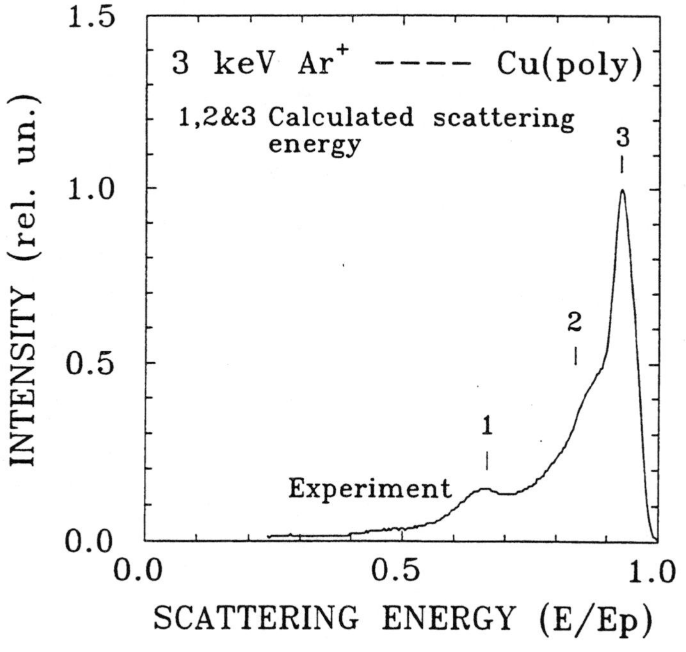

Figure 4: Energy distribution of Ar^+ ions scattered from polycrystalline Cu surface. The scattering angle is $\theta = 20°$, the incidence angle is $\alpha = 10°$. The incidence angle α is measured from the Cu surface. The markers **1**, **2** and **3** correspond to the calculated positions of the single scattered Ar^+ ions from Oxygen contaminants, Cu recoils and single scattered Ar^+ ions from target Cu atoms, respectively.

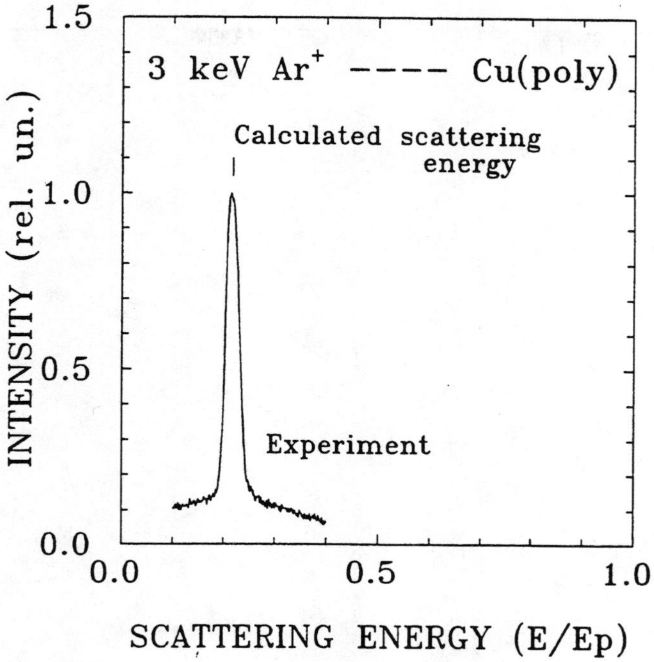

Figure 5: Energy distribution of Ar^+ ions scattered from polycrystalline Cu surface. The scattering angle is $\theta = 90°$, the incidence angle is $\alpha = 50°$. The angle α is measured from the Cu surface. The marker shows the calculated value of the scattering energy of the single scattered Ar^+ ions from Cu surface.

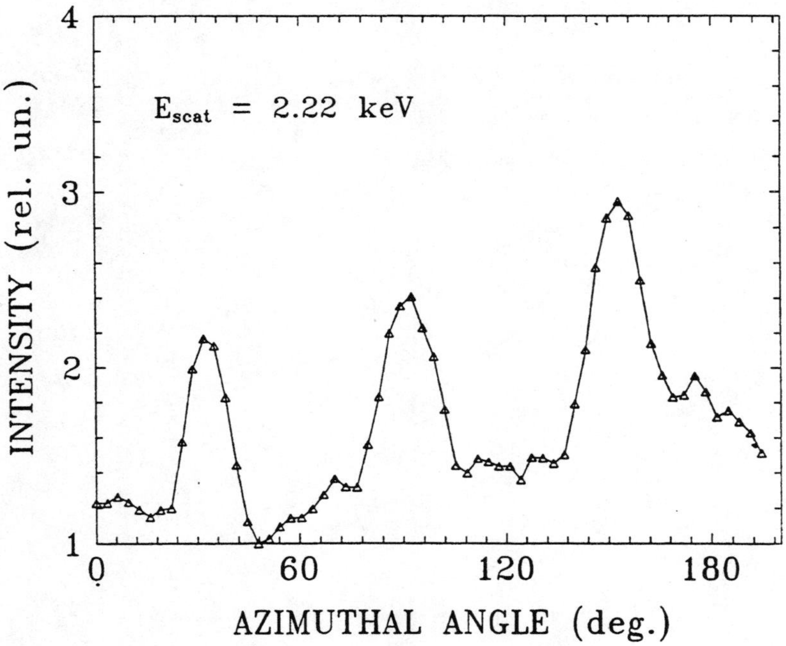

Figure 6: Azimuthal distributions of 3 keV Ar^+ scattered from $Ni(111)$ surface. The scattering energy is fixed and is chosen to be $E_{scat} = 2.33$ keV. The Ar^+ ions incident angle $\alpha = 5°$ is measured from the target face. The scattering angle is $\theta = 40°$. The working pressure is $p = 8 \times 10^{-10}$ $torr$.

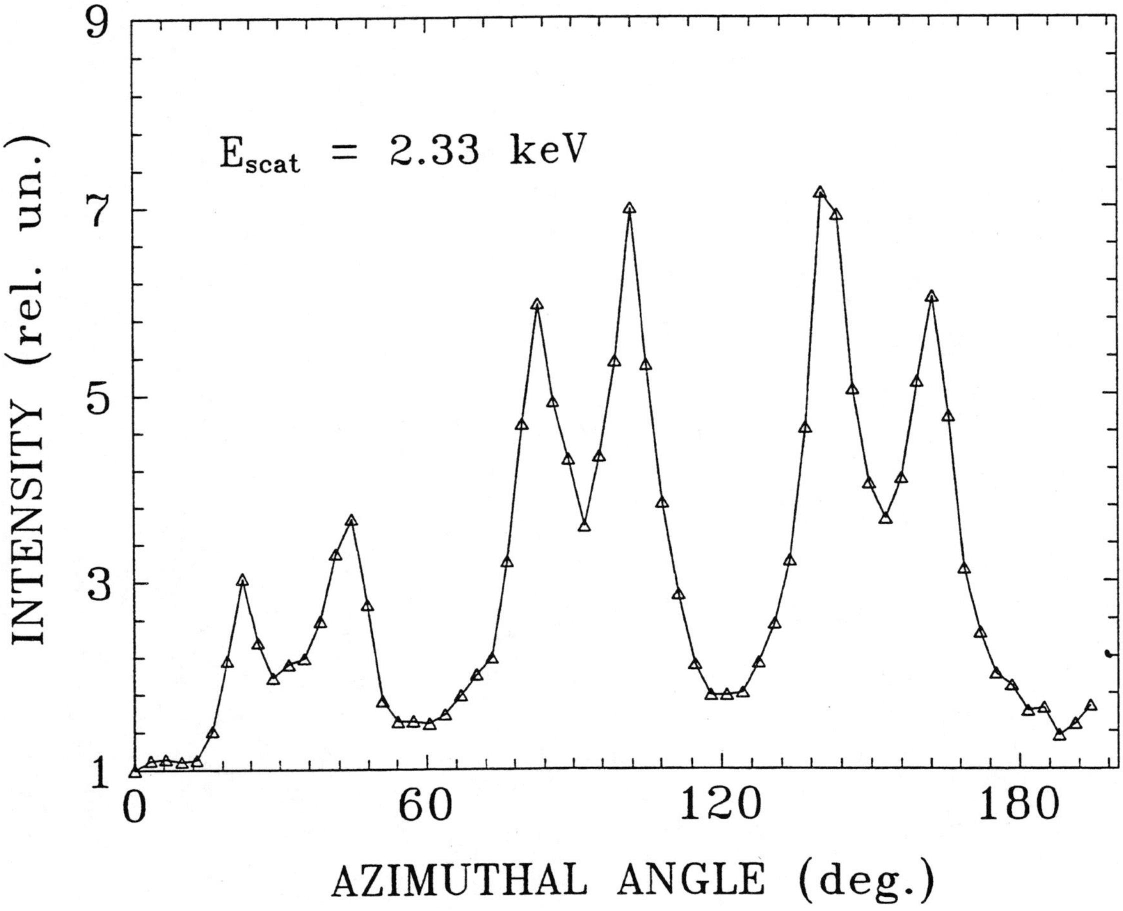

Figure 7: Azimuthal distributions of 3 keV Ar^+ scattered from $Ni(111)$ surface. The scattering energy is fixed and is choosen to be $E_{scat} = 2.22$ keV. The Ar^+ ions incident angle $\alpha = 5°$ is measured from the target face. The scattering angle $\theta = 40°$. The working pressure is $p = 8 \times 10^{-10}$ *torr*.

Vacuum Electron and Ion Technologies
D. S. Karpuzov (Ed.)

ISBN 1-56072-242-8

SURFACE COMPOSITION CHARGES OF Ni-Cu and Ni-Cr ALLOYS: EXPERIMENTAL AND COMPUTER SIMULATION STUDY

V. Miteva, I. Chakarov, D. Karpuzov, *P. Ivanov, *St. Angelova
Institute of Electronics, Bulgarian Academy of Sciences
72, Tzarigradsko Chaussee Blvd., 1784 Sofia. Bulgaria
*Surface Analysis Laboratory, 10, Nicola Petkov Blvd.,
DZU - DMON, Stara Zagora

ABSTRACT

An XPS study of the surface composition changes of $Ni_{0.56}Cu_{0.44}$ and $Ni_{0.53}Cr_{0.47}$ alloys under 5 keV Ar ion bombardment at room temperature is presented.

A Monte-Carlo computer simulation code was developed to include the surface effects of both radiation-induced segregation and radiation-enhanced diffusion, as well as preferential sputtering phenomena. Appropriate values of the diffusion and segregation constants were selected in terms of best fitting to the experimental results.

1. Introduction

The alteration of the near-surface composition that has often been observed in ion-bombarded alloys and compounds results from the dynamic interplay of several processes. In addition to displacement mixing, which is dominant at relatively low temperatures, and preferential loss of alloying elements by sputtering, many thermally-activated processes, including radiation-enhanced diffusion (RED), radiation-induced segregation (RIS) and Gibbsian adsorption, also play important roles.

Preferential sputtering of any of the surface constituents occurs if the ratio of the partial sputtering yields differs from that of the surface concentrations. It can be ascribed to large mass differences [1, 2], or differences in the surface binding energies [1, 3]. In addition, RED or RIS [4, 5] may occur due to atomic migration caused by the production and mobility of point defects, such as vacancies and interstitials. Their influence on the composition profile of the altered layer in the subsurface region has indicated that the preferential sputtering results in enrichment of the surface by constituent atoms of lower sputtering yields so that this composition gradually approaches the bulk concentration, mainly driven by RED. Improved approaches (e.g., Kelly's model [6, 7], Kurokawa et al. [8]) were proposed including RIS which controls the mass transport from the second atomic layer to the first, while RED is responsible for the relaxation of the composition dip caused by RIS, expanding the altered layer region much deeper into the sample.

In experiments, the chemical composition of the surface layers can be determined by analyzing the exposed surface after a certain sputtering time using, for example, AES, ISS, or XPS. Unfortunately, even given a well-calibrated analysis technique and a good deconvolution procedure, a number of factors can affect the determination of the depth profile. Computer simulation is the most powerful approach for solving this problem, since it allows one to describe the kinetic process of energy transfer through atomic collisions, projectile movement, particle ejection (sputtering) from each atomic layer beneatn the surface, Frenkel pair formation, etc.

Computer simulations based on a Monte Carlo method have already been applied to multi-component sputtering [10, 11] demonstrating their ability to describe the formation of the in-depth composition profile of the altered layer including even such transport processes as RIS and RED [8, 12].

A computer-simulation approach, based on a conventional Monte Carlo code and including RIS, RED, and an in-depth integration procedure, is presented in this work. The aim is to achieve good description of the near-surface in-depth changes of Cu-Ni and Ni-Cr alloys with the ion-bombardment dose.

2. Theoretical Grounds and Simulation Model

Since the basics of the present simulation have already been described [13] in what concerns the collision-determined preferential sputtering, we will only illustrate here the physical concept underlying the inclusion of RED and RIS.

All calculations were carried out with the following values of the parameters defining the surface dynamics: surface binding energy E_{Ni} = 4.46 eV, E_{Cu} = 3.5 eV, and E_{Cr} = 4.1 eV; cut-off energy (the energy of the pseudoparticle when it comes to rest) slightly below the surface binding energies, E_{cutoff} = 2.0 eV for all elements; the displacement energy E_d = 8 eV for all elements. The diffusion coefficient D participates in the calculations through the mean free path $\langle \Delta r^2 \rangle$ = 10 Å for NiCu and $\langle \Delta r^2 \rangle$ = 20 Å for NiCr (see below). The rest of the model parameters were chosen according to the experimental conditions: E_p = 5 keV for Ar^+ primary ions, incident angle $\alpha = 32^{\circ}$. For sufficiently good statistics, the number of pseudoparticles was set at 40000.

2.1. Modeling of RED

Let us consider the problem of diffusion of interstitials in an infinite target with initial concentration $c(x,0) = c_i(x)$. Then, the diffusion equation (the so-called Cauchy problem) can be written as:

$$\partial_t c = D\partial_x^2 c, \qquad -\infty < x < \infty,\ 0 < t < \infty$$

$$c(x,0) = c_i(x) \qquad -\infty < x < \infty$$

The solution is given by the expression

$$(3.1) \qquad c(x,t) = \frac{1}{\sqrt{\pi}} \frac{1}{\sqrt{4Dt}} \int_{-\infty}^{\infty} c_i(\xi)\exp\left[-\frac{(x-\xi)^2}{4Dt}\right] d\xi$$

where $c_i(x)$ is the initial distribution of the interstitial concentration, and

$$(3.2) \qquad G(x,t) = \frac{1}{\sqrt{\pi}} \frac{1}{\sqrt{4Dt}} \exp\left[-\frac{(x-\xi)^2}{4Dt}\right] d\xi$$

is Green's function, or "source function" (a Gaussian).

It can be shown that the source function $G(x,t)$ describes the system's response to a single initial pulse in the concentration distribution at $x = \xi$. It is clear that formula (3.2) can be interpreted as follows: the initial concentration of interstitial atoms $c(x,0) = c_i(x)$ can be expressed as a continuous set of point pulses, each having a value $c_i(\xi)$ at $x = \xi$. Each point concentration pulse generates the concentration distribution $c_i(\xi)G(x,t)$. The final distribution is obtained by summation of the point sources concentrations using (3.2). In this manner, one can represent the point diffusion of interstitial atoms after the ballistic cascade development. The time dependence is taken into account by using the mean-square free path $\langle \Delta r^2 \rangle = 4Dt$. In the specific case of modeling the diffusion of interstitial atoms, a surface is present at $x = 0$, defining the boundary condition. It is assumed that the diffusing elements have energy less than the surface binding energy E_{SB}, so that they cannot escape to the vacuum. If a particle reaches the surface, it is reflected back into the target.

The model is illustrated in Fig.1, where a simple case of interdiffusion at the interface of a two-layer single-metal target, namely a 50-Å Cu film on a semi-infinite Cu-substrate, is presented. The calculation is performed with 5 keV Ar ions at a dose of $1x10^{16}$ ions/cm^2 and different diffusion coefficients D for the two layers. The blurring of the interface is proportional to the diffusion constants. The surface erosion due to the ion bombardment is clearly seen.

2.2. Modeling of RIS

All recent experimental and theoretical treatments of the problem, e.g. [12], unambiguously indicate that RIS is expressed by preferential enrichment (depletion) of the first atomic layer at the expense of a mass transport from (to) the target bulk; the mass transport can be described by equations mathematically equivalent to the diffusion equation [14]. This led us to the idea of using the algorithm, outlined in the previous chapter, to describe the motion of atoms towards the surface. When modeling diffusion after the cascade's end, we redistribute each interstitial to a new slice with a probability given by a Gaussian centered at the old layer and having a half-width $\langle \Delta r^2 \rangle = 4Dt$. When modeling segregation, we execute the reverse procedure - knowing that the particle will inevitably enter the first atomic layer, we use a Gaussian having a half-width fitted to the experimentally observed width, and centered at the interface between the first and the second layer. Using this probability, we can determine the layer from which the atom has segregated. This approach is illustrated in Fig.2, which represents the depth distribution of the Cu concentration after bombardment by 5 keV Ar ions at a dose of 10^{15} ions/cm^2.

3. Experimental

The $Ni_{0.56}Cu_{0.44}$ and $Ni_{0.53}Cr_{0.47}$ alloys were investigated using XPS. The measurements were carried out in an ESCALAB - 2 (VG Scientific Ltd.) electron spectrometer at a residual gas pressure of $2x10^{-10}$ mbar. The photoelectron spectra were taken at normal emission with Mg K_{α} anode excitation ($h\nu$ = 1253.6 ± 0.7 eV) and an analyzer pass energy of 20 eV. The energy resolution over the whole spectral range was ≅ 1 eV. An Ar^{+}-ion beam with 5 keV energy and 0.5 $\mu A/cm^2$ ion-current density was used for sputtering of the top layers. The ion-beam incidence angle with respect to the sample normal was 32°. The energy analyzer was tilted at 15° from the normal to the surface. The areas of the $2p_{3/2}$ XPS peaks (852.7 eV for Ni, 932.7 eV for Cu, and 577 eV for Cr) were normalized to their relative sensitivity factor values (13.9, 15.9, and 7.8, respectively) for elemental concentration determination. To obtain the concentration profiles, we followed the changes of the XPS peak areas versus sputtering time (or ion fluence).

4. Results and Discussion

Figs.3 and 4 show results of the experimentally-obtained compositional changes at the surface of the alloys as a function of the ion dose. The existence of an altered-concentration layer at the surface is observed, namely Cu (respectively, Cr) atoms have segregated to the surface. A steady-state composition is reached at a fluence of $(6 - 8)x10^{15}$ ions/cm^2 for NiCu and $2x10^{15}$ ions/cm^2 for NiCr, when the measured elemental concentration of the temporary surface approaches that of the bulk of the alloy. One can see that the dynamics for the sécond sample is substantially different - the deviations are larger, while the saturation dose is lower.

It is difficult to make a direct comparison between the experimental and simulation results, since the XPS analysis yields an integral characteristics of the surface (the photoelectrons originate from a depth of 5 - 6 atomic layers).

This is why we executed an integration procedure following the formula [15]

$$I(\theta) = K\int_{0}^{x} n(x)\exp\left[\frac{-x}{\lambda\cos\theta}\right]dx,$$

where I is the intensity of the photoelectron flux at take-off angle θ, $n(x)$ is the atomic density at depth x, λ is the inelastic mean free path, K is a normalized proportionality coefficient. For λ we used the values quoted in [16], namely 8, 8, and 11.8 Å for Cu, Ni and Cr, respectively. We could select different depths of integration, which were estimated using the average atomic layer size. Figs.5 and 6 present a comparison of this integration with the experimental data for segregating components of the alloys, namely Cu and Cr.

5. Conclusion

Both radiation-enhanced diffusion and segregation effects were incorporated into a TRIM-based dynamic code described earlier. Comparisons with the XPS experimental data are encouraging.

REFERENCES

1. P. Sigmund, In Sputtering by Particle Bombardment I (Edited by R. Behrisch), Springer, Berlin, (1989), p.9
2. B. Baretsky, W. Möller, E. Taglauer, Vacuum, v.13, (1992), p.1207
3. H.H. Andersen, H.L. Bay, Radiat. Eff., v.13, (1972), p.67
4. Ri-Sheng Li, T. Koshikawa, K. Goro, Surf. Sci., v.121, (1982), p.L561
5. Ri-Sheng Li, T. Koshikawa, Surf. Sci., v.151 (1985), p.459
6. R. Kelly, Surf. & Interf. Anal., v.7, No.1, (1985)
7. R.Kelly, Nucl. Instrum. Methods, B39, (1989), p.43

8. A.Kurokawa, Hee Jae Kong, R. Shimizu, Surf. Sci., v.155, (1991), p.120

9. W. Möller, W. Ekstein, Nucl. Instrum. Methods, B718, (1985), p.727

10. O.F. Goktepe, M.L. Roush, Nucl. Instrum. Methods, B718, (1985), p.803

11. H.J. Kang, E. Kawatoh, R. Shimizu, Jap. Journ. Appl. Phys., v.24, No.11, (1985), p.1409

12. H.J. Kang, J.A. Kim, J.S. Kim, D.W. Moon, R. Shimizu, Surf. Sci., v.226, (1990), p.93

13. V. Miteva, D. Karpuzov, P. Ivanov, St. Angelova, accepted for publication in Nucl. Instrum. Methods B (1993)

14. R. Kelly, Surf. Interf. Analysis, v.7, No.1 (1985), p.1

15. R.S Yih, B.D. Ratner, Journ. El. Spectr. & Rel. Phen., v.43, (1987), p.61

16. S. Tanuma, C.J. Powell, D.R. Penn, Surf & Interf. Analysis, v.11, (1988), p.577

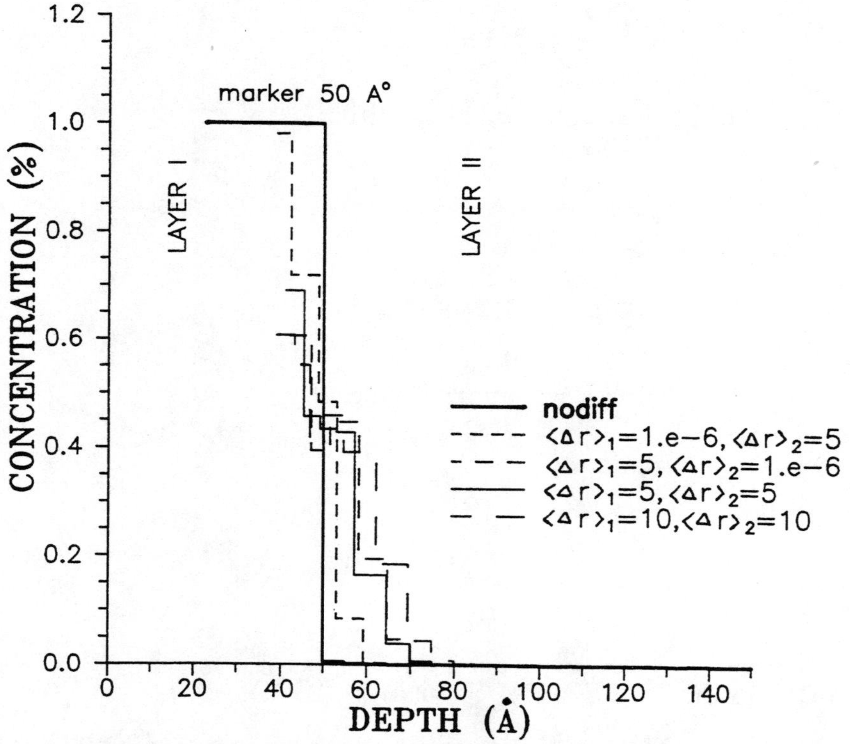

Fig.1 Computer simulation of the interdiffusion between a 50 −A copper film and a semi-infinite Cu substrate bombarded by 5 keV Ar ions.

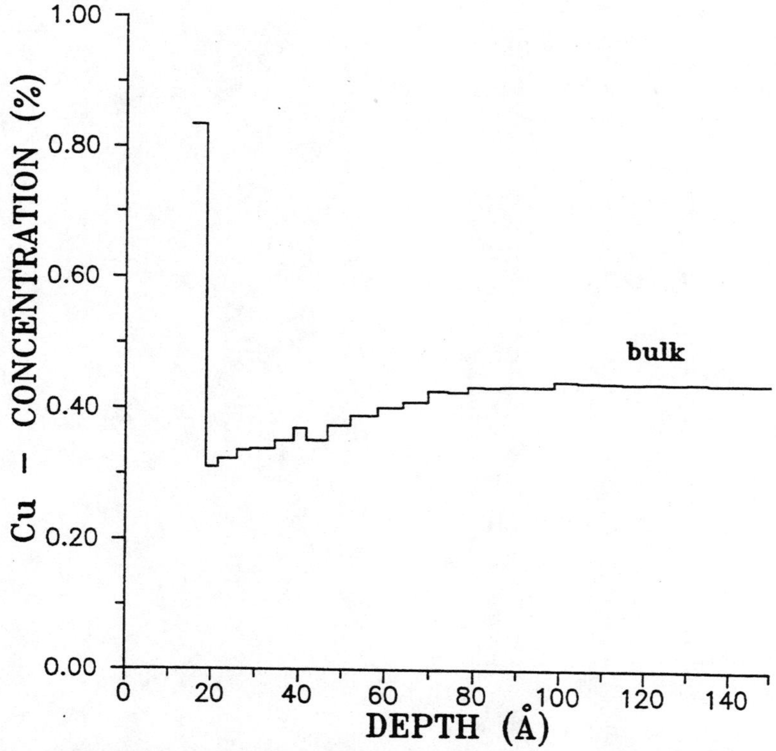

Fig.2 RIS−driven depth distribution of the Cu concentration in Ni−Cu target bombarded by 5 keV Ar ions at a dose of 10^{16} ions/cm^2

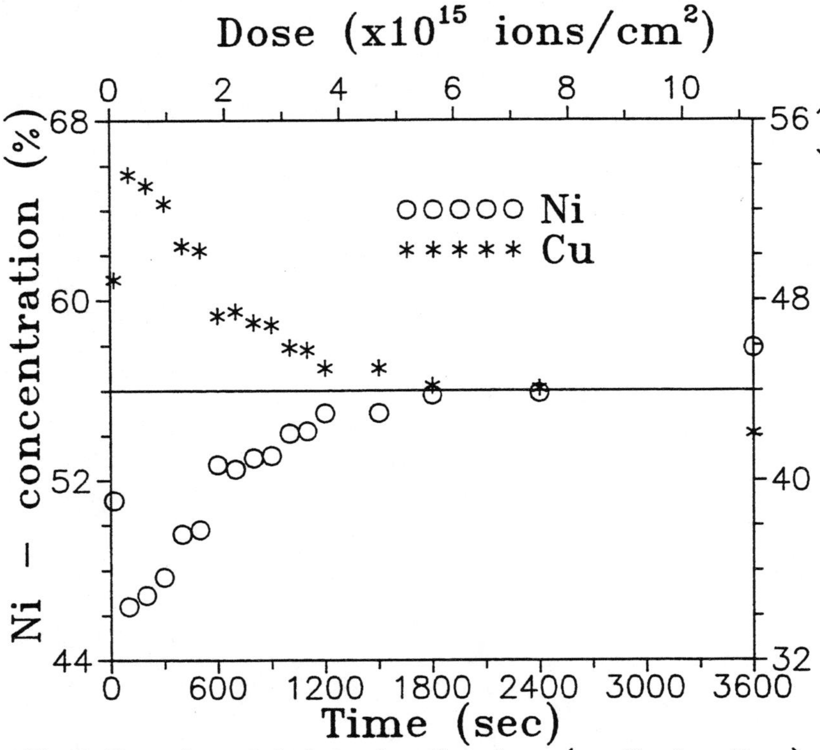

Fig.3 Experimental data for the dose (sputtering time) dependence of the surface concentration of a $Ni_{0.56}Cu_{0.44}$ alloy.

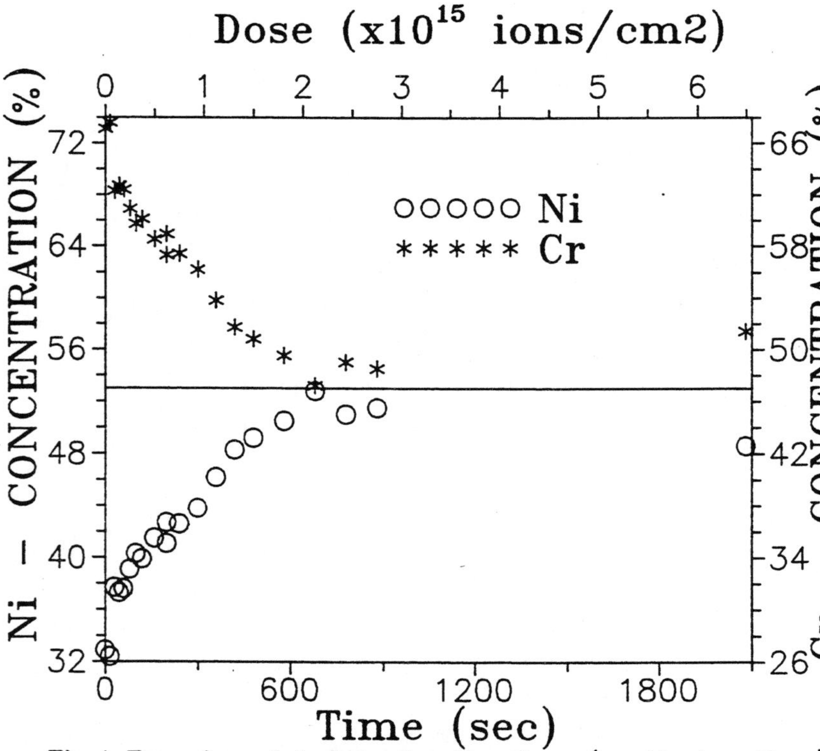

Fig.4 Experimental data for the dose (sputtering time) dependence of surface concentration of a $Ni_{0.53}Cr_{0.47}$ alloy.

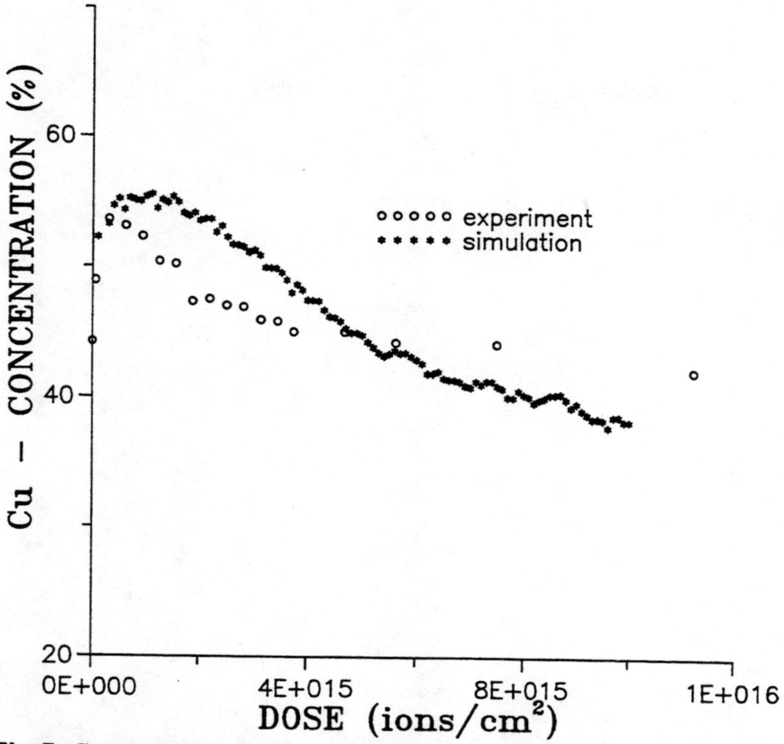

Fig.5 Comparison between the experimental data and the integrated computer-simulated data for Cu-atoms (incl. RIS and RED) in Ni-Cu.

Fig.6 Comparison between the experimental data and the integrated computer-simulated data for Cr-atoms (incl. RIS and RED) in Ni-Cr.

Vacuum Electron and Ion Technologies
D. S. Karpuzov (Ed.)

ISBN 1-56072-242-8

ON THE COMPUTER SIMULATED SPATIAL DISTRIBUTIONS OF EMITTED PARTICLES FROM SOLID SURFACES

I. Neshev
Physics Department, Illinois Institute of Technology
3301 S Dearborn Street, Chicago, Il, USA 60616
S. Tzanev and D. Karpuzov
Institute of Electronics, Bulgarian Academy of Sciences
blvd. Tzarigradsko Chaussée 72. 1784 Sofia. Bulgaria

Abstract

A new procedure for obtaining computer simulated, directly comparable to the experimental data spatial distributions of sputtered and reflected atomic particles from solid surfaces is proposed. It is based on a correct treatment of the constant size and shape of the simulated "detectors'" solid angle. Comparison with experimental data shows that the new procedure reproduces more adequately the experimental results. The differences between the existing extraction procedures and the one proposed here are discussed.

1 INTRODUCTION

For a correct quantitative description of ion beam interaction with solids it is important to obtain computer simulation results that are directly comparable with the experimental data. Also, recently, the importance of the information contained in the angular distributions of the sputtered and reflected atomic particles from solid surfaces has become increasingly recognized. There are some widely spread computer codes for simulation the interaction between impinging ions and solid materials that are used for such studies.

There are only a few experiments investigating the angular distributions of the sputtered constituents [1], [2], [3], [4], [5], although their importance for understanding the fundamentals of the interaction between the impinging ions and solid surfaces is doubtless. The scarceness of such experiments is probably related to the difficulties of differentiating the sputtered particle flux on energy E, polar θ and azimuthal φ angle of ejection and mass M. Thus, it is appropriate to investigate these differential characteristics of sputtering events by computer modeling.

Usually, most of the existing versions of the extraction procedures produce polar and azimuthal angular distributions of the particles emitted from single crystals. In the general case, care is taken for the preservation of the solid angle's size, but this solely does not lead to corresponding and could not be directly compared to the ones registered in actual experiments. These limitations will be discussed below.

It is the purpose of this paper to present an extraction procedure to be used with different computer codes that produces more adequate and directly comparable results with those obtained experimentally. The procedure can be applied to any computer program for simulation the interaction between accelerated ions and solids that gives as a result for the spatial distributions of emitted and/or reflected particles from solid surfaces the number of these particles together with their spatial direction of movement. We have chosen the MARLOWE binary computer code (BCC) [6] to carry out the simulation. It is one of the most widespread simulation programs and may be applied to modeling of different tasks and experimental conditions, including the above spatial distributions.

2 OBTAINING SPATIAL DISTRIBUTIONS OF EMITTED PARTICLES

In this section, we comment on the drawbacks of some existing and often used methods [7] for registering the spatial distributions of particles emerging from solid surfaces. A new method for registering the spatial distributions is proposed and compared with the one originally used in some already existing computer programs.

2.1 Main characteristics of the existing spatial distribution construction procedure

Usually, every computer program for simulating the interaction between the incoming ions and solids has as a goal producing adequate data for the number and the spatial movement directions of the emitted or reflected particles from the solid surfaces. The real directions and the number of the particles are the most important data for the simulated angular distributions but it could not be directly compared in this form to the experimentally obtained data. One should convert the simulation data in such a way, the results of which would enable a direct comparison with the experiment.

2.1.1 Construction of the azimuthal×polar angular distribution histogram

For the purposes of the calculations, the hemisphere above the target surface is usually overlaid by a grid that divides the azimuthal and polar directions into a set of channels. The azimuthal channels number could be varied independently on the number of polar channels (see fig. 1).

Consider a particle of mass M and energy E which is sputtered or reflected from the solid surface and is moving away from it along a direction determined by the polar angle θ and azimuthal angle φ (usually measured from the surface "*zero*"-direction). This emitted particle is registered (counted) in a grid cell corresponding to the relevant polar and azimuthal histogram channels, thus producing the spatial distributions of the emitted particles.

When simulating the spatial distributions, care is usually taken to maintain the solid angle of registration constant, i.e.

$$d\Omega = \sin\theta d\theta d\varphi = const,$$

where Ω, θ and φ are the solid angle, the polar angle, and the azimuthal angle of registration, respectively.

The azimuthal sector of interest is considered divided into NPH equidistant channels. An internal orthogonal coordinate system is chosen with the $\mathcal{X}$ and $\mathcal{Y}$ axes lying in the target surface plane and with the $\mathcal{Z}$ axis – along the normal $\vec{n}$, as shown in fig. 1. PHA is the angle in degrees, measured from the $\mathcal{X}$ axis and defines the beginning of the azimuthal histogram. The maximum angle in degrees measured from the $\mathcal{X}$ axis and corresponding to the end of the histogram is PHB ($PHB \leq 360°$).

Respectively, the polar histogram starts at $\theta = 0°$ and ends at $\theta = 90°$ and is divided into NTH channels. The requirement for maintaining the solid angle of registration constant is fulfilled by dividing the polar sector into equal $\Delta\cos\theta = const$, where $\Delta\cos\theta = \cos\theta_{k-1} - \cos\theta_k$ (θ_{k-1} and θ_k determine the "$k-th$" polar channel), i.e.,

$$\Delta\Omega = \Delta\varphi\Delta\cos\theta = const\ ,$$

where $\Delta\varphi = (PHB - PHA)/(NPH)$ is the width of the azimuthal channel and $\Delta\Omega$ is the solid angle of registration. This method, as we mentioned above, is often used, however, it leads to variable width (θ-limits) of the polar channels, e.g. for $NTH = 20$, the first channel (measured from the surface normal) lies between $\theta_0 = 0^o$ and $\theta_1 = 18^o$; the second channel — between $\theta_1 = 18^o$ and $\theta_2 = 25^o$, etc. Also, the shape of this constant-size solid angle is different for different polar angles of ejection.

On the other hand, practically all experiments measuring the spatial distributions of the emitted particles use a fixed entrance aperture of the detection system that often has the shape of a circular diaphragm. This fixed aperture determines the shape and size of the solid angle of registration. In experiments in which the spatial distribution of the emitted particles is investigated by collecting them onto a foil and subsequent analysis by a surface sensitive technique (e.g., Rutherford Backscattering Spectroscopy (RBS) [8], [2], X–ray analysis [3], Auger Electron Spectroscopy (AES) [4], Nuclear Reaction Analysis (NRA) [9]), one also registers particle flux within a constant solid angle Ω_{exp} due to the constancy of the size of the analyzing beam.

Thus, the above described procedure for registration the spatial distributions of the emitted particles does not adequately correspond to most experimental geometries and the obtained simulation results could and should not by directly compared with the experimental data.

2.1.2 Transformation of the azimuthal angle

Another problem in obtaining the spatial particle distributions could be connected to the transformation of the azimuthal angle in order to improve the statistics of the emitted particles. Let the maximum number of azimuthal channels is NPH. Their width then will be $\Delta\varphi = (360°)/(NPH)$ if all azimuthal angles of ejection are considered. Thus, to carry out precise simulations, e.g. of fine-structured anisotropic effects in sputtering or scattering one needs as many channels as possible. This will affect the memory required, the statistics, i.e., the computational time, etc. So, there should be a compromise between the above requirements. One possible solution for improving the insufficient angular resolution is to choose a smaller azimuthal sector of interest, which then is divided into NPH channels of smaller width.

In case the azimuthal sector of interest is less than 360°, there are always emitted particle directions that do not lie between the limits of the chosen sector. Some or all these directions could be properly transformed so that they lie within these limits, thus improving the statistics of the emitted particles without increasing the computational time. A possible transformation is supplied as an option of the MARLOWE program package and is performed as follows:

A check is made whether the particle's azimuthal direction of motion (measured from the X axis) lies within the doubled azimuthal sector of interest (between PHA and $2\times(PHB-PHA)$. In case it does not, it is rotated by a value equal to the doubled azimuthal sector of interest until it is put within it. Finally, it is reflected about PHB direction to make it lie between PHA and PHB. This procedure is repeated for all emitted particles with azimuthal directions not lying within the specified azimuthal sector.

This procedure, however, is not valid for solving the general case. It could be used in some cases, namely when the axes limiting the azimuthal sector of interest coincide with the crystallographic axes of symmetry characterized by both rotation and mirror symmetry. One example of validity of this method would be the (100) surface of a FCC or a BCC crystal when the chosen azimuthal sector of interest is equal to 90°. For example, however, the (111) surfaces of the FCC or BCC crystals are invariant to azimuthal rotations by 120° they do not possess mirror symmetry for this azimuthal sector of interest for which the above described extraction procedure could produce incorrect predictions. However, a 60° azimuthal sectors of interest could be chosen for the (111) face of FCC or BCC crystals to represent the whole azimuthal distribution, although this also not solving the general problem. That is why one should take special care to adequately represent the real angular distribution of reflected or sputtered particles. Thus, the existing procedures prove to be inapplicable in the general case of the particles' spatial distributions investigations and comparison with the experimental data.

2.2 New procedure for constructing the spatial distribution of the emitted particles

To make the registration procedure directly comparable to experiments, we propose a new method for constructing the grid and for counting the sputtered particles.

In both computer simulation and experiments a constant solid angle of registration is maintained. As discussed above, most often for obtaining the results by computer simulation constant $\Delta \cos\theta$ steps are used, whereas the experiments use a fixed aperture and therefore constant $\Delta\theta$.

Let us consider a fixed registration geometry with a circular entrance diaphragm. The solid angle of registration $\Delta\Omega_s$ is a cone with an opening angle $\Delta\theta_s$, see fig.2. The polar and azimuthal sectors of interest are divided into equidistant (on the respective angles) channels. The number of channels in the polar and azimuthal directions may be different. Let the polar sector of interest lie between the directions THA and THB ($0 \leq THA < THB \leq 90°$) and, also, let NTH be the number of polar channels. Then $\Delta\theta = (THB - THA)/(NTH)$ is the width of each polar channel. Similarly, if the azimuthal sector of interest lies between PHA and PHB ($0 \leq PHA < PHB \leq 360°$) and NPH is the number of azimuthal channels, the width of the azimuthal channel is $\Delta\varphi = (PHB - PHA)/(NPH)$. It is important to note that in this treatment the size of the solid angle of registration $\Delta\Omega_s$ is independent of the widths of the polar and azimuthal channels. The former corresponds to the size of the detector aperture and the latter — to the distance between successive positioning of the detector. In this manner, certain improvement of the polar angular resolution can be achieved without increasing the number of polar $\times$ azimuthal channels.

Let us examine the "i"-th azimuthal and "k"-th polar channels. We have the following for θ_k and φ_i (which determine the direction of the channel axis):

$$\theta_k = THA + (k - 1/2)\Delta\theta, k = 1, \ldots, NTH$$
$$\varphi_i = PHA + (i - 1/2)\Delta\varphi, i = 1, \ldots, NPH$$

A particle emitted with polar and azimuthal angles θ and φ will be counted within this grid cell only if its direction of motion lies within the solid angle $\Delta\Omega_s$ centered on the axis of the corresponding azimuthal $\times$ polar channel, as shown in fig. 2. This means that the following condition must be fulfilled:

$$\cos|\varphi - \varphi_i| \geq \frac{\cos\left(\frac{\Delta\theta_s}{2}\right) - \cos\theta\cos\theta_k}{\sin\theta\sin\theta_k}, \text{ [1]} \tag{1}$$

Thus, registration of the escaping particle within a given azimuthal $\times$ polar channel depends strongly on the solid angle of registration $\Delta\Omega_s$. For large $\Delta\Omega_s$, the particle may also be counted in other neighboring polar $\times$ azimuthal channels for which eq. (1) is also valid. Also, it is obvious that for directions near the normal, several neighboring azimuthal channels will be covered by even a small solid angle of registration. In the special case of placing the solid angle of registration normal to the surface, all emitted particles for which the polar angle is less than $(\Delta\theta_s)/(2)$ irrespective of their azimuthal angle will be registered.

This method of maintaining the solid angle of registration $\Delta\Omega_s = const$ more closely models the experimental situation and its results are directly comparable with the experimental data. It could also be easily applied to the case of a non circular entrance diaphragm.

Concerning the statistics enlargement we propose the following procedure:

The azimuthal sector of interest can be chosen as wanted. The statistics can be improved by summing only identical crystallographic sectors not lying within the *a priori* stated sector of interest. The azimuths of the particles' directions not lying within the desired sector are rotated by an angle corresponding to the crystallographic symmetry of the surface until they lie within the sector of interest. This ensured the applicability of the new procedure to the general case of simulating the spatial distributions of sputtered or scattered particles.

3 Comparison with experiment and simulation

The applicability of the proposed procedure was tested by comparison with experimental data [10] also with simulation data obtained by using the MARLOWE BCC [11] and using its original version of the extraction procedure.

[1]The derivation is made in the Appendix

Simulation of 600 eV Xe^+ ion beam bombardment of a semi-infinite $Au(111)$ crystal as a target surface was carried out. A 60° azimuthal sector of interest was used for the calculations. This was chosen concerning the *a priori* expected identity between the simulation results obtained by using the original version of the extraction procedure and the one proposed in this paper. Twenty thousand primaries were simulated.

The calculation used the Molière approximation to the Thomas-Fermi potential [12]. The default screening length used for the noble gas homonuclear pairs is assumed to be correctly predicted by the Firsov formula [13]. The screening lengths for the heteronuclear pairs also for the $Au - Au$ pairs are taken the same as those found to correctly describe the Wehner spot positions [11]. The cutoff energy and the surface binding energy were chosen be equal to 3.78 eV. The bulk binding energy for replacement sequences was 0.2 eV that has been shown to be the energy loss in such events via molecular dynamics simulations [14].

Equipartition between local and nonlocal inelastic energy losses to electron excitation is assumed.

The results of the simulation by both the existing and the newly proposed extraction procedures are shown in fig. 3. As expected, the polar positions of the spots calculated using the original procedure (fig. 3a) do not exhibit the experimentally observed characteristics, i.e. the peaks of the spots associated with the (110) and (100) directions can hardly be distinguished by their polar positions, although the associated low index directions are at $\theta \approx 35°$ and $\theta \approx 54°$, respectively. This is the result of the improper summing of one of the other two azimuthal sectors, i.e. the (110) peak is reflected onto the azimuthal position of the (100) peak and vice versa. The third sector is summed correctly, i.e. (110) peak onto the (110) peak of the sector of interest. Thus, the total number of particles in every grid cell always consists of a part of properly summed and a part of improperly summed particles. This makes the obtained spatial distribution inadequate. The height and the slope of the peaks are also affected by the existing procedure.

On the other hand, the spatial distributions calculated by the procedure proposed here using the same solid angle (fig. 3b) do exhibit the general characteristics of the experimentally acquired data concerning distinguishing the Wehner spots by their relative polar positions. The simulated peaks of the (100) and (110) Wehner spots are at polar angles of approximately 48° and 35°, respectively.

The resolution on the polar angular scale is better than the one using existing procedure with regard to angles near the surface normal.

Finally, fig. 3c shows the results calculated using the new procedure with the same number of polar and azimuthal channels but with a larger solid angle. It is seen that the obtained spatial resolution is determined not so much by the number of azimuthal $\times$ polar channels but by the value of the solid angle of registration that in our procedure may be independently varied.

4 Conclusions

A new procedure for obtaining the computer simulated polar and azimuthal distributions of particles emitted from solid surfaces that adequately describes the experimental situation is developed. The spatial distribution differentiated on various parameters (e.g. the energy of emission E, polar θ and azimuthal φ angles of emission, type or the species, etc.) or integrated on some of these parameters may be obtained by the proposed new procedure. This gives an additional opportunity for experimentally adequate computer simulations.

The proposed procedure allows improvement of the preciseness in determining the variable simulation parameters, and especially these affecting the particles' spatial distributions.

5 Acknowledgement

Thanks are due to S. Todorov for the help in preparing this paper.

This work was partially financed by contract MT-2 of the Bulgarian National Research Fund.

6 Appendix

In this paragraph, the condition for registering an escaping particle in a relevant azimuthal×polar channel (inequality 1 on page 8) is derived.

Consider a sphere with radius unity centered at the origin O fig.2. Let us intersect the sphere by a plane at a distance $p = \cos(\Delta\theta_s)/(2)$ from the center of the sphere. For easier calculation, the coordinate system $\mathcal{OXYZ}$ is rotated around the $\mathcal{Z}$ axis by the azimuthal angle φ_i till the $\mathcal{X}$ axis matches the φ_i direction. Thus, the relationship between the azimuthal angles φ and Φ measured in the $\mathcal{OXYZ}$ and the rotated coordinate systems, respectively, becomes $\Phi = \varphi - \varphi_i$. The polar angle θ remains unchanged in both the systems. The directional cosines of "p" in this case will be

$$\left\| \begin{array}{l} \cos\alpha = \sin\theta_k \\ \cos\beta = 0 \\ \cos\gamma = \cos\theta_k, \end{array} \right.$$

respectively.

The equation of a plane at a distance "p" from the beginning of the coordinate system has the form

$$x\cos\alpha + y\cos\beta + z\cos\gamma - p = 0$$

Substituting the relevant values for $\cos\alpha$, $\cos\beta$, $\cos\gamma$ and "p" in this equation we receive:

$$x\sin\theta_k + z\cos\theta_k - \cos\left(\frac{\Delta\theta_s}{2}\right) = 0$$

or in spherical coordinates:

$$r\sin\theta\cos\Phi\sin\theta_k + r\cos\theta\cos\theta_k - \cos\left(\frac{\Delta\theta_s}{2}\right) = 0$$

Solving the above equation together with the equation $r = 1$ describing the sphere we obtain the equation for the intersection of the plane with the sphere:

$$\sin\theta\cos\Phi\sin\theta_k + \cos\theta\cos\theta_k - \cos\left(\frac{\Delta\theta_s}{2}\right) = 0$$

Following some simple transformations we obtain:

$$\cos\Phi = \frac{\cos\left(\frac{\Delta\theta_s}{2}\right) - \cos\theta\cos\theta_k}{\sin\theta\sin\theta_k},$$

where

$$\theta_k - \frac{\Delta\theta_s}{2} \leq \theta \leq \theta_k + \frac{\Delta\theta_s}{2}$$

This dependence is exactly inequality (1) in the text. It states the condition for registering a particle: a particle ejected with a polar angle of emission "θ" (within the above interval) will be registered within the solid angle $\Delta\Omega_s$ if its azimuthal angle φ lies within the interval:

$$|\varphi - \varphi_i| \leq \arccos\left(\frac{\cos(\frac{\Delta\theta_s}{2}) - \cos\theta\cos\theta_k}{\sin\theta\sin\theta_k}\right), \qquad (2)$$

The definition of φ and φ_i is explained in fig. 2.

References

[1] J. P. Baxter, G. A. Schick, J. Singh, P. H. Kobrin and N. Winograd, J. Vac. Sci. Technol. **3** (1986) 1218.

[2] H. H. Andersen, B. Stenum, T. Sørensen, H. J. Whitlow, Nucl. Instr. Meth. **209/210** (1983) 39.

[3] S. Ichimura, H. Shimizu, H. Murakami, Y. Ishida, J. Nucl. Mater. **128/129** (1984) 601.

[4] I. Neshev, V. Hamishkeev, V. S. Chernysh, S. Postnikov and B. Mamaev, to be published in Vacuum.

[5] A. Johansen, E. Johnson, L. Sarholt-Kristensen, S. Steenstrup, H. H. Andersen, V. M. Buhanov, V. S. Chernysh, I. N. Ivanov, K. F. Minnebaev, Nucl. Instr. Meth. **B61** (1991) 21.

[6] M. T. Robinson, Phys. Rev. **179** (1967) 327.

[7] W. Eckstein, Computer Simulation of Ion-Solid Interactions, Springer, Heidelberg, 1991.

[8] H. H. Andersen, V. Chernysh, B. Stenum, T. Sørensen, H.J.Whitlow, Surf. Sci., **123** (1982) 39.

[9] J. Roth, J. Bohdansky, J. Nucl. Mater. **103** (1981) 339.

[10] W. Szymczak, PhD Thesis, GSF-report 18/85, Munchen (1985).

[11] W. Eckstein and M. Hou, Nucl. Instr. Meth. **B31** (1988) 386.

[12] G. Molière, Z. Naturforsch. **2A** (1947) 133.

[13] O. B. Firsov, Sov. Phys. JETP **6** (1958) 534.

[14] M. T. Robinson, in: *Sputtering by Particle Bombardment, Physics and Applications*, ed. R. Behrisch, Topics in Applied Physics (Springer, Berlin, Heidelberg, New York, 1981).

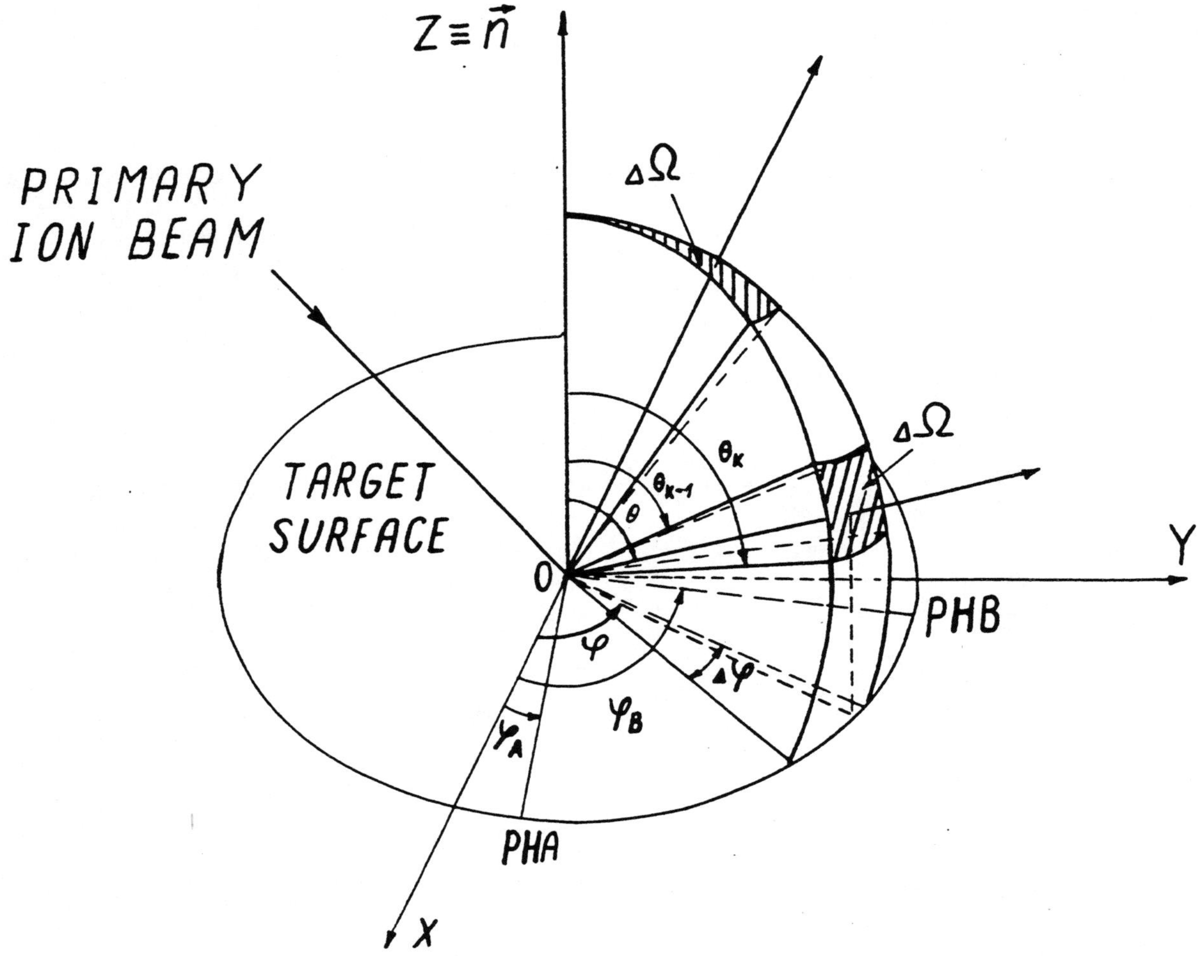

Figure 1: Geometry of the currently used by most of simultanionists extraction procedure. φ_A and φ_B are the angles of the beginning and the end of the azimuthal histogram, $\Delta\varphi = (\varphi_A - \varphi_B)/(NPH)$ is the azimuthal channel width, NPH is the number of the azimuthal channels, θ_{k-1} and θ_k determine the "k" - th polar channel. The number of the polar channels is 20. $\Delta\Omega$ is the solid angle of registration as determined by the θ and φ boundaries.

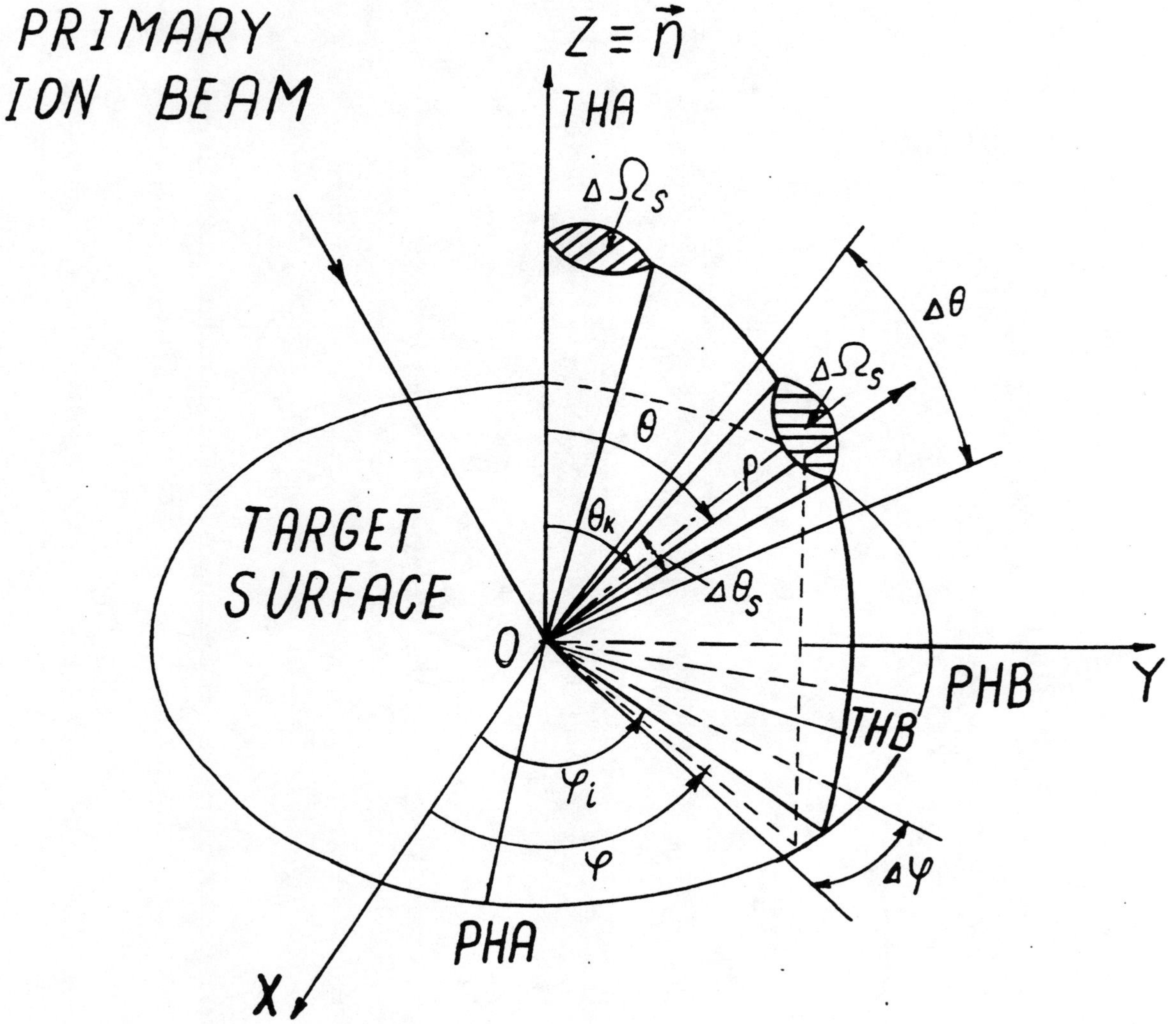

Figure 2: Geometry of the proposed extraction procedure. PHA and PHB determine the angles of the beginning and the end of the azimuthal histogram, respectively. THA and THB determine the beginning and the end of the polar histogram, respectively. $\Delta\varphi = (PHB - PHA)/(NPH)$ is the width of the azimuthal channel, NPH - the number of azimuthal channels. $\Delta\theta = (THB - THA)/(NTH)$ is the width of the polar channel, NTH - the number of the polar channels. $\Delta\Omega_s$ is the solid angle of registration whose shape and size is independent of the number of the azimuthal $\times$ polar channels.

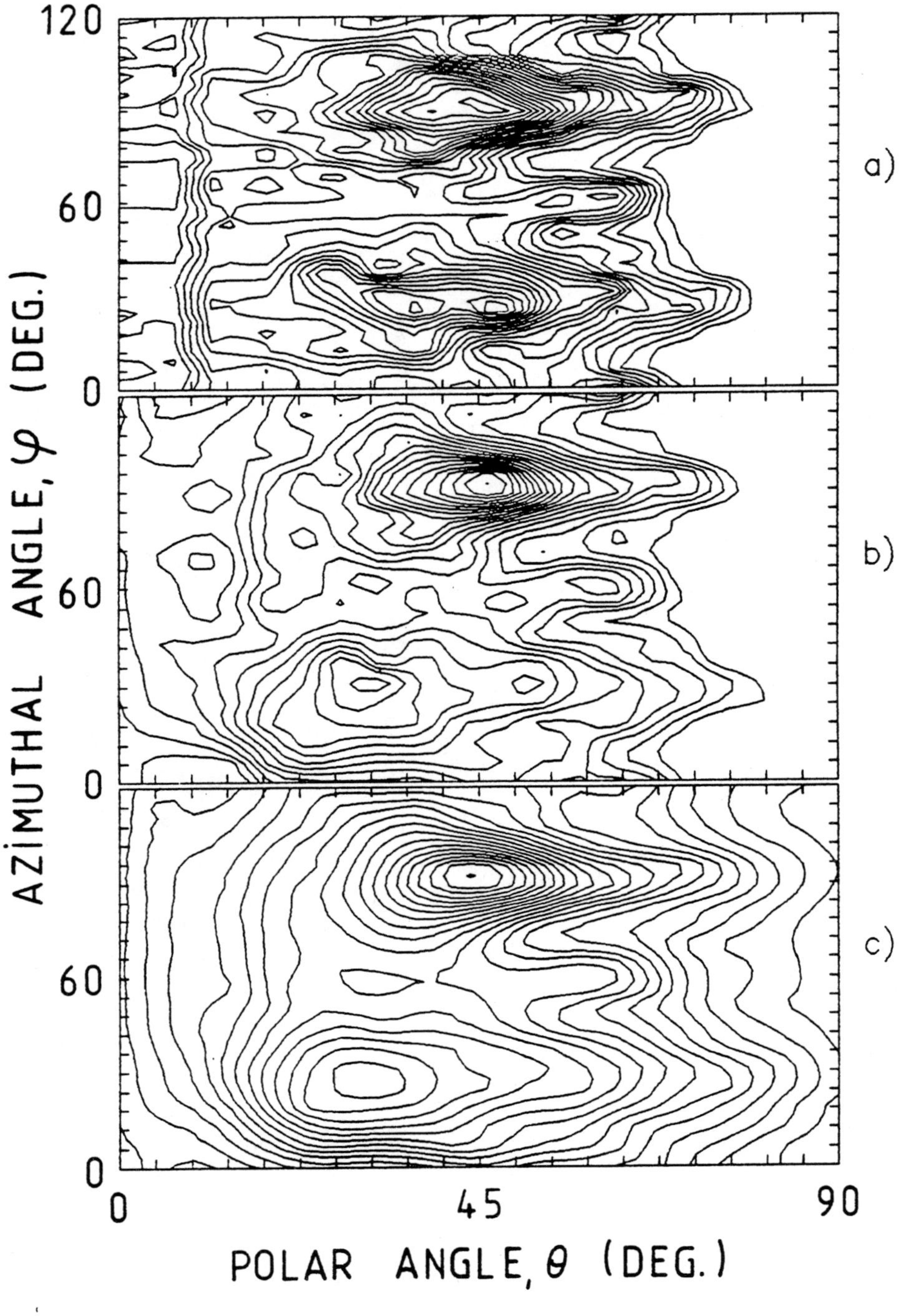

Figure 3: Sputtering direction distributions obtained by MARLOWE BCC: (a) using the existing extraction procedure, (b) using the proposed extraction procedure for the same value of the solid angle of registration $\Delta\Omega = 0.005\ sr$, and (c) — same as (b) for $\Delta\Omega = 0.024\ sr$. The contour lines represent 20 equidistant equal density levels. In all simulations, the input model parameters are the same as those used in ref. [10] determined by matching the simulation results to the experimental data of ref. [9]

AUTHOR INDEX

SUBJECT INDEX